Cinnamon

Ranjith Senaratne • Ranjith Pathirana
Editors

Cinnamon

Botany, Agronomy, Chemistry and Industrial Applications

Springer

Editors
Ranjith Senaratne
Department of Crop Science
University of Ruhuna
Mapalana, Kamburupitiya, Sri Lanka

Ranjith Pathirana
Plant & Food Research Australia Pty Ltd
Waite Campus Research Precinct
Urrbrae, SA, Australia

ISBN 978-3-030-54425-6 ISBN 978-3-030-54426-3 (eBook)
https://doi.org/10.1007/978-3-030-54426-3

This Springer imprint is published by the registered company Springer Nature Switzerland AG
The registered company address is: Gewerbestrasse 11, 6330 Cham, Switzerland

The editors dedicate this book to the people of Ceylon (Sri Lanka) who introduced cinnamon to the world, dedicated their lives to cinnamon and made it their livelihood; to those who were forced to harvest and process cinnamon under Sinhalese kings' royal prerogative, first for the kingdom and then to supply the foreign invaders; and to those who managed and sustained the industry despite change in priorities of the foreign powers and amidst untold obstacles and formidable challenges since independence in 1948 to date.

Peeling cinnamon in Ceylon circa 1672 from Atlas of Mutual Heritage; A collection of images related to Dutch East India Company

Preface

From time immemorial, cinnamon has been valued not only as a spice and an incense but also as an antiseptic. The history of cinnamon dates back to about 2800 B.C. and it is even mentioned in the Bible. Ancient Egyptians used cinnamon in mummification because of its antibacterial properties and fragrance, and the Romans used it in perfumes and fragrances and to flavor wines. Around the fourteenth century, Europeans started using cinnamon to improve the keeping quality of meat. It was the quest for spices that led European nations to explore the world in the fifteenth century.

The island nation of Sri Lanka had been a hub and a prominent cradle of spice trade in the ancient world, in which true cinnamon (*Cinnamomum zeylanicum* Blume) occupied an exalted position. It was the key spice that attracted Europeans to Sri Lanka (then Ceylon). The Portuguese, who conquered Sri Lanka in the early sixteenth century, wrested the cinnamon trade from the Arabs who had kept it as a jealously guarded secret for many centuries. Following the Portuguese takeover of the cinnamon trade, growing demand for cinnamon throughout the world created a lucrative market; this led the Dutch into battle with the Portuguese and gain control of the cinnamon industry and Sri Lanka in the mid-seventeenth century. In 1953, in his treatise titled "Ceylon under the British Occupation," Dr. Colvin R de Silva, a former minister and a legal colossus in Sri Lanka, recorded "If the vagaries of wind and wave brought the Portuguese to Ceylon, the lure of cinnamon kept them in the Island," and according to Fr. Philippus Baldaeus (1732), a Dutch minister, cinnamon was the "Helen or Bride" for whom the Dutch and Portuguese contended for many years. Such was the esteem in which cinnamon was held by Europeans.

C. zeylanicum, belonging to the Family Lauraceae, is indigenous to Sri Lanka and contributes around 90% of the global trade of true cinnamon. It is a unique and versatile plant that has oil in its leaves, bark, and roots, the chemical composition of which is completely different. Both bark and leaves contain essential oils, with cinnamaldehyde being predominant in the bark and eugenol in leaves; root-bark oil contains camphor. These substances have a wide range of industrial applications. Cinnamaldehyde is a proven natural bactericide widely used in the food and beverage industry, being particularly effective against *Salmonella* spp. and *Escherichia coli*. Given its organoleptic, medicinal, germicidal, and carminative properties, cinnamon has found applications in a wide range of industries, not only in food and beverage, but also liqueur, perfumery, nutraceutical, cosmeceutical and oral care, and traditional medicine.

Cinnamon, which has a highly fragrant aroma and a subtle, delicate, pungent, and sweet taste, contains insignificant traces of coumarin – a hepatotoxic hazardous substance. On the other hand, cassia – a cheap and inferior substitute for cinnamon – has a bitter, coarse, and strong flavor and contains appreciable quantities of coumarin, that is, 5–7 g kg^{-1} of bark and ~ 8% in oil. Yet no distinction has hitherto been made between cinnamon and cassia in global trade, and cinnamon is often adulterated with cheap cassia and sold under the label of cinnamon. In some countries like the USA, cassia is sold as cinnamon. Given the proven health implications of cassia, it is inconceivable why the European Union, North America, and other OECD countries, which have laid down stringent regulations and fastidious requirements to ensure food safety, have not enacted legislation to separate labeling of cassia from true cinnamon.

Today, markets are becoming increasingly globalized, sophisticated, and dynamic and consumers are increasingly health conscious. Therefore, there is growing demand for healthy natural foods and beverages, additives and flavors, pharmaceuticals, nutraceuticals, cosmeceuticals, and perfumes in the global market which has fueled the growth of the green economy. In view of this change in consumer behavior, preferences, and food habits, there has been a steady increase in the demand for cinnamon by the consumer as well as by a multitude of industries enumerated above. This has resulted in a surge of publications on various aspect of this agro-industrial crop ranging from ethnobotany, genetics, and agronomy to processing, value creation, and new product development.

To date, there has not been a comprehensive treatise covering the latest advances in research related to true cinnamon. This book, with chapters written by experts in diverse fields including ethnobotany, ecology, genetics, biotechnology, chemistry, pharmacology, agronomy, and value creation, has distilled and condensed current knowledge into a single source and contributes to the advancement and dissemination of knowledge and technology.

Inclusion of seminal work done and published recently related to molecular characterization, genetic barcoding, and chemical profiling of the germplasm of cinnamon and its pharmacological, nutraceutical, and cosmeceutical efficacy and industrial applications constitutes a salient feature of the book. Moreover, it has made a clear distinction between true cinnamon (Ceylon cinnamon) and cassia cinnamon (Chinese and Indonesian cassia) based on their biochemical profiles, highlighting the presence of a high content of coumarin, which is carcinogenic and hepatotoxic, in cassia. At present, both cinnamon and cassia are sold as cinnamon in the global market, and consumers are not aware of the health hazards associated with regular consumption of cassia. This book will create much needed global awareness about this issue, paving the way for formulating necessary food regulations and legislation in the interests of consumers.

Contributors to the book constitute internationally renowned senior scientists and academics with hands-on experience as well as movers and shakers in the cinnamon industry, thereby striking a right balance between theory and practice. Therefore, this will be a valuable source book for students, teachers, scientists, planners, policy makers, practicing agriculturists, and industrialists, and a prized acquisition to any library in higher education institutions, R & D institutions, and public and private sector institutions in agriculture and allied fields.

It is our fervent hope that this book will stimulate strategic research on cinnamon, unleashing and harnessing the potential of this wonder plant for the benefits of mankind while providing necessary knowledge and technology for the benefit of consumers and other stakeholders in the cinnamon industry.

Ranjith Senaratne
Mapalana, Sri Lanka

Ranjith Pathirana
Adelaide, SA, Australia

Acknowledgements

Cinnamomum zeylanicum Blume (true or Ceylon cinnamon) is indigenous to Sri Lanka and has a wide range of food and industrial applications. It has potential to be a multi-billion dollar industry, but operates far below its potential. Therefore, the Ministry of Primary Industries (MoPI) in Sri Lanka, the predecessor of the current Ministry of Plantation Industries and Export Agriculture, provided a special grant to carry out results-oriented, high-impact research with an aim to expand the cinnamon industry in Sri Lanka during 2017–2019. The project constructed a platform enabling senior scientists and technologists in academia, R&D institutions, and industry to work together to achieve the above objective. The project was administered by the Technology Division of the National Science Foundation (NSF), Sri Lanka.

The desire to produce a book on Ceylon cinnamon has been oscillating in us for quite some time, largely due to the lack of a comprehensive treatise on true cinnamon that covers the latest findings on ethnobotanical, biological, and agronomic aspects as well as its food, industrial, and medicinal applications. The project, funded by MoPI, gave a powerful stimulus that made us move forward with this idea. Findings emanating from research funded by the project constitute a sizable proportion of the book chapters. Therefore, we wish to record our sincere appreciation to Mr. Daya Gamage, former Minister of Primary Industries, for his initiative in allocating substantial funds for strategic research on cinnamon; chairman, director general, and staff of the Division of Technology of NSF during the said period; and members of the project steering committee and the project coordinator for the efficient administration, monitoring, and coordination of the project, which contributed in no small measure to the production of the book.

Contributions to this book have been made by renowned scientists and technical experts with considerable knowledge and experience in various aspects of cinnamon ranging from history, botany, and genetics to agronomy, medicinal properties, processing, and marketing. We wish to express our profound gratitude to them for their valuable contributions amidst busy schedules and heavy preoccupation.

The editors offer their heartfelt thanks to Dr. S. Sivasegaram, retired professor of mechanical engineering, University of Peradeniya, Sri Lanka; Prof. David Cooper, retired professor of philosophy, University of Durham, UK; and Drs. Andrew Granger, Vincent Bus, Michael Lay-Yee, Cath Kingston, Bruce Smallfield, and Ed Morgan and editorial staff at the New Zealand Institute for Plant and Food Research Limited for their valuable comments and for kindly editing some chapters of the book despite their heavy professional commitments. The editors wish to record their sincere gratitude to Springer for the meticulous care with which the book has been produced.

Last but not least, the editors wish to thank their spouses, Tamara and Sriya, and families for their understanding, support, love, and care during the period of editing the book.

Ranjith Senaratne and Ranjith Pathirana, May 11, 2020

Contents

Contributors

W. P. K. M. Abeysekera Herbal Technology Section, Modern Research & Development Center, Industrial Technology Institute, Halbarawa, Malabe, Sri Lanka

Pushpa Damayanthi Abeysinghe Department of Botany, Faculty of Science, University of Ruhuna, Matara, Sri Lanka

N. K. B. Adikaram National Institute of Fundamental Studies, Kandy, Sri Lanka

Pradeepa C. G. Bandaranayake Agricultural Biotechnology Centre, Faculty of Agriculture, University of Peradeniya, Peradeniya, Sri Lanka

K. R. Dayananda Retired Senior Research Officer, Industrial Technology Institute (successor to CISIR), Colombo, Sri Lanka

Achini M. De Silva Department of Agribusiness Management, Faculty of Agricultural Sciences, Sabaragamuwa University, Belihuloya, Sri Lanka

Ranil De Silva Interdisciplinary Centre for Innovation in Biotechnology and Neuroscience, University of Sri Jayewardenepura, Nugegoda, Sri Lanka

Sarada de Silva B. Darsin De Silva & Sons Private Limited, Colombo, Sri Lanka

Tuley De Silva R&D Center, Link Natural Products (Pvt) Ltd, Malinda, Kapugoda, Sri Lanka

Nirmal R. Dewasiri Department of History, University of Colombo, Colombo, Sri Lanka

A. P. P. Disna Department of Export Agriculture, Getambe, Peradeniya, Sri Lanka

Mohamed Esham Department of Agribusiness Management, Faculty of Agricultural Sciences, Sabaragamuwa University, Belihuloya, Sri Lanka

Lakmal Gonawala Interdisciplinary Centre for Innovation in Biotechnology and Neuroscience, University of Sri Jayewardenepura, Nugegoda, Sri Lanka

Kumudini Gunasekare Agriculture Sector Modernization Project, Ministry of Plantation Industries & Export Agriculture, Level 16, "Suhurupaya", Battaramulla, Sri Lanka

Jinasena W. Hewage Department of Chemistry, Faculty of Science, University of Ruhuna, Matara, Sri Lanka

Yoonus Imran Interdisciplinary Centre for Innovation in Biotechnology and Neuroscience, University of Sri Jayewardenepura, Nugegoda, Sri Lanka

Hanouf Ismail Interdisciplinary Centre for Innovation in Biotechnology and Neuroscience, University of Sri Jayewardenepura, Nugegoda, Sri Lanka

G. G. Jayasinghe National Cinnamon Research & Training Center, Department of Export Agriculture, Palolpitiya, Sri Lanka

Renda K. C. Jeewanthi Faculty of Agricultural Sciences, Sabaragamuwa University, Belihuloya, Sri Lanka

A. S. Karunaratne Faculty of Agricultural Sciences, Sabaragamuwa University of Sri Lanka, Sabaragamuwa, Sri Lanka

Liraj Kodithuwakku Interdisciplinary Centre for Innovation in Biotechnology and Neuroscience, University of Sri Jayewardenepura, Nugegoda, Sri Lanka

Nuwani Kodithuwakku Interdisciplinary Centre for Innovation in Biotechnology and Neuroscience, University of Sri Jayewardenepura, Nugegoda, Sri Lanka

K. L. Wasantha Kumara Department of Agricultural Biology, University of Ruhuna, Faculty of Agriculture, Mapalana, Kamburupitiya, Sri Lanka

Buddhi Marambe Department of Crop Science, Faculty of Agriculture, University of Peradeniya, Peradeniya, Sri Lanka

P. A. Paranagama Department of Chemistry, University of Kelaniya, Kelaniya, Sri Lanka

Ranjith Pathirana Plant & Food Research Australia Pty Ltd., Waite Campus Research Precinct, Urrbrae, SA, Australia

Stephanie Perera Interdisciplinary Centre for Innovation in Biotechnology and Neuroscience, University of Sri Jayewardenepura, Nugegoda, Sri Lanka

G. A. S. Premakumara Department of Basic Sciences and Social Sciences, Faculty of Nursing, University of Colombo, Colombo, Sri Lanka

D. K. N. G. Pushpakumara Department of Crop Science, Faculty of Agriculture , University of Peradeniya, Peradeniya, Sri Lanka

N. P. G. Pushpitha Ocean University of Sri Lanka, Crow Island, Mattakkuliya, Colombo, Sri Lanka

R. H. S. Rajapakse Department of Agricultural Biology, University of Ruhuna, Faculty of Agriculture, Mapalana, Kamburupitiya, Sri Lanka

D. Ratnasekera Department of Agricultural Biology, University of Ruhuna, Faculty of Agriculture, Mapalana, Kamburupitiya, Sri Lanka

Ravindu Runage Rathna Producers, Kamburupitiya, Sri Lanka

D. N. Samaraweera National Cinnamon Research and Training Center, Department of Export Agriculture, Palolpitiya, Thihagoda, Sri Lanka

Ranjith Senaratne Department of Crop Science, University of Ruhuna, Mapalana, Kamburipititya, Sri Lanka

S. Subasinghe Faculty of Agriculture, University of Ruhuna, Mapalana, Kamburipititya, Sri Lanka

Jayadeva Uyangoda Department of Political Science, University of Colombo, Colombo, Sri Lanka

Lodewijk J. Wagenaar Department of History, University of Amsterdam, East India House, Amsterdam, The Netherlands

Shamali Wasala Interdisciplinary Centre for Innovation in Biotechnology and Neuroscience, University of Sri Jayewardenepura, Nugegoda, Sri Lanka

Tharaka Weddagala Faculty of Agricultural Sciences, Sabaragamuwa University, Belihuloya, Sri Lanka

Jeevika Weerahewa Department of Agricultural Economics and Business Management, Faculty of Agriculture, University of Peradeniya, Peradeniya, Sri Lanka

K. D. N. Weerasinghe Faculty of Agriculture, University of Ruhuna, Kamburupitiya, Sri Lanka

Sheron N. Weerasuriya National Cinnamon Research and Training Center, Department of Export Agriculture, Palolpitiya, Thihagoda, Sri Lanka

Nalaka Wijekoon Interdisciplinary Centre for Innovation in Biotechnology and Neuroscience, University of Sri Jayewardenepura, Nugegoda, Sri Lanka

About the Editors

Ranjith Senaratne, PhD professor and chair, Department of Crop Science, Faculty of Agriculture, University of Ruhuna, Sri Lanka, has over 40 years of experience in higher education, including teaching, research, and administration and community development. He has held several senior administrative positions with distinction for a period of over 20 years, including the posts of dean, Faculty of Agriculture, and vice chancellor, University of Ruhuna, chairman of the Ocean University, and the vice chair of University Grants Commission of Sri Lanka. Currently he is Chairman of the National Science Foundation in Sri Lanka.

Prof. Senaratne is a visionary leader and an institution builder; moreover, in recognition of his outstanding contribution in education, science, community development, and international cooperation, the University of Durham, UK, conferred an honorary doctorate (*honoris causa*) on him in 2007. He has been the recipient of several internationally competitive and prestigious research grants and has held a number of coveted fellowships. He has over 100 research communications and papers to his credit and has written and edited around 10 books related to agriculture, higher education, science and technology, and national development.

Ranjith Pathirana, PhD began his career as a research officer in the Department of Agriculture, Sri Lanka, where he initiated the National Coordinated Oilseeds Programme. He was coordinator of the program before joining the University of Ruhuna, Sri Lanka, where he was the chair and professor of agricultural biology. There he initiated plant biotechnology programs and was the founding editor-in-chief of *Tropical Agricultural Research and Extension*, a peer-reviewed international journal. Dr. Pathirana has served as a FAO/IAEA expert in several developing countries and also served as a FAO/IAEA resource person for developing and implementing plant mutation breeding programs and conducting training courses.

In 2002, he joined the New Zealand Institute for Plant and Food Research Limited, where he initiated germplasm preservation research. Dr. Pathirana pioneered cryopreservation and cryotherapy of horticultural species in New Zealand and established the cryo-genebank in Palmerston North where he was the curator for 5 years and the principal investigator of several government-funded projects. He is a consultant to Acadia University, Nova Scotia, Canada, and the Kentville Research and Development Centre of Agriculture and Agri-Food, Canada, in cryopreservation and cryotherapy, as well as a reviewer of the Biological Sciences section of the Czech Academy of Sciences. His principal interests are in applying in vitro cell and plant biotechnologies for crop improvement, conservation, and elucidating biochemical pathways, particularly in plant stress response.

Chapter 1
An Introduction to Sri Lanka and Its Cinnamon Industry

Ranjith Pathirana and Ranjith Senaratne

1.1 Introduction

The cinnamon of commerce comes from the genus *Cinnamomum* Schaeffer belonging to the family Lauraceae, a large family of mostly evergreen woody trees or shrubs (except for the herbaceous hemiparasite *Cassytha*) consisting of about 53 genera and 2500–3000 species distributed throughout tropical and subtropical latitudes (Chanderbali et al. 2001; Kostermans 1957; Rohwer 1993). *Cinnamomum zeylanicum* Blume (syn. *Cinnamomum verum* Berchthold & Presl), the true or Ceylon cinnamon, and seven other species are endemic to Sri Lanka. Sri Lanka had different names in ancient times such as Lanka (in the Ramayana; seventh to fourth centuries BCE), Lankadeepa or Lakdiva (in Buddhist writings; fifth century BCE), Tambapanni (named by the first Sinhalese King of Sri Lanka, Vijaya; fifth century BCE), Taprobane (the name used by Megasthenes, the Greek ambassador to the court of the Mauryan King, Chandragupta, and Eratosthenes, one of the first Greek geographers; second century AD), Serendib (most of the Arab travellers and writers used this name; Taprobane, Sigaldip and Saheelan were also used by early Arab writers), Zeylan or Seylan (first European travellers used this name, derived from the Arab word Saheelan) and Ceylon—the name used by the colonizing British and the official name from that time until 1972 when it was changed to the present name. Sri Lanka was the major supplier of true cinnamon to the world for over seven centuries and continues to dominate the market with a 90% share. Originally harvested from the rainforests of the south-western region of Sri Lanka where

R. Pathirana (✉)
Plant & Food Research Australia Pty Ltd, Waite Campus Research Precinct,
Urrbrae, SA, Australia
e-mail: Ranjith.Pathirana@plantandfood.co.nz

R. Senaratne
Department of Crop Science, Faculty of Agriculture, University of Ruhuna, Mapalana,
Kamburupitiya, Sri Lanka

R. Senaratne, R. Pathirana (eds.), *Cinnamon*,
https://doi.org/10.1007/978-3-030-54426-3_1

C. zeylanicum originated, it was domesticated during the Dutch rule of the coastal areas of the country in the seventeenth century. By 2018, the area under cinnamon in Sri Lanka was up to 33,000 ha producing 24,000 MT (FAOSTAT 2018). Although low-quality cassia (mainly the product from *Cinnamomum cassia*, syn. *Cinnamomum aromaticum*, grown in China and Vietnam, and *Cinnamomum burmannii* grown in Indonesia) exports have increased much faster and exceed true cinnamon exports in quantity, Sri Lanka still ranks first in value of exports, with a US$259 million turnover in 2018. Over 90% of this is exported in the form of quills without value addition. This introductory chapter outlines the ecological and edaphic conditions of the cinnamon-growing area in Sri Lanka, a brief history of Sri Lankan cinnamon production and trade and a vision for the future.

1.2 Origin and a Brief Biogeography of the Rainforest Region of Sri Lanka, Home to Eight Endemic *Cinnamomum* Species

1.2.1 Origin and Geography of Sri Lanka

The island of Sri Lanka has a total area of 65,610 km^2 and is located southeast of the southernmost tip of peninsular India, between 50°53′ and 9°51′ North and 79°43′ and 81°53′ East. North to south the island is c. 435 km at its greatest length, and c. 240 km at its greatest width, east to west. Eighty per cent of the land mass consists of coastal plains and hills to an elevation of 900 m, and the remainder is a south-central mountainous region reaching 2527 m at Pidurutalagala. The physiography of the island consists of three peneplains: the lowest at 75 m above sea level (average 30 m), the second at 125–750 m (average 500 m) and the third at 750–2500 m (mostly 1500–1800 m) above sea level (Fig. 1.1; Ashton and Gunatilleke 1987; Cooray 1984; Erdelen 1988; Werner and Balasubramaniam 1992).

Sri Lanka was part of the ancient supercontinent of Pangea during the Mesozoic Era (c. 250 million years ago [mya]) when it was geologically connected to Madagascar, Africa, southern India and Antarctica (Dissanayake and Chandrajith 1999). During the second segment of the Mesozoic Era, the Jurassic period (c. 150 mya), it rifted from Africa of the Gondwana megacontinent as the Antarctica-India-Seychelles-Madagascar plate (Briggs 2000; Raven and Axelrod 1974). Thereafter, India-Madagascar separated from Antarctica c. 130 mya, India-Seychelles from Madagascar c. 84–96 mya and India from Seychelles c. 65 mya (Briggs 2000; Gunatilleke et al. 2017; Raven and Axelrod 1974). Post-Jurassic tectonic events shaped the geology of the mountains of Sri Lanka. The island consists mainly of Palaeozoic granitic rocks of the Deccan Plate of Gondwanan origin, uplifted in the post-Miocene period (Cooray 1967). Nevertheless, Gunatilleke et al. (2017) argue that there is no evidence for a mountain bridge between India and Sri Lanka since the Miocene, and possibly since the Jurassic period. Similarly, there is no reason to

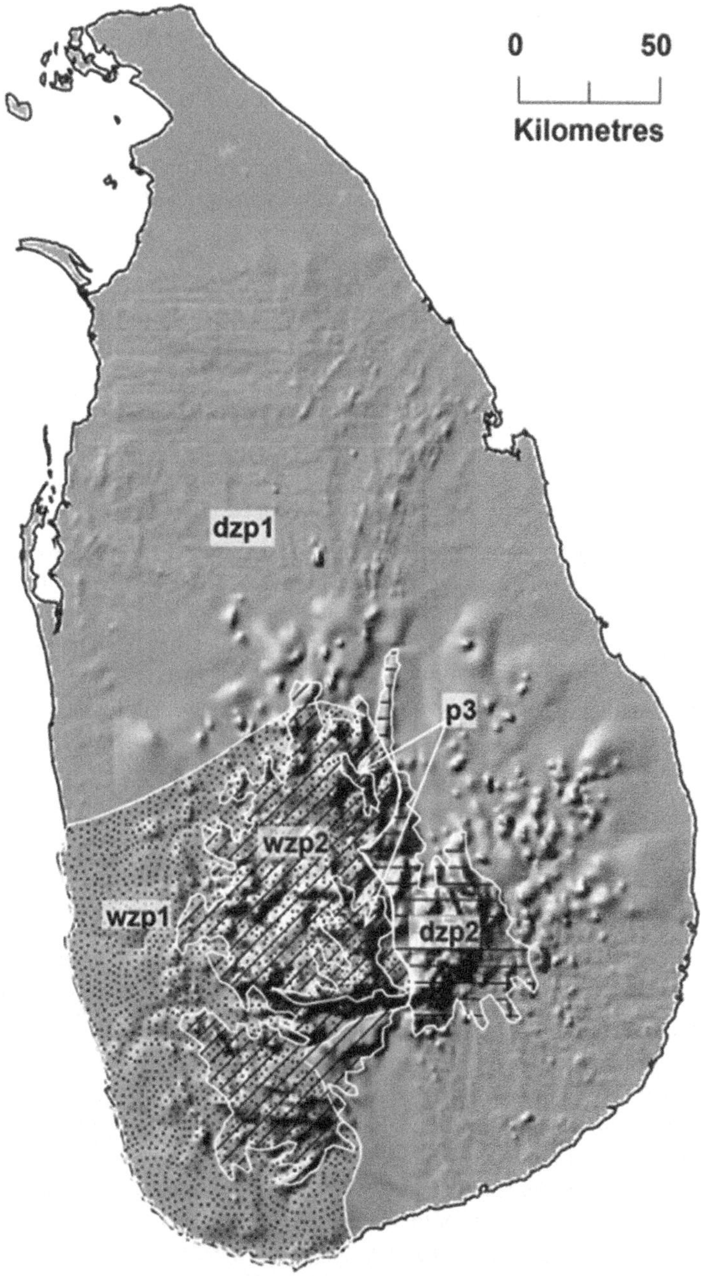

Fig. 1.1 The three peneplains of Sri Lanka and the main climatic zones: dzp1, dry zone peneplain 1; dzp2, dry zone peneplain 2; wzp1, wet zone peneplain 1; wzp2, wet zone peneplain 2; p3, peneplain 3. (Adapted from Biswas and Pawar (2006))

believe that there has been a direct mountain bridge between the newly uplifted Hindu Kush and Himalaya in the past either (Gunatilleke et al. 2017). Based on geological, mineralogical and isotopic data, Dissanayake and Chandrajith (1999) suggest that Sri Lanka is more closely associated with the south-eastern parts of Madagascar than with the Archean granulites of southern and eastern parts of India as has long been postulated. It is part of the same belt connecting to the Lützow-Holm Bay area in Antarctica (Dissanayake and Chandrajith 1999).

1.2.2 Biogeography of the Rainforest Region

South Asia to which Sri Lanka belongs lies on the Indian plate that has traversed from warm temperate through dry, then seasonally wet and equatorial tropical climates (Dittus 2017; Gunatilleke et al. 2017). Fossil pollen and wood reveal a southern retreat of the rich forest tree flora, which is now confined to a tiny, 10,000 km^2 area in southwest Sri Lanka (Gunatilleke et al. 2017), including seven wild *Cinnamomum* spp. In the lower parts of the south and south-western 'wet zone', the natural vegetation comprises evergreen rainforest while in the western slopes of the highlands, between 600 and 900 m, the lowland rainforest gradually changes into the lower montane rainforest (Gunatilleke and Gunatilleke 1991; Werner and Balasubramaniam 1992).

The lower montane rainforests are similar to lowland rainforests in terms of floristics and structure, and these rainforest floras must have survived the partly drier conditions of the Pleistocene (Werner and Balasubramaniam 1992). The 1500 m elevation forms the lower limit of a frequent and dense cloud cover, thus marking a sharp boundary between the lower and upper montane rainforests. Temperature in the montane forest region ranges between 15 and 25 °C (Ranawana 2014; Weerawardhena and Russell 2012; Wickramasinghe et al. 2008).

The mountain ranges in the upper part of the montane rainforest resemble an anchor set from west to east with a slight curve to the south. Peak Wilderness Sanctuary forms the western part and Haputale ranges in the east. The highest peaks (1800–2100 m) are in the centre, with another mountain chain that runs northwest (Fig. 1.1). Further to the north of this massif is the relatively isolated Knuckles range formed in a northwest to southeast direction, exactly perpendicular to the two principal wind currents (southwest and northeast monsoons) that bring a substantial proportion of the rainfall to the island. Being perpendicular to the wind currents, it also acts as a climatic barrier, influencing climatic conditions on its two sides. Thus, the highland and western parts of the Knuckles range are extremely wet throughout the year, with an annual rainfall of about 5000 mm, whereas the lower eastern slopes are drier with less than 2500 mm annual rainfall (DMSL 2019).

Unlike in many montane forests, those in Sri Lanka lack conifers and Fagaceae, which form a major component of the diversity in montane forests elsewhere in Southeast Asia (Manos and Stanford 2001). In the upper montane rainforests of Sri Lanka, their ecological role has been taken over by other families, with Lauraceae,

Myrtaceae, Clusiaceae and Symplocaceae having an exceptionally high frequency (Erdelen 1988; Werner and Balasubramaniam 1992). This composition obviously has developed from the lowland flora during the gradual uplift of the mountains during the Tertiary period. The lower montane and lowland rainforests are a mixed dipterocarp type (Ashton et al. 2001; Ashton and Gunatilleke 1987; Gunatilleke and Ashton 1987).

Perhumid southwest Sri Lanka has an extraordinary degree of endemism and species richness. For example, in Sinharaja Forest Reserve, a UNESCO Heritage site, 60% of tree species are endemic and many are rare. This refuge is the last extensive patch of tropical rainforest in Sri Lanka, covering an area of 8864 ha with a range in altitude from 300 to 1700 m. It is home to at least 139 endemic plant species (UNESCO 2020), including several endemic *Cinnamomum* species (details in Chap. 4 in this book). The cinnamon species endemic to Sri Lanka are dispersed in all parts of the country except the vast dry zone peneplain 1 shown in Fig. 1.1. However, some species are limited to just one zone (details in Chap. 4 of this book).

1.3 Cinnamon Trade in Ancient Ceylon (Sri Lanka): A Historical Sketch

1.3.1 Pre-colonial Period

According to many authors, cinnamon has been a prized commodity since the time of early civilizations of the Middle East. Cinnamon was used in ancient Egypt for mummification and as a medicine. It was an expensive commodity in ancient times and was available only to monarchs and the elite. Cinnamon is also mentioned in the Bible, and in ancient Greek and Latin texts (Abdel-Maksouda and El-Amin 2011; Braun 2006; David 1992). Using gas chromatography–mass spectrometry, Namdar et al. (2013) provide archaeological evidence for storage of cinnamon oil in small Phoenician flasks discovered in the store rooms of temples and treasuries of the southern Levant, dating back to the eleventh and late tenth centuries BCE. The organic residue in these clay flasks retained cinnamaldehyde, the major component of the oil of some of the *Cinnamomum* spp., including bark oil of *C. zeylanicum* from Sri Lanka (Ariyarathne et al. 2018; Senanayake and Wijesekara 2004; Sritharan 1984). Cinnamaldehyde is a terpenoid apparently exclusive to *Cinnamomum* spp. (Joshi et al. 2009; Simic et al. 2004; Senanayake and Wijesekara 2004; Shylaja et al. 2004). Furthermore, the sixth-century merchant and traveller, Cosmas Indicopleustes, the first westerner to visit the Island, mentioned in his book, titled *Topographia Christiana*, the importance of the spice trade in Ceylon (Ginigaddara 2018). There is evidence that between A.D. 1273 and A.D. 1284, the Sinhalese kings endeavoured to enter into a trade agreement with the Sultan of Egypt, to supply him with cinnamon, precious stones and elephants. The Sinhalese emissaries travelled by sea up the Persian Gulf and thence by land to Cairo, passing through Baghdad and the desert of Syria. Foreign trade had really become a matter of considerable importance during this period to the Sinhalese kingdom (Jayasuriya 1949). Although evidence indicates that trade in this region with Asia, including cinnamon, took place much earlier than hitherto suspected (Gilboa and Namdar 2016), specific mention of cinnamon of Sri Lankan origin is unavailable in literature until about the tenth century. However, Haw (2017) argues that the reference to cinnamon in the Mediaeval period is related to some other taxa from the African continent. For more details on trade between south Asia and the Mediterranean in the Iron and Bronze Ages, the reader is referred to Gilboa and Namdar (2016) and Jayasuriya (1949).

Cinnamon was one of the important types of merchandise travelling via Indian Ocean trade routes since the tenth century (Gunaratna 2002; Paranavitana 1960; Biedermann and Strathern 2017). The Sri Lankan origin of quality cinnamon was a well-kept secret among the Muslim merchants who made fortunes by shipping the product to the Red Sea and then transferring it to the Mediterranean region. International trade was welcomed by Sinhalese kings as an important factor in economic prosperity. Arab traders established themselves in small communities along the Sri Lankan west coast near many ports with the king's patronage, operating as autonomous clusters and paying port dues (Paranavitana 1960; Somaratne 1975). Some historians argue that the royal centre Kotte was located close to Colombo to take advantage of the developing cinnamon trade (Somaratne 1975; Strathern 2004; de Silva 2000; Halikowski Smith 2001).

1.3.2 Colonial Period

1.3.2.1 The Portuguese Period

In the pre-colonial period, cinnamon peelers worked for the king in return for the crown land they cultivated (land tenure) under a system called *Rajakariya*. This system was later to be used by the colonial powers to exploit labour for harvesting and peeling cinnamon. The arrival of Lourenço de Almeida, the son of the first Portuguese Viceroy of India, in Colombo in 1505 or 1506 led to the establishment of Portuguese diplomatic ties with the Kingdom of Kotte ruled by King Parakramabahu VI, with the Portuguese eventually taking control of the cinnamon trade from the Arab merchants. King Vijayabahu of Kotte in 1518 allowed the construction of a fort in Colombo to protect the Portuguese trading interests. The king even allowed the Portuguese to establish a cinnamon processing factory (Gunaratna 2002) and the whole island was brought into suzerainty and required to pay an annual tribute payment of a certain quantity of cinnamon under the watchful eye of the Portuguese governor (Halikowski Smith 2001; de Silva 1989). This alliance continued until 1540, but due to emerging conflicts the Kotte Kingdom was abandoned by 1597 (Gunaratna 2002; Strathern 2004; de Silva 2000; Halikowski Smith 2001). In the period between the 1590s and the 1640s, the Portuguese came close to controlling the production and world trade of cinnamon, but this was challenged by cinnamon from other Asian countries (Halikowski Smith 2001).

During the Portuguese period, harvesting of cinnamon was mainly concentrated in the forests of the lower slopes of western and south-western parts of Sri Lanka (coastal plains of wet zone peneplain 1—Fig. 1.1), where the Kotte Kingdom operated. However, vast swathes of the rainforest under the Kandyan and Sitawaka Kingdoms did not have agreements with the Portuguese: these Kingdoms were in constant rivalry with the Kotte Kingdom and the Portuguese were not welcome. Within their jurisdiction, the Portuguese ruthlessly exploited cinnamon forests by changing the services (*Rajakariya*) due from peelers. This was earlier based on extent and quality of land held by the local population. The Portuguese later associated it with the person rather than the land and by the mid-seventeenth century it became the standard for every person of *Salagama* caste over 12 years to supply a fixed quantity of cinnamon to the 'state', thus establishing a system of slavery (de Silva 1981; Ginigaddara 2018; Halikowski Smith 2001; Somaratne 1975). The peeling and the preparation of cinnamon for the market was a skilled profession and had become the caste occupation of an important segment of the population, the *Salagama* or the *Chalea* caste. The *Salagamas* were immigrants from South India (Biedermann and Strathern 2017; de Silva 1993). The Portuguese, and thereafter the Dutch, managed to take advantage of the system established by Sinhalese Kings, keeping the *Rajakariya* (free labour for the king) intact. For further details of the Portuguese period and cinnamon trade, see Chap. 2 of this book and de Silva (1993).

1.3.2.2 The Dutch Period

Access to cinnamon was the reason for the Dutch arrival in Sri Lanka in the fifteenth century (de Silva 1981, 1993; Devasiri 2008). Early in the seventeenth century, several competing Dutch trading companies had merged into a new commercial enterprise: the Dutch East India Company (Dutch: *Vereenigde Oostindische Compagnie*—VOC), whose primary goal was to gain a trade monopoly of East Asian spices. In order to protect trade, the VOC was given a mandate by the government of the Netherlands to conquer land, build forts and maintain an army (Jayasena 2013).

By the time the Dutch arrived in Sri Lanka in the 1630s, the Kingdom of Kandy, which was an accessory kingdom of the Kingdom of Kotte, had established itself as the sole independent polity on the island. King Rajasinghe II of Kandy signed a treaty in 1638 (Hanguranketha Treaty, also called Batticaloa treaty) with the VOC that secured the terms under which the two parties would cooperate in defending the Kingdom of Kandy from the Portuguese; the Dutch were appointed as the sole protector of the country. In exchange, the VOC secured the lucrative cinnamon trade. It took another two decades for the Dutch to drive the last Portuguese forces out of Sri Lanka. The VOC controlled the territories of southern and western Sri Lanka and continued to use forced labour for harvesting and peeling of cinnamon (Biedermann and Strathern 2017; van Meersbergen 2017).

Within the 5-year period from 1665 to 1670, the area occupied by the Dutch increased twofold and more people of other castes were thrust into cinnamon peeling (de Silva 1981), the most arduous task in the industry. Slaves of Indian origin were also imported. In the period around 1660, over 10,000 Indian slaves were transported by the VOC (Schrikker and Ekama. 2017). The price of cinnamon in European markets doubled in the mid-seventeenth century, at which time relocation of villages from cinnamon lands to marginal areas and further pressure on the cinnamon lands and on peelers resulted in discontent among the population. Cinnamon prices remained steady until the end of Dutch rule in Sri Lanka near the end of the eighteenth century. The Dutch managed to meet this demand and export sufficient quantities of cinnamon by improving relations with the Kandyan Kingdom where more cinnamon was available for harvest. At the same time, the peelers were unable to escape to Kandyan Kingdom.

To meet the export demand and compensate for the loss of cinnamon trees due to population growth in the coastal areas of the southwest, the Dutch started establishing cinnamon plantations during 1767–1770 (Weiss 2002). A report on cinnamon plantations in 1786 and a series of maps from 1794 provide a vivid picture of the expansion of cinnamon plantations along the western coastal line. By the late eighteenth century, cinnamon plantations occupied a large proportion of land from Kalutara, in the south of Colombo, to Negombo, in the north (Dewasiri 2007). In the years to come, this decision to grow cinnamon helped to contain further erosion of genetic resources in the rainforest (Pathirana 2000; Wijesinghe and Pathirana 2000), albeit for a short period until the British transformed the landscape into coffee and then to tea and rubber plantations. The cinnamon industry and its rise during the Dutch period are discussed in more detail in Chap. 2 of this book and in de Silva (1993) and Dewasiri (2007).

1.3.2.3 The British Period

With the surrender of the Dutch forces to the British military on 17th February 1796 and the capture of the last King of Kandy, Sri Wickrama Rajasinghe, in January 1816, the whole island came under British rule. The British succeeded to a very rich legacy in cinnamon and were determined to preserve it. The British East India Company enjoyed a monopoly of export of cinnamon from 1802 until it was abolished and taken over by the Crown in 1821 (Mendis 1952). The period of British rule had a more significant impact on the rainforest region (where high endemism, including that of cinnamon exists) of Sri Lanka than the Portuguese or the Dutch. In parallel with cinnamon production and exports, they increased land under coffee and rubber, resulting in deforestation of large swathes of rainforest.

A regulation instituted in the very early days of British rule prohibiting Europeans from acquiring land in the island was repealed, and Europeans were given grants of land free of tax. Local civil servants were also encouraged to take to planting, and very soon tens of thousands of hectares were brought under coffee cultivation. Nevertheless, the annual revenue of Ceylon in and round 1830 did not exceed

£333,000, of which nearly half was derived from the cinnamon and salt monopolies. The coffee peak was reached in 1845 by which time most of the hill country forests of the island colony had been converted to plantations. European settlers were attracted in increasing numbers, and followed by cheap labour obtained in the form of workers from South India who were settled in plantation areas (Jayasuriya 1949). By 1936, Indian Tamils numbered over 1.1 million, more than 15% of the total population of the country, the majority of whom did not have citizenship of either country until in the post-Independence Indo-Ceylon pact of 30th October 1964 (MEA 1964). The area under coffee plantations doubled between 1845 and 1847 to 50,071 acres (20,263 ha; de Silva 1981). Cultivation of coffee had to be abandoned gradually in the 1870s after the incursion of coffee rust (*Hemileia vastatrix*) in 1869 and the decline of the crop as a result (Waller 1982). Planters then took to the cultivation of tea, mainly in the second and third peneplains, and rubber in the first peneplain of the wet zone (Fig. 1.1).

The British continued to derive considerable profit from the sale of cinnamon until 1835, but from that year its price began to fall owing to the competition from cassia. The Dutch, who had been deprived of the high-quality cinnamon from Ceylon by the British, carried away 3000 cinnamon plants and seeds as well as a number of cinnamon peelers from Ceylon in 1825 in one of their ships. Before long the Dutch were selling quality cinnamon from Java at a lower price than British product from Ceylon and the Dutch began to re-capture the market (Mendis 1952). The higher price of true cinnamon from Ceylon was because of a high export duty that remained in place until 1853. By then it was too late for the British to recover trade and capture the market, resulting in conversion of large tracts of cinnamon plantations to rubber and coconut, and some at higher elevations to tea (de Silva 1981; Peiris 1981).

Peiris (1981) considered the period between 1830 and 1930 as the beginning and the end of major landmarks of Sri Lanka's political and economic history and suggests that it be treated as a distinct phase. The Portuguese and the Dutch who ruled the maritime provinces of Sri Lanka in the two centuries that preceded the British regime did not attempt to initiate basic structural changes in the traditional system of land tenure. Their overall impact on the economy was slight and, by and large, they remained content with extracting services, revenue and products for trade through the media of existing tenure relationships. In most parts of their operations, the existing service obligations of the land holders were retained with only minor modification (de Silva 1981; Peiris 1981).

In terms of economic development, the period of British rule was characterized by the increasing importance of systematic and large-scale expansion of plantation crops for export. Plantation agriculture, which was confined to a few hundred acres of coffee at the commencement of the period, soon became the dominant sector of the colony's economy. By the end of this period, plantation crops covered over two million acres (ca. 405,000 ha; Peiris 1981). Within a period of less than half a century, most of the forests in the hill country were cleared for plantation crops. For example, 2796 ha of ecologically fragile Knuckles alpine forest with unique biodiversity (Werner 1995) was deforested to plant tea within 2 years (1874–1875) for the establishment of the Kallebokka estate (Forrest 1967). Between 1830 and 1930, nearly one million ha of land was sold, most of which was in hill country forests. In 1875, a total of 191,351 ha had been bought by plantation owners in the hill country. However, only 52%, or 99,079 ha, of the land bought was under coffee cultivation; most of these lands were later planted with tea. An overwhelming proportion of the land that was under forest cover at the turn of the nineteenth century was covered with plantation crops by the end of this period.

Development of the plantation industry in Sri Lanka by the colonial administration became possible because of the appropriation and sale of land to which no proof of ownership could be presented by the local population. Most of the land in the two upper peneplains fell into this category and the ruling British elite as well as new European arrivals in the island were the main buyers of land. Under the Sinhalese kings, land was given for services provided, and most of the land used by the villagers did not have titles. The Crown Lands Encroachment Ordinance no. 12 of 1840 opened the way for the acquisition and sale of the King's forests, unoccupied lands, waste lands and the seasonally cultivated lands including communal forests. Introduction of land sales by the colonial government enabled the Europeans and locals loyal to the administration to start coffee plantations in the island (Bandarage 2019; Mendis 1952; Meyer 2008; Peiris 1981). At the time of Independence in 1948, 90% of Sri Lanka's foreign exchange earnings were from tea, rubber and coconut—all these crops occupying the former rainforest regions of western and south-western wetlands and the central hills; and a country that had been self-sufficient in its main staple rice at the beginning of the colonial era, was spending 25% of its foreign exchange to import rice.

1.4 Agro-Ecology of Cultivated Cinnamon in Sri Lanka

1.4.1 Impact of Deforestation to Establish Plantation Agriculture

Unlike in many other European colonies, the primary agricultural strategy adopted by the British in Sri Lanka was not peasant or small-holder cultivation, but large-scale plantation agriculture. The effects of expansion of the plantation industry on the wet zone forest areas and the resulting population increase had long-term consequences on the ecology of the central hills and the southwest montane and rainforest regions of Sri Lanka. It even had effects on the drier areas because of silting of river beds resulting in frequent flooding (Bandarage 2019; Peiris 1981) as most of the dry zone rivers, including the country's largest and longest, Mahaweli River, have their headwaters in the wet montane area. Unlike in many Asian countries, 'population explosion' in Sri Lanka is not a twentieth-century phenomenon. It started in the mid-nineteenth century. Within a 70-year period, it rose from 1.72 to reach 4.50 million by 1921. The population density was highest in the strip extending about 12 km into the interior, that covered the coastline from Negombo to Dondra Head—the area that was used by the Dutch to plant cinnamon in the west and southwest. This trend of population growth continued into the post-Independence period and most of the districts where cinnamon is grown now have a population density of over 600 persons per square kilometre compared with the national average of 310 persons (BM 2015; DCS 2006b; de Silva 1981). Overall, Sri Lanka, the 25th largest island in the world, now has the 5th largest island population density after Java, Honshu, Luzon and Mindanao (BM 2015).

Although the present level of forest cover in Sri Lanka as a whole is around 20%, only Matale and Nuwara Eliya among the plantation districts have a forest cover at or above this national average. This is due to the banning of the sale of land above 1524 m (5000 ft) and the declaration of two areas, Horton Plains-Peak Wilderness region and Hakgala, as nature reserves in the later phase of the colonial government (Wickramagamage 1998) and declaration of more forest reserves under protection post-Independence. The change from forest cover to over 2,000,000 acres (ca. 810,000 ha) of plantations has had its effects on the climates of the cinnamon-growing regions, with lower and more erratic rainfall. The stability of the eco-system and soil has suffered irreparable damage as immediate changes occurred in the pattern of soil formation (Wickramagamage 1998). As Peiris (1981) recorded, the edaphic effects of the plantation practices that were adopted in Sri Lanka during the colonial period can be summarized as follows:

- Direct loss of soil through removal by erosion.
- Formation of a hard-baked upper soil surface lacking in adequate moisture.
- Lowering of the potential nitrogen and organic matter content of the soil.
- Widening of the soil temperature range resulting in the enhancement of soil activity.
- Enhanced siltation of river beds due to forest clearing resulting in increased frequency and intensity of flooding and silting of irrigation tanks, destroying irrigation systems for paddy and other crops.
- In extreme cases, excessive siltation resulting in the conversion of paddy lands into uncultivable wasteland.

1.4.2 Socio-economic Structure

It was in the 'cinnamon belt' (from Negombo to Dondra Head) that most of the churches were built and missionary schools started during the colonial period, resulting in a population of educated youth who were not too inclined to work in agriculture, particularly in the arduous and monotonous task of cinnamon peeling. Moreover, as discussed before, there was social stratification in different occupations during the times of kings, described as the caste system. Thus, cinnamon peeling was the 'duty' of the *Salagama* caste. People of this caste are descendants of workers from South India, brought into Sri Lanka before the arrival of the Portuguese

(Biedermann and Strathern 2017; de Silva 1993; Dewasiri 2007). Thus, even today there is reluctance from people of other castes to participate in cinnamon peeling as rural populations hold on to cultural traditions more strongly than city populations. Hence, even to this date the labour shortage in the cinnamon industry, particularly of peelers, is felt despite the high population density in the areas of cinnamon cultivation. Attempts to mechanize the peeling process are underway and highlighted in Chap. 9 of this book. Breeding programmes should go hand in hand with this, to select genotypes more amenable for mechanization of peeling and processing. Another aspect highlighted later in this book (Chaps. 15 and 16) is the value addition before export, so that the technique of peeling does not have to be 'perfect' to fit the stringent requirements of quality of the bales as demonstrated in https://www.srilankanspices.com/sl_spices_cinnamon.html (CCE 2020).

1.5 Progress and Challenges of the Sri Lankan Cinnamon Industry

1.5.1 Trends in Cinnamon Production in Post-Independence Sri Lanka

Since Independence in 1948, the cinnamon industry was neglected as preference was given to three major plantation crops, tea, rubber and coconut, that were earning enough to support the import of rice and other food, production of which was neglected during the colonial period. In 1972, the Government of Sri Lanka created a separate department (Department of Minor Export Crops) targeting spices and other neglected crops with potential. In the 1960s, annual cinnamon production averaged 6275 MT from about 13,200 ha (Table 1.1; range from 5500 to 8250 MT). There was an expansion in cinnamon lands in the 1970s but thereafter the area of cinnamon cultivation remained static until 1992. The production increased to a modest ~10,000 MT in the 1970s, and stagnated at that level until 1992.

Table 1.1 Trends in cinnamon production in Sri Lanka in the last five decades

Period	Area, ha	Yield, kg/ha	Production, MT
1961	11,000	4727	5200
1961–70	13,177	4760	6275
1971–80	21,628	4773	10,324
1981–90	21,088	4752	10,016
1991–00	23,558	4967	11,695
2001–10	26,157	5155	13,492
2011–18	30,353	6513	19,797
2018	33,094	7258	24,020

Note: Values calculated from FAOSTAT (2018); the values are averages for the period indicated

In the early 1990s the Government of Sri Lanka started taking other export crops more seriously and in 1992 even changed the name of the government department that used to handle spice crops from the Department of Minor Export Crops to the Department of Export Agriculture. Since then, successive governments have implemented different programmes including incentives to ensure the growth of the cinnamon industry. The land area under cinnamon increased from 24,510 ha in 1998 to 33,094 ha in 2018, within a 20-year period. This was also accompanied by an increase in yield from 4790 kg/ha to 7258 kg/ha for the same period. As a result, the production doubled in the last 20 years, from 11,740 MT in 1998 to 24,020 MT in 2018 (FAOSTAT 2018). An increasing trend in market prices in the same period ensured good profit for growers and generated foreign exchange for the country. As a result, US$132.28 million of export earnings from cinnamon bales in 2014 increased to US$213.25 million by 2018. When the other cinnamon products for which records are available such as oleoresins and essential oils are added to the above figures, the export value of US$166.7 million in 2014 increased to US$259 million in 2018 (SLEDB 2020). More details about the trends in world trade of cinnamon products can be found in Chap. 3 of this book.

1.5.2 Status of Endemic Cinnamomum *Species in Sri Lanka*

The humid, tropical evergreen forests of the southwest and the montane rain forests of the central hills of Sri Lanka are home to eight *Cinnamomum* species, including the cultivated *C. zeylanicum* Blume. Their distribution and conservation status are discussed in detail in Chap. 4 of this book. Different species occur from 30 m to 2400 m altitude from tropical rain forest to semi-dry areas with annual rainfall ranging from 1100–4450 mm. The distribution of *Cinnamomum* species covers all the three peneplains (Fig. 1.1) of the country. All the endemic species except *Cinnamomum dubium* are rare and threatened to different extents in the wild (Table 1.2).

Table 1.2 *Cinnamomum* species endemic to Sri Lanka and their vulnerability status

Cinnamomum species	Vernacular name in Sinhalese	Global Red Listing category[a]	National Red Listing category[b]
C. capparu-coronde	Kapuru Kurundu	Endangered	Vulnerable
C. citriodorum	Pengiri Kurundu	Critically endangered	Vulnerable
C. dubium	Sevel Kurundu	Least concern	Vulnerable
C. litseaefolium	Kudu Kurundu	Critically endangered	Endangered
C. ovalifolium	Bola Kurundu	Vulnerable	Vulnerable
C. rivulorum	Wal Kurundu	Critically endangered	Endangered
C. sinharajaense	Sinharaja Kurundu	Critically endangered	Vulnerable
C. zeylanicum	Kurundu	Not listed	Vulnerable

[a]IUCN (1998)
[b]MOE (2012)

The threats to the endemic species include fragmentation of forests and over harvesting of most species for use in local medicine. The bark, leaf and root of these species have different major constituents as described in Chaps. 4 and 10 of this book. Thus, different parts of the same species are used for different ailments. Genetic variation in phytochemicals among species provides opportunity for further improving cultivated cinnamon or direct use of endemic species after introducing to cultivation. In addition to the well-known uses in food, nutraceutical and cosmeceutical industries (described in detail in Chap. 13 of this book), and in the pharmaceutical industry (described in Chaps. 11 and 12), some of the chemicals identified in these species are well-known anti-fungal agents (e.g. cinnamaldehyde, eugenol and β-caryophyllene; Ranasinghe et al. 2002; Simic et al. 2004), insecticides (e.g. linalool; Praveena and Sanjayan 2011), insect repellents (geraniol, citral, citronellol, eugenol, limonene, benzyl benzoate and linalool; Li et al. 2010), acaricides, scabicides and topical treatments for human scabies (benzyl benzoate; Hamid et al. 2015). Therefore, utilizing wild species for the development of natural therapies, including natural pesticides and topical medications, thereby replacing synthetic chemicals, will facilitate their cultivation and utilization in crop improvement programmes (as discussed in Chap. 6 of this book) and ensure their conservation as well.

Cinnamomum capparu-coronde has the highest eugenol content in its leaf and bark oil among the endemic *Cinnamomum* species and Kumarathilake (2012) has proposed this as a strong candidate for cultivation. Similarly, *Cinnamomum citriodorum* leaf and bark oil contain nearly 50% citronellol, which is a valuable component in cosmetics, flavourings and fragrances and as an insect repellent. The major chemical in *Cinnamomum dubium* leaf oil is citral-b and in bark oil it is linalool (Kumarathilake 2012); both compounds are used widely in food and cosmetic industries. The *Cinnamomum* species *C. zeylanicum* and *Cinnamomum sinharajaense* occur as sister groups in phylogenies with good support (Bootstrap 94%; Kumarathilake 2012). Furthermore, the oil profile of *C. sinharajaense* is closest to that of *C. zeylanicum*—the cultivated species—with high cinnamaldehyde content in bark oil and high eugenol in leaf oil (Ariyarathne et al. 2018; Kumarathilake 2012). Thus, these two species appear to be closely related and hybridization may be successful without intervening biotechnologies such as embryo rescue. Furthermore, all the *Cinnamomum* species of Sri Lankan origin have evolved at diploid level ($2x = 2n = 24$; Sritharan et al. 1993; Wijesinghe and Pathirana 2000) and hence there is a greater chance of success in inter-specific hybridization.

Considering medicinal value and the possibility of development of varieties with new chemical traits through inter-specific hybridization with cultivated cinnamon, these species need to be conserved ex situ and utilized. This will also ensure conservation of the species. Ex situ conservation of cinnamon genetic resources in Sri Lanka is a matter of urgency as the forest cover has been decreasing rapidly. The forest cover estimated by aerial photography in 1956 was 44% and had reduced to 26.6% by 1983. A forest map based on 1992 satellite remote sensing data supplemented by field survey revealed only 23.9% cover. The data indicate that the rate of deforestation has been 42,000 ha/year in the period between 1956 and 1983.

This has increased to 54,000 ha/year in the period from 1983 to 1992 (Yamamoto 2000). Of these forests, 70% are in the dry zone where there are no endemic *Cinnamomum* spp. Southwest lowland forests and the montane forests in the central hills are fragmented due to high population and plantation agriculture. Hence the threat to endemic cinnamon species in this zone is very high.

The changes in climate and edaphic factors enumerated in the previous sections call for the development of climate resilient cultivars. This requirement is further felt as Sri Lanka is changing cinnamon plantations from seedling populations to clonal since the release of two clonal varieties 'Sri Gemunu' and 'Sri Vijaya' (details of these in Chaps. 4 and 6 of this book). Furthermore, with less use of herbicides and other agrochemicals to meet quality standards in world markets, pest and disease resistant cultivars are also sought. In addition, there is an urgent need to breed cultivars high in identified bioactives for specialized markets and for this the endemic species with different chemical profiles will be of immense value. Finally, varieties suitable for mechanized peeling, such as those with straight stems and easily peelable bark, are required as machinery is already available (details in Chap. 9 of this book). The possible approaches to incorporate these traits are enumerated in Chap. 6.

1.5.3 Avenues for Expanding Cultivation and Increasing Production

Expansion of land under cinnamon in the areas most suited for cinnamon cultivation is hindered by land fragmentation due to population growth and the presence of existing coconut and rubber plantations. Considering that the ideal agro-ecological zones for cinnamon cultivation are occupied by coconut and rubber in Sri Lanka, the possibility of multiple cropping systems involving cinnamon has been investigated. For example, Liyanage et al. (1984) showed that among six coconut intercropping systems, intercropping with cinnamon produced the second highest yield in coconut after coffee. However, it was more economical to grow cinnamon than coffee because the cinnamon intercrop had less than half the fertilizer requirements of the coconut/coffee intercrop (Liyanage et al. 1984). Furthermore, fertilizer recommendations for intercropping cinnamon with coconut is available (Gunathilake and Manjula 2006). Further coconut yield benefits can be obtained by composting cinnamon leaves, which have high potassium—an important nutrient that is deficient in coconut plantations (Gunathilake and Manjula 2006). Similar positive outcomes have been recorded in rubber/cinnamon intercropping (Pathiratna 2006). Pathiratna and Perera (2006) have shown that high bark yield of cinnamon can be obtained by denser planting (17,500 bushes ha^{-1}) of cinnamon intercropped with rubber. Cinnamon as an intercrop over an 11-year study period has shown no competition with rubber (Seneviratne and Perera 2011). Thus, expansion of the cinnamon industry through intercropping can increase, stabilize and diversify farm income, particularly for small holders, and also can help in optimizing labour use. Growing cinnamon as an intercrop with rubber can facilitate organic matter build up, better soil retention in lands prone to erosion and an improved microclimate (Liyanage et al. 1984; Gunathilake and Manjula 2006; Mapa 1995; Seneviratne and Perera 2011; Pathiratna 2006).

Currently, there are 116,500 ha of rubber plantations in Sri Lanka (DCS 2006a) and we estimate that 67,300 ha of this area is suitable for intercropping with cinnamon. Coconut is cultivated in close to 400,000 ha in Sri Lanka (Pathiraja et al. 2015) and 15–20% of these coconut plantations are suitable for intercropping with cinnamon.

1.5.4 Value Addition, Product Diversification and Compliance with Quality Standards

Value addition to cinnamon to capture better revenue is another aspect that has been given priority in the last 20–30 years by successive Sri Lankan governments, and progress has been slow but steady. Even to this day the largest earnings from cinnamon come from export of the raw quills. Several reasons have been shown for this lack of pace in value addition: poor knowledge of the requirements of the international markets including market intelligence; poor understanding of quality standards of importing countries; lack of equipment, skills and training in value addition; and, of course, constraints in funds for upgrading processing facilities. Research in cinnamon was also lagging for many years until the Ministry of Primary Industries launched a research fund focused solely on the cinnamon industry that resulted in the formation of focused research groups within academia, research organizations and industry, which partly inspired this book. It is hoped that the momentum will continue so that the work begun will continue to bring benefit to the industry and result in increased revenue for the country.

1.5.5 International Cinnamon Trade

After black pepper, cinnamon is the most important spice traded in European and North American markets (Ravindran and Nirmal Babu 2004; Woehrlin et al. 2010). There are two types of cinnamon in the world market. The first is the inner bark of *C. zeylanicum* Blume (syn. *C. verum* Berchthold & Presl), which is called Ceylon cinnamon, Sri Lankan cinnamon or true cinnamon and is grown only in Sri Lanka, Madagascar and the Seychelles. The other type is generally known as cassia or cassia cinnamon, and is derived mainly from two other species of the genus *Cinnamomum: C. cassia* (L.) Berchthold & Presl. (Syn. *C. aromaticum* Nees) from China and Vietnam, which is called Chinese cassia, and *C. burmannii* C.G. Th.

Nees, which is called Indonesian or Java cassia (Dao 2004; WHO 1999). Due to a different processing technique, Vietnam product, also called Saigon cassia, had been thought to come from *Cinnamomum loureiroi* Nees. This is a rare species, and what is exported from Vietnam has now been confirmed as coming from *C. cassia* (Dao 2004).

Both bark and leaf oils of Indonesian cassia and Chinese cassia have very high cinnamaldehyde content, giving them a very strong spicy flavour. The other major constituent in cassia is benzaldehyde (Senanayake and Wijesekara 2004). The bark of these species has a dark brown-red colour (Fig. 1.2), has a rough texture and is thicker than bark of *C. zeylanicam* (true cinnamon), which is thin and papery and forms multiple layers when rolled (Fig. 1.2). Thus, true cinnamon is easier to grind than cassia and has a delicate and mildly sweet flavour, making it ideal for desserts as well as for spice mixtures. Other than cinnamaldehyde (50–75%), the main compounds (with 4–9% abundance) in *C. zeylanicum* bark oil are linalool, β-caryophyllene, 1,8-cineole and cinnamyl acetate (Paranagama et al. 2001; Senanayake and Wijesekara 2004).

True cinnamon from *C. zeylanicum* is much more valuable than cassia, as reflected in the export value of the products from the four main suppliers to the world market. From the values reported by different countries to the Food and Agriculture Organization of the United Nations (Table 1.3), it can be calculated that a metric ton of true cinnamon was sold at US$12,186 in 2017 whereas the value of a ton of cinnamon from Indonesia, China and Vietnam sold in the same year was US$3043, 1924 and 7812, respectively. Thus, it is a common practice to adulterate true cinnamon with cassia cinnamon, and also to sell cassia cinnamon under the name cinnamon, implying it is true cinnamon. Furthermore, in many parts of the USA, the largest importer of cinnamon in the world (Table 1.3), true cinnamon is unknown and both *C. cassia* and *C. burmannii* are simply labelled as cinnamon (Ravindran and Nirmal Babu 2004). More than 90% of 'cinnamon' imported into

Fig. 1.2 Appearance of cinnamon (*C. zeylanicum*) (left) and cassia (*C. cassia*) bark (right)

Table 1.3 Cinnamon production by the five major producers, and main importing and exporting countries (FAOSTAT 2018)

Countries	Production, MT		Exports, MT	Imports, MT	Export value (in million US$)
	2018	2017	2017	2017	2017
Indonesia	83,734	86,246	48,632		148.08
China	81,545	79,225	57,184	2224[a]	110.09[b]
Vietnam	29,053	29,388	13,184	1918	103.13
Sri Lanka	24,020	24,680	16,617	62	202.52
Madagascar	3113	2823	3313		5.84
United States			2096	34,185	10.60
India	80[c]	150[c]	1280	27,634	5.43
Bangladesh			8	7962	0.03
Netherlands			6176	7441	21.03
Mexico			470	7308	1.31
Saudi Arabia				5255	
United Arab Emirates			1346	5004	1.86
Pakistan				4986	
Germany			1631	4845	10.1
Iran (Islamic Republic of)			10	4739	0.01
Brazil			7	3104	0.03
Sudan				2647	
United Kingdom			342	2572	2.34
Canada			188	2520	0.68
Turkey			34	2281	0.019
Malaysia			226	2251	1.23
Japan				2239	
Republic of Korea			4	2177	
Thailand			34	1923	
Spain			569	1700	
Iraq				1450	
France			690	1431	5.5

[a]492 MT imported by mainland China, balance by Taiwan (1592 MT), Hong Kong (135 MT) and Macao (5 MT)
[b]Data from SBI (2019)
[c]Includes mainland China (109.74), Hong Kong (0.23) and Taiwan (0.11)

the USA is *C. burmannii* (Dinesh et al. 2015). It is the same situation in Canada, with the largest suppliers of cinnamon being Vietnam, Indonesia, India and China; Sri Lanka contributes just 3.3% of imports of cinnamon (Trendeconomy 2020). Due to the lower price, cassia from Indonesia, China and Vietnam is replacing true cinnamon from Sri Lanka in many European countries as well (Blahova and Svobodova 2012; Piyasiri and Wijeratne 2016; Dinesh et al. 2015).

In many European countries including the UK, the term 'cinnamon' is applied to *C. zeylanicum* and all other cinnamon products are labelled either 'cassia' or 'cassia cinnamon'. However, the irony is that in a world where quality products are sought, particularly in the food industry where there is increased interest in health foods, cinnamon products coming from different species and countries with altogether different chemical profiles are bunched together as 'cinnamon' in many other markets. Only specialized spice traders indicate the geographic origin of cinnamon in many markets. In contrast, beverages like tea and coffee are identified by their country of origin, the method of processing and the species (such as Arabica and Robusta coffee). For example, many coffee companies around the world advertise their coffee as sustainably produced and have Fair Trade certified products, but consumers have no clue where the cinnamon in their cappuccino comes from.

1.6 The Tale of Four C's: Cinnamon, Cassia and Coumarin Conundrum

1.6.1 Regulations Pertaining to Coumarin Due to Hepatotoxic and Carcinogenic Properties

Coumarins are compounds that contain a 1,2-benzopyrone skeleton. These are widespread in vegetables, spices, fruit and medicinal plants (Murray 2002). Coumarin (2*H*-chromen-2-one or 1,2-benzopyrone; Fig. 1.3), the simplest member of this class, is present in a variety of plants including some *Cinnamomum* species. It was first isolated from tonka beans (*Dipteryx odorata* (Aubl.) Willd. Fabaceae) in 1822 and chemically synthesized in 1868 (Bruneton and Hatton 1999) and was a popular flavouring agent in the food industry due to its pleasant spicy odour of fresh hay, woodruff or vanilla (Abraham et al. 2010). After about a century of use, it was revealed that coumarin has hepatotoxic effects in laboratory animals (Hazleton et al. 1956). As a result, its use as a flavouring agent was banned in the USA in 1954 by the U.S. Food & Drug Administration (FDA), including the use of tonka beans and its extract as a food (FDA 1977). Later, in the 1960s coumarin was shown to have carcinogenic properties in long-term animal studies. In addition to adenomas and carcinomas of the liver and bile ducts and adenomas of the kidneys in rats, adenomas and carcinomas in the lungs and adenomas in the liver of mice were reported (EFSA 2004; Lake 1999). These observations resulted in the European Food Safety Authority (EFSA) setting a limit in 1988 of 2 mg/kg coumarin in food derived from natural spices and herbs (Abraham et al. 2010). However, genotoxicity of coumarin was not established (Lake 1999). It was shown that coumarin does not bind to DNA covalently. With a genotoxic mode of action of coumarin discounted, based on animal hepatotoxicity data, a tolerable daily intake (TDI) value of ≤0.1 mg/kg body weight was determined in 2004 by the EFSA (EFSA 2004). This value was confirmed by the German Federal Institute for Risk Assessment (BfR) in 2006, which also considered human data from administration of coumarin as a drug (Abraham et al. 2010). This was made possible because it has been shown that

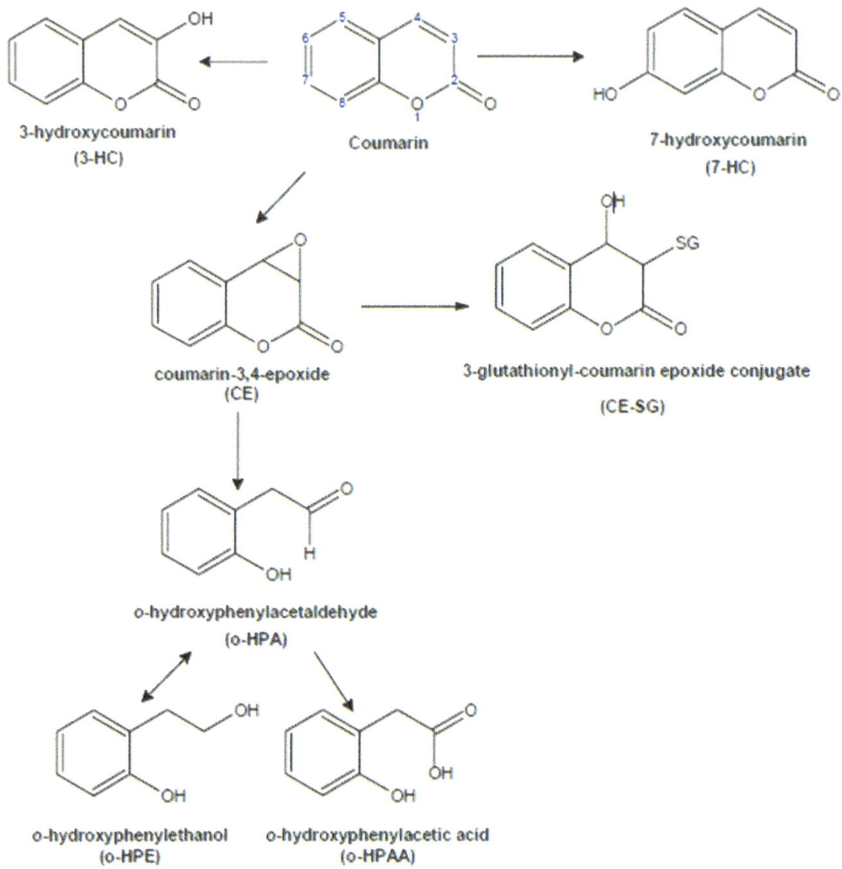

Fig. 1.3 Two main metabolic pathways of coumarin (1,2-benzopyrone) metabolism. (Adapted from EFSA (2004)). 7-Hydroxycoumarin is the predominant metabolite in primates. In rodents and some humans, the metabolism of the lactone ring of coumarin leads to the intermediate coumarin 3,4-coumarin epoxide (CE). Spontaneous loss of CO_2 or conjugation of CE with glutathione produces *o*-hydroxyphenylacetaldehyde (*o*-HPA), a hepatotoxic metabolite

coumarin absorption from powder of cassia cinnamon is similar to that of isolated coumarin used in drugs (Abraham et al. 2011).

It is now known that there are significant differences in the metabolism of coumarin among species, particularly between primates and rodents. Human clinical data indicated that the majority of patients were more tolerant to coumarin-induced hepatotoxicity than rodent models. High-performance liquid chromatography (HPLC) studies using liver microsomes demonstrated that in humans the major detoxifying metabolite of coumarin is 7-hydroxycoumarin (Fig. 1.3), mediated by the enzyme CYP2A6, whereas in rats coumarin is metabolized to *o*-hydroxyphenylacetaldehyde (*o*-HPA), the hepatotoxic intermediate of epoxidation of coumarin (Rietjens et al. 2010; Fig. 1.3). However, a subpopulation of humans is much more susceptible to coumarin-induced hepatotoxicity than rodents (EFSA 2008; Wang et al. 2013). It is postulated that these individuals are polymorphic for CYP2A6, leading to an increased formation of 3,4-coumarin epoxide and *o*-HPA as demonstrated by Hadidi et al. (1997) in an individual homozygous for an inactivating CYP2A6*2 allele. However, Rietjens et al. (2010) argue that a higher proportion of the general population has elevated toxic metabolites of coumarin in liver than are homozygous for the defective CYP2A6 allele, and this suggests that hepatotoxicity may be due to other compounding factors. For example, under normal conditions the transcription factor (TF) Nrf2, which regulates antioxidant and cytoprotective genes, is sequestered in the cytoplasm by Kelch-like ECH-associated protein 1 (Keap1) that stimulates its proteosomal degradation (Itoh et al. 1997). During oxidative stress Nrf2 dissociates from Keap1 and accumulates in nucleus as it is a major TF, regulating oxidative stress-related genes (Hakooz and Hamdan 2007). Yokota et al. (2011) showed that human CYP2A6 is regulated via Nrf2, suggesting its induction as a result of oxidative stress, partly explaining the high amount of *o*-HPA found in individuals with homozygous CYP2A6 allele. An increase in the 3-hydroxycoumarin is associated with an increased production of the significant cytotoxic product *o*-HPA, suggesting that a shunting of coumarin metabolism away from 7-hydroxylation is the cause of the toxicity. Hence, poor CYP2A6 metabolizers are more likely to metabolize coumarin via the cytotoxic pathway. As a result of reports of patients developing signs of hepatotoxicity with the use of coumarin, the drug was discontinued in France and Australia (Casley-Smith and Casley-Smith 1995; Wang et al. 2013). After reviewing accumulated human data and the possibility that mechanisms other than CYP2A6 polymorphism can play a role in the toxicity of coumarin in humans, EFSA maintained the TDI recommendation of 0.1 mg/kg body weight for coumarin in its latest review (EFSA 2008). Based on even more recent studies conducted by BfR (Abraham et al. 2011), it reaffirmed the TDI of ≤0.1 mg/kg body weight in 2012 (BfR 2012b).

Codex Alimentarius, which provides internationally recognized standards and general guidelines for natural flavourings, recommended a maximum level of coumarin of 2 mg/kg (exceptions were alcoholic beverages and caramels—10 mg/kg maximum) in 1985 (Sproll et al. 2008). These standards were subsequently introduced into European law in 1988. It is therefore of considerable importance to assess market products for coumarin content. The final part of this section deals with coumarin content in *Cinnamomum* species and foods containing cinnamon in retail markets of different countries.

1.6.2 Coumarin Content in Cinnamon, Cassia and Food Products

Although cinnamaldehyde is the main constituent in all the commercially important *Cinnamomum* species, they differ in coumarin content. According to BfR, 'cassia cinnamon on average contains approximately 3000 mg (3 g) of coumarin per kg of cinnamon. However, the highest measured levels were as high as 10000 mg (10 g) of coumarin per kg of cassia cinnamon' (BfR 2012a). BfR in the same document suggests 'Consumers who frequently use large amounts of cinnamon as spice in their home cooking, for example for rice pudding with sugar and cinnamon, should use Ceylon cinnamon which contains low levels of coumarin'. BfR maintains a TDI of 0.1 mg/kg body weight limit of intake in its updated document (BfR 2012a). As cinnamon is used in large quantities in various food products, it is important that countries enact legislation for manufacturers to label the product, so that consumers can decide on the type of cinnamon they consume. However, to this date, there are no guidelines by the FDA for the use of cinnamon in food, except for the prohibition of direct addition of coumarin, tonka bean or its extract, instituted back in 1954 (FDA 1977). The Canadian Food Inspection Agency under its Food Safety Action Plan conducted a survey of coumarin in cinnamon products and cinnamon-containing foods in the retail market in 2010/11 in 11 cities. In the 193 samples analysed, coumarin was detected in 98% of the samples. The highest concentrations of coumarin were observed in ground cinnamon (7816 mg/kg), cinnamon bark (6823 mg/kg) and followed by spice mixes (2014 mg/kg). Baby food samples recorded 14.9 mg/kg of coumarin. As the Canadian authorities were looking for signs of added coumarin, they concluded the samples have not been adulterated with coumarin and hence 'do not pose an unacceptable concern to human health' (CFIA 2018). It is ironic that most of the cassia produced in Indonesia and China is imported to North America either directly or through intermediate markets, and true cinnamon from Sri Lanka with undetectable or very low concentrations of coumarin is virtually unknown there.

In contrast to North America, the European Union and individual countries within the EU are actively monitoring coumarin content in food products and cinnamon in trade. Regulatory bodies in the EU emphasize that coumarin is well absorbed by the body from a plant matrix of cinnamon and the TDI value, which was derived from data with isolated coumarin, can therefore also be applied to coumarin in cinnamon-containing foods (BfR 2012b; EFSA 2008). Yet, in many European countries marketed cinnamon products need not be identified as derived from cassia or true cinnamon. For example, in Czech supermarkets cassia is sold under the name of cinnamon (Blahova and Svobodova 2012). In Germany, after analyses revealed 22 mg/kg of coumarin in popular star cookies (BfR 2006), more analytical work was commissioned (Abraham et al. 2010, 2011; Sproll et al. 2008).

Because of the more serious consideration of coumarin effects on human health by the European Union and countries in Europe compared to other parts of the world, the majority of literature available for coumarin content in cinnamon and food sold in retail is from this region. It is ideal to compare analytical results from the same lab in the same experiment for comparative purposes as there is much variation in methods of analysis between labs.

Wang et al. (2013) recently analysed authentic samples of four *Cinnamomum* species for coumarin content using a fingerprinting ultra performance liquid chromatography–UV/mass spectrometry (UPLC–UV/MS) method and compared results with those from commercial sources in the US market (Table 1.4). As several

Table 1.4 Coumarin content in four *Cinnamomum* species from authentic and commercial samples analysed by fingerprinting UPLC-UV/MS method, adapted from Wang et al. (2013)

Cinnamomum species	Source	Coumarin (mg/g)
C. verum	Authentic sample	0.017
C. verum	Bark/Sri Lanka commercial	0.005–0.025 ($n = 15$)
C. verum	Bark/US commercial	0.013
C. burmannii	Authentic sample	2.14
C. burmannii	Bark/US commercial	2.37–9.30 ($n = 7$)
C. loureiroi	Authentic sample	6.97
C. loureiroi	Bark/Vietnam commercial	1.06
C. cassia	Authentic sample	0.310
C. cassia	Bark/China commercial	0.145
C. cassia	Bark/US commercial	0.085–0.262 ($n = 2$)
Cinnamomum spp.	Bark/US commercial	2.00–5.79 ($n = 7$)
Cinnamomum spp.	Powder/US commercial	2.06–6.16 ($n = 6$)

authors have reported, the variation in coumarin content between samples within a species was high. Nevertheless, the sample of Ceylon cinnamon with the highest coumarin content (0.025 mg/g) was lower by 3.4-fold than the lowest coumarin content recorded in any of the samples from the other three species (0.085 mg/g in one of the two *C. cassia* commercial samples). Indonesian cassia *C. burmannii* has the largest market share in the USA, and commercial bark samples of this species in the US market recorded up to 9.3 mg/g coumarin, 372 times higher than the Ceylon cinnamon sample that recorded the highest coumarin content among Ceylon cinnamon. As there are no restrictions or even guidelines for sale and consumption of Indonesian and Chinese cassia in the USA, consumers there are unaware of the potential harm caused if they are regular users of cinnamon in their diet. Wang et al. (2013) noted that the cinnamaldehyde to coumarin ratio in Indonesian cassia and Ceylon cinnamon is 8.5 and 805, respectively, and is a good marker to differentiate the two species.

Blahova and Svobodova (2012) using an HPLC system with a UV detector analysed 60 ground samples of cinnamon from 12 brands (five samples for each brand) available in Czech supermarkets for coumarin content. None of the packets indicated whether the product was cassia or cinnamon (Blahova and Svobodova 2012). The authors imported a *C. verum* sample directly from a plantation in Sri Lanka as a control. The mean coumarin content of each brand ranged from 2650 to 7017 mg/kg, whereas in the Sri Lankan cinnamon sample coumarin was below the limit of detection. The authors concluded that all the supermarkets were selling cassia and not true cinnamon.

To understand the basis of variation in coumarin content within species, Woehrlin et al. (2010) analysed bark samples from different parts of Indonesian cassia plant and from plants of different age using HPLC. They also studied cinnamon bark and powder from the German retail market. Coumarin concentrations of powdered cassia cinnamon varied between 1740 and 7670 mg/kg ($n = 40$). A much greater variation was found in the 29 cassia bark samples, from limit of detection to 9900 mg/kg. The mean value of coumarin for powders (4020 mg/kg) was not significantly different from the mean value for bark (3250 mg/kg). Again, as in previously described two studies, coumarin concentrations in Ceylon cinnamon varied between lower than detection level and 297 mg/kg for powder and lower than detection level and 486 mg/kg for bark (Woehrlin et al. 2010). With regard to the variation of coumarin concentration within Indonesian cassia trees, Woehrlin et al. (2010) found that bark from the middle part of the Indonesian cassia tree has the highest concentration at 6180 mg/kg. Furthermore, both younger and older trees recorded high levels of coumarin in their bark, in the range of 2240–6660 mg/kg (Woehrlin et al. 2010). Bark samples of true cinnamon and cassia cinnamon are easy to differentiate by their folding characteristics, thickness and colour (Fig. 1.2). However, once ground they both look similar. To identify the origin of cinnamon marketed in powder form, Woehrlin et al. (2010) proposed that any cinnamon powder sample with over 1700 mg/kg of coumarin can be regarded as of cassia origin.

As already mentioned, the BfR in Germany conducts its own research on food safety. In 170 cinnamon samples sourced from the German retail market, the mean coumarin content was found to be 2680 mg/kg (Abraham et al. 2010). Several other cinnamon-containing food products also had appreciable quantities of coumarin. Among these, cinnamon tea (231.3 mg/kg; n = 16), cinnamon star cookies called *zimtsterne* (37.7 mg/kg, n = 218), almond cookies called *spekulatius* (16.2 mg/kg, n = 40), gingerbread cake called *lebkuchen* (10.3 mg/kg, n = 80), desserts with cinnamon (10.2 mg/kg, n = 29) and chocolate with cinnamon (9.4 mg/kg, n = 25) are noteworthy. Based on their results, Abraham et al. (2010) estimated that a 4-year-old child weighing 15 kg will be ingesting 0.13 mg/kg body weight daily if he/she is consuming three cinnamon star cookies weighing about 6 g. This value is already above the TDI, and the authors noted that there may be other coumarin-containing foods being consumed simultaneously. Considering the high amount of coumarin in some food in the German market, the authors from BfR entrusted an independent company specializing in market surveys to carry out a survey involving 1012 persons over the age of 14 years (Abraham et al. 2010). Coumarin intake of individuals was estimated by adding up the intake data from the ten different foods. It was estimated that the heaviest consumers' intake exceeds 35 mg of coumarin per week (maximum 47.5 mg/week), thus exceeding the 0.1 mg/kg body weight TDI. The contribution of star cookies (one of the highest coumarin contents among the foods) to the total intake was minimal (5.2%) and 80% of the interviewees had not eaten these cookies as they had seen warnings in the media against the consumption of too much 'cinnamon' and 'cinnamon products' (Abraham et al. 2010).

Sproll et al. (2008) developed a simple HPLC method with diode array detection (DAD) for the analysis of coumarin and used it to determine the concentration of coumarin in *Cinnamomum* species, flavourings and bakery products submitted to the Chemical and Veterinary Investigation Office in Karlsruhe, Germany. They did not detect coumarin in any of the five samples of Ceylon (Sri Lankan) cinnamon analysed. The mean coumarin content in cassia was 3612 mg/kg with a range of

2880–4820 mg/kg (n = 5; Sproll et al. 2008). Appreciable amounts of coumarin were found in cereals and bakery products. Germany's most popular Christmas star cookies had an average of 25 mg/kg of coumarin, more than 12 times over the *Codex Alimentarius* recommended maximum level of 2 mg/kg. Some of these cookies had 88 mg/kg of coumarin. By their calculations, the TDI of coumarin by eating these star cookies alone was in agreement with Abraham et al. (2010) as three to four star cookies for a child and ten for an adult would be sufficient to reach 0.1 mg/kg body weight. Analysing their results, Sproll et al. (2008) came to the conclusion that flavouring a food with 0.1% (w/w) of cassia containing 3000 mg/kg would lead to a concentration above the limit of 2 mg/kg in the food. It should be noted that the value of 3000 mg/kg used is less than even the average of 3612 mg/kg coumarin in cassia cinnamon in the German retail market in the study. Furthermore, as there are no maximum levels for coumarin in flavourings, even the products with unusually high coumarin concentrations of up to 8790 mg/kg found in their study would not be rejected by food safety authorities. The authors argued that **'because of the different flavouring characteristics, including the differences in coumarin content, a change in food policy that demands the different species, cinnamon and cassia, be labelled with their specific names appears to be required'** (Pg. 466) (Sproll et al. 2008). As for now, the German word *Zimt* includes both cinnamon and cassia (Abraham et al. 2010; Sproll et al. 2008).

In an attempt to improve extraction of coumarin and other compounds in cinnamon for analysis, Miller et al. (1995) demonstrated that solvent-assisted supercritical fluid extraction is superior to other methods tested. In 24 cinnamon and cassia samples obtained from Sri Lanka, Korea, the UK and the USA, they found coumarin to be a major compound in cassia (0.7–12.2 mg/kg), whereas it is a trace compound in true cinnamon (0.00–0.19 mg/kg), and concluded that true cinnamon is easily distinguished from cassia by the presence of eugenol, absence of 6-cadinine, much lower amounts of coumarin and larger amounts of benzyl benzoate (Miller et al. 1995).

Lungarini et al. (2008) performed HPLC-DAD analysis of 14 samples of cinnamon quills, 20 of cinnamon powder and 50 samples of cinnamon-containing foods from the Italian market for coumarin and cinnamaldehyde contents. The quill samples were segregated into Ceylon cinnamon or cassia by the large difference in coumarin content and the corresponding appearance. With regard to ground cinnamon samples tested, 70% were derived from cassia with high coumarin content and hence their use for flavouring in large quantities (cheaper in the market) can pose a health risk for the consumers, above all for children (Lungarini et al. 2008). Furthermore, based on the average daily consumption of biscuits, cakes, deserts, chocolate and confectionery in Italy and the content of coumarin found in the study, the authors calculated that the intake of coumarin was 0.21 mg/kg body weight for a 6-year-old child weighing 20 kg, twice the TDI value for coumarin (0.1 mg/kg body weight) established by EFSA (Lungarini et al. 2008).

In a government inspection programme, the Danish Veterinary and Food Administration (DVFA) investigated food products in the Danish market using a UPLC-photodiode array analysis for coumarin content. In the fine bakery category,

49% of the samples exceeded the EU limits for coumarin, with one sample exceeding it by a factor of three (Ballin and Sørensen 2014). Considering that small bakery owners are unaware of how much cassia can be added to their products without exceeding the EU limit for coumarin, the authors prepared a practical guide for bakeries. They also suggested a strategy of replacing cassia with true (Ceylon) cinnamon (Ballin and Sørensen 2014) as suggested by several other authors.

In another European study, Fotland et al. (2012) re-assessed coumarin toxicity using a benchmark dose approach and revised the TDI for coumarin at 0.07 mg/kg body weight/day. Based on analyses of coumarin in Norwegian foods, they showed that children eating oatmeal porridge sprinkled with cinnamon a few days a week could have an intake of 1.63 mg/kg body weight/day. With other foods, these children will exceed the TDI by several folds and the authors warned that this could cause serious health issues. Similarly, adults consuming cinnamon tea and health supplements may exceed the TDI by 7–20 folds (Fotland et al. 2012). These and many other publications from different groups have shown that the high content of coumarin in cassia is a serious issue in food and the replacement of true cinnamon with cheap cassia is a dangerous trend that needs to be addressed by food and health authorities.

1.7 Way Forward for Sri Lankan Cinnamon Industry

1.7.1 True Ceylon Cinnamon Is Like No Other Spice; Unique to Sri Lanka

Cinnamon, which is indigenous to Sri Lanka, possesses superior organoleptic, medicinal and anti-microbial properties and has found wide applications in a range of industries, including food and beverage, pharmaceutical, nutraceutical, cosmeceutical, liqueur, perfumery and oral care. Sri Lanka practically commands a monopoly of true cinnamon in the global market, accounting for over 90% of its trade (FAOSTAT 2018; SLEDB 2020). However, given the economic potential of the crop, growth of the local cinnamon industry in Sri Lanka has been slow, with the present area under cinnamon totalling only around 33,000 ha (FAOSTAT 2018). Cinnamon is mainly grown on small holdings of about 0.5 ha, with over 70,000 small holders engaged in its cultivation. Only 5–10% of land for cinnamon cultivation includes larger plantations in the range of 5–20 ha.

Currently, cinnamon produces an average bark yield of around 700 kg/ha/year, although the potential yield is about 1500 kg/ha/year under good management with continuous harvesting, that is, harvesting of stems as and when they mature rather than just harvesting twice a year. Conventional processing technology accounts for around 60% of the cost of production, which has not only made the crop rather unprofitable to the grower, but also much more expensive than its substitutes in the global market. Sri Lanka annually produces around 17,000 tons of cinnamon bark of which nearly 90% is exported mainly in the form of quills with hardly any value addition. A few countries in South and North America are the main importers with Mexico accounting for 70% and such market concentration is undesirable. Moreover, most of the cinnamon plantations in Sri Lanka were established from seeds and, as cinnamon is a cross-pollinating species, there is considerable variation in yield and quality of cinnamon between plants. This poses a challenge in producing particularly drugs based on pharmacological efficacy of cinnamon as consistency and reproducibility cannot be guaranteed. In addition, cassia—a cheap substitute of cinnamon—is produced in large quantities by countries such as China, Vietnam and Indonesia. Cassia contains appreciable amounts of coumarin, which is hepatotoxic. However, in the global market, no distinction is hitherto made between cassia and cinnamon despite a hazardous substance being present in the former, and both go as 'cinnamon' in the global market where the proportion of cassia:cinnamon is around 9:1. Since true cinnamon is about three to four times more expensive than cassia and as no trade restrictions have thus far been imposed on cassia even by the EU and the USA, despite the presence of appreciable quantities of coumarin, it is a formidable challenge to enhance the market share of true cinnamon in the global 'cinnamon' trade unless strategic interventions are made through appropriate awareness and education programmes with the support of foreign missions of Sri Lanka.

Today, markets are increasingly globalized, sophisticated and dynamic; consumers are increasingly becoming health-conscious and there is growing interest in natural food additives and flavours. Therefore, consumers are prepared to pay a premium price for natural spices and healthy and wholesome food and beverages (Shirai 2010; Thapa et al. 2020; Papacharalampous 2017). Foods fortified with true cinnamon because of its medicinal, nutraceutical and anti-microbial properties with undetectable coumarin levels are no exception.

Sri Lanka possesses core competencies and competitive advantage over other countries that produce cinnamon and its substitutes due to its outstandingly rich ecological, edaphic and genetic resources. This includes several wild relatives of cinnamon, and a competent and dedicated manpower base in the country that has sustained this industry over several centuries amidst manifold obstacles and formidable challenges, particularly during the British rule.

In view of the above, cinnamon, a unique and versatile spice, has an immense potential in the global market. However, intervention by way of genetic, agronomic, technological, advisory, quality assurance and accreditation, and marketing reinforced by a consistent, coherent and rational policy package is needed to harness the economic potential of this crop.

1.7.2 Development of Promising Clonal Cultivars

Development of improved genotypes with high yield, quality and resistance to biotic and abiotic stresses is important, and requires the identification and location of relevant genetics. In this regard, the genetic repository established at the National

Cinnamon Research and Training Centre of the Department of Export Agriculture, with over 550 accessions, including landraces, will prove invaluable. Cinnamon is indigenous to Sri Lanka, and this being the only repository in the country makes it a global treasure. Hence, this rich gene bank, DNA library and biorepository should, with the support of a dedicated fund, be enhanced, systematized and comprehensively characterized in line with international standards to realize the potential benefits. Future research on chemo-profiling and identification of new flavours, fragrances, bioactive compounds and nutraceuticals is important in producing functional food with bio-fortification and new perfumes with unique fragrances. Moreover 'Omic' technologies that are primarily aimed at the universal detection of genes (genomics), mRNA (transcriptomics), proteins (proteomics) and metabolites (metabolomics) offer new research directions.

Alongside, it is apposite to develop a repository of indigenous knowledge (conventional wisdom) related to various aspects of production and processing of cinnamon in the older generation before it is lost, which is also important in developing geographical indications.

1.7.3 Production of Elite Planting Material

Production of quality and disease-free planting material is a prerequisite for any production programme to be successful. In this connection, establishment and maintenance of accredited and certified nurseries and development of protocols for micro-propagation are important. Cinnamon is difficult to multiply by micro-propagation. However, as discussed in Chap. 7 of this book, Subasinghe et al. have developed a protocol for micro-propagation. Recently two promising varieties of cinnamon, namely 'Sri Gemunu' and 'Sri Vijaya', have been developed and released for cultivation by the Department of Export Agriculture in Sri Lanka. Therefore, micro-propagation techniques should be used for rapid multiplication of these promising varieties, as well as those in the pipeline for release in future.

1.7.4 Expansion of Cinnamon Cultivation

Though Sri Lanka accounts for around 90% of the global trade of true cinnamon, its annual production is only around 17,000 MT, which is not sufficient to enhance its global footprint. The present area under cinnamon cultivation is only around 33,000 ha whereas the area under paddy is over 750,000 ha and under plantation crops—tea, rubber and coconut combined—exceeds 700,000 ha. Therefore, there are prospects for establishing cinnamon as an intercrop under coconut, rubber and in abandoned agricultural fields. Hence, steps should be taken to expand the cultivation of cinnamon into climatically and edaphically suitable areas to meet the growing global demand and consolidate the position of Sri Lanka as the principal supplier of premium quality cinnamon in the world. In this connection, establishment and management of cinnamon in a matched crop-climate-land resources scenario through land suitability analysis based on the Ecocrop model (FAO 2012) is an urgent and integral need. Such suitability assessment is important in identifying potential growing areas, formulating appropriate management decisions, recommending ameliorative measures and adopting best practices for increasing productivity.

1.7.5 Agro-Technology

Present-day crop production should be eco-friendly, resource-efficient, tech-savvy and knowledge-based in order to be globally competitive and sustainable. Besides, the mode of production the world over is being transformed from linear to circular, producing zero waste. This demands confluence of technologies, particularly biotech, nano-tech, electronics and information and communications technology (ICT), and consequently the technological landscape of agriculture is rapidly changing.

Development of cost-effective microbial technologies for increased assimilation of nitrogen and phosphorous to improve crop production with reduced fertilizer usage is important to promote eco-friendly farming. Use of advanced drones to collect real-time data on crop health, abiotic stresses, soil analysis and weather patterns is useful to optimize the use of fertilizer, pesticides, water etc., thereby enhancing resource use efficiency and reducing ecological footprint.

1.7.6 Land Reforms and Digital Intervention for Improved Productivity and Profitability

Despite bright prospects for the cinnamon industry, it is beset with a myriad of issues and challenges. The scattered nature and small size of holdings, lack of timely access to information such as weather, market and service providers, low productivity, inefficient processing technologies, unsatisfactory extension service, inequitable distribution of profits among the key stakeholders and lack of linkage to the global value chain are key issues affecting the growth of the Sri Lankan cinnamon industry. While land reforms are urgently needed to consolidate holdings to overcome land tenure issues, promotion of investment to derive economies of scale is also important. ICT can offer valuable inputs as well as solutions to many issues and problems encountered from establishment, management, harvesting, processing, value addition, branding and marketing of cinnamon globally while harnessing opportunities. ICT should be leveraged to enhance productivity, profitability and competitiveness of the cinnamon industry in a globalized environment as is done in developed as well as in many developing countries.

1.7.7 Studies on Factors Affecting Chemical Profile and Organoleptic Properties

Cinnamon has unique chemical, medicinal and organoleptic properties with manifold industrial applications and its organoleptic properties such as flavour, aroma and taste are affected by the duration and conditions of storage, as shown for example in encapsulated Indonesian cassia oil (Pratiwi et al. 2016). However, hardly any studies have been conducted to ascertain the effect of duration and conditions of storage on the time course of variation in chemical profile and organoleptic properties of true cinnamon bark or leaves. Moreover, organoleptic properties of cinnamon can be affected by genotype, maturity of stem, physiological state of the plant (i.e. flushing, flowering, seed ripening) and soil and climatic conditions. As described in Chap. 7 of this book, climatic factors have a marked effect on quality and flavour development in tea, coffee and grapes. Studies carried out to determine the effect of climatic, soil and plant factors on the organoleptic properties of crops are few and far between. Such studies on cinnamon will help identify if any flavour season exists as with tea in Yunnan Province in China (Larson 2015) and Dimbula in Nuwara Eliya, Sri Lanka (SLTB 2014).

1.7.8 Development of Cost-Effective and Time-Efficient Processing Technologies

Processing technology for cinnamon introduced by the Dutch about 350 years ago is still being used, which is time-consuming and labour-intensive, and accounts for around 60% of the cost of production. It demands skilled labour, and every year some plantations are not harvested regularly due to a dearth of experienced peelers resulting in a considerable loss of yield. Moreover, the traditional processing technology takes more than 2 months to complete the process during which there could be loss of organoleptic and other properties. To overcome these problems, a cinnamon rubbing machine RUWEKA-CG has been introduced to mechanize one of the most time-consuming operations in cinnamon processing, including a bench suitable for all the processing operations (details in Chap. 9 of this book). It is being introduced to processing centres. In contrast to true cinnamon, cassia is marketed in the form of powder or bark pieces. Therefore, it is important to develop a cost-effective and time-efficient processing technology for cinnamon that will appreciably reduce the cost of production, and consequently the price difference between cinnamon and cassia, increasing the demand for the former.

1.7.9 Value Addition and New Product Development

In view of the medicinal properties of cinnamon as described in Chaps. 11 and 12, novel products of cinnamon in the form of medicines, therapeutics and nutraceuticals can be developed. Additionally, its fragrant, volatile oils can be used to produce green perfumery and air-fresheners. These are in addition to applications in food and beverage, liqueur, oral care and cosmeceutical industries. At present, cinnamon is mainly exported in its primary form and value addition takes place in importing countries, meaning that Sri Lanka loses a great deal of potential export revenue. Therefore, policy intervention to develop robust public–private partnerships to restrict export of cinnamon in primary form and to promote value creation is of crucial importance.

1.7.10 Compliance with Stringent Requirements of Importing Countries

With the imposition of stringent laws and regulations by importing countries for food products, particularly in Europe and North America, which jointly account for over 50% of Sri Lanka's cinnamon export market, it is of utmost importance to build capacity in terms of compliance infrastructure and competencies for international trade so that all exports, especially food commodities, conform to the stringent requirements (including traceability in certain instances) and guidelines laid down by importing countries. This will require compliance with good agricultural practices (GAP), good manufacturing practices (GMP), good hygiene practices (GHP), safe quality food (SQF), hazard analysis and critical control points (HACCP) and ISO 22000. This demands new investment from public as well as private sector institutions, including producers, processors, accreditation and certification authorities and importers. Return or rejection of a consignment for non-compliance will affect not only the credibility of the exporting company, but also the image of the country of origin. Therefore, it is of utmost importance to ensure that goods failing to conform to the specific requirements of the buyer do not leave the port, and there is stringent enforcement to ensure this. Biosensors and DNA bar coding technologies can be developed to detect adulteration of true cinnamon with cheap and inferior substitutes and to authenticate genuine products.

1.7.11 Establishment of a Global Market Research Facility

A Global Market Research Centre is a prerequisite for any country interested in promoting its footprint in the fiercely competitive global market. Research conducted in Sri Lanka for export promotion is woefully inadequate and there is no institutional mechanism to harness Sri Lankan foreign missions to this end. Today consumers are health-conscious, their food habits are changing and socio-economic standards are improving. Consequently, global markets are dynamic and sophisticated and are in a state of constant flux with some developed markets becoming unstable while new markets, both mainstream and niche, emerge with new competitors entering. Therefore, establishment of a Global Market Research Facility with a clear mandate and the requisite powers and resources is a high priority to enhance global market penetration of Sri Lankan exports, including cinnamon. Such a facility could collect, collate, record, analyse, interpret and report to relevant institutions and companies to support decision-making and make timely intervention to contend with challenges and to consolidate the existing markets while harnessing emerging opportunities. Creating a brand value for 'Cinnamon of Origin' with geographical indications followed by a vigorous and aggressive marketing drive based on its unique intrinsic qualities will enhance global market penetration.

In summary, Ceylon cinnamon is a unique plant that has oil in its leaves, bark and roots, but the chemical composition of oil from each plant part is completely different. Both bark and leaves contain essential oils with cinnamaldehyde being predominant in the bark and eugenol in leaves; root-bark oil contains camphor. These substances have wide applications in industry. Cinnamon is a versatile spice that can be added to many food items such as confectionaries, beverages, salads, baking, soups and sauces. Additionally, its bark oil possesses antiseptic properties that can improve the hygiene and keeping quality of food products, one of its first uses in Europe. Cinnamon thus has promise as an agro-industrial crop with immense potential. However, in order to harness this potential, a range of interventions by way of genetic, agronomic, technological, institutional, quality assurance and accreditation, marketing and policy is needed to transform cinnamon into a multi-billion-dollar industry for Sri Lanka.

Acknowledgements The authors thank Ms Sheron Weerasuriya of the National Cinnamon Research and Training Center, Department of Export Agriculture, Thihagoda, Sri Lanka, for providing some of the information contained in Sect. 1.5.1. We also acknowledge Bruce Smallfield, Cath Kingston and Ed Morgan for their valuable comments on the manuscript.

References

Abdel-Maksouda G, El-Amin A (2011) A review on the materials used during the mummification processes in ancient Egypt. Mediter Archaeol Archaeom 11:129–150

Abraham K, Wohrlin F, Lindtner O, Heinemeyer G, Lampen A (2010) Toxicology and risk assessment of coumarin: focus on human data. Mol Nutr Food Res 54:228–239

Abraham K, Pfister M, Wöhrlin F, Lampen A (2011) Relative bioavailability of coumarin from cinnamon and cinnamon-containing foods compared to isolated coumarin: a four-way crossover study in human volunteers. Mol Nutr Food Res 55:644–653

Ariyarathne HBMA, Weerasuriya SN, Senarath WTPSK (2018) Comparison of morphological and chemical characteristics of two selected accessions and six wild species of genus *Cinnamomum* Schaeff. Sri Lankan J Biol 3:11–23

Ashton PS, Gunatilleke CVS (1987) New light on the plant geography of Ceylon I. Historical plant geography. J Biogeogr 14:249–285

Ashton MS, Gunatilleke CVS, Singhakumara BMP, Gunatilleke IAUN (2001) Restoration pathways for rain forest in southwest Sri Lanka: a review of concepts and models. For Ecol Manag 154:409–430

Ballin NZ, Sørensen AT (2014) Coumarin content in cinnamon containing food products on the Danish market. Food Control 38:198–203

Bandarage A (2019) The establishment and consolidation of the plantation economy in Sri Lanka. Bull Concerned Asian Sch 14:2–22

BfR (2006) High daily intakes of cinnamon: health risk cannot be ruled out, BfR Health Assessment No. 044/2006, 18 August 2006. German Federal Institute for Risk Assessment. Available from: https://mobil.bfr.bund.de/cm/349/high_daily_intakes_of_cinnamon_health_risk_cannot_be_ruled_out.pdf. Accessed 3 May 2020

BfR (2012a) FAQ on coumarin in cinnamon and other foods. German Federal Institute for Risk Assessment (BfR). Available from: https://www.bfr.bund.de/cm/349/faq-on-coumarin-in-cinnamon-and-other-foods.pdf. Accessed 3 May 2020

BfR (2012b) New insights into coumarin contained in cinnamon. Federal Institute for Risk Assessment. Available from: https://www.bfr.bund.de/cm/349/new-insights-into-coumarin-contained-in-cinnamon.pdf. Accessed 3 May 2012

Biedermann Z, Strathern A (2017) Sri Lanka at the crossroads of history. UCL Press, London. https://doi.org/10.14324/111.9781911307822

Biswas S, Pawar SS (2006) Phylogenetic tests of distribution patterns in South Asia: towards an integrative approach. J Biosci 31:95–113

Blahova J, Svobodova Z (2012) Assessment of coumarin levels in ground cinnamon available in the Czech retail market. Sci World J 2012:263851

BM (2015) Brilliant Maps: the 26 largest islands in the world compared. Available from: https://brilliantmaps.com/largest-islands/. Accessed 13 Apr 2020

Braun L (2006) Cinnamon. J Complement Med 5:67–70

Briggs JC (2000) The biogeographic and tectonic history of India. J Biogeogr 30:381–388

Bruneton J, Hatton CK (1999) Pharmacognosy, phytochemistry, medicinal plants, 2nd edn. Intercept, Hampshire

Casley-Smith JR, Casley-Smith JR (1995) Frequency of coumarin hepatotoxicity. Med J Aust 162:391–391

CCE (2020) Cinnamon. Sri Lanka Spice Council. Available from: http://www.ceyloncinnamonexports.com/en/process. Accessed 13 December 2020

CFIA (2018) 2011–2012 Coumarin in cinnamon and cinnamon-containing products. Canadian Food Inspection Agency. Available from: https://www.inspection.gc.ca/food-safety-for-industry/chemical-residues-microbiology/chemical-residues/cinnamon-and-cinnamon-containing-products/eng/1430827809220/1430827809954. Accessed 3 May 2020

Chanderbali AS, van der Werff H, Renner SS (2001) Phylogeny and historical biogeography of Lauraceae: evidence from the chloroplast and nuclear genomes. Ann Mo Bot Gard 88:104–134

Cooray PG (1967) An introduction to the geology of Ceylon. National Museum, Colombo

Cooray PG (1984) Geology, with special reference to the Precambrian. In: Fernando CH (ed) Ecology and biogeography in Sri Lanka, Monographiae biologicae, vol 57. Martinus Nijhoff Publishers, The Hague, pp 1–34. https://doi.org/10.1007/978-94-009-6545-4_1

Dao NK (2004) Chinese cassia. In: Ravindran P, Babu K, Shylaja M (eds) Cinnamon and Cassia the genus *Cinnamomum*. CRC Press, Boca Raten, pp 156–184

David AR (1992) Plants and plant-products used in mummification. In: Nigg HN, Seigler D (eds) Phytochemical resources for medicine and agriculture. Plenum Press, New York, pp 15–33. https://doi.org/10.1007/978-1-4899-2584-8_2

DCS (2006a) Detailed information on rubber in Sri Lanka. Department of Census and Statistics. Available from: http://www.statistics.gov.lk/agriculture/rubber/all.pdf. Accessed 14 Feb 2020

DCS (2006b) Brief analysis of population and housing characteristics. Department of Censes and Statistics, Government of Sri Lanka. Available from: http://www.statistics.gov.lk/PopHouSat/PDF/p7%20population%20and%20Housing%20Text-11-12-06.pdf. Accessed 12 Apr 2020
de Silva KM (1981) A history of Sri Lanka. University of California Press, Berkeley
de Silva CR (1989) The Portuguese impact on the production and trade in Sri Lanka cinnamon in Asia in the sixteenth and seventeenth centuries. Indicator 26:25–38
de Silva MU (1993) Caste feudalism under slave masters – a review of change in Sri Lanka with special reference to mahabadda of the Cinnamon Department, 1597–1832. Rohana 4:77–120
de Silva CR (2000) Chapter 7. Islands and beaches: indigenous relations with the Portuguese in Sri Lanka after Vasco da Gama. In: Disney A, Booth E (eds) Vasco da Gama and the linking of Europe and Asia. Oxford University Press, New Delhi, New York, p 2
Devasiri NR (2008) The adaptable peasant: Agrarian Society in Western Sri Lanka under Dutch rule, 1740–1800. Brill, Leiden, Boston
Dewasiri NR (2007) Peasant in transition: Agrarian Society in Western Sri Lanka under Dutch rule, 1740–1800. Leiden University, The Netherlands
Dinesh R, Leela NK, John Zachariah T, Anandaraj M (2015) Controversies surrounding coumarin in cassia: the good, the bad and the not so ugly. Curr Sci 108:482–484
Dissanayake CB, Chandrajith R (1999) Sri Lanka–Madagascar Gondwana linkage: evidence for a Pan-African mineral belt. J Geol 107:223–235
Dittus WPJ (2017) The biogeography and ecology of Sri Lankan mammals point to conservation priorities. Ceylon J Sci 46:33
DMSL (2019) Climate of Sri Lanka. Sri Lanka: Department of Meteorology. Available from: http://www.meteo.gov.lk/index.php?option=com_content&view=article&id=94&Itemid=310&lang=en. Accessed 3 May 2020
EFSA (2004) EFSA, opinion of the scientific panel on food additives, flavourings, processing aids and materials in contacts with food (AFC) on a request from the commission related to coumarin; adopted on 6 October 2004. EFSA J 104:1–36
EFSA (2008) European Food Safety Authority. Coumarin in flavourings and other food ingredients with flavouring properties. EFSA J 793:1–15
Erdelen W (1988) Forest ecosystems and nature conservation in Sri Lanka. Biol Conserv 43:115–135
FAO (2012) Assessing suitable land: the methodological framework and its implementation. p 19
FAOSTAT (2018) Food and Agriculture Organisation of the United Nations. http://www.fao.org/faostat/en/#data. Accessed 19 Apr 2020
FDA (1977) Code of federal regulations. Title 21, Part 189- substances prohibited from use in human food: Sec. 189.130 Coumarin. (21 CFR 189.130). U.S. Food & Drug Administration, US Department of Health and Human Services. Available from: https://www.accessdata.fda.gov/scripts/cdrh/cfdocs/cfcfr/CFRSearch.cfm?CFRPart=189&showFR=1. Accessed 2 May 2020
Forrest DM (1967) A hundred years of Ceylon tea: 1867–1967. Chatto and Windus, London
Fotland TØ, Paulsen JE, Sanner T, Alexander J, Husøy T (2012) Risk assessment of coumarin using the bench mark dose (BMD) approach: children in Norway which regularly eat oatmeal porridge with cinnamon may exceed the TDI for coumarin with several folds. Food Chem Toxicol 50:903–912
Gilboa A, Namdar D (2016) On the beginnings of South Asian spice trade with the Mediterranean region: a review. Radiocarbon 57:265–283
Ginigaddara GAS (2018) Wonders of spices in Sri Lanka. In: Sharangi AB (ed) Indian spices: the legacy, production and processing of India's treasured export. Springer International Publishing, Cham, pp 63–93. https://doi.org/10.1007/978-3-319-75016-3_5
Gunaratna KL (2002) The spatial evolution of Colombo and needed interventions for the future. J R Asiat Soc Sri Lanka 47:1–34
Gunathilake HAG, Manjula BAS (2006) Balanced fertilization for sustainable coconut and coconut intercrop systems in Sri Lanka. Paper presented at the Balanced fertilization for sustaining crop productivity, Ludhiana, India, 22–25 November 2006

Gunatilleke CVS, Ashton PS (1987) New light on the plant geography of Ceylon. II. The ecological biogeography of the lowland endemic tree flora. J Biogeogr 14:295–327
Gunatilleke IAUN, Gunatilleke CVS (1991) Threatened woody endemics of the wet lowlands of Sri Lanka and their conservation. Biol Conserv 55:17–36
Gunatilleke N, Gunatilleke S, Ashton PS (2017) South-west Sri Lanka: a floristic refugium in South Asia. Ceylon J Sci 46:65–78
Hadidi H, Zahlsen K, Idle JR, Cholerton S (1997) A single amino acid substitution (Leu160His) in cytochrome P450 CYP2A6 causes switching from 7-hydroxylation to 3-hydroxylation of coumarin. Food Chem Toxicol 35:903–907
Hakooz N, Hamdan I (2007) Effects of dietary broccoli on human in vivo caffeine metabolism: a pilot study on a group of Jordanian volunteers. Curr Drug Metab 8:9–15
Halikowski Smith S (2001) Portugal and the European spice trade, 1480–1580. European University Institute, Fiesole, Italy
Hamid KA, Ibrahim SI, Hashim MA, Salama M (2015) Formulation and evaluation of benzyl benzoate emulgel. J Pharm Biol Sci 10:6–9
Haw SG (2017) Cinnamon, cassia and ancient trade. J Ancient Hist Archaeol 4:5–18
Hazleton LW, Tusing TW, Zeitlin BR, Thiessen R, Murer HK (1956) Toxicity of coumarin. J Pharmacol Exp Ther 118:348–358
Itoh K, Chiba T, Takahashi S, Ishii T, Igarashi K, Katoh Y, Oyake T, Hayashi N, Satoh K, Hatayama I, Yamamoto M, Nabeshima Y-i (1997) An Nrf2/Small Maf heterodimer mediates the induction of phase II detoxifying enzyme genes through antioxidant response elements. Biochem Biophys Res Commun 236:313–322
Jayasena RM (2013) The historical archaeology of Katuwana, a Dutch East India Company fort in Sri Lanka. Post Mediev Archaeol 40:111–128
Jayasuriya CEP (1949) A historical survey of Ceylon trade. J R Soc Arts 97:757–768
Joshi SC, Padalia RC, Bisht DS, Mathela CS (2009) Terpenoid diversity in the leaf essential oils of Himalayan Lauraceae species. Chem Biodivers 6:1364–1373
Kostermans AJGH (1957) Lauraceae. Reinwardtia 4:193–256
Kumarathilake DMHC (2012) Study on the extinction risks, conservation and domestication of endemic wild *Cinnamomum* species in Sri Lanka. M.Phil, Faculty of Agricuture, University of Ruhuna, Sri Lanka
Lake BG (1999) Coumarin metabolism, toxicity and carcinogenicity: relevance for human risk assessment. Food Chem Toxicol 37:423–453
Larson C (2015) Reading the tea leaves for effects of climate change. Science 348:953–954
Li Y, Xiang S, Zhu W (2010) Field experiment by using methyl eugenol to trap and kill *Dacus dorsalis* (male). Pest Sci Adm 31:49–50
Liyanage MDS, Tejwani KG, Nair PKR (1984) Intercropping under coconuts in Sri Lanka. Agrofor Syst 2:215–228
Lungarini S, Aureli F, Coni E (2008) Coumarin and cinnamaldehyde in cinnamon marketed in Italy: a natural chemical hazard? Food Addit Contam Part A 25:1297–1305
Manos PS, Stanford AM (2001) The historical biogeography of Fagaceae: tracking the tertiary history of temperate and subtropical forests of the northern hemisphere. Int J Plant Sci 162:S77–S93
Mapa RB (1995) Effect of intercropping coconut lands on soil water retention. Biol Agric Hortic 12:173–183
MEA (1964) Agreement on persons of Indian origin in Ceylon. Ministry of External Affairs, Government of India. Available from: https://mea.gov.in/bilateral-documents.htm?dtl/6426/Agreement+on+Persons+of+Indian+Origin+in+Ceylon. Accessed 11 Apr 2020
Mendis GC (1952) Ceylon under the British, 3rd Revised edn. The Colombo Apothecaries' Co. Ltd., Educational Publishers, Colombo, Sri Lanka
Meyer E (2008) From landgrabbing to landhunger: high land appropriation in the plantation areas of Sri Lanka during the British period. Mod Asian Stud 26:321–361

Miller KG, Poole CF, Chichila TMP (1995) Solvent-assisted supercritical fluid extraction for the isolation of semivolatile flavor compounds from the cinnamons of commerce and their separation by series-coupled column gas chromatography. J High Resolut Chromatogr 18:461–471

MOE (2012) The National Red List 2012 of Sri Lanka; conservation status of the Fauna and Flora, vol viii. Ministry of Environment, Colombo, Sri Lanka, p 476

Murray RDH (2002) The naturally occurring coumarins. In: Herz W, Falk H, Kirby GW, Moore RE (eds) Progress in the chemistry of organic natural products. Springer Vienna, Vienna, pp 1–619. https://doi.org/10.1007/978-3-7091-6172-2_1

Namdar D, Gilboa A, Neumann R, Finkelstein I, Weiner S (2013) Cinnamaldehyde in early Iron Age Phoenician flasks raises the possibility of Levantine trade with South East Asia. Mediter Archaeol Archaeom 12:1–19

Papacharalampous M (2017) A business opportunity in food industry: natural food products that boost health and have healing attributes. International Hellenic University, Thessaloniki

Paranagama PA, Wimalasena S, Jayathilake GS, Jayawardena AL, Senanayake UM, Mubarak AM (2001) A comparison of essential oil constituents of bark, leaf, root and fruit of Cinnamon (*Cinnamomum zeylanicum* (Blum.)) grown in Sri Lanka. J Natl Sci Found Sri 29:147–153

Paranavitana S (1960) The Kotte Kingdom up to 1505. In: History of Ceylon, Book V, Chapter III. Ceylon University Press, Colombo, pp 669–671

Pathiraja PMEK, Griffith GR, Farquharson RJ, Faggian R (2015) The Sri Lankan coconut industry: current status and future prospects in a changing climate. Australas Agribus Perspect Paper 106:23

Pathirana R (2000) A phytosociological study of angiosperms in lowland wet tropical forests of Sri Lanka, Final report submitted to the World Wildlife Fund, Faculty of Agriculture, University of Ruhuna, Sri Lanka

Pathiratna LSS (2006) Management of intercrops under rubber: implications of competition and possibilities for improvement. Bull Rubber Res Inst Sri Lanka 47:8–16

Pathiratna LSS, Perera MKP (2006) Effect of plant density on bark yield of cinnamon intercropped under mature rubber. Agrofor Syst 68:123–131

Peiris GH (1981) Agrarian transformations in British Sri Lanka. Sri Lanka J Agrar Stud 2:1–26

Piyasiri KHKL, Wijeratne M (2016) Comparison of the cultivated area and the production trends of Ceylon cinnamon with the main competitors in the worlds' total cinnamon market. Int J Sci Res Publ 6:476–480

Pratiwi IY, Darmadji P, Hastuti P (2016) Effect of storage temperature on the stability of microencapsulated essential oil from cinnamon *(Cinnamomum burmanii)* 1741:130014

Praveena A, Sanjayan KP (2011) Inhibition of acetylcholinesterase in three insects of economic importance by linalool, a monoterpene phytochemical. Insect Pest Manag 1:340–345

Ranasinghe L, Jayawardena B, Abeywickrama K (2002) Fungicidal activity of essential oils of *Cinnamomum zeylanicum* (L.) and *Syzygium aromaticum* (L.) Merr et L.M.Perry against crown rot and anthracnose pathogens isolated from banana. Lett Appl Microbiol 35:208–211

Ranawana KB (2014) Forest dieback in Horton Plains National Park. J Dep Wildl Conserv 2:103–112

Raven PH, Axelrod DI (1974) Angiosperm biogeography and past continental movements. Ann Mo Bot Gard 61:539–637

Ravindran PN, Nirmal Babu K (2004) Introduction. In: Ravindran PN, Nirmal Babu K, Shylaja M (eds) Cinnamon and Cassia: the genus *Cinnamomum*. CRC Press, Boca Raton, pp 1–13

Rietjens IMCM, Punt A, Schilter B, Scholz G, Delatour T, van Bladeren PJ (2010) In silico methods for physiologically based biokinetic models describing bioactivation and detoxification of coumarin and estragole: implications for risk assessment. Mol Nutr Food Res 54:195–207

Rohwer JG (1993) Lauraceae. In: Kubitzki K, Rohwer JG, Bittrich V (eds) Flowering plants dicotyledons: magnoliid, hamamelid and caryophyllid families. Springer Berlin Heidelberg, Berlin, Heidelberg, pp 366–391. https://doi.org/10.1007/978-3-662-02899-5_46

SBI (2019) Spice Board of India, Spice Wise Area and Production. Ministry of Commerce & Industry, Government of India. Available from: http://www.indianspices.com/sites/default/files/Major%20spice%20wise%20area%20and%20production_2018.pdf. Accessed 26 Apr 2020

Schrikker A, Ekama KJ (2017) Through the lens of slavery: Dutch Sri Lanka in the eighteenth century. In: Biedermann Z, Strathern A (eds) Sri Lanka at the cross roads of history. UCL Press, London, pp 178–193. https://doi.org/10.14324/111.9781911307822

Senanayake UM, Wijesekara ROB (2004) Chemistry of cinnamon and cassia. In: Ravindran PN, Babu KN, Shylaja M (eds) Cinnamon and cassia: the genus *Cinnamomum*. Medicinal and aromatic plants — industrial profiles. CRC Press, London, pp 81–120

Seneviratne P, Perera MKP (2011) Cinnamon as an intercrop for rubber plantations. Bull Rubber Res Inst Sri Lanka 52:12–17

Shirai M (2010) Analyzing price premiums for foods in Japan: measuring consumers' willingness to pay for quality-related attributes. J Food Prod Mark 16:184–198

Shylaja M, Ravindran PN, Nirmal Babu K (2004) Other useful species of *Cinnamomum*. In: Ravindran PN, Nirmal Babu K, Shylaja M (eds) Cinnamon and Cassia: the genus *Cinnamomum*. Medicinal and aromatic plants — industrial profiles. CRC Press, Boca Raton, pp 330–355

Simic A, Sokovic MD, Ristic M, Grujic-Jovanovic S, Vukojevic J, Marin PD (2004) The chemical composition of some Lauraceae essential oils and their antifungal activities. Phytother Res 18:713–717

SLEDB (2020) Export performance indicators. Sri Lanka Export Development Board. Available from: https://www.srilankabusiness.com/pdf/Export_Performance_Indicators_of_Sri_Lanka_2009_2018.pdf. Accessed 23 Apr 2020

SLTB (2014) Sri Lanka Tea Board: Dimbula. Sri Lanka Tea Board. Available from: http://www.srilankateaboard.lk/index.php/dimbula. Accessed 2 May 2020

Somaratne GPV (1975) Political history of the Kingdom of Kotte: 1400–1521. Nugegoda, Colombo

Sproll C, Ruge W, Andlauer C, Godelmann R, Lachenmeier DW (2008) HPLC analysis and safety assessment of coumarin in foods. Food Chem 109:462–469

Sritharan R (1984) Studies of the genus *Cinnamomum*. University of Peradeniya, Peradeniya

Sritharan R, Jacob VJ, Balasubramaniam S, Bavappa KVA (1993) Palynological and cytological studies of the genus *Cinnamomum*. Acta Hortic 330:107–113

Strathern A (2004) Theoretical approaches to Sri Lankan history and the early Portuguese period. Mod Asian Stud 38:191–226

Thapa YB, Sharma MD, Regmi PP, Kattel RR (2020) Consumers' preference and willingness to pay for Nepalese large cardamom in the global market. Int J Soc Sci Manag 7:55–69

The IUCN Red List of Threatened Species (1998) World Conservation Monitoring Centre. www.iucnredlist.org. Accessed 27 Feb 2020

Trendeconomy (2020) Canada imports: cinnamon. TrendEconomy.com. Available from: https://trendeconomy.com/data/import_h2?time_period=2019&reporter=Canada&trade_flow=Import&commodity=0906. Accessed 2 May 2020

UNESCO (2020) Sinharaja forest reserve. UNESCO. Available from: https://whc.unesco.org/en/list/405/. Accessed 30 Mar 2020

van Meersbergen G (2017) Cinnamon and Elephants: Sri Lanka and the Netherlands from 1600. Mariner's Mirror 103:227–228

Waller JM (1982) Coffee rust—epidemiology and control. Crop Prot 1:385–404

Wang Y-H, Avula B, Nanayakkara NPD, Zhao J, Khan IA (2013) Cassia cinnamon as a source of coumarin in cinnamon-flavored food and food supplements in the United States. J Agric Food Chem 61:4470–4476

Weerawardhena SR, Russell AP (2012) Historical land-use patterns in relation to conservation strategies for the Riverstone area, the Knuckles massif, Sri Lanka: insights gained from the recovery of anuran communities. Taprobonica 4:92–102

Weiss EA (2002) Spice crops. CABI Publ, Oxford

Werner WL (1995) Biogeography and ecology of the upper montane rain forest of Sri Lanka (Ceylon). In: Hamilton LS, Juvik JO, Scatena FN (eds) Tropical montane cloud forests, Ecological studies (Analysis and synthesis), vol 110. Springer, New York, pp 343–352

Werner WL, Balasubramaniam S (1992) Structure and dynamics of the upper montane rain forests of Sri Lanka. In: Goldammer JG (ed) Tropical forests in transition. Advances in life sciences. Birkhäuse, Basel. https://doi.org/10.1007/978-3-0348-7256-0_11

WHO (1999) WHO monograph on selected medicinal plants, vol 1. World Health Organisation, Geneva

Wickramagamage P (1998) Large-scale deforestation for plantation agriculture in the hill country of Sri Lanka and its impacts. Hydrol Process 12:2015–2028

Wickramasinghe K, Steele P, Senaratne A (2008) Socio-economic impacts of forest conservation on peripheral communities: case of Knuckles National Wilderness Heritage of Sri Lanka. Paper presented at the governing shared resources: connecting local experience to global challenges, the Twelfth Biennial Conference of the International Association for the Study of Commons, Cheltenham, England

Wijesinghe KGG, Pathirana R (2000) Germplasm screening and varietal development in cinnamon. In: Senaratne R, Dayatilake GA (eds) Revitalisation of the cinnamon industry in Sri Lanka: constaints and remadies. Faculty of Agriculture, University of Ruhuna, Matara, pp 15–25

Woehrlin F, Fry H, Abraham K, Preiss-Weigert A (2010) Quantification of flavoring constituents in cinnamon: high variation of coumarin in cassia bark from the German retail market and in authentic samples from Indonesia. J Agric Food Chem 58:10568–10575

Yamamoto W (2000) Participatory forest conservation in southwest lowland rainforests of Sri Lanka. J For Res 5:195–199

Yokota S, Higashi E, Fukami T, Yokoi T, Nakajima M (2011) Human CYP2A6 is regulated by nuclear factor-erythroid 2 related factor 2. Biochem Pharmacol 81:289–294

Chapter 2
Historical, Ethno-Botanical and Social Aspects of Cinnamon Cultivation in Sri Lanka

Nirmal R. Dewasiri, Lodewijk J. Wagenaar, and Jayadeva Uyangoda

2.1 Introduction

Cinnamon was an important commodity in the economic life of Sri Lanka from the thirteenth to nineteenth centuries. It had also been a commodity central to the growing Indian Ocean trade. As Sri Lanka was the sole supplier of quality cinnamon for the global market, it became a determinant of the domestic life in the country's cinnamon growing region. Although the majority of the people had no special interest in cinnamon as a commodity, it made a lasting impact on their lives over the centuries. Most importantly, it attracted commercial and military powers that were establishing themselves as the dominant players in the Indian Ocean. Against this backdrop, the central problem that this chapter will attempt to address is how cinnamon shaped the historical destinies of the island of Ceylon and also the process of social change it set in motion in the cinnamon growing region.

Cinnamon grew wild in the Western and Southern parts of Sri Lanka and remained so until the Dutch administration began cinnamon plantations in the late eighteenth century. The earliest mention of Sri Lankan cinnamon is in the works of the thirteenth century Arab writer Kazwini. Ibn Battuta, a fourteenth century traveller who visited Sri Lanka, has left a detailed account on cinnamon in Sri Lanka (Kanapathypillai 1969: 280).

Under the indigenous kings, the collection of cinnamon was a royal prerogative. At the time when the Portuguese involved themselves in the island's affairs during

N. R. Dewasiri (✉)
Department of History, University of Colombo, Colombo, Sri Lanka

L. J. Wagenaar
Department of History, University of Amsterdam, East India House, Amsterdam, The Netherlands

J. Uyangoda
Department of Political Science, University of Colombo, Colombo, Sri Lanka

R. Senaratne, R. Pathirana (eds.), *Cinnamon*,
https://doi.org/10.1007/978-3-030-54426-3_2

Fig. 2.1 Cinnamon peelers c. 1672. (Wagenaar 2017: 146)

the early sixteenth century, with cinnamon as one of their major interests, peeling and delivering cinnamon had evolved into a well-organized state function (Fig. 2.1). A separate department called *mahabadda* (the great department) was in operation and this nomenclature remained under successive European colonial powers as well. Until drastic economic changes occurred in the nineteenth century under the British colonial domination, cinnamon remained the most important commercial crop in Sri Lanka.

The discussion developed in the chapter is organized around eight subthemes. The first subtheme is on the growth of the Indian Ocean trade since the tenth century and its impact on the economic and political changes in Sri Lanka, leading to a situation in which cinnamon emerged as the primary commercial attraction to the Europeans. The second subtheme extends the discussion of the historical context to developments after the thirteenth century during which the erstwhile 'Rajarata Civilization' came to an end and a new socio-economic formation began to evolve in the Southwest of the island. The discussion will also show that it was in this new Southwestern socio-economic formation that cinnamon emerged as a historically important merchandise. The discussion under the third subtheme explores the significance of cinnamon in the economic relations developed in the Southwestern region. The focus of the fourth subtheme is on Sri Lanka's early encounters with European trading powers and the role that cinnamon played in that process. Under the fifth subtheme, the focus is on the production and trade in cinnamon under the Dutch and key changes that occurred in labour relations, land use and political relations in the broad context of Dutch cinnamon monopoly. Cinnamon was also a major factor in the relations between the Dutch and the Kandyan Kings and that constitutes the chapter's sixth subtheme. The cinnamon economy under the Dutch also had a significant impact on the peasant economy in a variety of ways and that constitutes the seventh subtheme of the chapter. The eighth, which is the final subtheme, is on the beginning of organized cinnamon plantations during the latter part of the Dutch rule and the subsequent decline of the commercial and economic significance of cinnamon after Sri Lanka came under the British rule. The chapter ends with a brief section devoted to concluding remarks.

2.2 Growth of the Indian Ocean Trade and Sri Lanka: A Brief Historical Sketch

The importance of cinnamon as a merchandise was closely associated with the growth of the Indian Ocean trade since around the tenth century. Although the Indian Ocean has a very long history in the long-distance trade, the period following the rise of Islam is considered to be most important. It coincided with the emergence of Europe from the so-called 'Dark Age' to become a dominant player in world history since the end of the fifteenth century.

Parallel to the growth of Indian Ocean trade, important political changes were taking place in the island of Ceylon too. The overseas expansion of the South Indian Chola Empire during 993–1017 CE resulted in the decline and fall of Anuradhapura, the capital of the kingdom. Polonnaruwa then became the provincial Chola capital. The situation changed in 1076 after its conquest by Vijayabahu, who established a new kingdom in Polonnaruwa. Continued Tamil immigration from Southern India weakened Polonnaruwa and consequently the city was abandoned.

Manthai—the ancient port Mahatittha ('Great Port')—for one and a half millennia played a major role as a trade centre halfway on the long route between China and the Middle East/East Africa. It was an important port for the merchants who traded between Sri Lanka and India (Carswell et al. 2013). The heyday of the harbour of Manthai was between the first and eleventh centuries CE. The collapse of Anuradhapura around 1000 CE put an end to this once flourishing hub of regional and international trade.

This, however, didn't cause the maritime isolation of Sri Lanka. Contacts with faraway places located east and west of Sri Lanka continued (Perera 1952: 14–22).

In the process, in Sri Lanka, the trade centres had changed: in the South of Kotte was the most important harbour. In the north, it was Jaffna.

The sources before the thirteenth century praise the island for its precious stones, such as rubies and sapphires, and also mention enthusiastically about the splendid pearls as the best found in the whole world. However, nowhere could we find references made to cinnamon until the late thirteenth century (Perera 1952: 17).

The famous traveller Marco Polo visited 'Seilan' around that time and travelled all the way (possibly from Jaffna through the interior) to Adam's Peak and even climbed it. Marco Polo speaks amply about the precious stones: 'the Island produces more beautiful and valuable rubies than are found in any other part of the world, and likewise sapphires and topazes and amethysts, garnets, and many other precious and costly stones (Travels of Marco Polo 1997: 224)'. We may assume that Marco Polo knew about the spice cinnamon, famous from antiquity, and also well known in his time. But he didn't mention it, understandably so, for he had not visited the maritime provinces of southwest Sri Lanka. He was perhaps a victim of the secrecy of the Arab traders, who kept the origin of the produce a secret in order to protect their monopoly: only Arab merchants shipped the cinnamon to the Red Sea, where colleagues took over the costly merchandise and transferred it to the Mediterranean.

The secret was disclosed when Lourenço de Almeida, son of the Portuguese Viceroy in Goa, accidently arrived at the harbour of Kotte in November 1506. He was commissioned to intercept Arabian competitors and destroy their vessels. Strong winds forced his flotilla to land in Sri Lanka, and there he was met with the animosity of Arabian merchants. They complotted with the king of Kotte and presented the Portuguese with 400 bahars (1 bahar = 80 pounds) of cinnamon in the hope that they would never return to the island. We know this was a miscalculation, for in 1517 Lopo Soarez de Albergara arrived with a fleet of 17 ships and 1000 soldiers, commissioned to set up business. Auxiliary forces helped to force the king to accept a treaty, permitting the Portuguese to build a fortified factory and for the king to pay an annual tribute of 300 bahars of cinnamon, 12 ruby and sapphire rings and 6 elephants. In return, the Portuguese promised to protect the king against all possible enemies. With that gesture, the Portuguese effectively eliminated their Muslim competitors.

Vasco da Gama revolutionized the existing maritime systems, since his first voyage in 1497/1498 by the direct route between India and Europe proved to be a success with no rival to challenge the Portuguese. Earlier, produce from the East arrived at the Mediterranean by the southern route (South India) or by the northern route (from Bharuch/Surat) via the Red Sea or the Persian Gulf, where the goods were further transferred over land to—mostly—the coastal ports of Lebanon or to Egypt. To a great extent, the old combined maritime–land routes had become obsolete, and thereby monsoon dependent merchants had to reorganize and focus on regional trade—that, for the greater part, they did successfully. However, the direct connection between Europe and Asia had a tremendous, even fatal, effect: Spain and Portugal were able to occupy enormous pieces of land, even complete kingdoms. Spain focused especially on the Philippines (since 1565), while the colonial activities of the Portuguese were expanded in South India (from 1505) and Sri

Lanka (from 1517). They occupied Malacca in 1512 and extended their influence further East until the Dutch came into the picture.

As for Sri Lanka, it can be said without exaggeration that it was cinnamon that attracted the Europeans—first the Portuguese, then the Dutch and then the British—and they stayed in the maritime province engaging in cinnamon trade. In the process, the Arabian trade was cut short. The local Muslim and Chetty traders engaged only in the regional commercial exchange between South India and Sri Lanka. The enterprise of Muslims, generally called 'Moors', was obstructed to a maximum till the British started a more liberal policy. The trade in gems, for which Serendib/Taprobane/Ceilan/Ceylon had become famous for more than two millennia, was kept solidly in Muslim hands. Cinnamon, once the principal produce of Sri Lanka, lost its place of pride under the British. Colonial exploitation that had been greatly expanded by the skilful Dutch was even more professionally developed by the British. After the incorporation of the kingdom of Kandy in 1815, Sri Lanka's economy was fully exploited to serve the British coffers. However, cinnamon played only a very modest role in that process.

2.3 Historical Developments in Sri Lanka Since Thirteenth Century and Cinnamon Trade

The thirteenth century was a turning point in the historical evolution of Sri Lanka. It marked the end of the erstwhile Rajarata civilization, which had lasted more than a millennium in a developed form. The end of the Rajarata civilization saw the emergence of new socio-economic formations, which were alien to the former, in the Western and Southern parts of the island. It was in this socio-economic formation that cinnamon became a historically important merchandise. During the post-thirteenth century period, this cinnamon-growing zone of the island emerged politically, economically and demographically as the most significant region.

The post-thirteenth century Western and Southern socio-economic formation of the island (WS formation hereinafter) could be distinguished temporally from the pre-thirteenth century Rajarata social formation and spatially from other socio-economic formations that emerged after the thirteenth century. The characteristic feature of the Rajarata formation was that its economic life was largely organized around a complex but centrally managed irrigation system. The WS formation, in contrast, was located in the region where the annual rainfall was higher and it was possible for the peasantry to engage in agriculture with the rainwater. This, coupled with other factors, led to a socio-economic formation different from the previous one. In the Rajarata civilization, the major concentrations of population had been located in northern, eastern and south-eastern parts of the island, now known as the 'dry zone'. The post-thirteenth century period saw new centres of population developing outside the areas of the Rajarata civilization. They included the Northern peninsular region, which remained, hitherto, at the periphery of the Rajarata system, the

Western and Southern regions and the central highlands. These three regions were subjected to separate state formation processes in the post-thirteenth century period with three monarchical centres, namely, Jaffna, Kotte and Kandy (*Kandaudarata*). What is noteworthy is the fact that the areas that came under Kotte and Kandy monarchical centres were those with direct association with the cinnamon industry.

In terms of the economic structure of the region, it had two interrelated characteristics. The first was that the majority of the population, which was thinly distributed in a number of sub-regions, was engaged in peasant agriculture as their main economic activity. Cultivation was enabled by conducive ecological factors, particularly the annual rainfall from the two monsoon systems. The surplus generation capacity of the peasant agriculture was, however, significantly lower than that of the Rajatara system. The second was the presence of non-agricultural economic activities in the coastal belt, where trade occupied the central place. The importance of various trading communities and their economic, political and cultural significance was strongly felt in the region. The strength of the political elites was heavily dependent upon its ability to control the trade that was carried out from a large number of harbours along the Western coastline. There was stiff competition between the rulers of the Western coastal region and those of Jaffna to control this trade. The long drawn military confrontation between Aryachakravartis of Jaffna and Alakeshwaras of Raigama bears witness to this conflict over trade and sea routes.[1] Both these elite groups had strong connections to the Indian Ocean trade. They established themselves as rulers in Jaffna and on the Western coast, respectively, and controlling of trade became a major priority as it was the chief source of income.

When Aryachakravartis sought to extend their dominance Southward to gain full control of the brisk Western coastal trade and to extract land revenue, Alakeshwaras mobilized themselves to check the advance of the former with Kotte as the seat of their power. The ascendency of Kotte did not, however, last long. Centrifugal forces were more powerful than the centripetal ones. When the Portuguese landed in the island, besides Kotte and Jaffna, a separate political entity had emerged in *Kandaudarata* as well. Sooner than later Kotte itself came to be divided into two centres of authority, namely, Kotte and Sitavaka, following the assassination of the old king by his three sons in 1521.

2.4 Cinnamon in the Economic Relations in the WS Region

Cinnamon happened to be the most significant commodity to enter the brisk Indian Ocean trade from this region. There were, of course, other commodities, namely, elephants, pearls, precious stones and areca nuts. None of them overcame the importance of cinnamon in its vital role in world trade in centuries to come.

Local rulers could benefit from their ability to mobilize the corvée labour (*rajakariya*) that was the dominant form of the 'non-economic' transfer of value from the

[1] For Aryachakravartis of Jaffna, see De Silva and Pathmanathan (1995). For Alakeshwaras, see Liyanagamage (2001).

subjects to the ruling elite. The *rajakariya* system, as a 'rent-in-labour', had become the dominant form of payment of the king's due because the other two forms of payment of the king's due in high productivity situations—'rent-in-kind' and 'rent-in-money'—had become less attractive as the level of production was significantly low, especially among the peasantry. Therefore, the kings and their subordinates in the hierarchy of the ruling elite transformed this unpaid (or sometimes underpaid) labour of *rajakariya* into exchange value by various means. A classic example was the transformation of the labour of *Salagama* caste into exchange value. This was enabled by two factors. First, the cinnamon yielded a high level of income in overseas trade and, therefore, the kings were able to sell the cinnamon that they collected at minimal cost. Second, the people of the *Salagama* caste were less integrated to the agrarian system, and thus drawing their corvée labour to the arduous task of peeling cinnamon was perhaps relatively easy. Although this hypothesis requires substantive verification, there is evidence to support this claim.

The 'caste system' as existed in this period could be identified as a mode of mediating the relationship between various historically evolved communities and integrating them into a ritually sanctioned hierarchical edifice.[2] As the *Salagama* community is known to have been formed by recent South Indian immigration, a process that was frequent, especially in the post-thirteenth century period, they were less integrated into the agrarian system in the interior. Partially mythological accounts claim that ancient kings brought people of this community from South India as weavers and thereafter used them as cinnamon peelers. They remained an immigrant community with only weak links with the peasant economy and the Sinhalese social system.

Furthermore, historians rank this caste group along with two others, namely, *Karava* and *Durava* (Roberts 1982). The *Karava*, numerically the largest among the three caste communities, showed a high degree of social mobility, both horizontally and vertically. In this context, using people of the *Salagama* caste as cinnamon peelers made sense.

Evidence does not suggest that cinnamon was a major factor in the economic life of the peasant society at the time of the Portuguese arrival. However, it was with the Portuguese that cinnamon as a commodity began to play a crucial role as a catalyst for social transformation. Consolidation of the caste structure in the Southern and the Western Sinhalese society and the emergence of cinnamon peeling labour as a distinct caste group are the key facets of this transformation.

2.5 Early European Encounter and the Role of Cinnamon

For the Portuguese, having a strong foothold for controlling commercial and military affairs in the Indian Ocean was a sine qua non. Although the appearance of a Portuguese fleet at the shores of the Western coast of the island in November 1506[3]

---

[2] See Dewasiri (2008: 185–219) for this approach to the caste system in Sri Lanka.

[3] 1505 was known to be the year of first arrival of Portuguese on the island. However, it has been established now that this occurred in 1506.

seems unintended, it would have happened at any time sooner than later. The Portuguese returned after 12 years, in 1517, with a definite purpose. Initially, they got the consent of the King of Kotte to build a small fort near the Colombo harbour, perhaps, to the fury of the Muslim traders, who were dominating commerce at the Colombo harbour at the time. The Muslims tried to convince the king that it was an unwise decision.[4] Following the involvement of the Muslim traders, a conflict emerged between the Portuguese and the king. However, the serious internal political crisis of the declining Kotte kingdom allowed the Portuguese access to the internal affairs of the kingdom and thereby consolidate themselves at the expense of the Muslims as well as the local ruling elite (Dewaraja 1994).

Portuguese interests in the island could be seen in terms of two objectives. One was to use this strategically located island to achieve their main goal, namely, ousting of Muslims from the Indian Ocean. The second was to obtain the monopoly over cinnamon trade. Their initial interest was to get monopsony right for cinnamon from the Kotte King. The opportunity to be a party to the internal power struggles of the Kotte kingdom provided ideal conditions for them to achieve both these objectives. First and foremost, the Portuguese used this opportunity to obtain as much cinnamon as possible from the King of Kotte in return for their military support.

The absolute right for obtaining cinnamon was in the hand of the king, who mobilized the people of the *Salagama* caste group to peel cinnamon which grew in the wild. The *Salagama* community was forced to perform this annual duty free of remuneration as their duty to the king (*Rajakariya*). A separate department called *Mahabadda* (great department) was set up for organizing the labour of the *Salagama* caste for cinnamon peeling.

When the Portuguese inherited the administration of Kotte from the indigenous kings, they continued with the existing system, probably with greater vigour. However, it is difficult to imagine whether any major changes occurred in the way cinnamon was peeled and delivered to the Colombo harbour under the Portuguese. Real changes came about when the Dutch dominated the region.

2.6 Cinnamon and the Dutch

A fundamental feature of the Dutch overseas presence that began in the early sixteenth century was its commercial interests. Unlike the Portuguese, the Dutch overseas exploration was carried out by trading companies and commercial interests. The fundamental objective of Dutch trade with Asia, carried out by Verniegde Oostindiesche Compagnie (VOC), was to obtain the merchandise from the East, which were in high demand in Europe, from the places of their origin. Therefore, getting direct access to the cinnamon growing region in Sri Lanka, identified in the

[4] See Dewaraja (1994) for a detailed account of this event.

VOC maps as *'Kaneel Land'* (the country of cinnamon), occupied a high priority.[5] While the Portuguese managed to occupy the entire cinnamon growing area, the territory that the VOC occupied was much smaller, with many cinnamon growing areas left as the territory of the Kandyan kings. However, the Dutch administration struck a deal with the Kandyan kings whereby they could send their peelers to Kandyan territory to peel cinnamon there.

The Dutch achieved great success in their quest to exploit the supply of cinnamon in the island. Van den Belt has calculated that, from 1700 to 1760, staggering volume of 36,700,000 pounds of cinnamon was exported from the island towards the Netherlands. This amount was 71.3% of the total cinnamon export, of which 27.3% was sent to Batavia (present Jakarta). The remaining 1.4% was sent to the Dutch factories in the Indian mainland. Cinnamon peelers of the *Salagama* caste peeled and delivered altogether 466,000 bales (37.28 million pounds). A share of 98.5% of this quantity was exported and about 560,000 pounds were consumed locally (Van den Belt 2008: 54).

There was also a gradual increase in the relative importance of cinnamon export. In the first decade of the eighteenth century, the share of cinnamon export out of the total export to the Netherlands was 5%. By the 1760s, this amount had increased to 13.2% (Van den Belt 2008: 56).

The Dutch followed an extremely systematic and cautious approach to the collection of cinnamon. There were several objectives they wanted to achieve in connection with cinnamon. The most important one was, of course, to ensure that no party other than the VOC had access to Sri Lankan cinnamon. Strict control over coastal trade was maintained for this end. Particular attention was paid to ensure that the Kandyan kings would not trade cinnamon with non-VOC parties (Arasaratnam 1958: 181–193).

Ensuring the availability of sufficient labour for the collection and delivery of cinnamon was also a major concern. The main structural obstruction to overcome in this connection was that the mobilization of labour to collect cinnamon was organized within the framework of caste-based division of labour. It was not easy to change this system. The Dutch attempted some alterations to the system of caste-based mobilization of labour. For example, they tried to employ people of the *Wahumpura* caste in the Matara region to collect cinnamon, without much success. Moreover, people of the *Karava* caste were heavily utilized as boatmen on interior waterways to transport cinnamon. It was not easy to extract labour utilization out of the existing system of labour mobilization. Thus, the Dutch administration had to be content with the limited supply of labour of the *Salagama* caste for the most arduous task of the cinnamon industry, namely, peeling of the cinnamon bark.

Mobilizing the *Salagama* labour for the cinnamon peeling was a challenging task for the Dutch in the context of a growing demand for cinnamon as the population of the *Salagama* caste did not grow at a pace to match the growing demand for

[5] See Wagenaar (2017: 149–155) for a broader perspective of the place of cinnamon in the Dutch administration in Sri Lanka.

Table 2.1 Annual cinnamon collection from 1700 to 1750 and 1764 to 1793

Year	Bales	Year	Bales	Year	Bales	Year	Bales
1700/1701	7290	1724/1725	8515	1748/1749	5673	1772/1773	5316
1701/1702	7866	1725/1726	8673	1749/1750	6692	1773/1774	5837
1702/1703	6849	1726/1727	8889	1750/1751	N/A	1774/1775	4690
1703/1704	5883	1727/1728	9065	1751/1752	N/A	1775/1776	4156
1704/1705	5878	1728/1729	9224	1752/1753	N/A	1776/1777	5097
1705/1706	5868	1729/1730	9041	1753/1754	N/A	1777/1778	3129
1706/1707	6179	1730/1731	9722	1754/1755	N/A	1778/1779	5150
1707/1708	6018	1731/1732	9763	1755/1756	N/A	1779/1780	N/A
1708/1709	6685	1732/1733	9423	1756/1757	N/A	1780/1781	N/A
1709/1710	6393	1733/1734	9425	1757/1758	N/A	1781/1782	4476
1710/1711	6138	1734/1735	8843	1758/1759	N/A	1782/1783	4333
1711/1712	6011	1735/1736	2888	1759/1760	N/A	1783/1784	3999
1712/1713	7285	1736/1737	6546	1760/1761	N/A	1784/1785	5674
1713/1714	4337	1737/1738	9137	1761/1762	N/A	1785/1786	5750
1714/1715	13,158	1738/1739	8721	1762/1763	N/A	1786/1787	5522
1715/1716	8413	1739/1740	9736	1763/1764	N/A	1787/1788	5081
1716/1717	8449	1740/1741	5439	1764/1765	5000	1788/1789	4716
1717/1718	11,300	1741/1742	9182	1765/1766	5315	1789/1790	5143
1718/1719	7685	1742/1743	8736	1766/1767	10,009	1790/1791	4290
1719/1720	8108	1743/1744	9112	1767/1768	7724	1791/1792	5585
1720/1721	8820	1744/1745	8248	1768/1769	5655	1792/1793	5360
1721/1722	8881	1745/1746	8144	1769/1770	5542		
1722/1723	7724	1746/1747	8028	1770/1771	5348		
1723/1724[a]	N/A	1747/1748	7964	1771/1772	5355		

Source: Kanapathypillai (1969: 290)
[a]Data not available

cinnamon. The heavy burden on cinnamon peelers to meet the rising demand for cinnamon had been felt even during the Portuguese times as well. Queyroz states that the cinnamon peelers of the *Salagama* caste petitioned to the Portuguese captain of Colombo asking for relief from the burden of cinnamon peeling (Arasaratnam 1958: 186). There seems to be a significant mismatch between the demand for cinnamon and the available level of labour. As the available data shows, there were significant fluctuations in the supply of cinnamon (Table 2.1). Meeting the demand for cinnamon was heavily dependent upon the unhindered supply of labour. This was, however, not the case, at least until the latter part of the eighteenth century. Dearth of the cinnamon peelers was a constant complaint. Measures taken for the effective mobilization of the *Salagama* people for cinnamon peeling and their effects are well known (Kotelawele 1995: 248). The following remarks made by Governor Gollenesse is pertinent to this discussion:

> ...the proposal which I have made to their Excellencies by letter of 10th January, 1748 to abolish the money cinnamon on account of the considerable increase of the number of cinnamon peelers as shown above from 968 to 2924 men, should be postponed until their

Excellencies' further orders as many of these men died in the past year and the things have also changed considerably since (Gollenesse 1974: 74).

It is possible to surmise that the labour shortage may have forced the limited number of cinnamon peelers to spend more time on cinnamon peeling. This would have affected their life in a number of ways. It reduced the time available for the peelers to engage in paddy cultivation (and also slash-and-burn cultivation), which was necessary for their sustenance. Arasaratnam (1958, 185–186) states that the cinnamon peelers were forced to remain in the woods around 8 months every year. The traditional system ensured that the usual involvement of peasant agriculture of cinnamon peelers would not be disturbed as a result of their engagement in cinnamon peeling. This systemic balance obviously changed under the Dutch, with more labour time for cinnamon peeling, resulting in less labour time for peasant agriculture.

The heavy involvement of cinnamon peelers in peasant rebellions in the eighteenth century was a remarkable phenomenon (Jayawardena 2010). Moreover, cinnamon peelers openly expressed their displeasure over the overburden in their customary annual audience to the governor (Dewasiri 2017). The net result was that the Dutch administration was forced to partially remunerate cinnamon peelers, whereas traditionally it was an unpaid labour.[6]

One way in which *Salagama* people resorted to evade this burdensome task was to use some of the limited avenues for changing their caste identity. Some would get their children married to persons of 'lower castes' so that their children may be listed under the latter caste and thus be relieved from the burden of cinnamon peeling. However, to prevent this manipulation of the traditional caste norms, the Dutch administration introduced a new law preventing cross-caste marriages among the *Salagama* people. The law would further say that even when such marriages occur, children of such unions would be listed as cinnamon peelers, thus making the move of cinnamon peelers futile.[7]

The other challenge that the Dutch faced in ensuring the maximum supply of cinnamon was the protection of cinnamon trees. This was a major issue because peasants would pay little attention to cinnamon trees, especially when clearing the lands for slash-and-burn cultivation. This became necessary at a time when the population began to increase slightly in the cinnamon growing regions following the end of the incessant warfare that characterized most of the sixteenth and seventeenth centuries (Kotelawele 1968: 21; Paranavitana 2001: 47). Probably slash-and-burn cultivation became widespread and the destruction of cinnamon trees in the process became a concern for the Dutch administrations (Loten 1935: 38–39).

---

[6] 'For one man: 7 1/2 Fanams [12 fanam = 1 rix – dollar], one parrah of rice monthly [1 parrah = 40 Dutch lbs.] and 2 pieces of ordinary Salampores annually' (Pybus 1958).

[7] There is a series of proclamation issued in this respect. The title of the proclamation issued on 23rd March 1753 reads: 'Mandate-Ola forbidding various sub-castes of the Chalias to marry mutually, and prescribing that a child from such mix marriage shall always belong to the proper cinnamon peelers' (Hovy 1991: 570–571).

Strict regularization of *chena* cultivation was thus maintained in order to protect cinnamon trees. Chena cultivation was beneficial for the growth of cinnamon trees as long as the cinnamon trees and saplings were not harmed while land was being cleared, since it would create an environment conducive for cinnamon trees to grow (Loten 1935: 39).

2.7 Cinnamon in Kandyan Lands

A good part of the cinnamon growing areas was still lying in the territory of the King of Kandy. Although the strategic objective of the Dutch was to occupy the entire cinnamon growing region, they could not achieve it. However, the access to the cinnamon in the king's territory was essential to meet the annual demand for cinnamon. When, during the early 1680s, the relationship between the Dutch and Kandy calmed down after a long war between them, particularly after Rajasingha II's decision to pursue peace with the Dutch, the latter succeeded in securing exclusive rights for peeling cinnamon in the Kandyan territory.

An annual ceremonial visit to the Court of Kandy was organized with gifts to the king in order to obtain formal consent for peeling cinnamon in the king's territory (Fig. 2.2). This annual visit had a broader meaning in Kandy-Dutch relations, the symbolic aspect of it being linked to the assertion of sovereign right of the King of Kandy to the Dutch occupied territory.

Following the circumstances of the Dutch conquest, which was linked to the Kandy–Dutch treaty of 1638, the Dutch could effectively occupy parts of the former Portuguese territory. Although the Dutch may have probably misinterpreted the treaty so that they could retain the areas they occupied after wars with the Portuguese, they could not claim absolute sovereign power to the territory. They were compelled to accept, at least formally, the claim of sovereign power of the King of Kandy to the entire island. Given the fact that there were no other indigenous centres of power, the claim of the King of Kandy went unchallenged. The Dutch also could not challenge this claim as their intervention in the affairs of the island was solely based on the 1638 treaty.[8]

The ambiguity regarding the legality of the Dutch occupation was ultimately resolved by the Kandy-Dutch Treaty of 1766, by which the Dutch gained considerably, including the acceptance of the full sovereign power of the Dutch over the territory occupied by them. However, the Dutch made use of this ambiguity to their advantage. The Dutch formally addressed the King of Kandy as 'Your Majesty the Emperor of the Island of Ceylon', and loyalty to the king was pledged at the annual visit to the Kandyan Court when permission for peeling cinnamon in the Kandyan territory was granted.

[8] For more details about the Kandy-Dutch Treaty of 1638 and the controversy around the sovereignty of Dutch occupied areas, see Arasarathnam (1958) and Gunawardana (1958).

Fig. 2.2 Visit of Dutch envoy Daniel Agreen to Kandy in 1744. (Source: Wagenaar 2017: 108)

Dutch officials resented the annual visit and certain court rituals that the Dutch embassy was forced to perform, including kneeling before the king (Wagenaar 1996). The 1766 Treaty removed this burden, without, of course, compromising Dutch interests. Besides the major territorial gains, including the entire coastline of the island and the acceptance of the legitimacy of Dutch occupation of territory, access to Kandyan territory was granted to the VOC's cinnamon peelers.

2.8 Cinnamon and the Socio-economic Life of the Peasantry

Although only the people of the *Salagama* caste were directly involved in the cinnamon industry, Dutch cinnamon policies significantly affected the peasantry in general. For the *Salagama* people, the impact of cinnamon was seemingly twofold.

As evident from their rebellious behaviour, the impact was seemingly negative in the early eighteenth century. However, the situation may have changed gradually and they seemed to have economically benefitted from their engagement in cinnamon collection in the latter part of the eighteenth century.

The most compelling testimony to the negative impact of cinnamon on the life of ordinary peasants is the character of peasant revolts in the Dutch administered territory in the WS region in the eighteenth century. Both the *Salagama* people and the peasantry in general participated in the widespread rebellious activities in the early and mid-eighteenth century. The way in which the Dutch administered cinnamon was a major factor behind these revolts.[9]

As will be discussed in the next section, the initial resistance from the *Salagama* community came for the cinnamon plantations, probably based on the assumption that the plantations may deprive them of a source of income. This suggests a major shift in the way in which cinnamon peeling conditioned the lives of *Salagama* people.

Before this shift in attitude, cinnamon peeling had created major discontent among members of the *Salagama* community. Much of the cinnamon growing region was in a state of strife for long and there apparently was a serious shortage of labour for cinnamon peeling. Heavy control over the life and labour of the *Salagama* people was evident in the early part of the eighteenth century, and as already mentioned the Dutch administration took tough measures to protect the limited supply of *Salagama* labour for cinnamon peeling as well as to protect cinnamon trees.

The *Salagama* people resorted to various means to avoid the responsibility of peeling cinnamon. Apart from revolting, many of them fled to the Kandyan lands to avoid cinnamon peeling. This annoyed the Dutch administration very much, and bringing back absconding cinnamon peelers was a key issue in the Dutch-Kandy relations during the first half of the eighteenth century.

The primary reason for the tension between *Salagama* people and the Dutch administration in the first half of the eighteenth century was arguably the widening gap between the demand for cinnamon and the availability of labour for cinnamon peeling. While there was no major increase in the demand for cinnamon, especially in the European market, there was a decline in labour supply, possibly due to the general decrease of the population. Incessant conflict in the region for more than one and half centuries would have taken its toll on the population of the region (Arasaratnam 1996). This situation had begun to change in the early eighteenth century (Paranavitana 2001: 47).

The wave of riots by cinnamon peelers in the first half of the eighteenth century was probably a manifestation of this demographic change. It might have taken some time for the administration to come to terms with the demographic change. Governor Van Gollenesse in 1748 reported an increase in the number of cinnamon peelers from 968 to 2924 men (Gollenesse 1974: 74). Cinnamon workers escaping to Kandyan territory would have further aggravated the problem.

[9] For more details about these revolts, see Jayawardana (2010).

The increase in population seems to have had a dual impact on the cinnamon industry. It had a positive impact by way of the increase in labour supply for cinnamon. However, the population increase also had a negative impact as it demanded more land for *chena* cultivation. The administration perceived this as detrimental to the protection of cinnamon trees since clearing of fresh land for settlement and agricultural production purposes caused the destruction of cinnamon trees.

Another social conflict that emerged out of this problem is the growing conflict between cinnamon peelers and other ordinary peasants. Ordinary peasants regularly complained against cinnamon peelers forcefully entering their land to peel cinnamon trees. This appears to be the result of the difficulties faced by cinnamon peelers in finding cinnamon trees in a sufficient number, leading to a problem for both peelers and other peasants.

The crisis of the first half of the eighteenth century could be understood in this context. When Governor Diederik van Domburg died suddenly in 1736, the entire Dutch territory in the West coast was in a state of insurgency. The reason for the turmoil is generally attributed to the mishandling of the situation by Governor Domburg.[10] This allegation was in particular levelled by his successor Van Imhoff (1736–1739) (Imhoff 1910). While the allegation of mishandling might have been justified, one cannot ignore the larger context of the background to the turmoil.

As discussed in the previous section, while the burden of collection of cinnamon fell on the people of the *Salagama* caste, that burden also fell on the entire peasant population owing to the harsh measures introduced by the Dutch administration to protect cinnamon trees from destruction. The peasant population suffered most by restrictions imposed on *chena* cultivation. Instructions issued in connection with *chena* clearance required a formal request by aspiring cultivators, whereupon commissioners may be appointed by the government to visit the lands intended for clearing. Applicants were expected to produce a detailed report on cinnamon trees found in those lands. Needless to say, this was an entirely intimidating procedure for the peasants in the eighteenth century.

The situation seemed to have changed with the setting up of cinnamon plantations in the 1770s (Fig. 2.3). Even by the time plantations were inaugurated, cinnamon peelers were probably benefitting from their role in this economic activity. General unrest among the peasantry also let up, at least temporarily.

2.9 Cinnamon Industry After Plantations

The VOC administration's decision to start cinnamon plantations was a crucial decision in many respects. It affected the entire organization of the cinnamon production. The organization of labour, the mode of land utilization and transportation

[10] For more details, see Kotelawele (1968: 177–222).

Fig. 2.3 Governor Iman Willem Falck (1765–1785) under whom cinnamon plantations were inaugurated. (Source: Wikipedia)

of cinnamon from peeling centres to the Colombo harbour were some of the important aspects that were affected. Most of the problems faced by the Dutch administration in the early and middle parts of the eighteenth century were associated with cinnamon collection. The success of cinnamon plantations seems to have solved many of these problems.

It is true that the cinnamon plantations were inaugurated at the last phase of the Dutch administration of Sri Lanka. It is also important to note that this was a time that the VOC was in a state of deep crisis as a commercial venture. None of these factors, however, belittle the importance of cinnamon plantations in relation to the social and economic structure of the country. Cinnamon plantations were inaugurated in the year 1770. The establishment of plantations was followed by successful experiment under Governor Willem Falck (1765–1785) (Fig. 2.3). When Governor Falck successfully cultivated cinnamon in his garden, the established belief that cinnamon cannot be planted was debunked.

The lands that were brought under cinnamon plantations were comparatively large. A series of maps prepared in 1794 shows the massive extent of lands that were brought under cinnamon cultivation in the present-day Colombo and Gampaha districts (Fig. 2.4). While the government was by far the main owner of the plantations, a sizable amount of plantations was owned by private individuals, particularly indigenous chiefs. The government encouraged indigenous chiefs to establish cinnamon plantations. Evidence shows that the chiefs responded positively, and large tracts of new lands were made available to start new cultivations (Table 2.2).

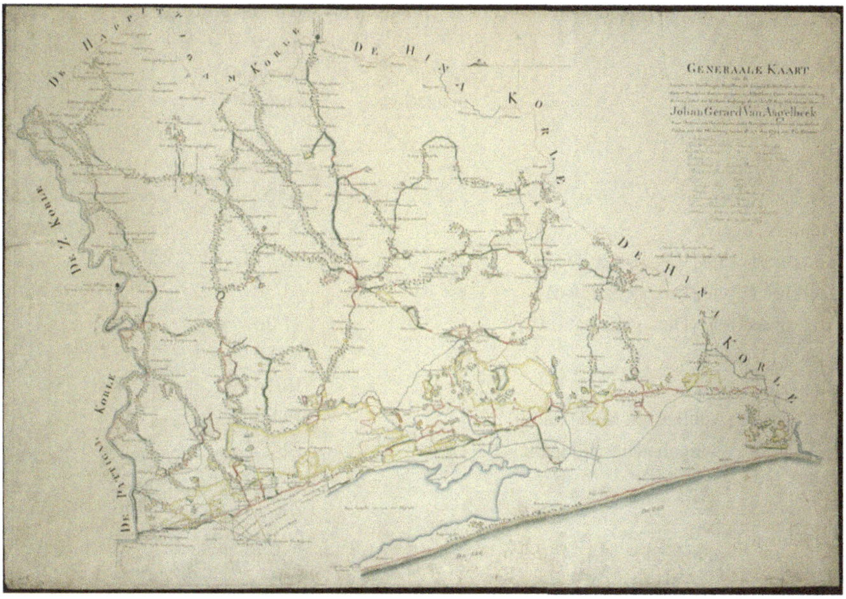

Fig. 2.4 A map of Aluthkuru Korale (prepared in 1794). Cinnamon plantations are shown in yellow. (Source: National Archives, The Hague)

The labour needs were also diversified. As Table 2.2 shows, a large number of overseers were needed to look after plantations. There is evidence of the lack of proper maintenance of the plantations (Dewasiri 2008: 99–100). It could be anticipated that the labour conditions of the peelers also became much favourable owing to the plantations. No major resistance has been reported from the cinnamon peelers in the second half of the eighteenth century. It has also been shown that there was a considerable degree of upward social mobility among members of the *Salagama* caste. This was an obvious sign of the changing circumstances of the lives of cinnamon peelers compared to the appalling conditions of their life in the early eighteenth century (Table 2.3).

Kanapathypillai maintains a rather critical view towards the cinnamon plantations. He argues that the plantations failed to meet the expectations and concludes that the whole exercise was a failure (Kanapathypillai 1969: 280–307). Schrikker, with the backing of a more comprehensive perusal of the contemporary documentation of the VOC administration of Sri Lanka, challenges Kanapathypillai's conclusions. Schrikker has shown that at the time that English captured the island from the Dutch, cinnamon stocks sufficient for the European market were found in the stores (Schrikker 2007: 56).

It is important to place the inauguration of plantations and its accomplishments within the overall policy framework of the Dutch administration, which for long had been following a policy of encouraging the local population to bring more land under commercial crops. Large tracts of lands were made available for aspiring

Table 2.2 Regional distribution of cinnamon gardens in the Colombo disavany[a]

Geographical area	No. of plantations	Size (Amunu/ kuruni)[b]	Overseers needed
Colombo four Gravets	23	1056/06	274
Vidane Kotte	6	11/20	18
Vidane Wattala	4	9/30	9
Vidane Kaleniya	4	2/30	4
Alutkuru korale (Ragam pattuwa)	37	789/13	50
Alutkuru korale (Dasiya pattuwa)	25	83/20	66
Alutkuru korale (Dunagaha pattuwa)	13	41/26	38
Negombo district	12	60/30	46
Hewagam korale	9	18/16	28
Hina korale (Adhikari & Meda pattu)	14	132/30	58
Hina korale (Gangaboda & Udugaha pattu)	13	22/23	28
Salpity korale	19	202/14	146
Moratuwa	22	201/25	105
Raigam korale	18	67	67
Panadura district	9	23/03	30
Kalutara district	29	76/19	107
Hapitigam korale	36	56/03	86
Pasdun korale	13	22/29	50
Walallawiti korale	7	8	6

Source: Dewasiri (2008: 80)

[a]The term that the Portuguese and the Dutch used for the province administered by a *disave*. *Disave* is a position existed during the pre-colonial kings too. The Portuguese as well as the Dutch retained the nomenclature

[b]A unit of measure. In the case of paddy and other grains, it is based on sowing capacity, but it varied from region to region. Forty kuruni = one amuna, or approximately two acres of paddy lands. The same units of measures were used for measuring other types of lands too

cultivators. Initially, popular cash crops such as coconut, pepper and coffee, as well as paddy cultivation, were also encouraged (Kotelawela 1968: 109176). The availability of land in large measure for cultivation was enabled by the ways in which the term ‘Dutch land’ was interpreted and put into practice in such a way that the VOC got hold of large extents of land at the expense of the ordinary peasants (Dewasiri 2008: 105–140). The administration made use of the unpaid, or partially paid, labour of ordinary peasants to cultivate these lands.

When the British occupied the Dutch possessions in the island in 1796, they decided to revisit the whole issue of cinnamon plantations, which ‘lay dispersed in great numbers on the south and south-west coasts of the island, between Matura and Chilaw (Bertolacci 1983, sic.). The distribution of cinnamon plantations was to be rationalized following new needs of land utilization. For example, the plantations in Maradana (now within the Colombo municipality limits) was considered to be the best one (Fig. 2.5). It was difficult for this to be expanded because more lands were

Table 2.3 Ownership of cinnamon plantations in 1786

Owner	No. of plantations owned	Amount (*amunu/ kuruni*)
Company	38	2022/02
Dissave[a]	3	41/00
Mudliyars[b]	8	387/13
Muhandirams[c]	51	89/25
Mahavidane[d]	14	118/26
Korale[e]	5	4/15
Athukorala[f]	2	00/14
Vidane[g]	33	75/26
Arachchi[h]	18	22/30
Wibadde (vidane)[i]	2	1/10
School master	3	3/05
Kangaan[j]	8	2/35
Mahabadderala[k]	2	24/00
Majoraal[l]	1	00/05
Baas	1	2/00
Lascarine[m]	2	00/20
Nainde[n]	1	00/06
Gamekeeper	1	00/10
Interpreter	1	00/03
Mohottirala[o]	1	00/15
Clerk	1	00/10
Others	33	72/04

Source: Dewasiri (2008: 82)
[a]Administrator of a province
[b]Highest rank of indigenous chiefs under the Dutch
[c]High-ranking indigenous chief below the rank of *Mudaliyar*
[d]An indigenous chief who is responsible for mobilizing corvée labour
[e]Chief of a sub-unit of a province bearing the same name
[f]A chief below the rank of *Korala*
[g]A minor indigenous chief
[h]A local level chief, below *Korala*
[i]A village-level headman responsible for collecting paddy revenue from the cultivator on behalf of the VOC
[j]Overseer
[k]An indigenous chief of the cinnamon department
[l]An indigenous chief who is responsible for mobilizing corvee labour
[m]A term used primarily for indigenous soldiers who also served as messengers and guards
[n]Generally translated as husbandman. They performed many sorts of menial work as their corvée labour. They were used mostly for work on plantations
[o]A clerk or secretary

Fig. 2.5 Cinnamon processing sheds at Maradana. c. 1785. (Source: Wagenaar 2017: 154)

needed to expand the city of Colombo. 'Further, the plantation being close to a garrison and surrounded by a numerous population, much destruction was caused by multitude of people who daily crossed it, and by the depredations of cattle and those searching for firewood' (De Silva 1995a, b, c: 418).

Colvin R. de Silva states that '[w]hen the Maritime Provinces were captured, East India Company .... purchased for £180,000 the 732,885 lbs. of the price cinnamon which had been found in the Colombo warehouses and in the Dutch ships captured in 1795' (De Silva 1995b,: 414).

At the time of the East India Company's monopoly on the cinnamon trade was abolished by the Colebrooke-Cameron Commission in 1932, Sri Lanka was still supplying best quality cinnamon to the European market. There were, however, two factors that affected the central place that cinnamon occupied hitherto in the economy of Sri Lanka. First, there was a growing competition from the cheaper variety of cassia from Indonesia and the Indian mainland. Although the export of cinnamon had been nearly doubled in 1857 compared to the situation in 1841, gross value had decreased (Tennent 1999: 165).

The other factor is the growth of significance of other plantation crops, such as coffee and cocoa and later tea, rubber and coconut. These cash crops were becoming more important in the global trade following the changes brought about by the industrialization of the West. While cinnamon remained, and still remains, an important commodity from Sri Lanka in the international market, its relative importance and social impact have drastically changed.

2.10 Concluding Remarks

Philippus Baldaeus, writing in the late seventeenth century, described Sri Lankan cinnamon as the 'Helen or Bride in contest in this Isle' (De Silva 1995a, b, c: 414). This portrayal was in fact to designate the importance of cinnamon as a vital commodity of international trade at the time. Sri Lanka had the reputation as the place of origin of the best quality cinnamon. When the global trade revived in the early second millennium, the economic social and political significance of Southern and Western parts of the island also increased.

While Portuguese were more concerned about the strategic location of Sri Lanka in their quest for expelling Muslims from the Indian Ocean, they had an interest in the flourishing cinnamon trade in Sri Lanka too. The transformative effect of cinnamon in the social formation was, however, clearly evident during the long presence of the Dutch in the cinnamon growing parts of the country. The land and labour policies of the Dutch administration were largely shaped by its interests in cinnamon. The Dutch, as well as the Portuguese, made full use of the corvée labour of the *Salagama* caste to peel cinnamon. Significant changes were affected in the system under the Dutch.

The biggest challenge that the Dutch confronted was how to make use of the limited supply of labour of the *Salagama* caste to meet the full demand for cinnamon. Moreover, the problem of labour mobilization for the cinnamon production was supplemented by the issue of the administration of cinnamon growing terrain. Cinnamon trees had to be protected, practically, from the ordinary peasants, measures taken for which were major nuisance for them. This situation created much tension between the administration and the people and it caused a wave of peasant rebellions. There were of course other contributory factors for this tension and the peasant riots that followed. Cinnamon was, however, a major factor.

The cinnamon policy was linked with many aspects of the policies related to land tenure and the mobilization of labour. The eighteenth century saw major changes being brought about in relation to these aspects of the social formation.

While cinnamon plantations could resolve some of the major problems related to the cinnamon industry, they also contributed to other important developments, such as the creation of new avenues for upward social mobility, especially among the *Salagama* caste people. It also significantly contributed to the change in the socio-economic landscape of the WS region. These changes altogether restructured the traditional agrarian system.

While the British continued to utilize the infrastructure of the cinnamon industry as developed by the Dutch, new developments in the nineteenth century deprived the central position that cinnamon industry enjoyed hitherto. Increasing demand for inferior quality cinnamon in the European market and the higher focus on other cash crops became major factors.

References

Arasaratnam S (1958) Dutch power in Ceylon, 1658–1687. Djambatan, Amsterdam

Arasaratnam S (1996) Ceylon and the Dutch, 1600–1800. Variorum, Aldershot

Bertolacci, A (1983) A view of the agricultural, commercial, and financial interests of Ceylon, Thisara Publishers, Dehiwala

Carswell J, Deraniyagala S, Graham A (2013) Mantai: city by the sea. Cambridge University Press, Cambridge

De Silva CR (1995a) Ceylon under the British occupation, 1795–1833: its political, administrative and economic development, II. Navrang, New Delhi

De Silva CR (1995b) The rise and fall of the kingdom of Sitavaka. In: De Silva KM (ed) History of Sri Lanka, II. University of Peradeniya, Peradeniya, pp 61–104

De Silva CR (1995c) Sri Lanka in the early sixteenth century: economic and social conditions. In: De Silva KM (ed) History of Sri Lanka, II. University of Peradeniya, Peradeniya, pp 37–60

De Silva CR, Pathmanathan S (1995) The kingdom of Jaffna up to 1620. In: De Silva KM (ed) History of Sri Lanka, II. University of Peradeniya, Peradeniya, pp 105–121

Dewaraja LS (1994) The Muslims of Sri Lanka: one thousand years of ethnic harmony, 900–1915. Lanka Islamic Foundation, Colombo

Dewasiri NR (2008) The adaptable peasant: Agrarian society in Western Sri Lanka under Dutch rule, 1740–1800. Brill, Leiden

Dewasiri NR (2017) Dutch and cinnamon peelers: a critical evaluation of the relations between the VOC and cinnamon peeling Chalias community in the mid-eighteenth century Sri Lanka. In: A felicitation volume in honour of Professor Indrani Munasinghe Professor Emeritus of History. Department of History, University of Colombo, Colombo, pp 32–56

Gunawardana KW (1958) Foundation of the Dutch power in Ceylon. Djambatan, Amsterdam

Hovy L (1991) Ceylonees Plakkaatoek: plakkaten en andere wetten uitgevaardigd door het Nederlandse estuur op Ceylon, 1638–1796, Deel II. Verloren, Hilversum

Jayawardana K (2010) Perpetual ferment: popular revolts in Sri Lanka in the 18th and 19th centuries. Social Scientist Association, Colombo

Kanapathypillai V (1969) Dutch rule in maritime Ceylon. Dissertation, University of London

Kotelawele DA (1968) The Dutch in Ceylon, 1743–1766. Dissertation, University of London

Kotelawele DA (1995) The VOC in Sri Lanka 1658–1796: problems and policies. In: History of Sri Lanka, II. University of Peradeniya, Peradeniya, pp 417–451

Liyanagamage A (2001) Keralas in medieval Sri Lankan history: a study of two contrasting roles. In: Liyanagamage A (ed) Society, state, and religion in premodern Sri Lanka. University of Kelaniya, Kelaniya

Loten JG (1935) Memoir of Joan Gideon Loten Governor of Ceylon delivered to his successor Jan Schreuder on February 28, 1757 (trans: Reimers E). Ceylon Government Press, Colombo

Paranavitana KD (2001) Land for money: Dutch land registration in Sri Lanka. Author, Colombo

Perera BJ (1952) The foreign trade and commerce of ancient Ceylon, IV – exports and imports of ancient Ceylon. Ceylon Hist J II(1 & 2):14–22

Pybus J (1958) The Pybus embassy to Kandy, 1762 (trans: Raven-Hart R). The National Museums of Ceylon, Colombo

Roberts M (1982) Caste conflict and elite formation: the rise of a Karava Elite in Sri Lanka, 1500–1931. Cambridge University Press, Cambridge

Schrikker A (2007) Dutch and British colonial intervention in Sri Lanka, 1780–1815: expansion and reform. Brill, Leiden

Tennent JE (1999) Ceylon: an account of the island physical, historical and topographical with notice of its natural history, antiquities and productions. Asian Educational Service, New Delhi

The Travels of Marco Polo (1997). Wordsworth Classics of World Literature, Hertfordshire

Van den Belt A (2008) Het VOC-bedrijf op Ceylon, Een voorname vestiging van de Oost-Indische Compagnie in de 18de eeuw. Walburg Pers, Zutphen

van Gollenesse JS (1974) Memoir of Julius Stein van Gollenesse, Governor of Ceylon, 1743–1751 (trans: Arasaratnam S). Department of National Archives, Colombo

Van Imhoff B (1910) Memoir of Gustaaf Willem Baron Van Imhoff Governor of Ceylon delivered to his successor 1740 (trans: Pieters S). Ceylon Government Press, Colombo

Wagenaar LJ (1996) Knielen of buigen? De gezantschappen van de Compagnie naar Kandy na het vredesverdrag van 1766. In: Davids CA, Fritschy W, van der Valk LA (eds) Kapitaal, ondernemerschap en beleid: studies over economie en politiek in Nederland, Europa en Azië van 1500 tot heden — Afscheidsbundel voor prof. dr. P.W. Klein. NEHA, Amsterdam, pp 441–466

Wagenaar LJ (2017) Cinnamon and elephants: Sri Lanka and the Netherlands from 1600. Rijk Museum, Vantilt Publishers, Amsterdam

Chapter 3
Ceylon Cinnamon Production and Markets

Achini M. De Silva and Mohamed Esham

3.1 Introduction

Ceylon cinnamon is known as the miracle spice of Sri Lanka. Botanically known as *Cinnamomum zelanicum* Blume, it is indigenous to Sri Lanka. It is one of the first spices known to the human race since the early cultivation. In ancient times Ceylon cinnamon was the most important spice exported from the island. Sri Lankan cinnamon has established its identity in the global market place as "Ceylon cinnamon." Sri Lanka produces and exports about 90% of Sri Lankan true cinnamon to the world annually (Lankage 2017). Cinnamon at present is the fourth most important export agricultural crop and the main spice exported in terms of foreign exchange earnings. It is popularly known as "True cinnamon" or "Pure cinnamon" and it is the only trade in which Sri Lanka holds a monopoly in the world market.

3.2 Production

Ceylon cinnamon is predominantly grown in the intermediate and wet zones of the country. The main production areas of cinnamon are Galle, Matara, Kalutara, Ratnapura, and Hambantota districts with a total annual production of 24,000 MT covering an extent of 33,000 ha (Central Bank of Sri Lanka 2019a). The production has doubled since 2001 while the extent of cultivation has expanded by 35%, implying that yield has increased by 42% over the same period (Fig. 3.1).

Globally cinnamon production is dominated by four countries, namely Indonesia, China, Sri Lanka, and Vietnam (Table 3.1). Among the four countries except for Sri

A. M. De Silva · M. Esham (✉)
Department of Agribusiness Management, Faculty of Agricultural Sciences, Sabaragamuwa University, Belihuloya, Sri Lanka
e-mail: esham@agri.sab.ac.lk

R. Senaratne, R. Pathirana (eds.), *Cinnamon*,
https://doi.org/10.1007/978-3-030-54426-3_3

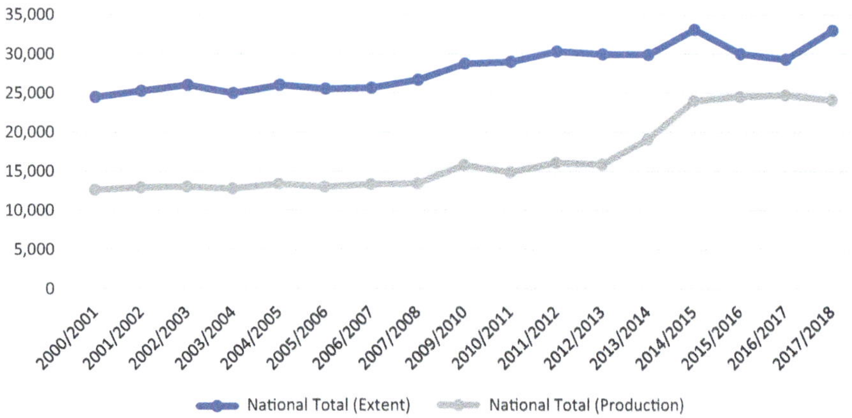

Fig. 3.1 Cinnamon extent and production in Sri Lanka from 2000 to 2018

Table 3.1 Production of cinnamon: top 10 countries (average production 1994–2014)

Country	Production (tones)
1. Indonesia	70,861
2. China, mainland	47,549
3. Sri Lanka	13,520
4. Vietnam	12,855
5. Madagascar	1779
6. Seychelles	239
7. Timor-Leste	92
8. Grenada	69
9. Sao Tome and Principe	59
10. Dominica	52

Source: FAOSTAT (2018)

Lanka, other countries produce cassia. Madagascar and Seychelles are the other two countries producing true cinnamon. As shown in Figs. 3.2 and 3.3, cassia production and the extent of cultivation have grown significantly over the past three decades. For instance, in Vietnam extent cultivated and production has grown at an unprecedented rate of 4700% and 3800%, respectively. A similar trend can be observed in Indonesia and China. However, in Sri Lanka both extents cultivated, and production has shown mediocre growth over the same period, despite the country enjoying a competitive advantage and being the leading exporter of Ceylon cinnamon to the global market.

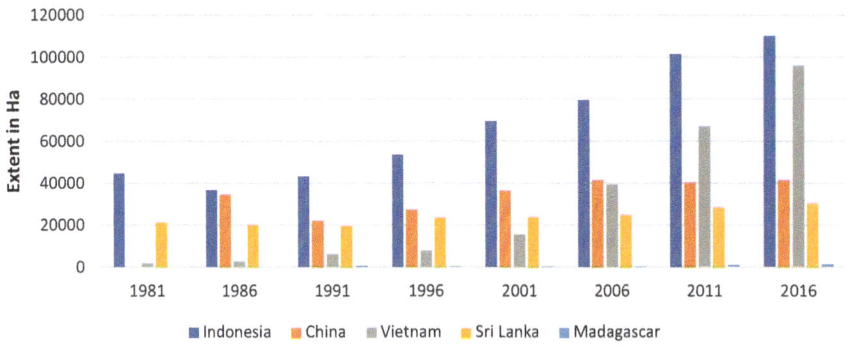

Fig. 3.2 Country-wise cinnamon cultivation trend. (Source: FAOSTAT (2018))

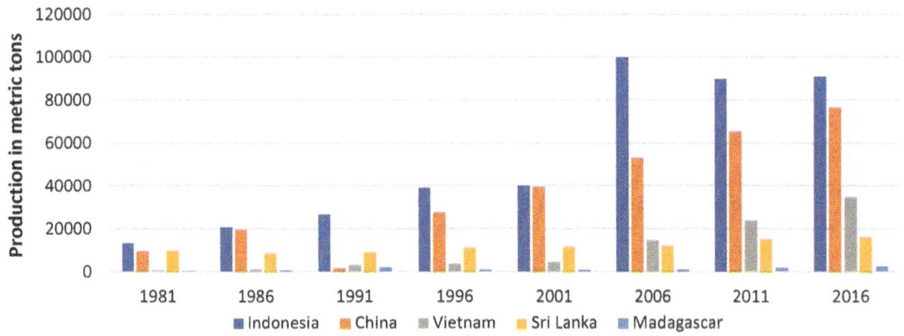

Fig. 3.3 Country-wise cinnamon production trend. (Source: FAOSTAT (2018))

3.3 Cinnamon Products

Sri Lanka is the largest true cinnamon producer. Among true cinnamon produce, 90% is in the form of quills. However, now there is a growing interest in value-added forms such as bark powder and leaf oil. Cinnamon is exported in three main forms, namely non-value-added bulk, value-added bulk, and value-added consumer packs. A survey of 43 exporters revealed that 71% of exports are in bulk form and 29% are exported in value-added packs (Harindra 2017).

In the Cinnamon industry, bulk products are described as packed products in large quantity, which are transported by cargos or air freight to international business to business (B2B) buyer. Mainly the following cinnamon products are bulked:

- Cinnamon bales: Bundle of quills maximum length is 42 inches and each bundle weighs about 25–45 kg, wrapped with gunny bags
- Quillings: Pieces of quills resulting from the baling process
- Chips: Dried bark of unpeeled cinnamon stems, branches, and trimmings inclusive of the outer bark

- Cut cinnamon: Perfectly cut cinnamon quills to the required size and any desired grade
- Powder: Ceylon cinnamon of any grade, finely powdered to any required mesh size
- Leaf oil: Essential oil extracted from cinnamon leaves, containing a high percentage of Eugenol
- Bark oil: Essential oil extracted from cinnamon quillings, containing a high percentage of cinnamaldehyde

3.3.1 Harmonized System Code-Based Cinnamon Product Categories

According to the designated harmonized system (HS) codes, there are 12 different product categories being exported as bulk forms. Table 3.2 provides an HS code-wise segregation list of products intended for the bulk market.

3.3.2 Ceylon Cinnamon and Cassia

Cassia or *Cinnamomum aromaticum* syn *C. cassia* is the main competitive product for Ceylon cinnamon. It is sold in the world market with the label of cinnamon. China, Indonesia, and Vietnam are the dominant producers and exporters of cassia. Because of the cheaper price, it is increasingly replacing the Ceylon cinnamon in the global market. This is despite the warning of leading health agencies about its

Table 3.2 List of harmonized system (HS) code-wise segregated cinnamon products for exports

HS code	Product type
09061090	Cinnamon and cinnamon tree flowers
09061110	Quills cut, in retail packs of 1 kg or less
09061120	Other cinnamon quills
09061130	Cinnamon quillings
09061140	Cinnamon featherings
09061150	Chips
09061190	Other
090619	Other
09062010	Cinnamon (*Cinnamomum zeylanicum* Blume) crushed
09062020	Cinnamon (*Cinnamomum zeylanicum* Blume) ground
09062090	Other

Source: Sri Lanka Customs, Colombo (2019)

Table 3.3 Ceylon cinnamon vs. cassia

	Ceylon cinnamon	Cassia
Botanical name	*Cinnamomum zeylanicum*	*Cinnamomum aromaticum; Cinnamomum cassia*
Other names	Sweet cinnamon, true cinnamon, Mexican cinnamon	Chinese cinnamon, Tung Hing
Country of origin	Sri Lanka	China, Indonesia, Vietnam
Flavor	A mild, soft, and sweet aromatic flavor	Intense flavor, pungent, and very spicy flavor
Coumarin content	0.0004%	5%
Health effect	Generally safe	Known to cause liver and kidney damage
Price	Relatively expensive	Relatively cheap

Sources: Piyasiri and Wijeratne (2016), and Lungarini et al. (2008)

negative impact on health due to the high content of coumarin. Ceylon cinnamon compared to cassia is lighter brown in color, softer, and sweeter. Although it is in quill form, it is easy to distinguish cassia from cinnamon. However, in powdered form, it is not easy to distinguish cassia from cinnamon. Hence, Ceylon cinnamon is often adulterated with cassia, especially when it is in powdered form. A study based on cinnamon samples obtained from the Italian market revealed that about 51% of cinnamon samples consisted of cassia, 10% were a blend of cassia and Ceylon cinnamon, whereas only 39% were actually Ceylon cinnamon (Lungarini et al. 2008). Table 3.3 shows a comparison between Ceylon cinnamon and cassia.

Ceylon cinnamon has a distinct competitive advantage over cassia due to its superior chemical and physical properties. Moreover, Ceylon cinnamon is the only product exported in the form of cinnamon quills. Preparation of the cinnamon quill is an art unique to Sri Lanka, which is carried over from generation to generation.

3.4 Compliance on Food Safety and Quality Standards

Although Sri Lanka has a competitive edge over Ceylon cinnamon in the world market so far, the country has failed to capitalize on this due to the failure to adopt proper food safety and quality management systems especially at the upstream of the value chain. Some of the cinnamon value chain actors follow poor production and processing practices compromising the food safety and quality of the entire value chain. This has challenged the country's international market share and its ability to compete in the global market.

Deterioration of quality of cinnamon happens at every level of the cinnamon value chain from farmer to the table. Most of the value chain actors are less conscious about the quality of the cinnamon due to the lack of awareness of international market requirements, standards, and specifications of cinnamon.

3.4.1 Product and Process Standards for Cinnamon

Standards are defined by ISO as "documented agreements containing the technical specification or other precise criteria to be used consistently as rules, guidelines or definitions, to ensure that materials, products, processes, and services are fit for this purpose."

Product standards are a set of specifications and criteria that lay down the properties of the product. Process standards are a set of criteria to be followed in the process of making the product. Social and environmental standards in agriculture are means of commitment to social and environmental sustainability and they are essentially process standards. Product standards pertinent to cinnamon include the following:

1. Sri Lanka standard specification for Ceylon cinnamon—SLS81:2010
2. Code of hygienic practice for spices and other dried aromatic plants—SLS 1327:2008
3. Code of practice for general principle's food hygiene—SLS 143:1999
4. Classifications for quills—ISO 6535:1997 SLS 81:2000

3.4.2 Ceylon Cinnamon Grades

Quills are graded on the basis of the diameter of the quill and the level of foxing. There are four main grades and accompanying subgrades. The main grades are Alba, Continental (C grade), Mexican (M Grade), and Hamburg (H Grade) (TSP 2010) (Table 3.4, Fig. 3.4).

3.4.2.1 Foxing

The occurrence of reddish-brown patches on the surface of the quills, which may become dark brown with time, is known as foxing. According to the Spice Council of Sri Lanka (2019), foxing can be of two types:

(a) Superficial patches (*"malkorahedi"*): appearing on the surface of the quills
(b) Heavy patches (*"korahedi"*): resulting in damage to the surface of the quills and making the quills uneven

Common process standards relevant to cinnamon are included in the following sections.

Table 3.4 Classification for quills (ISO 6535:1997) (SLS 81:2000)

Grade	Diameter of quill (max. mm)	Min. no. of 42″ long quills/ kg	% rough quills/ kg	Min. length of quills/ bail	Max. % of single quality quills/ bail	Min permissible overall extend of foxing	Max. % weight of cinnamon pieces in a bail
Alba	6	45	None	200	1	0	1
Continental (*C*)							
C5 Special	6	35	10	200	1	10	1
C5	10	31	10			10	1
C4	13	24	10			10	1
C3	16	22	15			15	1
C2	17	20	20			20	1
C1	19	18	25			25	1
Mexican (*M*)							
M Special	16	22	50	200	2	50	2
M5	16	22	60			60	2
M4	19	18	60			60	2
Hamburg (*H*)							
H1	23	11	25	150	3	25	3
H2	25	9	40			40	3
H3	38	7	60			65	3

Source: Samarawickrema (2015)

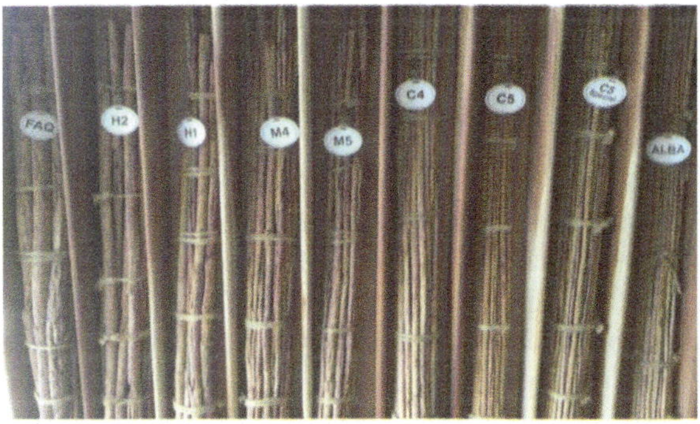

Fig. 3.4 Quality grades of cinnamon. (Note: FAQ is a H2 grade cinnamon quill)

3.4.2.2 Good Agricultural Practices

Good Agricultural Practices (GAPs) are a production- and farm-level approach to ensure the safety of fresh produce for human consumption. It is a voluntary audit that verifies agricultural products are produced, packed, handled, and stored as safely as possible to minimize risks of microbial food safety hazards. GAPs are practices that address the environmental, economic, and social sustainability of on-farm processes and result in safe and quality food and nonfood agricultural products (FAO 2013). GAP becomes highly relevant in the context of stringent maximum residue level (MRL) fixed by the European Commission for all foodstuff to ensure amounts of residues found in food are safe for consumers. MRL is the highest level of a pesticide residue that is legally allowed in food when pesticides are applied correctly following GAPs. There are high chances that the MRL would be made zero by prohibiting the presence of any traces of pesticides in the near future. This would demand the adoption of GAPs in all cinnamon plantations as well as setting up stringent quality control mechanisms to prevent substandard products from reaching the international market.

3.4.2.3 Good Manufacturing Practice

Good manufacturing practices (GMPs) are a system of processes, procedures, and documentation required to ensure *that* products are consistently produced and controlled adhering to quality standards. GMP guidelines and regulations address many issues that can influence the safety and quality of a product such as design and facilities of the establishment, clean and hygienic processing area, the establishment of hygiene in the premises, personnel hygiene and health of employees, transportation, and records of manufacturing and food safety training. GMPs in cinnamon cover the basic conditions and activities that need to be established in cinnamon processing, thus maintaining a hygienic environment suitable for the production, handling, and provision of safe end-product for human consumption (Mendis 2016).

3.4.2.4 ISO 22000:2005

ISO 22000:2005 specifies requirements for a food safety management system where an organization in the food industry has to demonstrate its ability to control food safety hazards to ensure that food is safe at all times until human consumption. This standard integrates the requirements defined by ISO 9001 and the methodology used by Hazard Analysis and Critical Control Points (HACCP) management systems (Teixeira and Sampaio 2011).

3.4.2.5 Food Safety System Certification 22000

Food Safety System Certification (FSSC) 22000 is a Food Safety Management System (FSMS) certification scheme based on ISO 22000. FSSC 22000 ensures food safety by defining, evaluating, and controlling risks and hazards in processing, manufacturing, packaging, storage activities transport, and distribution in the food value chain.

3.4.2.6 Hazard Analysis and Critical Control Points

Hazard Analysis and Critical Control Points (HACCP) is a food safety management system, in which food safety is addressed through the analysis and control of biological, chemical, and physical hazards starting from raw material until the finished product reaches the consumer. It is based on seven principles to ensure that potentially hazardous products do not reach the end customer (FDA 2014).

3.4.2.7 British Retail Consortium

The British Retail Consortium (BRC) certification is a global standard for food safety that has been developed to specify the safety, quality, and operational criteria required to be in a working area of the food industry. These standards guarantee the standardization of quality, safety, and operational criteria and ensure that manufacturers fulfill their legal obligations and provide protection for the end consumer. BRC global standards are now often a fundamental requirement of leading retailers, manufacturers, and food service organizations (BRC 2019).

3.4.3 Compliance, Awareness, and Perception of Food Safety and Quality Standards

Based on a survey of 328 value chain actors including growers, peelers, processors, traders, and exporters, it was found that food safety and quality compliance levels varied significantly among different levels of the cinnamon value chain. Compliance levels were low at the upstream compared with the downstream of the value chain. It was found that 58% and 48% of the exporters were GMP and ISO certified, respectively. Considering the level of awareness, exporters had a higher level of awareness on all standard and certification requirements. Other players had average knowledge about organic, GAP, and GMP certifications. However, their awareness about ISO, HACCP, and FSSC process standards was poor or negligible. The majority of exporters sourced information about food safety and quality requirements mainly from the suppliers and buyers, and industry literature, and by attending trade fairs (Madushani 2018).

Survey results revealed that the grower's peelers and traders showed negative attitudes towards food safety and quality standards and certifications. This can be attributed to a lack of returns or premium prices for products produced adhering to food safety and quality standards. Adoption of food safety and quality practices among the value chain actors is constrained by lack of awareness on consumer behavior in international markets, inadequate availability of local institutional services, cost of services, and lack of knowledge to assess the cost and benefits of adoption of certification system as against conventional products.

In cinnamon processing, more attention should be given to processing methods and hygienic conditions in the processing facilities. The traditional method of processing of cinnamon fails to meet the food safety requirements of high-end markets. The cinnamon processing is generally carried out on the floor, which leads to higher contamination as well as health hazards for processors (Weerasinghe et al. 2010). As the majority of cinnamon in Sri Lanka is produced by smallholder operators, quality assurance is challenging. Moreover, the majority of small producers are not trained in maintaining hygienic standards and food safety.

3.5 Marketing of Ceylon Cinnamon

Today, marketing is not just a function. It is a way of doing business giving priority to customer needs. Marketing helps to decide what business will sell, to whom, when, and how. According to the famous marketing guru, Philip Kotler, marketing is the science and art of exploring, creating, and delivering value to satisfy the needs of a target market at a profit. Where do Sri Lanka and other true cinnamon producing countries stand with respect to true cinnamon, for example, Sri Lanka's fourth most important exporting agricultural crop and the predominant spice exported in terms of foreign exchange earnings? Sadly, the performance in the international market appears to be poor. According to available literature, Sri Lanka has lost about 10–5% of its market share in value and volume, respectively, in the world market during the past 5 years (Piyasiri and Wijeratne 2016). This can be attributed to gains made by major cassia exporters: Indonesia, China, and Vietnam (Figs. 3.2 and 3.3).

3.5.1 Marketing Mix-Product

Cinnamon marketing mix is the combination of the marketing elements that need to be used effectively in promoting cinnamon and deliver a high-quality product to the customer. It is important to have a clear understanding of the marketing elements to craft marking strategies. Cinnamon products mainly consist of quills, quislings, featherings, and chips. The commercial parameters of quills include color, odor, flavor, moisture, volatile oil, shelf life, and packing (Table 3.5). There are four main grades of quills in the cinnamon industry, namely Alba, Continental, Hamburg, and Mexican. These grades are based on the diameter of quills, a number of whole quills per kg, the extent of foxing, minimum length of quills in a bale, etc.

Table 3.5 Commercial specifications of cinnamon

Character	Specification
Color	Pale brown to slightly reddish color of ground cinnamon—yellowish to reddish-brown in color
Odor	Characteristic fresh aroma
Flavor	Delicate and sweet flavor characteristic to Ceylon cinnamon. It shall be free from the foreign flavors including mustiness.
Moisture	Not more than 15% for quills and 12% for other grades
Volatile oil	Minimum 1% for quills and 0.7% for other grades on a dry basis.
Shelf life	Minimum of 1 year
Packing	Packaged in clean, sound, dry packages, made of jute, cloth, paper, or polyethylene bags.

Source: IPS (2017)

Figure 3.5 shows the percentage of quantities exported by sampled 45 traders/exporters in Sri Lanka in 2016/2017 (Warnakulasooriya 2017).[1] It is clear that the premium-quality grade Alba is hardly exported. According to exporters, producing Alba is time-consuming and needs meticulous efforts and skills (Warnakulasooriya 2017). Therefore, most of the processors tend to produce other grades instead of Alba, leading to a low volume of exports of Alba. Among all grades, the export basket consists of a higher volume of C5 and C5 special grades.

3.5.1.1 Major Ceylon Cinnamon Product Forms Exported Under Private Labeling

A survey of 43 exporters[2] from Sri Lanka by Harindra (2017) enabled to identify nine products exported under private labeling. These include bales, quillings, cut cinnamon, chips, powder, leaf oil, bark oil, oleoresin, and cinnamon tea. As shown in Fig. 3.6, the major export product is cut cinnamon. Cut cinnamon significantly contributes to bulk export volume. Cut cinnamon is a minimum value-added product and has a growing demand in the international market. Cut cinnamon was available in the export basket as wrapped individual sticks or unwrapped bundles of sticks. Further, most of the traditional exporters moved to this product form instead of exporting bales though it generates a waste (offcuts). Other than these major

[1] This study was conducted to identify the present status, issues, and marketing trends in the Ceylon cinnamon industry to explore marketing strategy concerning the entire value chain actors and marketing mix. Both quantitative and qualitative data were gathered from 45 exporters of cinnamon products using interviewer-administered questionnaires and in-depth interviews from August to October 2017 (Warnakulasooriya 2017).

[2] This study was undertaken to determine the impact of country of origin image over brand "Ceylon Cinnamon" to use as a secure vehicle to win the international market. Primary data were collected from 43 Ceylon cinnamon exporters using interviewer-administered questionnaires and focus group discussions. The study was conducted from July to November 2017 (Harindra 2017).

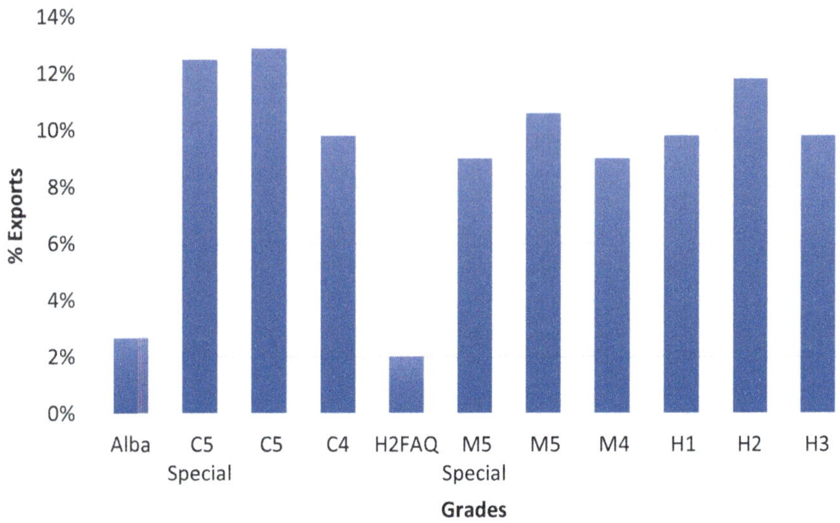

Fig. 3.5 Export volumes of cinnamon grades by 45 Sri Lankan exporters in 2016/2017

Fig. 3.6 Composition of different cinnamon-based product forms exported from Sri Lanka by 43 exporters in 2016/2017

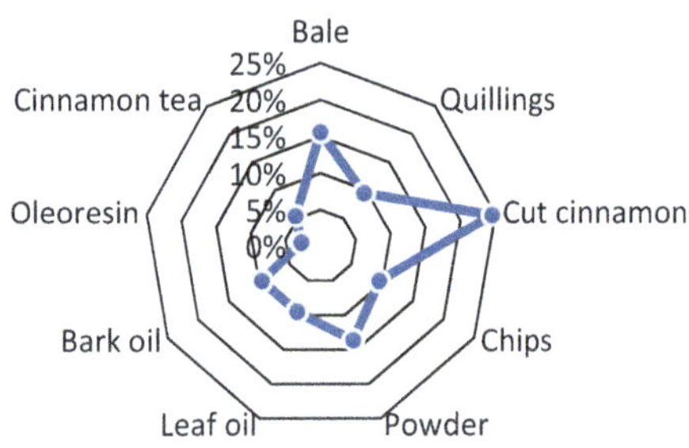

products, there were other value-added products exported in small quantities consisting of cinnamon tablets, personal care products, gift packs, scented candles, cinnamon with bee honey, cinnamon toothpicks, cinnamon spread, cinnamon grinder, cake spices, cinnamon snakes, cinnamon leaves, fractionates, cocktail infusion, and cinnamon cigar (Harindra 2017).

3.5.2 *Marketing Mix-Price*

The price of cinnamon to a larger extent depends on the grade. The highest premium price is fetched by Alba (Table 3.6). However, processors prefer to produce other grades despite low prices due to the difficulty of producing Alba.

According to the processors, the number of quills needed to make one bale, or 1 kg of cinnamon is a critical factor in determining the profitability. To produce 1 kg

Table 3.6 Producers' prices (farm gate) of cinnamon in Sri Lanka in 2019 (on December 17, 2019)

Grade of cinnamon	Highest price (Rs./kg)	Average price (Rs./kg)
Alba	2600	2513
C-5 Sp	2500	2280
C-5	2300	2053
C-4	2050	1816
M-5	1750	1672
M-4	1700	1633
H-1	1650	1537
H-2	1600	1441
H-Faq	1400	1250

Source: Department of Export Agriculture (2019)

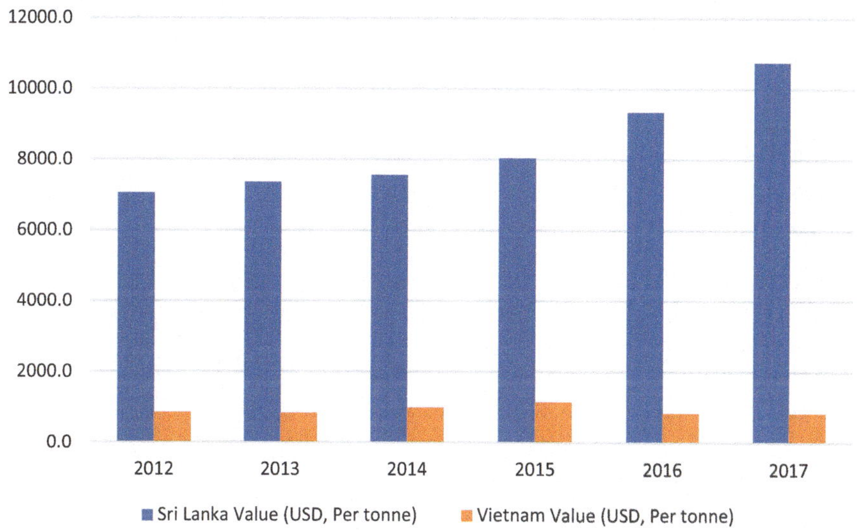

Fig. 3.7 Comparison of producer prices of Ceylon cinnamon (Sri Lanka) and cassia (Vietnam). (Source: FAOSTAT (2018))

of Alba, at least 45 Alba quills are needed while for other grades a lesser number of quills are required (Warnakulasooriya 2017).

An industry survey of 45 exporters[1] revealed that exporters always fetched a higher price for Ceylon cinnamon compared to the substitute cassia (Warnakulasooriya 2017). Figure 3.7 shows a comparison of producer prices of Ceylon cinnamon and cassia produced in Sri Lanka and Vietnam, respectively. It clearly shows a significant difference in producer prices between Ceylon cinnamon and cassia. The prices of cinnamon in the export market have shown a positive trend over the past years with significant price increases except during the 2008/2009 financial crisis (Fig. 3.8).

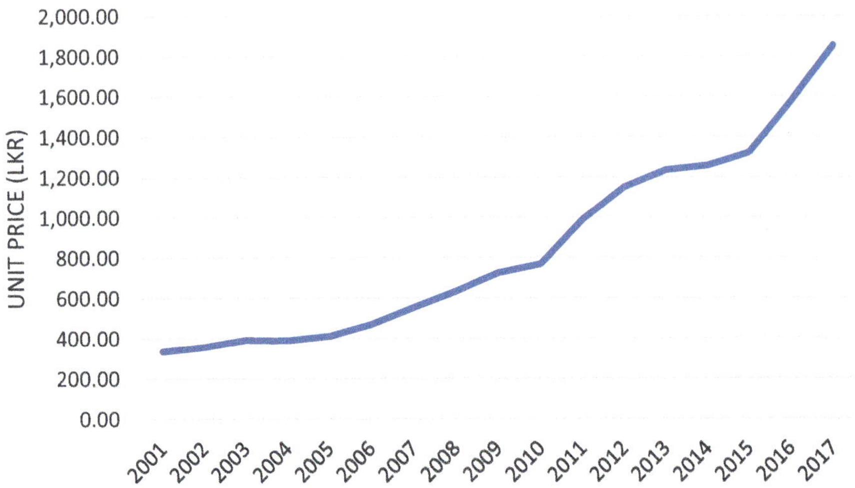

Fig. 3.8 Cinnamon export average price trend (price/kg). (Source: EDB (2019))

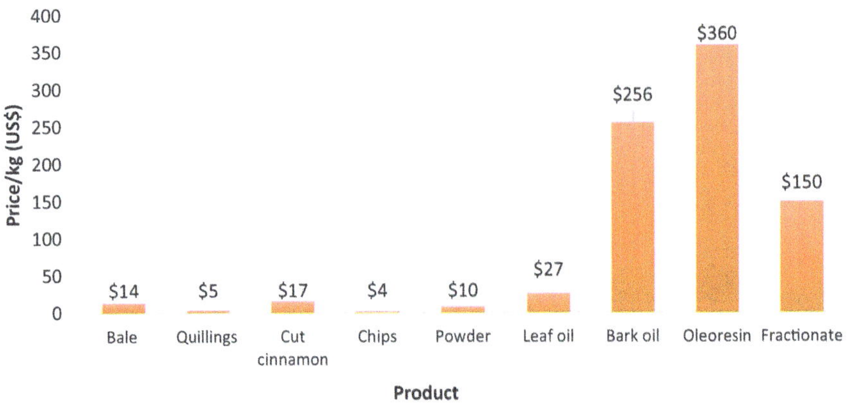

Fig. 3.9 Mean price for major Ceylon cinnamon product exports to the world market in 2017. (Source: Harindra (2017))

3.5.2.1 Export Prices for Ceylon Cinnamon Products Under Private Labeling

The variability of prices of private-labeled value-added cinnamon products is depicted in Fig. 3.9. Oleoresins received the highest price followed by bark oil, fractionates, and other products. This price variability indicates that minimally processed products receive significantly low prices compared to value-added products. The need of the hour is to produce value-added products and adopt diversification strategies to match the global customer needs.

The main reasons behind the reluctance of processors to add value to the cinnamon product were lack of technical know-how, the need for high-tech equipment, and the knowledge gap on end consumers' product preferences (Harindra 2017). The technical know-how of producing value-added products like oleoresins and fractionates is available with a handful of market leaders in the essential oil industry, which is not shared with others.

3.5.3 Marketing Mix-Place

The export destination of private-labeled products was identified on the basis of a survey of 43 exporters from Sri Lanka (Harindra 2017). Export market destinations were grouped into four categories. The categorization was done on the basis of the percentage volume of private-labeled products exported by an individual exporter to each of their market destinations. The categories were less than 25% (low), 25–50% (medium), 50–75% (high), and more than 75% (very high).

Figure 3.10 shows the countries where cinnamon exports were less than 25% of the total exports of individual exporters. Japan was the leading market destination for less than 25% category. This implies that Japan imported relatively smaller quantities of private-labeled products from individual exporters from Sri Lanka viz. 18.2% of the exporters export less than 25% in terms of volume of their private-labeled cinnamon products to Japan. According to EDB, Japan ranked in the 21st position with a market share of 0.49% and a market value of 1.05 US$ million (EDB 2019). Exporters revealed that Japan is a valuable niche market for Ceylon cinnamon. Value-added consumer packs are the most common value-added product exported to Japan.

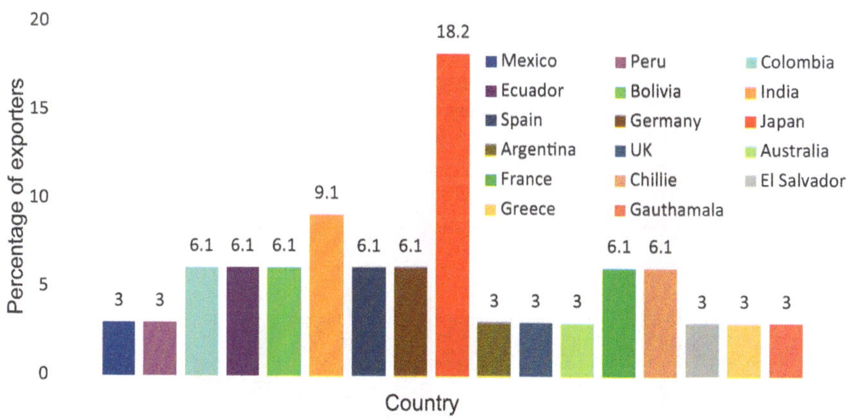

Fig. 3.10 Export market destinations of Sri Lankan cinnamon in the less than 25% exports category. (Source: Harindra (2017))

In the 25–50% category, Germany is the leading market destination for Ceylon cinnamon (Fig. 3.11). This implies that Germany imports 25–50% in terms of volume of private-labeled cinnamon products from individual Sri Lankan exporters. Germany ranks in the 11th position with a market share of 1.48% and a market value of 3.16 US$ million (EDB 2019).

The USA is the leading market destination for 50–75% category (Fig. 3.12). The individual Sri Lankan exporter exports 50–75% in terms of volume of their private-labeled cinnamon products to the USA. The USA is the second-largest market for Ceylon cinnamon with a market share of 11.9% and a market value of 25.32 US$ million (EDB 2019).

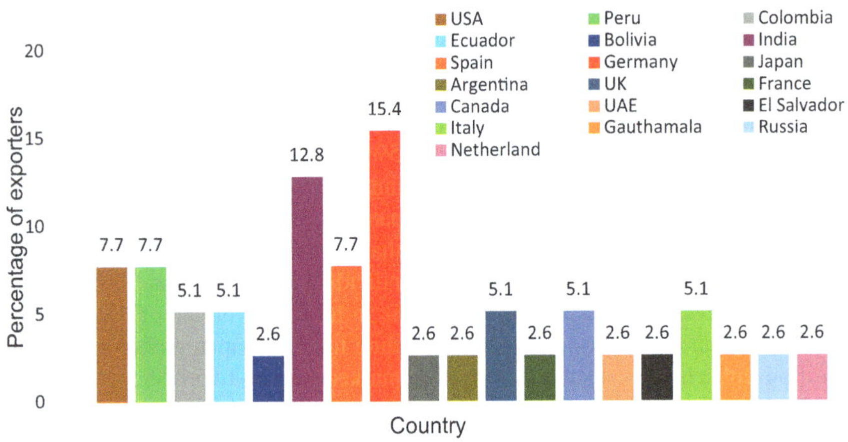

Fig. 3.11 Export market destinations of Sri Lankan cinnamon in the 25–50% exports category. (Source: Harindra (2017))

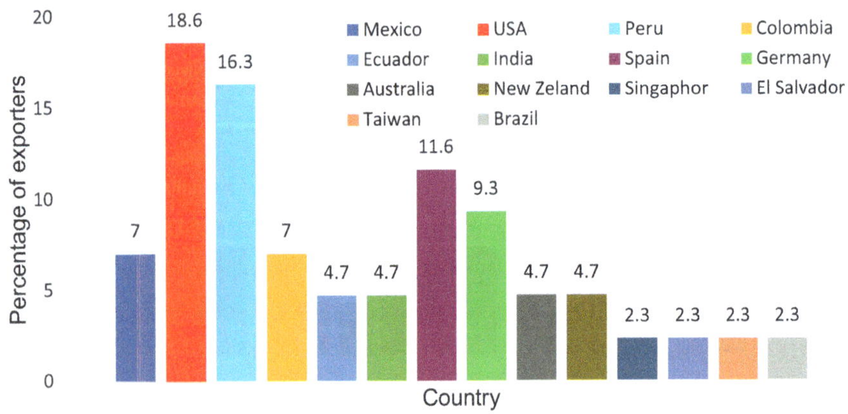

Fig. 3.12 Export market destinations of Sri Lankan cinnamon in the 50–75% exports category. (Source: Harindra (2017))

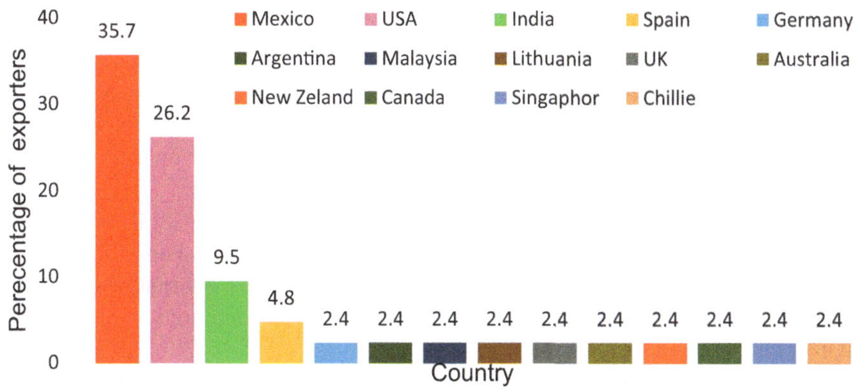

Fig. 3.13 Export market destinations of Sri Lankan cinnamon in the more than 75% exports category. (Source: Harindra (2017))

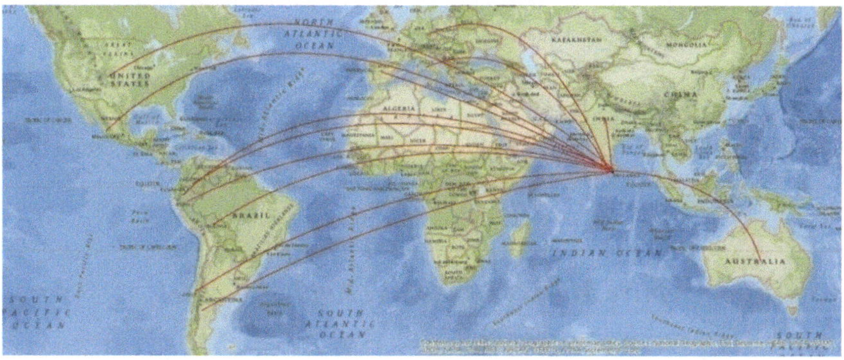

Fig. 3.14 Global Ceylon cinnamon export market destinations

In the more than 75% export category, Mexico is the leading market destination (Fig. 3.13). Mexico imports the bulk of private-labeled cinnamon products of individual exporters from Sri Lanka. Mexico, known as the hub of cinnamon, imported 94.11 US$ million worth of cinnamon from Sri Lanka in 2018. Mexico is the leading market for Ceylon cinnamon with a market share of 44.1% (EDB 2019). Mostly cinnamon is exported to Mexico in bulk form. A significant volume of this cinnamon is reexported to other South American countries. Overall Sri Lanka has more than 90% market share in the global marketplace (UNIDO 2016) with captive markets placed in the South and Central American regions. Figure 3.14 depicts the global Ceylon cinnamon routes.

3.5.4 Marketing Mix-Promotion

Promotional tools play a vital role in marketing. Ceylon cinnamon is a unique product of Sri Lanka and needs extensive promotion to convey its superiority over other cinnamon products. Ceylon cinnamon exporters are having their own promotional strategies to promote Ceylon cinnamon internationally.

The industry survey[1] revealed that the exporters were adopting all forms of market promotion strategies on an individual and limited scale. The budgets allocated for sales promotion were inadequate to make an impact on the international market. The common approaches consisted of advertising, sales promotions, social media, public relation, and sales organizations.

The promotional strategies and tools commonly used were as follows:

- Advertising: Web pages, magazines, newspapers, and email campaigns
- Sales promotions: Exhibitions, trade fairs, conferences, and forums
- Social media: Facebook and Twitter
- Public relations: International delegates, buyers, and business partners
- Sales organization: Market leaders such as Laksala and Pearl of Asia

Figure 3.15 shows the weightage given to each promotional strategy by the exporters. The most prominent are advertising, public relations, and sales promotion. It is interesting to observe that less prominence is given to social media in an era where it is used extensively for sales promotion (Warnakulasooriya 2017).

3.5.5 Competitive Market Landscape Analysis

Competitive market landscape analysis enables to identify direct and indirect competitors at the same time. Based on a survey of 124 spice traders[3] from Sri Lanka in 2017, five main categories of traders/exporters were identified by considering their

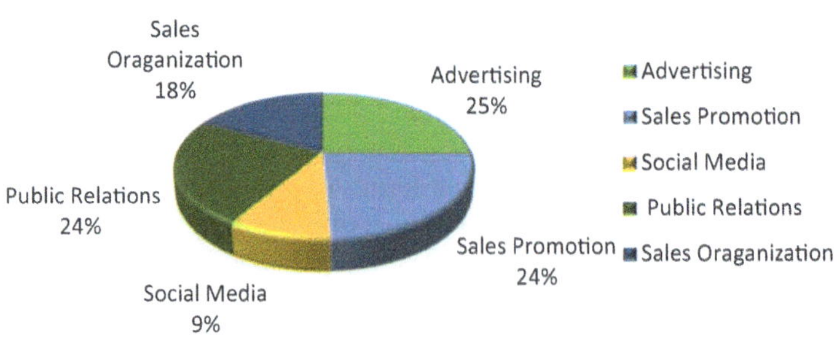

Fig. 3.15 Distribution of promotional methods of Ceylon cinnamon by Sri Lankan exporters in 2017. (Source: Warnakulasooriya (2017))

[3] Business profiles of 124 spice traders who have been involved in cinnamon trading and registered at the Export Development Board of Sri Lanka were considered when developing the "Competitive Market Landscape" (Hettiarchchi et al. 2018).

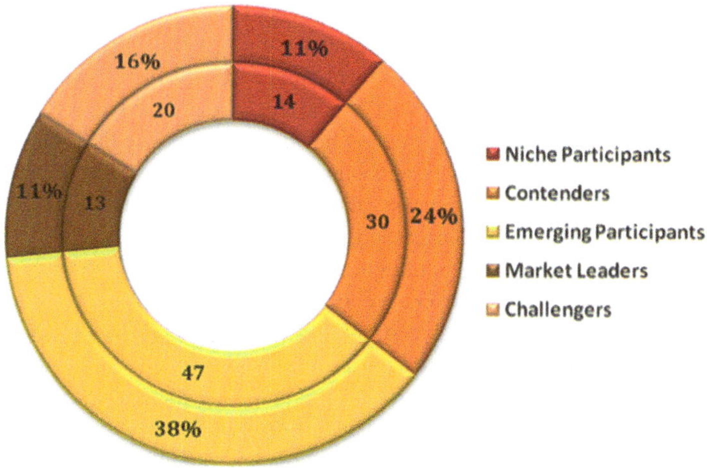

Fig. 3.16 Competitive market landscape analysis of Ceylon cinnamon exporters in Sri Lanka in 2017

export product baskets, export destinations, product prices, and export quantities. These categories included niche participants, contenders, emerging participants, market leaders, and challengers (Fig. 3.16).

A niche participant was defined as "a participant who aims to satisfy the needs of a specific market, in view of the product quality, price range and the demographics that is intended to impact."[3] Eleven percent of spice traders were identified as "*Niche participants.*" Of the sample, 24% spice traders were recognized as "*Contenders*" defined as "a participant who is a traditional processor, with no/less product diversification and different from a competitor who takes on every unexpected challenge."[3] Thirty-eight percent of spice traders who displayed blend of characteristics of "niche participants" and "market leaders," but did not meet the standards to become a "niche participant" or a "market leader," were grouped in to "*Emerging participants*" category while 16% of spice traders fell in to "*Market challengers*" category as a result of their efforts to expand the market share by aggressively flooding the market with products at competitive prices under their brand names. Eleven percent of spice traders were distinguished as "*Market Leaders*" with their largest market share in the industry and with the capacity to affect the competitive landscape and market direction (Hettiarchchi et al. 2018) (Fig. 3.16).

3.5.6 Brand Landscape for Ceylon Cinnamon

Having the monopoly for supply of "True cinnamon" to the world market, Sri Lanka annually earns more than 228 million US$ in foreign currency (Central Bank of Sri Lanka 2019b). From the global market perspective, Sri Lanka holds a dominating position with a 90% market share (Seychelles and Madagascar collectively hold only 10% market share) of the true cinnamon trade (UNIDO 2016). According to the Central Bank of Sri Lanka (2019b), Sri Lanka exported 16,967 tons and 17,860 tons of cinnamon during the years of 2017 and 2018, respectively.

The share of cinnamon, on an average, has been just above 60% of total export earnings from spices. Despite this increasing trend in the annual turnover from cinnamon exports, for the past 5 years, the annual export volumes of cinnamon remained almost constant at an average of 12,336 tons (UNIDO 2016). The USA, Mexico, Peru, and Colombia remain as the main buyers of Ceylon cinnamon, while Sri Lanka exports cinnamon to over seventy (70) countries, spreading across almost all continents. Unlike the strongly established bulk market, Sri Lanka only has a 10% share in the global value-added cinnamon markets, where the trade for these products was limited to the USA, Europe, Japan, and Mexico (UNIDO 2016). The industry stakeholders believe that this market position is grossly inadequate, compared to the potential of the industry, particularly the product development potentials in oils, oleoresins, gift packs, and packaging. Since the fragile global market linkages seem to negatively affect the market expansion and recognition of global consumer demands, a brand landscape analysis was performed on the basis of web analysis of relevant websites of seven major Ceylon cinnamon consuming countries. Industry intelligence was gathered by studying the stakeholder websites, online shopping sites, competitor product websites, and following the stakeholders on social media during the period from July to November 2017. It was possible to develop a brand landscape for Ceylon cinnamon as depicted in Fig. 3.17, identifying Ceylon cinnamon brands popular in the seven identified countries.

Fig. 3.17 Brand landscape for Ceylon cinnamon

3.6 Unutilized Potentials of Value Addition in Ceylon Cinnamon

It is evident that Sri Lanka is mainly exporting cinnamon quills in bulk, not focusing much on value addition. However, the value addition of raw spices is essential to increase competitiveness in international markets. Adding value to cinnamon in Sri Lanka itself can be a viable option to increase the income levels of industry stakeholders with sales to niche markets in Europe, Japan, and the USA. Besides the aforementioned products, the following additional potentials were identified and few of them can be mentioned as follows: *guided tours;* visits to cinnamon factories, estates, and cinnamon smallholders can be included into touristic programs, *peeling knife;* similar to the "Swiss Army Knife," the small, handy versions of the peeling knife (a unique knife developed by Sri Lankan cinnamon peelers) can be crafted and sold to tourists as souvenirs, *spice gift packs;* gift packs containing a variety of spices packaged in attractive packages can be produced and sold as souvenirs, *crafts;* walking sticks, furniture, or picture frames can be made out from peeled cinnamon sticks for sale, *cinnamon paper;* Cinnamon wood paper can be used for gift cards, letters, serviettes, corporate gifts, or notebooks, *filling materials*; remaining scrapples could be utilized to fill pillows, mattresses, and similar items that would emit a benevolent smell of cinnamon, *cleaner production*; peeled cinnamon sticks as biomass, to generate electricity or hot water. Although some of these value additions are already practiced, they are on a very small scale. Therefore, government agencies and stakeholders need to drive toward these untouched potentials to thrive in a highly competitive market environment.

Acknowledgments We are grateful to Mr. Sarada De Silva, Mr. Vijith Jayathilake, and Mr. D.A. Perera for their dedicated and expeditious help in arranging filed work as well as linking up the research team with overseas markets. We like to thank the participants in our focus groups for their time and willingness to share their experiences with us. We thank the project coordinators of special project on cinnamon, Prof. Ranjith Senaratne, Prof. K.D.N. Weerasinghe, and our contact persons at the National Science Foundation for their enormous efforts in the development of cinnamon research. Finally, but not the least, we wish to covey our sincere thanks to Ministry of Primary Industries, for opening up the research grant scheme on Ceylon cinnamon and the grant for value chain development (SP/CIN/2016/05) under which this research work was accomplished.

References

BRC (2019) BRC global standards. Retrieved from https://www.brcglobalstandards.com

Central Bank of Sri Lanka (2019a) Annual report, Central Bank of Sri Lanka

Central Bank of Sri Lanka (2019b) Economic and social statistics of Sri Lanka

Department of Export Agriculture (2019) Market information. Retrieved from http://exagri.info/mkt/2019/17.12.2019.html

EDB (2019) Export performance indicators. Export Development Board, Colombo

FAO (2013) Good agricultural practices. Food and Agriculture Organization of the United Nations. FAO, New York

FAOSTAT (2018). Retrieved from http://www.fao.org/faostat/en/#data/QC/visualize
FDA (2014) Hazard analysis and critical control point principles and application guidelines
Harindra WAM (2017) Branding as a secure vehicle for Ceylon cinnamon to win the international market (Unpublished thesis). Sabaragamuwa University of Sri Lanka
Hettiarachchi IC, De Silva DAM, Esham M, Liyanagamage TM, Abeysinghe AMIP, Warnakulasooriya S, Harindra WAM (2018) Ceylon cinnamon market landscape analysis for broader strategies. Unpublished manuscript
IPS Institute of Policy Studies of Sri Lanka (2017) Analysis of cinnamon, pepper and cardamom value chains in Sri Lanka. Institute of Policy Studies of Sri Lanka, Colombo
Lankage J (2017) How to identify Ceylon cinnamon and why cinnamon should be consumed instead of cassia. Sri Lanka Exporter
Lungarini S, Aureli F, Coni E (2008) Coumarin and cinnamaldehyde in cinnamon marketed in Italy: a natural chemical hazard? Food Addit Contam A 25(11):1297–1305
Madushani KP (2018) Standards and certifications as a means for competitive advantage in marketing- a case of Ceylon cinnamon industry (Unpublished thesis). Sabaragamuwa University of Sri Lanka
Mendis E (2016) Guide to good manufacturing practice (GMP) requirements for cinnamon processors. United Nations Industrial Development Organization (UNIDO) and The Spice Council (TSC) of Sri Lanka
Piyasiri KH, Wijeratne M (2016) Comparison of the cultivated area and the production trends of Ceylon cinnamon with the main competitors in the worlds' total cinnamon market. Int J Sci Res Publ 6(1):476–480
Samarawickrema S (2015) Assessment of the *kalli*- and line system in the Sri Lankan cinnamon sector. United Nations Industrial Development Organization (UNIDO). Retrieved from https://www.standardsfacility.org/sites/default/files/PG_343_Kalli_Line_System.pdf
Sri Lanka Customs (2019) Colombo, Sri Lanka
Teixeira S, Sampaio P (2011) An analysis of food safety management systems certification: the Portuguese case. Sinergie J 85:53–68
The Spice Council (2019) Cinnamon. Retrieved from https://www.srilankanspices.com/sl_spices_cinnamon.html
UNIDO (2016) Enhancing the compliance and productive capacities and competitiveness of the cinnamon value chain in Sri Lanka. United Nations International Development Organization, Geneva
Warnakulasooriya WMS (2017) Marketing of Ceylon cinnamon: managing tangible and intangible assets (Unpublished thesis). Sabaragamuwa University of Sri Lanka
Weerasinghe KDN, Gunasena CP, Karunarathne G (2010) Technological aspects and mechanization of cinnamon peeling. Faculty of Agriculture, University of Ruhuna, Sri Lanka

Chapter 4
Botany of Endemic *Cinnamomum* Species of Sri Lanka

Pushpa Damayanthi Abeysinghe, Pradeepa C. G. Bandaranayake, and Ranjith Pathirana

4.1 Introduction

Cinnamon belongs to the family Lauraceae, a large family of mostly evergreen woody trees or shrubs (except for the herbaceous parasite *Cassytha*) consisting of about 53 genera and 2500–3000 species distributed throughout tropical and subtropical latitudes (Chanderbali et al. 2001; Kostermans 1957; Rohwer 1993). Botanical description of Lauraceae has been dealt with in detail by Kostermans (1957) and Rohwer (1993), with the detailed description of specific characteristics such as pollen (van der Merwe et al. 1990), embryology and phylogeny (Sastri 1963), flower morphology (Sastri 1965) etc. by other authors. Recent findings from sequence variation in the chloroplast and nuclear genomes have confirmed the neotropical representation of the Lauraceae, primarily derived from Early Miocene distribution (Chanderbali et al. 2001). Earlier thought to exist only in the Asia-Pacific region, *Cinnamomum* now has a pantropical distribution except Africa after the revision of the genus with the transfer of neotropical *Phobe* to *Cinnamomum*

P. D. Abeysinghe
Department of Botany, Faculty of Science, University of Ruhuna, Matara, Sri Lanka
e-mail: pushpa@bot.ruh.ac.lk

P. C. G. Bandaranayake
Agricultural Biotechnology Centre, Faculty of Agriculture, University of Peradeniya, Peradeniya, Sri Lanka
e-mail: pradeepag@agri.pdn.ac.lk

R. Pathirana (✉)
Plant & Food Research Australia Pty Ltd, Waite Campus Research Precinct, Urrbrae, SA, Australia
e-mail: Ranjith.Pathirana@plantandfood.co.nz

R. Senaratne, R. Pathirana (eds.), *Cinnamon*,
https://doi.org/10.1007/978-3-030-54426-3_4

Schaeffer (Kostermans 1957, 1961). The name of the genus *Cinnamomum* is derived from the Greek root *kinnamon* or *kinnamomum* (Ravindran et al. 2004), and the Arabic and Hebraic term "Amomon" means aromatic spice tree.

Cinnamon is well known for its aromatic bark used as a spice, and for its various products used in pharmaceutical, cosmeceutical (Nabavi et al. 2015), and food (Wang et al. 2013) and beverage industries (Helal et al. 2014). Cinnamon as a spice comes mainly from four different species: *Cinnamomum zeylanicum* Blume (syn. *C. verum* Berchthold & Presl) called Ceylon cinnamon, Sri Lankan cinnamon or true cinnamon, *C. cassia* (L.) Berchthold & Presl. (Syn. C. aromaticum Nees)—Chinese cassia, *C. burmannii* C.G. Th. Nees—Indonesian or Java cassia and *C. tamala* Nees—Indian cassia. *C. loureiroi* Nees, once considered to be Saigon cassia exported in large quantities from Vietnam, is a rare species, and what is exported from Vietnam is now considered as *C. cassia* (Dao 2004). Because of different harvesting and processing practices in Vietnam to that in China, the product is distinctly different and hence considered to be derived from a different species (Dao 2004; Ravindran et al. 2004). However, Wang et al. (2013) sourced commercial samples of *C. loureiroi* from Vietnam to compare with other species of commercially important *Cinnamomum* for a study on coumarin content and other marker compounds where they showed that true cinnamon (*C. zeylanicum*) has only traces whereas other three species have high amounts of coumarin. Ravindran et al. (2004) have described the botany of these species in detail. Therefore, in this chapter we present the botanical descriptions and conservation status of *Cinnamomum* species endemic to Sri Lanka with emphasis on morphological features. Undocumented so far, detailed anatomical features of *C. zeylanicum*, the true cinnamon of Sri Lankan origin, are also presented.

4.2 *Cinnamomum* Species Endemic to Sri Lanka and Their Features

4.2.1 *Identification of Cinnamomum Species Endemic to Sri Lanka*

In addition to *C. zeylanicum* Blume, the True or Ceylon cinnamon, Sri Lanka is home to seven other *Cinnamomum* species: *C. capparu-coronde* Blume, *C. citriodorum* Thw., *C. dubium* Nees, *C. litseaefolium* Thw., *C. ovalifolium ovalifolium* Weight, *C. rivulorum* Kosterm, and *C. sinharajaense* Kosterm (Ariyarathne et al. 2018; Kumarathilake 2012; Liyanage 2010). *C. zeylanicum* found in South India is introduced from Sri Lanka (Kostermans 1983). In addition, *Cinnamomum camphora* (L.) J. Presl (camphor tree, camphor laurel) has been introduced. Recently Ariyarathne et al. (2018) examined morphological characters of flower, leaf, seed and bark, and bark and leaf oil content, as well as the chemical characters of leaf and bark oil of all the endemic species, including two *C. zeylanicum* cultivars released for cultivation in Sri Lanka— "Sri Wijaya" and "Sri Gemunu". A dendrogram was constructed, and a diagnostic key for identification of the species was developed on the basis of oil yield, morphological and chemical characters. These are presented in Chap. 6 of this book. This key uses chemical constituents and is not helpful for field identification. Similarly, a key developed by Kostermans (1964a) uses floral characteristics, again not available throughout the year under field conditions. To overcome these issues in field identification, Kumarathilake (2012) developed a key for the identification of Sri Lankan *Cinnamomum* species based on morphological characteristics avoiding floral characteristics. It is given in the following.

1. Leaf shape oval to elliptic **_Cinnamomum ovalifolium_**
1. Leaf shape ovate-lanceolate or lanceolate
    2. Pinnate venation present (three veins not prominent) **_Cinnamomum citriodorum_**
    2. Mainly three venations (midrib and two lateral veins prominent)
        3. Lateral veins visible up to near the apex
            4. Large (length; 35–40 cm) thick leaves, fresh leaves odourles **_Cinnamomum sinharajaense_**
            4. Small- to medium-size (7–17 cm long) thin leaves; fresh leaves aromatic
5. Immature leaves densely sericeous, bark not slimy **_Cinnamomum capparu-coronde_**
5. Immature leaves not sericeous, bark slimy
    6. Secondary veins parallel, immature bark reddish to orange, strong pungent, erect and less branched plant **_Cinnamomum zeylanicum_**
    6. Secondary veins reticulate, immature bark greenish to brown, not pungent, stem much branched and not erect **_Cinnamomum dubium_**
        3. Lateral veins visible only up to the middle of the leaf
            7. Leaves curled upward, margin entire, apex acute, flush reddish to yellowish **_Cinnamomum litseaefolium_**
            7. Leaves not curled upward, margin undulate, apex narrowly acuminate, flush greenish **_Cinnamomum rivulorum_**

Adapted from Kumarathilake (2012)

Abeysinghe and Scharaschkin (2019) studied the petiolar anatomy of all the eight endemic Sri Lankan *Cinnamomum* species and found major differences that helped them to develop a key for identification of these species purely on the basis of these characters. Main characters that turned out to be useful for this differentiation were presence or absence of trichomes, their shape, vascular bundle shape, stone cell shape, and symmetry of the petiolar axis and pore size or their absence.

Traditionally, identification of cinnamon in Sri Lanka has been based more on organoleptic characteristics than on morphological characters. The vernacular names in Sinhalese identify different types as *Peni Miris Kurundu* (sweet and hot cinnamon), *Thitta Kurundu* (bitter cinnamon), *Peni Kurundu* (sweet cinnamon), *Naga Kurundu* (snake cinnamon), *Veli Kurundu* (sandy cinnamon implicating the granular nature), *Sevel kurundu* (slimy cinnamon—botanically *C. dubium*), and *Kahata Kurundu* (acrid cinnamon) (Wijesinghe and Pathirana 2000). For example, *C. dubium* has a slimy bark, so *Sevel Kurundu* (slimy cinnamon) is the Sinhalese vernacular name for this species. Similarly, *C. citriodorum* has a citronella smell in bark due to a high amount of citronellol (50%; Kumarathilake 2012) and is called *Pengiri Kurundu* meaning citronella cinnamon.

All the species of *Cinnamomum* endemic to Sri Lanka have spheroidal and apolar pollen grains (Sritharan et al. 1993), similar to other *Cinnamomum* species (Pal 1976). In fact, tribes Perseeae, Cinnamomeae, Litseeae, and Hypodaphneae have spheroidal, apolar, and spinulose pollen (van der Merwe et al. 1990). All the species of *Cinnamomum* found in Sri Lanka have a diploid chromosome number of $2x = 2n = 24$ (Sritharan et al. 1993; Wijesinghe and Pathirana 2000).

4.2.2 Differences in Leaf and Bark Oil Content and Composition in *Cinnamomum* Species Endemic to Sri Lanka

Among the species, the leaf oil content is highest in *C. zeylanicum* (3–3.2%) followed by *C. sinharajaense* (2.6%) (Ariyarathne et al. 2018). Contrastingly, Liyanage et al. (2017) reported significantly higher bark oil in *C. sinharajaense* (3.53%) than in *C. zeylanicum* (3.06%). However, in agreement with Ariyarathne et al. (2018), *C. zeylanicum* (3.26%) leaf oil content in their studies was significantly higher than that in *C. sinharajaense* (2.41%). *C. capparu coronde* recorded 1.49% essential oil in leaf and *C. dubium*—1.55% in the bark, whereas *C. rivulorum* and *C. citriodorum* recorded less than 0.9% bark and leaf oil (Liyanage et al. 2017). The lowest leaf oil content among all the endemic Sri Lankan *Cinnamomum* species is in *C. revulorum* (0.4%) and *C. citriodorum* (0.6%) and the lowest bark oil is in *C. dubium* (0.4%) and *C. capparu coronde* (0.48%) (Ariyarathne et al. 2018). In a review of composition of essential oils of *C. zeylanicum*, Mallavarapur and Rao (2007) state that the leaf oil content of *C. zeylanicum* can vary from 0.4 to 4.7%, whereas leaf of this species from Benin had 1.48% oil (Philippe et al. 2012).

Several endemic *Cinnamomum* species are used in traditional Ayurveda system of medicine, employing different parts of the plant for different ailments. In all the species studied, the composition of essential oil varies within a plant depending on the source used to distil oil viz. bark, leaf, bud, fruit, flower, or root (Ariyarathne et al. 2018; Kumarathilake 2012; Liyanage et al. 2017; Mallavarapur and Rajeswara Rao 2007; Senanayake and Wijesekara 2004). Much information on the oil composition of C. *zeylanicum* is available, but only a limited number of publications are devoted to this subject in endemic wild species. Eugenol and cinnamaldehyde are the two most important constituents in cinnamon oils from an industrial perspective. Ariyarathne et al. (2018) analyzed these two chemicals and benzyl benzoate concentrations in all indigenous *Cinnamomum* species from Sri Lanka except *C. ovalifolium*. On average *C. zeylanicum* leaf and bark oil contained 83.4 and 34.3% eugenol, respectively. Results of Liyanage et al. (2017) are in agreement, with 85.7% and 16.3%, respectively, and also similar contents of eugenol were reported from *C. verum* grown in Kerala, India (Krishnamoorthy et al. 1996). Senanayake and Wijesekara (2004) reported 2.2% eugenol in stem bark oil and 70.1% in leaf oil of 2½-year-old plants. These results are in agreement with Paranagama et al. (2001) with 4.15% and 76.74%, respectively. Among the seven species studied by Ariyarathne et al. (2018), the highest cinnamaldehyde content was in *C. zeylanicum* bark oil (74.1%). Bark oil cinnamaldehyde content in Sri Lankan-grown *C. zeylanicum* has been reported to vary from 50.5% (Paranagama et al. 2001) to 75% (Senanayake and Wijesekara 2004). Interestingly *C. sinharajaense* recorded the closest profiles of these chemicals to *C. zeylanicum* among the seven species studied by Ariyarathne et al. (2018), with leaf oil having 77% eugenol and bark oil with 15.3% eugenol and 51.4% cinnamaldehyde. Other noteworthy sources of eugenol were bark oils of *C. capparu coronde* (31.1%) and *C. liseaefolium* (25.5%). *C. capparu coronde* recorded high bark oil cinnamaldehyde (58%) content as well as relatively high leaf oil cinnamaldehyde (19.3%) and benzyl benzoate (16.4%) contents (Ariyarathne et al. 2018). According to Kumarathilake (2012), *C. capparu corande* has 90.3% eugenol in leaf oil and 36.3–67% in bark oil, depending on the location. *C. liseaefolium* leaf oil had the highest benzyl benzoate content (19.7%) with *C. zeylanicum* bark oil recording an average of 14.1% (Ariyarathne et al. 2018). Bark oil of *C. capparu corande* has recorded a high content (11.9–29%) of linalool (Kumarathilake 2012; Wijesekera and Jayewardene 1974), widely used in the cosmeceutical industry. *C. citriodorum* leaf oil has citronellol (66.5–71.1%), citronellal (10.8–10.9%), and trans-geraniol (5.9–6.5%) (Leela et al. 2012) and could easily be used as an alternative to citronella oil distilled from the grass of *Cymbopogon* spp.

The chemical composition of the essential oil extracted from leaves of *C. zeyalanicum* from Benin showed ethyl cinnamate, cinnamaldehyde, and benzyl benzoate (39.9, 25.0, and 20.5%, respectively) as major compounds, different from what is reported for Sri Lankan-grown material as described earlier. *C. zeylanicum* fruit oil was reported to have high concentrations of δ- and γ-cadinene (36%), α-cadinene (5.6%), T-cadinol (7.7%), and β-caryophyllene (5.63%) with negligible amounts of cinnamaldehyde, eugenol, and camphor—the major constituents of bark, leaf, and root oils, respectively (Senanayake and Wijesekara 2004). Also, these profiles are different from those reported for fruit oil from *C. zeylanicum* from Kerala and Karnataka, India (Jayaprakasha et al. 1997; Mallavarapur and Rao 2007), indicating either environmental influence or a genotypic difference.

4.3 Distribution, Conservation Status, and General Features of *Cinnamomum* Species Endemic to Sri Lanka

Seven endemic wild *Cinnamomum* species have been recorded in 12 districts in seven provinces of Sri Lanka and are concentrated in and around Sinharaja forest reserve, Enasalwatte-Deniyaya, Kanneliya forest reserve, Walankanda forest reserve, Horton Plains National Park, Gilimale-Erathne proposed reserve, Peak-Wilderness sanctuary, Knuckles conservation area, and Haldummulla-Halpe regions (Gunatilleke and Gunatilleke 1991; Kostermans 1957, 1964b; Kumarathilake 2012; Kumarathilake et al. 2010; Liyanage 2010; Ranawana 2014; Sritharan 1984; Werner and Balasubramaniam 1992; Wickramasinghe et al. 2008; Wijeratne and Piyasiri 2015) (Fig. 4.1). The regions cover forest areas ranging from 30 to 2400 m in altitude from tropical rain forest to semi-dry areas with annual rainfall ranging from 1100 to 4450 mm. The distribution of *Cinnamomum* species covers all the three peneplains of the country—30 m, 500 m, and 1500–1800 m (Erdelen 1988) in the Southern, Western, North Western, Central, Sabaragamuwa, and Uva provinces

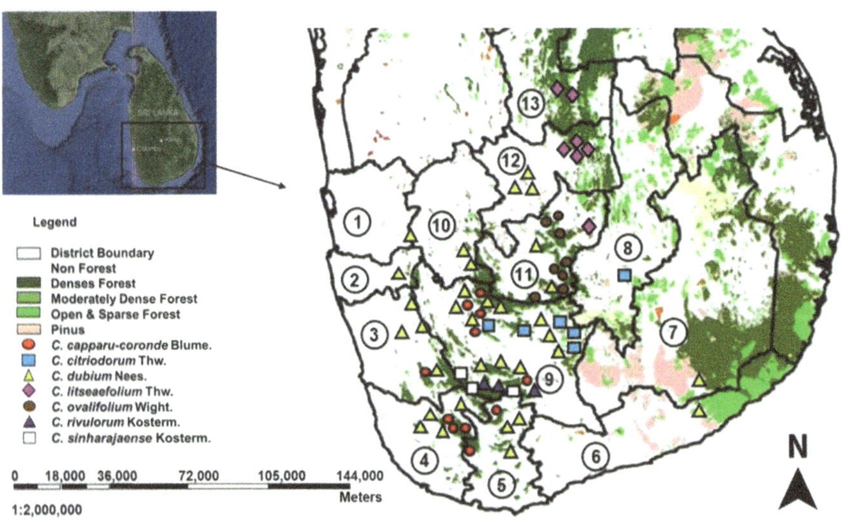

Fig. 4.1 Distribution of endemic wild *Cinnamomum* species in Sri Lanka with district boundaries where endemism occurs. Districts: (1) Gampaha, (2) Colombo, (3) Kalutara, (4) Galle, (5) Matara, (6) Hambantota, (7) Moneragala, (8) Badulla, (9) Ratnapura, (10) Kegalle, (11) Nuwara Eliya, (12) Kandy, and (13) Matale. (Courtesy: Bandusekara et al. (unpublished), adapted from Liyanage (2010))

Table 4.1 Vulnerability status of the wild *Cinnamomum* species endemic to Sri Lanka

Species	Vernacular name in Sinhalese	Global red listing category[a]	National red listing category[b]
C. capparu-coronde	Kapuru Kurundu	Endangered	Vulnerable
C. citriodorum	Pengiri Kurundu	Critically endangered	Vulnerable
C. dubium	Wal kurundu, Sevel kurundu	Least concern	Vulnerable
C. litseaefolium	Kudu Kurundu	Critically endangered	Endangered
C. ovalifolium	Bola Kurundu	Vulnerable	Vulnerable
C. rivulorum	Wal Kurundu	Critically endangered	Endangered
C. sinharajaense		Critically endangered	Vulnerable
C. zeylanicum	Kurundu		Vulnerable

[a]IUCN (1998)
[b]MOE (2012)

(Fig. 4.1). Considerable variation in morphology can be observed among plants of the same species in these locations, and Kumarathilake et al. (2010) attribute this to the diversity in climate in these habitats and it is described in detail by Erdelen (1988). The conservation status of the wild cinnamon species is a concern as all the endemic species except *C. dubium* are endangered or critically endangered, with *C. ovalifolium* vulnerable according to IUCN red listing criteria (IUCN 1998) (Table 4.1).

4.3.1 C. zeylanicum

C. zeylanicum—the cultivated species of true cinnamon originates in Sri Lanka (Kostermans 1983), but can now be found in 10 Asian countries, including eight states in India, 12 African countries, nine countries in the Central American and Caribbean Region, five South American countries, and almost all main islands of the Pacific, and has become an invasive species in many islands (CABI 2020). In the Seychelles, for example, it is the main invader of forest and intermediate areas threatening endemic species such as *Phoenicophorium borsigianum*, *Deckenia nobilis,* and *Roscheria melanochaetes* (Fleischmann 1997). Introduced in the year 1772 to Mahé Islands of the Seychelles, it dominates the canopy of Morne Seychellois National Park (Kueffer et al. 2008). Although the species was threatened in Sri Lanka due to extensive harvesting from the forests, establishment of plantations in the period 1767–1770 by the Dutch (Weiss 2002) helped to prevent further erosion of genetic resources through forest harvest (Wijesinghe and Pathirana 2000).

It is an evergreen broad-leaved perennial woody tree up to 18 m tall propagated by seeds. In cultivation it is usually a bushy plant about 2–3 m high. Bark is strongly aromatic, about 10 mm thick, smooth and pale brown in young branches, rough and dark in mature trees. Leaves ovate to elliptical, opposite, strongly aromatic with a petiole grooved on upper surface, 1–2 cm long, three or five veined, apex acuminate, glabrous, and shiny dark green. Inflorescence is a panicle of variable length and can be axillary or terminal with a softly hairy, creamy-white 5–7 cm long peduncle. Pale yellow flowers are small, about 3 mm in diameter subtended by a small, ovate, and hairy bract; perianth about 8 mm long, silky hairy, with short companulate tube and six persistent tepals about 3 mm long. Stamens are arranged in three whorls, fertile with two small glands at the base of the third whorl. A fourth innermost whorl consists of staminodes. Anthers 2- or 4-celled, carried on hairy stout filaments, with the largest pollen (26 μm diameter) among Sri Lankan *Cinnamomum* species (Sritharan et al. 1993); ovary is superior, 1-celled, with a single ovule and a short style. Fruit is an ellipsoidal drupe, black when ripe surrounded by the enlarged perianth at the base with a 1- to 2-cm-long single seed.

Although native habitat is tropical rainforest, *C. zeylanicum* has adapted to many more environments, both in cultivation and in the wild such as secondary forests, forest gaps, rock outcrops, and riparian regions. Genetic resources and diversity of the cultivated *C. zeylanicum* are described in Wijesinghe and Pathirana (2000) and in Chap. 6 of this book.

4.3.2 C. litseaefolium

C. litseaefolium is commonly known as *Kudu Kurundu* in Sinhalese (Liyanage 2010) and is found in the Knuckles conservation area and surrounding areas in the Kandy, Matale, and Nuwara Eliya districts (Kumarathilake 2012). Liyanage (2010) was able to find specimens in Badulla district as well, mostly at elevations 800–2100 m. Primary habitat of this species is up-country dry montane forest (Pigmy forest), but larger specimens are found in fertile soils of lower submontane and lowland forests (Fig. 4.1).

Trees, up to 20 m tall, mostly 10–15 m, are the tallest among the Sri Lankan cinnamons; leaves simple, subopposite to opposite, lanceolate, apex acute, base acute, leaf margin entire, surface coriaceous, smooth to rough texture, mainly three venation present, lateral vein seen only up to middle of the leaf, reddish to yellowish color flush, leaves normally curled upward, leaves and bark are less aromatic and branching of trunk is common; flower white to greenish, flowering season March to May. Bark is thick and difficult to peel (Kumarathilake 2012; Liyanage 2010).

According to the national red listing criteria, *C. litseaefolium* is considered Endangered (EN) (MOE 2012). The Knuckles Forest Range situated 7°21′ N, 81°45′ E in the Intermediate Zone of the Central Province of Sri Lanka, where the diversity of *C. liseaefolium* occurs is unique as it is perpendicular to the two principal wind currents that bring rains and acts as a climatic barrier. The highland and western areas of the Knuckles ranges are extremely wet throughout the year, with an annual rainfall of about 5000 mm, whereas the lower eastern slopes are drier with less than 2500 mm annual rainfall. Temperature in the forest region ranges between

15 and 25 °C (Weerawardhena and Russell 2012; Wickramasinghe et al. 2008). The survival of *C. litseaefolium* and many other endemic species in this forest region consisting of 210 km^2 was first threatened when commercial plantations were established during the British colonial period, with major estates such as Nichola Oya, Rangalle, and Kallebokka. For example, 2796 ha of this rainforest was deforested to plant tea within 2 years (1874–1875) for the establishment of Kallebokka estate (Forrest 1967). Thereafter legalizing cardamom (*Elettaria cardamomum*) cultivation and traditional shifting cultivation by farmers in the adjoining communities resulted in further decline of biodiversity in the region as forest had to be first cleared of undergrowth and forest-sourced firewood was used for curing cardamom (Gunatilake et al. 1993; Wickramasinghe et al. 2008). Although cardamom cultivation in the Knuckles has been practiced for over a century, it was intensified in 1960s with the local government encouraging the practice and leasing land in the forest area for cardamom cultivation (Wickramasinghe et al. 2008). In the year 2000, a conservation zone was declared for the Knuckles in areas above 3500 ft (1068 m) (Wickramasinghe et al. 2008) and cardamom cultivation and agricultural activities have since then ceased, bringing in more economic hardships to peripheral communities (Gunatilake et al. 1993). Currently the Knuckles represents a fragmented forest area with 21% of it heavily degraded and only 12% left as virgin forest (Wickramasinghe et al. 2008). Nevertheless, declaration of the conservation area gives an opportunity for the restoration of habitat and for in situ conservation of species such as *C. liseaefolium* and other endangered species. According to Liyanage (2010), the distribution of *C. litseaefolium* is in the Badulla, Nuwara Eliya, Kandy, and Rathnapura districts, with some specimens found outside the Knuckles forest conservation area.

4.3.3 C. dubium

Vernacular Sinhalese names for *C. dubium* are *Sevel Kurundu* (slimy cinnamon) or *Wal Kurundu* (wild cinnamon). It is widely distributed and can be found in 12 districts of Sri Lanka, from south-eastern Hambantota and Moneragala, southern Matara and Galle to Western Kalutara, Ratnapura, Colombo, Gampaha, mid-country Kegalla and Kandy, and in up-country Nuwara Eliya and Badulla districts between altitudes 20 and 2000 m (Kumarathilake 2012; Kumarathilake et al. 2010; Liyanage 2010) (Fig. 4.1). It is a common species in the wet rainforests and in adjacent areas, including abandoned lands. According to two independent geographical surveys, *C. dubium* was the most widespread species among the wild *Cinnamomum* spp. in Sri Lanka (Kumarathilake et al. 2010; Liyanage 2010) (Fig. 4.1), and according to national red listing criteria, it is considered Vulnerable (VU) (MOE 2012).

Long and narrow leaf with three veins is characteristic of *C. dubium* and is easily distinguished from other endemic *Cinnamomum* species of Sri Lanka by the shape of its leaves. The average height of this species is 6–8 m with rare specimens reaching 15 m (Liyanage 2010). *C. dubium* has the smallest pollen (17.6 μm) among endemic Sri Lankan *Cinnamomum* spp. (Sritharan et al. 1993). Slimy bark and low concentration of eugenol, cinnamaldehyde, and linalool in leaf and bark oil are other distinguishing features (Ariyarathne et al. 2018; Liyanage 2010). The leaves are simple, subopposite to opposite, lanceolate, acuminate or sometime narrowly acuminate, base acute, entire margin, surface coriaceous, smooth to rough texture, mainly three venations are present, lateral vein parallel to mid rib and prominent. Flush color yellowish to reddish and branching of trunk is common. Less aromatic smell and no particular taste of leaf and bark. Flowering season is from March to April, flower yellowish to white, perianth segment persistent, bark thin, smooth to rough texture, slimy and easy to peel (Ariyarathne et al. 2018; Kumarathilake 2012; Liyanage 2010). Bark oil of this species has ~55.2% linalool, 10.8% eugenol, and 8.5% myrcene, and the leaf oil has citral-b (18.6–35.8%, depending on the location), myrcene (16.5%), cinnamyl alcohol (16.1–28.8%), β-caryophyllene (16.7%), and 1–8 cineole (13.8%) (Kumarathilake 2012).

4.3.4 C. capparu coronde

The species name for *C. capparu coronde* was coined on the basis of Sinhalese vernacular name of *Kapuru Kurundu* (camphor cinnamon) because of its strong camphor smell. Bark has a nutmeg/clove smell and is sold in the market for medicinal purposes (Liyanage et al. 2017; Shylaja et al. 2004). Although this species has a wide adaptation and specimens found in Galle, Kandy, Ratnapura, and Nuwara Eliya districts (Fig. 4.1), the plants are sparse even within conservation areas because it is felled for its bark, which has medicinal value (Kostermanns 1973; Kumarathilake 2012; Kumarathilake et al. 2010; Liyanage 2010). Kostermanns (1973) warned that this species is becoming extinct and emphasized the need for its conservation. With an average score of 4.0 on Sri Lanka national red listing criteria, this species is one of two Highly Threatened (HT) *Cinnamomum* species in Sri Lanka (Kumarathilake et al. 2010). However, it is considered vulnerable in the National Red List of 2012 (MOE 2012).

C. capparu coronde is a medium-sized tree of about 20 m, but rare trees of 40 m with ~50 cm bole have also been found in forest areas of Kalutara district (Liyanage 2010). The bark of the tree is rough, light to dark brown with a very strong odor of clove and nutmeg. Leaves 7–14 × 2.5–5 cm, opposite, ovate-oblong to ovate, gradually acute or subacuminate with blunt tip, base short and rounded, three main veins, filiform, lateral veins prominent, terminating 1–2 cm below leaf tip. About 1 cm long petiole is slender and sericeous. Inflorescence is a densely sericeous axillary (or pseudoterminal) panicle bearing small (about 3 mm long) flowers, tepals 1.5–2 mm, ellipsoid, apiculate with resinous odor; cotyledons red. Fruit is a one-seeded ellipsoid apiculate drupe about 1 cm long and has a resinous odor (Kostermans 1964b; Kumarathilake 2012; Shylaja et al. 2004).

C. capparu coronde leaf oil consists of 31.1% eugenol and 16.4% benzyl benzoate, whereas bark oil is predominantly cinnamaldehyde (58%) (Ariyarathne et al.

2018). Wijesekera and Jayewardene (1974) analyzed this species and found the main chemical constituents to be 1,8-cineole (15%), linalool (29%), and eugenol (23%). Kumarathilake (2012) reports 90.1% eugenol in leaf oil and 36.3–67.1% in bark oil, other constituents in the latter being linalool (11.9–27.1%), 1–8 cineole (15%), myrcene, and geraniol (10.7 and 11.1%, respectively). Considering good adaptability to field conditions and the wide uses of this species having valuable chemical constituents in both leaf and bark oil, *C. capparu coronde* is a strong candidate for domestication as well as distant hybridization with cultivated cinnamon for selecting superior genotypes. This would allow better conservation prospects for this highly endangered species.

4.3.5 C. citriodorum

For its strong smell of citronella (*Cymbopogon nardus* (L.) Rendle, Syn *Andropogon nardus* L. family Poaceae) oil, the vernacular Sinhalese name for *C. citriodorum* is *Pengiri Kurundu*; *Pengiri* is the Sinhalese name for citronella. The primary habitat of *C. citriodorum* is tropical semidry forests and is distributed at elevations of 500–1000 m, with largest populations found in Haputale region of Badulla district and Halpe region of Balangoda in Rathnapura district of Sri Lanka (Kumarathilake 2012) (Fig. 4.1), and it also occurs in the south western Ghats of India (Leela et al. 2012; Shylaja et al. 2004). Because of continuous decline of forest habitat of the species in Sri Lanka, confining the species into small pockets of land that are unprotected, *C. citriodorum* is an Endangered (EN) *Cinnamomum* species (IUCN 1998). According to the National Red List of 2012, it is vulnerable (VU) (MOE 2012). In addition to these areas of distribution, Liyanage (2010) recorded this species in some parts of Kandy district and at elevations up to 1450 m.

C. citriodorum is a medium-sized tree growing to about 10–14 m with a thin, easily peelable bark and leaves having a characteristic citronella (lemongrass) odor due to the presence of 66.5–71.1% citronellol and 10.8–10.9% citronellal in leaf oil (Leela et al. 2012), a unique feature of this species. Kumarathilake (2012) reports slightly lower citronellol (52.8%) and higher citronellal (21.8%) concentration in leaves of this species of Sri Lankan origin, in addition to 14.4% citral-a. The bark oil is not much different from leaf oil in major constituents—citronellol 49.6%, citronellal 12%, and citral-a 14.4% (Kumarathilake 2012). It is used both as a spice and as a medicinal plant, mainly for stomach ailments. Leaves are lanceolate or ovate-lanceolate, with acuminate apex (Ariyarathne et al. 2018) and pinnate venation, often curled upward (Kumarathilake 2012). The new flush is yellowish to reddish; greenish white flowers are sericeous with a short tube and 3-mm-long sepals (Kostermans 1964b). Perianth segment is deciduous. Branching of trunk is common.

Because of its use as a spice as well as a medicine, the species is highly threatened in its natural habitat. The species may be useful in hybridization programs for improving cultivated cinnamon for alternate uses. Alternately *C. citriodorum* could be commercialized for the production of citronellol extensively used in the cosmetic industry. Its antimicrobial, insecticidal, and acaricidal properties warrant field trialing this species, thus minimizing forest harvesting.

4.3.6 C. ovalifolium

C. ovalifolium is called *Bola Kurundu* in Sinhalese probably because of its distinct oval-shaped leaves. It is distributed in the Nuwara Eliya, Kandy, Rathnapura, and Badulla districts of Sri Lanka in the wet montane habitat at elevations 1300–2200 m (Kumarathilake 2012; Liyanage 2010) (Fig. 4.1). Recently this species was found in Kerala part of Agasthyamala Biosphere Reserve of Thiruvananthapuram district, southern Western Ghats as well (Robi et al. 2018). In Sri Lanka it is mainly confined to few forest fragments such as Peak-Wilderness sanctuary, Sooriyakanda region, Knuckles conservation area, and Horton plain National Park, Haggala—Kandapola—Piduruthalagala—Mahakudugala range (Kumarathilake 2012; Liyanage 2010; Werner and Balasubramaniam 1992) (Fig. 4.1).

C. ovalifolium is comparatively smaller than other endemic *Cinnamomum* species, growing to a height of about 4–6 m (Liyanage 2010). Leaves broadly ovate to subrotundate (Kostermans 1964b), rigidly coriaceous, base obtuse, subchordate, and apex subcordate (Kumarathilake 2012). Two main lateral veins do not reach apex. Leaf and bark odorless and light brown, perianth segments deciduous (Kostermans 1964b). According to global red list criteria, *C. ovalifolium* is considered Endangered (EN) (IUCN 1998), but according to national red list criteria, it is considered as vulnerable (MOE 2012) and dominates the vegetation in some parts of Horton Plains National Park (Ranawana 2014). No uses have been reported for this species.

Sritharan (1984) analyzed *C. ovalifolium* bark, leaf and root oil by thin layer chromatography and found linalool and unidentified terpenes in bark oil, α-terpineol, and eugenol in leaf and root oil.

4.3.7 C. rivulorum

C. rivulorum is also called *Wal Kurundu* in Sinhalese, meaning wild cinnamon. It is found in Matara, Kaluara, and Rathnapura districts in Sinharaja forest reserve (Liyanage 2010), mainly in Ensalwatte Division (Kumarathilake 2012); the majority of these locations are at 300–1100 m elevation (Fig. 4.1). However, Liyanage (2010) reports two locations at less than 50 m elevation in Ratnapura district. It is adapted to moist rainforest habitat, but some specimens do survive in cleared areas. Out of 23 locations marked by Liyanage (2010), only 14 are in protected areas and most of the rest are at risk. This species has seen continued decline and now limited to only one forest fragment in Ensalwatte Division of Sinharaja range but not under the Forest Department ownership, and hence considered as one of the most endangered among the Sri Lankan endemic cinnamons (Kumarathilake 2012). According to global red list criteria, *C. rivulorum* is Critically Endangered (CR) (IUCN 1998), and according to National Red List criteria it is endangered (MOE 2012).

C. rivulorum is a small tree of 4 m tall, leaf lamina is lanceolate, gradually acuminate, base shortly acute, leaf margin undulate, apex narrowly acuminate, surface sub coriaceous, smooth to rough texture, flush greenish (Kostermans 1964b; Kumarathilake 2012). Leaves are not curled upward, and two main lateral veins do not reach apex, reticulation prominent on both surfaces. Flower greenish, densely sericeous (Kostermans 1964b), perianth segment persistent, flowering season March to April. Bark thin, smooth to rough, easy to peel. Leaf, petiole or bark have no particular pungency, non-aromatic. Branching is common and bark is slimy. *C. rivulorum* is a shade loving plant found mainly in lower canopy. According to Liyanage et al. (2017), the bark oil has both eugenol (22.3%) and cinnamaldehyde (31.6%), whereas leaf oil has only eugenol, but in higher concentration (63.5%) than bark oil. This species has no recorded uses.

4.3.8 C. sinharajaense

Kostermans (1980) describes this species as very rare as he could find only three saplings and one mature tree in the thick Sinharaja rainforest. According to Kostermans (1980), it is a tree, 6 m tall, bole grey, smooth, 30 cm in diameter. Live bark 3 mm, pale, the outside bright red, without odor or taste. Branchlets are glabrous. Leaves are glabrous, sub-opposite, coriaceous, elliptic to broadly elliptic, 6 × 14—12 × 20 cm, acuminate (acumen broad, 1.5–2 cm long), base contracted into the petiole, obtuse, the center very shortly cuneate; above dull, obscurely, minutely pitted to smooth, midrib and subbasal lateral nerves smoothly prominent, below paler, midrib prominent, the two subbasal lateral nerves prominent, ending 2–5 cm below the leaf apex; secondary nerves obscure, parallel, horizontal, widely spaced, also present at the outside of the lateral nerves, or invisible, reticulation dense, obscure, or invisible. Petioles are thick, 1.5–3 cm long. No aromatic smell in leaf or bark. Both Kumarathilake et al. (2010) and Liyanage (2010) did not find this species outside Sinharaja Man and Biosphere Reserve. Flowering season is from March to April, and flowers pale yellow with persistent perianth segment. Bark is easy to peel (Kumarathilake 2012). The species occurs at 500–1100 m elevation, closer to streams.

According to global red list criteria, *C. sinharajaense* is Critically Endangered (CR) (IUCN 1998) and according to national red list criteria, it is described as vulnerable (MOE 2012). *C. sinharajaense* has the highest leaf (2.64%) and bark oil (1.21–3.53%) contents among the wild species, very similar to cultivated varieties of *C. zeylanicum* (Ariyarathne et al. 2018; Liyanage et al. 2017). Sritharan (1984) analyzed *C. sinharajaense* bark and leaf oil by thin layer chromatography and found cinnamaldehyde, eugenol, and linalool to be present in bark oil, eugenol, linalool, and α-terpineol in leaf oil. This was later confirmed in the studies by Ariyarathne et al. (2018) and Liyanage et al. (2017), using gas-liquid chromatography. Liyanage et al. (2017) reported the highest leaf eugenol content (87.5%) and highest bark cinnamaldehyde content (57.5%) in *C. sinharajaense* among the endemic *Cinnamomum* species. In addition, both leaf (11.2%) and bark oil (8.1%) contain benzyl benzoate (Ariyarathne et al. 2018). Considering good peelability of bark, the presence of valuable chemical constituents similar to cultivated cinnamon and high bark and leaf oil contents, *C. sinharajaense* is considered a good candidate for domestication and research for interspecific hybridization with *C. zeylanicum*. Even petiolar anatomy of this species was similar to that of cultivated cinnamon (Abeysinghe and Scharaschkin 2019) and furthermore, when phylogenetic relationships were analyzed using cpDNA profiles, the two *C. sinharajaense* accessions clustered together with several *C. zeylanicum* accessions (Abeysinghe et al. 2009), thus showing close relationship among these two species.

4.4 Comparative Morphology of *Cinnamomum* Species Endemic to Sri Lanka

4.4.1 Leaf Morphology

Leaf morphology is highly diverse in the genus *Cinnamomum*. This variation is common at species and subspecies levels as well as within a species or subspecies. Leaf morphological characteristics are widely used in identification of species (Dassanayaka 1995; Brewer and Stott 2017; Baruah and Nath 2007; Wuu-Kuang 2011).

Main leaf morphological characters important in species identification include leaf size (small, medium, large), leaf length (cm), leaf width (cm), leaf arrangement (opposite, sub-opposite, opposite or sub-opposite in different branch but in same branch in same plant), leaf shape (elliptic, broadly elliptic, narrowly elliptic, ovate, broadly ovate, oval, lanceolate, ovate—lanceolate, oblong—lanceolate), leaf apex (acute, obtuse, acuminate, long acuminate, acuminate with broad acumen, other), and leaf base (acute, subacute, cuneate, rounded, subcordate, obtuse, contracted into petiole, then shortly cuneate, other) (TURIS 2013).

4.4.1.1 Leaf Size and Shape

Ravindran et al. (2004) reviewed literature on leaf morphological characters of *Cinnamomum* species in cultivation. *C. verum* leaf length may vary from 8.7 to 22.7 cm with the mean length of 13 cm whereas leaf width may vary 3.3–8.3 cm with the mean width of 5.1 cm (Krishnamoorthy et al., 1992). Interestingly *C. zeylanicum* leaves collected from Sri Lanka were larger than Indian samples (Shylaja 1984). Our recent analysis with more than 500 *C. zeylanicum* accessions grown in the same environmental conditions showed significant variation among leaf size with an average leaf length of 12.3 ± 2.3 cm and width of 5.8 ± 1.1 cm (Table 4.2). The leaf length-to-width ratio also varied considerably with a mean of 2.02 ± 1.02 cm (Liyanage et al. 2020). Similar variations in leaf length and width have been reported in *C. zeylanicum* germplasm in Matara district of Sri Lanka (Azad et al. 2016).

Variations in leaf length, leaf width, and length-to-width ratio are even more prominent in wild relatives of cultivated cinnamon. Among the wild relatives of *C. zeylanicum*, *C. sinhajaense* has the largest leaves while *C. ovalifolium* has the smallest leaves (Table 4.3). Therefore, as expected, the leaf weight, measured using a sample of 30 fully mature leaves, was highest in *C. sinhajaense* and lowest in *C. ovalifolium*.

Table 4.2 Variation in quantitative traits of leaf and plant in *C. zeylanicum* germplasm ($n = 515$)

Traits		Average ± SD	Maximum	Minimum
Leaf	Leaf petiole length (cm)	1.79 ± 0.73	3.32 (220)	0.62
	Leaf length (cm)	12.3 ± 2.3	22.8 (764)	3.0
	Leaf width (cm)	5.8 ± 1.1	9.0 (330)	3.0
Plant	Canopy spread (cm)	82.3 ± 55.3	325.0 (461)	10.0
	Tree height (m)	2.14 ± 0.78	5.1 (490)	0.6
Stem	Trunk circumference (cm)	9.6 ± 6.7	25.0 (747)	1.0

Note: From Liyanage et al. (2020). Values are mean ± SD. The accession recorded highest value is given in the brackets. $n = 515$

Table 4.3 Quantitative leaf traits of *Cinnamomum* species found in Sri Lanka. Values are mean ± SD, $n = 30$

Species	Leaf length (LL) (cm)	Leaf width (LW) (cm)	Petiole length (mm)	LL/LW	Leaf fresh weight (g)
C. ovalifolium	5.5 ± 0.6^{g}	3.3 ± 0.1^{e}	9.9 ± 1.1$^{f, e}$	1.6 ± 0.2^{c}	11.6 ± 2.1^{f}
C. litseaefolium	8.8 ± 0.81^{f}	3.9 ± 0.4$^{e, d}$	13.6 ± 1.6$^{d, e}$	2.2 ± 0.1^{c}	17.2 ± 3.2^{e}
C. dubium	12.8 ± 2.4$^{c, d}$	3.4 ± 0.51^{e}	13.1 ± 2.3$^{e, f}$	3.3 ± 0.5^{a}	23.5 ± 3.5^{d}
C. capparu-coronde	15.4 ± 1.6^{b}	6.1 ± 0.7^{b}	18.6 ± 2.1^{b}	2.5 ± 0.1^{c}	28.6 ± 2.4^{c}
C. sinharajaense	28.4 ± 7.6^{a}	11.1 ± 2.2^{a}	29.2 ± 6.1^{a}	2.7 ± 0.3^{b}	66.1 ± 28.3^{a}
C. rivulorum	11.5 ± 1.9^{d}	4.1 ± 0.6^{d}	13.4 ± 1.3$^{d, e}$	2.6 ± 0.3^{b}	15.7 ± 2.1^{e}
C. zeylenicum (wild)	13.5 ± 2.8^{c}	5.3 ± 0.5^{c}	17.2 ± 1.4^{b}	2.7 ± 0.2^{b}	30.7 ± 2.2^{c}
C. zeylanicum ("Sri Gemunu")	11.2 ± 1.2^{d}	5.1 ± 0.4^{c}	14.8 ± 1.1$^{c, d}$	2.2 ± 0.3^{c}	29.7 ± 2/2^{b}
C. zeylanicum ("Sri Wijaya")	13.5 ± 1.4^{c}	5.2 ± 0.3^{c}	15.3 ± 1.2$^{c, d}$	2.6 ± 0.1^{b}	33.1 ± 2.5^{c}
C. citriodorum	10.8 ± 2.9^{e}	4.2 ± 1.2^{d}	14.8 ± 2.5$^{c, d}$	2.5 ± 0.2$^{b, c}$	15.5 ± 3.6^{e}

From Baundusekera et al. (unpublished). Values represented by the same letter within each column are not significantly different according to Duncan's Multiple Range Test; $n = 30$

Leaf Length—along the mid rib to obtain maximum length, Width; maximum width along the laminar, Petiole length; from leaf detachment to starting point of mid rib

Fig. 4.2 Representative leaf images of *Cinnamomum* species found in Sri Lanka. From Bandusekera et al. (unpublished). (**A**) *C. camphora*, (**B**) *C. ovalifolium*, (**C**) *C. litseaefolium*, (**D**) *C. citriodorum*, (**E**) *C. capparu-coronde*, (**F**) *C. dubium*, (**G**) *C. revulorum*, (**H**) *C. zeylanicum*, (**I**) *C. sinharajaense*. Bar = 10 cm, leaf pictures were captured from the tenth leaf from tip of the first secondary matured branch. Where more than one sample per species, the sampling sites were (from left to right): (**C**) Haggala, Delpitiya, (**F**) Sinharaja rain forest, Delpitiya (**H**) Norwood, Matara, Bibila, and (**I**) Matara, Sinharaja rain forest

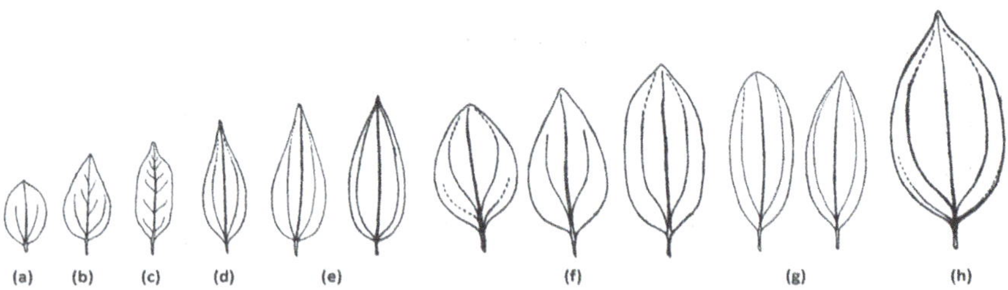

Fig. 4.3 Variation of leaf venation patterns of *Cinnamomum* species found in Sri Lanka. From Bandusekera et al. (unpublished). (**a**) *C. ovalifolium*, (**b**) *C. litseaefolium*, (**c**) *C. citriodorum*, (**d**) *C. rivulorum*, (**e**) *C. dubium*, (**f**) *C. zeylanicum*, (**g**) *C. capparu-coronde*, and (**h**) *C. sinharajaense*

4.4.1.2 Other Leaf Characters

Qualitative leaf characteristics such as leaf apex, shape, base, texture, color, margin, and flush color also vary considerably among cultivated cinnamon and its wild relatives in Sri Lanka (Figs. 4.2, 4.3, and 4.4 and Table 4.4). Leaf venation patterns also vary considerably among *Cinnamomum* species (Baruah and Nath 1997; Brewer and Stott 2017). Similarly, different venation patterns were recorded in wild relatives of cinnamon found in Sri Lankan forests (Figs. 4.2 and 4.3).

Leaf venation may have different levels, primary (1°), secondary (2°), tertiary (3°), and minor veins of 4° and 5° order. The origin, length, and arrangement vary considerably from species to species. Interestingly, *C. zeylanicum* and all the endemic wild relatives that exist in Sri Lanka can easily be identified by the leaf morphological characteristics, especially venation. For example, Bandusekera and

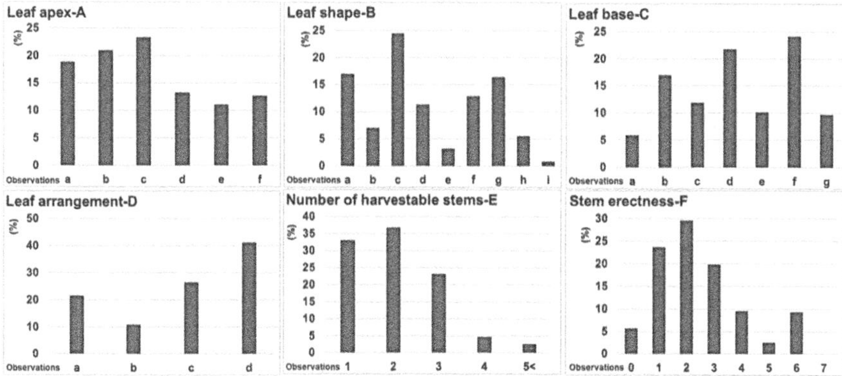

Fig. 4.4 Variation in qualitative leaf morphological traits of *Cinnamomum zeylenicum* germplasm in Sri Lanka ($n = 515$). From Liyanage et al. (2020). (**a**) leaf apex: a- acute, b- obtuse, c- acuminate, d- short acuminate, e- narrowly acuminate, f- acuminate with broad acumen; (**b**) leaf shape: a- elliptic, b- broadly elliptic, c- narrowly elliptic, d-ovate, e- broadly ovate, f- oval, g- lanceolate, h- ovate-lanceolate, i- oblong-lanceolate; (**c**) leaf base: a- acute, b- subacute, c- cuneate, d- rounded, e- subcordate, f- obtuse, g- obtuse; (**d**) leaf arrangements: a- Opposite, b- Subopposite, c- Opposite or subopposite in different branch but in same plant, d- Opposite to subopposite in the same branch in same plant; (**e**) number of stems 1, 2, 3, 4, and above 5; (**f**) "0"- no bends, "1"- one bend, "2"- two bends, "3"- three bends up to "7"- seven bends

colleagues recently developed a morphological index for field-level identification of *Cinnamomum* species in Sri Lanka (Bandusekara et al. 2020). Nevertheless, the number of 2° veins and the angle between 1° and 2° veins can vary from leaves to leaves in a single taxon (Baruah and Nath 1997).

The leaf shape and the related characteristics such as leaf apex, leaf base, leaf texture, leaf color, leaf arrangement, and flush color vary among species as well as within species. For example, Baruah and Nath (2007) have identified 14 taxa with variable phenotype and odoriferous characters found in Northeast India. Based on their overall morphological traits, these 14 taxa could be categorized into seven species. For example, *C. bejolghota* (Buch-Ham) consists of three clear variants while *C. cassia* consists of two variants. Two clear variants or phenotypic forms of *C. verum* were identified in the same study (Baruah and Nath 2007).

The leaf morphological characters, both qualitative and quantitative, vary considerably among *C. ovalifolium*, *C. litseaefolium*, *C. citriodorum*, *C. rivulorum*, *C. dubium*, *C. zeylanicum*, *C. capparu-coronde,* and *C. sinharajaense* (Tables 4.2 and 4.3, and Figs. 4.2, 4.3, and 4.4) (Bandusekara et al. unpublished). Furthermore, significant differences were found when more than one accession from the same wild species growing at the same location were assessed using morphological characters mentioned earlier (Bandusekara et al. unpublished). Similarly, Azad et al. (2019) found variation in leaf morphological characters among different accessions of *C. zeylanicum* to be more evident than the variation observed for stem and inflorescence characteristics. Such within-species diversity may be attributed to genetic diversity among them, environmental effect, or their interaction. To

Table 4.4 Qualitative leaf traits of genus *Cinnamomum* in Sri Lanka

Spp.	Leaf size	Apex	Shape	Base	Arrangement	Upper surface leaf texture	Leaf texture	Leaf color	Flush color	Leaf Margin
C1	Small	Obtuse^{1}/Blunt^{2}	Oval^{1} /Broadly eliptic^{2}	Acute	Opposite–subopposite	Glabrous, Glossy, Smooth	Rigidly coriaceous	137A	153A	Entire
C2	Small	Acute^{1}/Blunt^{2}	Ovate	Rounded^{1}/ Subacute^{2}	Opposite–subopposite	Glossy, Smooth	Coriaceous	137A	153A	Entire/ undulate
C3	Small to medium	Acuminate	Narrowly elliptic	Subacute	Opposite	Glabrous, Glossy, Smooth	Rigidly coriaceous	NN137	153A	Entire/ undulate
C4	Medium	Acute	Lanceolate	Subscute^{1} / Cunate^{2}	Opposite	Glabrous	Coriaceous	NN137D	47C	Entire
C5	Medium	Acute	Lanceolate	Obtuse	Opposite–subopposite	Glabrous	Coriaceous	NN137B	46B	Entire
C6	Medium	Acute	Ovate	Rounded^{1}/ Obtuse^{2}	Opposite–subopposite	Glabrous	Coriaceous	NN137B	46B	Entire
C7	Medium	Acute^{1} / Obtuse^{2}	Oval	Obtuse	Opposite–subopposite	Glabrous	Coriaceous	NN137B	46B	Entire
C8	Large	Narrowly acuminate	Elliptic	Obtuse	Subopposite	Glabrous	Coriaceous	NN137D	153A	Entire
C9	Medium	Short acuminate	Ovate lanceolate	Rounded	Opposite	Glabrous, Glossy	Sub coriaceous	NN137D	59B	Entire
C10	Medium	Long acuminate	Ovate lanceolate	Rounded	Opposite–subopposite	Glabrous, Glossy, Smooth	Sub coriaceous	137B	47A	Entire

Apex, shape, and base were selected on the basis of observation of 10 mature leaves, "1" indicates the prominent observation while "2" indicates second prominent observation

Leaf color, texture, and margin were recorded from the tenth leaf from tip of a first secondary matured branch. Leaf and flush colors were recorded on the basis of RHS sixth edition (2015). Leaf size: small (<10 cm length/<4 cm width), medium (10–16 cm length/4–8 cm width), large (> 16 cm length/>8 cm width (Source: Bandusekera et al. unpublished)

C1 C. ovalifolium, *C2 C. litseaefolium*, *C3 C. citriodorum*, *C4 C. capparu-coronde*, *C5 C. zeylenicum* (wild), *C6 C. zeylanicum* ("Sri Wijaya"), *C7 C. zeylanicum* ("Sri Gemunu"), *C8 C. revulorum*, *C9 C. dubium*, *C10 C. sinhajaense*

address this question, Liyanage and colleagues studied morphological diversity of a *C. zeylanicum* germplasm collection of more than 500 accessions growing in the same environmental conditions. There was a significant diversity among different accessions (Liyanage et al. 2020). Interestingly, the biochemical composition and yield parameters among them were also different (Liyanage et al. 2020). This suggests that the morphological diversity observed is a reflection of genetic diversity among them.

4.5 Plant Morphology

In the assessment of plant morphological characters in germplasm studies tree age and tree type (seedling, vegetative propagated plant, micro propagation) should be considered along with tree vigor (weak, intermediate, strong), tree spread (measured as the mean diameter using two directions), tree height (from ground level to the top of the tree), tree shape (columnar, pyramidal, obovate, rectangular, circular, semicircular, semi elliptic, irregular, other), trunk surface (smooth, rough, very rough), trunk circumstance (recorded at 30 cm above ground level), branching pattern (extensive, intensive, both patterns), distribution of branches (ascendant, irregular, verticillate, axial, horizontal), crotch angle of main branches (acute, obtuse, other), extension growth of twigs (measured after major growth flush following harvest), internode length of twigs (measured at the intermediate part of the twigs, after current season of growth has ceased), twig diameter (intermediate part of the twigs, after current season's growth has ceased), and surface of young twig (glabrous, pubescent, other) (TURIS 2013). Some of these traits are directly related to the yield, while others have indirect or no clear relationship with leaf or bark yield.

Some qualitative characters such as leaf shape, leaf base, and leaf apex are identified as environment independent characters (TURIS 2013). Other characters may vary significantly with the macro and micro environmental factors. These traits also vary among species and subspecies. Within-species diversity of such traits has been studied briefly in Northeast India (Baruah and Nath 2007), and in Matara district of Sri Lanka and in 269 accessions from six districts of Sri Lanka (Azad et al. 2019). Liyanage and colleagues have recently characterized the largest *C. zeylanicum* germplasm in Palolpitiya, Matara district, Sri Lanka (Liyanage et al. 2020). More than 500 accessions grown in the same environment were assessed for 25 traits including plant, leaf, and biochemical characters. There is a significant difference among them in plant morphological traits such as canopy spread, tree height, stem circumference, the number of harvestable stems, and erectness (Table 4.2, Fig. 4.4). Among them, tree height, the number of harvestable stems, and stem circumference are directly related to the yield. On the other hand, studying the plant characteristics of wild species is not practical since those change with the habitat and other environmental factors. For example, *C. ovalifolium* grows at high elevation forests and is adapted to the mountain environment and plants are short and bushy. Unfortunately, this species cannot even grow in germplasm collections maintained in other regions of the country. The natural habitat of *C. sinhajaense* is the Sinharaja rain forest. Both plant and leaf morphology of *C. sinhajaense* change drastically when it grows in different habitats (Bandusekara et al. unpublished).

4.6 Floral Morphology

Cinnamon flowers usually appear in long-pedunculate, axillary, and pseudoterminal panicles. Flowers are yellowish green with shallow flower-tube, six equal tepals, either completely deciduous, or deciduous upper part or indurate with fruit. Nine stamens are arranged in three whorls with distinct filaments. The anthers of outer six stamens are 4-celled introse; those of third whorl are often reduced to 2, extrose; upper pair of cells is much smaller than lower; basal glands are lateral to stamens of third whorl. Style is short, stigma conspicuous, and usually peltate (Dassanayaka 1995). Pollen grains are binucleate (Erdtman 1952; Brewbaker 1967). Cinnamon flowers are insect pollinated. Mohanakumar and colleagues observed 13 insect pollinators on cinnamon (Mohanakumar et al. 1985).

Flower color, number of flowers per panicle, length of panicle, pedicel length, petal length, and anther shape vary considerably among different species (Table 4.5 and Fig. 4.5) (Bandusekara et al. unpublished). Azad et al. (2018) studied the flower morphological diversity of *C. zeylanicum* in Matara district, Sri Lanka. They evaluated *C. zeylanicum* accessions collected from 15 locations for peduncle length, flower length, flower width, and floral tube length. The variation in tepal shape was clear, having two whorls of tepals of a single flower exhibiting two shapes.

Table 4.5 Qualitative and quantitative flower characters of genus *Cinnamomum* in Sri Lanka

Species	FC	MNFP	MLP	APL	ATL	TP
C. ovalifolium	Yellowish white	2–3	38.6	2.85	1.42 ± 0.21	Densely silvery
C. litseaefolium	Yellowish white	2–3	12.6	2.45	1.58 ± 0.15	Densely silvery
C. dubium	Greenish white	6–8	64.5	5.52 ± 0.94	2.65 ± 0.31	Densely white
C. capparu-coronde	White	3–5	128.6	5.45 ± 0.75	1.83 ± 0.23	Intermediate white
C. sinharajaense	White	3–5	110.8	5.51 ± 0.31	3.48 ± 0.28	Intermediate silvery gray
C. rivulorum	Greenish white	8–10	66.8	3.94 ± 0.85	2.43 ± 0.15	Densely white
C. zeylanicum	White	12–14	286.8	4.56 ± 0.89	2.83 ± 0.83	Densely white

Notes: *FC* Flower color, *MNFP* maximum number of fully open flowers per panicle, *MLP* maximum length of panicle (cm); from top most flower to attachment in branch, *APL* average pedicel length (cm); from beginning of pedicel to flower attachment, mean ± SD, *ATL* average tepal length (cm), mean ± SD, *TP* tepal pubescence; dense, intermediate, sparse; color of pubescence; silvery, white, yellowish

Fig. 4.5 Representative flower samples of wild *Cinnamomum* species and the cultivated *C. zeylanicum*. From Bandusekera et al. (unpublished). (**A**) *C. sinhajaense,* (**B**) *C. zeylenicum.* (**C**) *C. dubium,* (**D**) *C. capparu-coronde,* (**E**) *C. revulorum,* (**F**) *C. litseaefolium,* (**G**) *C. ovalifolium,* Bar = 1 cm, indicating flower width at fully open stage

Shivaprasad et al. (2015) studied the reproductive biology of *C. sulphuratum* Nees growing in wet evergreen forest of western Ghats in Karnataka.

While few early (Josy 1981; Kubitzki and Kurz1984; Mohanakumar et al. 1985) and recent (Hathurusinghe and Bandaranayake 2018, Bandusekera et al. unpublished) studies on floral biology of *Cinnamomum* are available, the information is still incomplete. Flowering season starts in November and lasts till early March. Duration from visible flower initiation to flowering is about 2 weeks. Cinnamon flowers are botanically identified as protogynous dichogamy where female and male phases are separated temporally. The female parts mature first and the male parts mature about a day later ensuring outcrossing.

When the flower opens in the stage one, its stigma is whitish and fresh and receptive. At that stage, there is no anther dehiscence, and the stamens of the first whorl and those of the third whorl are fused. Flower will open for about 5–6 hours and closes. In stage 2, the flower opens again in the next day. The stigma appears pale brown and shriveled and not receptive. The stamens of the third whorl are now separated from those of the first whorl and adhere to the pistil. The anthers dehisce 30–60 minutes after flower opening and behave as a male flower until it closes permanently after about 5 hours (Josy 1981). Josy (1981) reported variations in the time of occurrence and duration of first and second stages. Out of 55 plants studied, in 33 the first stage occurred in the morning while the rest opened at around noon. The opposite happened in their opening for the second time, where 22 plants among those that showed late opening in the first day had early opening in the second day morning while the rest of 33 plants had second opening around noon. These observations suggested complete cross-pollination in cinnamon.

Josy (1981) and Kubitzki and Kurz (1984) used the term synchronized dichogamy in relation to flowering behavior of cinnamon. Like in avocado, *Persea americana* (Sedgley and Griffin 2013), there are two flowering types: Type (a), with female phase in the morning and male phase next day afternoon, and Type (b), with female phase in the afternoon and male phase on the next day morning. Since both type (a) and (b) plants are mixed in any natural population, large numbers of functional female and male flowers are available during the flowering season. Because of the temporal separation, within flower pollination and allogamy between flower pollination of the same plant are prevented allowing pollination only between different trees (Kubitzki and Kurz 1984).

However, a recent study showed the possibility of autogamy in *C. zeylanicum*. A study conducted with two local varieties "Sri Gemunu" (Type a) and "Sri Wijaya" (Type b) showed that occurrence of an overlapping period of stage I female flowers (open for the first time) and stage II male flowers (open for the second time) in the same tree for about 1–2 hours. The overlapping period was longer in "Sri Gemunu" (nearly 2 hours) than in "Sri Wijaya" (30–45 minutes) (Hathurusinghe et al. unpublished). Insects, mainly ants, moved between flowers during this period, possibly moving the pollen around. Fruits were also set from those flowers. Furthermore, fruits were set normally in an isolated *C. zeylanicum* plant with no access to another plant. This observation also suggests allogamy in *C. zeylanicum*. This will be important in future breeding attempts under controlled pollination conditions.

4.7 Petiole, Leaf, Root, and Stem Anatomy of *C. zeylanicum* Blume from Sri Lanka: Taxonomic Implications

4.7.1 Petiole Anatomy

A comprehensive study of petiole anatomy of the *Cinnamomum* species showed marked variations (Abeysinghe and Scharaschkin 2019). Shape of the entire petiole of *C. zeylanicum* in cross section is roundish, bifacial, and vertically symmetrical (Fig. 4.6a). Tissue arrangement of *C. zeylanicum* consists of a cortex with fundamental parenchyma cells with discontinuous patches of sclerenchymatic cells that surround the vascular bundle (Fig. 4.6b, c) and open arch vascular bundle with collateral type xylem tissues (Fig. 4.6c). Similar tissue arrangement has been observed in *Ocotea duckei* (Coutinho et al. 2006). *C. zeylanicum* has slight concave-shaped petioles with a shallow groove on the adaxial side. Trichomes and upper surface wings are absent. Epidermal cells are observed to be mostly rectangular. Abeysinghe and Scharaschkin (2019) observed pigmented cells in ground tissues (Fig. 4.6c) that have been observed in worldwide genera of the family Lauraceae (Kamel and Loutfy 2001).

4.7.2 Leaf Anatomy: Cuticle Epidermis, Trichomes, and Stomatal Complex

According to Scanning Electron Micrography (SEM), a continuous thick film of epicuticular waxes is present on both surfaces of the leaf (Fig. 4.7a–d). No clear pattern of the wax deposition can be observed on the upper surface of *C. zeylanicum*

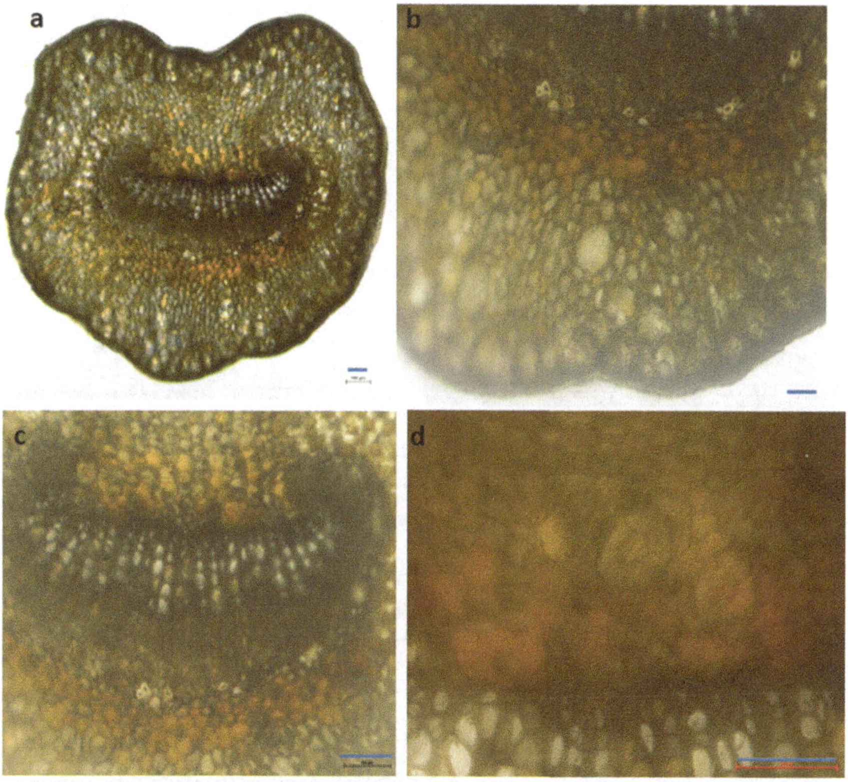

Fig. 4.6 Petiole anatomical structures of *C. zeylanicum*. (**a**) Cross section of the petiole; (**b**) lower surface of the petiole showing thick cuticle and the slight groove in the middle part of the petiole; (**c**) dark pigment mainly present around the vascular bundle, "C"-shaped vascular strand with slightly curved ends; (**d**) stone cells above the vascular strand. Scale bar 100 μm

leaf (Fig. 4.7a, b), but unique ultrastructure of the wax deposition and clear clusters of crusts are observed on lower surface (Fig. 4.7c, d).

C. zeylanicum leaf samples were collected from the fully sun-exposed plants. Therefore, thick cuticle (Fig. 4.8a, b) would help to prevent the water loss effectively, as well as the leaf from wilting through cell dehydration (Metcalfe and Chalk 1972). In *O. duckei*, thick and smooth cuticle has been observed by Coutinho et al. (2006). Structural differences of cuticle of family Lauraceae have been observed in *Beilschmiedia* and *Endiandra* (Christophel et al. 1996); *Apollonias* (Loutfy 2001); *O. duckei* and *Cinnamomum pauciflorum* (Baruah and Nath 2006; Coutinho et al. 2006); *Aspidostemon*, *Beilschmiedia*, *Cryptocarya*, and *Potameia* (Nishida and van der Werff 2007); *Syndiclis* (Yang et al. 2012); *Cryptocarya* (Nishida et al. 2016); and *Beilschmiedia* and *Litsea* (Pole 2019). Furthermore, cuticular materials from extinct plants such as *Machilus maomingensis* from the Eocene (Tang et al. 2016), foliar fossils from late Palaeocene (Carpenter et al. 2010), *Cinnamomum*, *Neolitsea*,

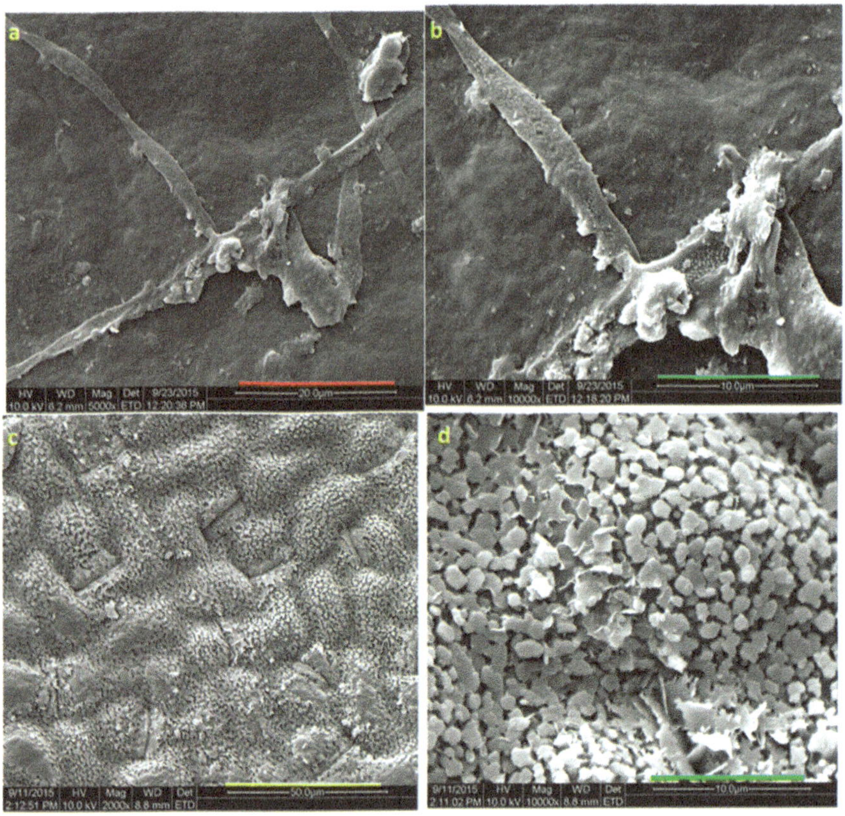

Fig. 4.7 Scanning electron micrography of wax deposition on abaxial and adaxial surfaces of *C. zeylanicum* leaf. (**a**) Irregularly deposited wax on the adaxial surface; (**b**) closer view of adaxial surface; (**c**) clusters of wax; (**d**) closer view of wax deposition on abaxial surface showing cuticular flakes/crusts. Scale bar 10 μm (green), 20 μm (red), and 50 μm (yellow)

Litsea, *Alseodaphne*, *Laurus*, and *Beilschmiedia* from the Oligocene period (Shi et al. 2014) have been studied.

On the upper leaf surface of *C. zeylanicum*, isodiametric epidermal cells form a reticulate pattern and is devoid of stomata (Fig. 4.8a, b, d), while irregular shape epidermal cells, small double semicircle, paracytic, and hypostomatic stomata are found only on abaxial surface (Fig. 4.8c). The epidermal cell walls are mucilaginous as revealed in test with ruthenium red. Unbranched trichomes (Bakker et al. 1992; Kamel and Loutfy 2001; Loutfy 2001; Bhatt and Pandya 2012) and glanduate and filliform trichomes have been observed in *Litsea chinensis* Lam. (Bhatt and Pandya 2012). Hypostomatic with small, sunken guard cells are also consistent with Christophel et al. (1996), Baruah and Nath (2006), Ceolin et al. (2009), Coutinho et al. (2006), Yang et al. (2012), Zeng et al. (2014), and Gonçalves et al. (2018). However, amphistomatic and anisocytic type stomata have been observed in *Litsea chinensis* Lam. (Bhatt and Pandya 2012). As we observed, Bakker et al. (1992) have examined the palisade and spongy parenchyma with idioblasts, oil, and mucilage

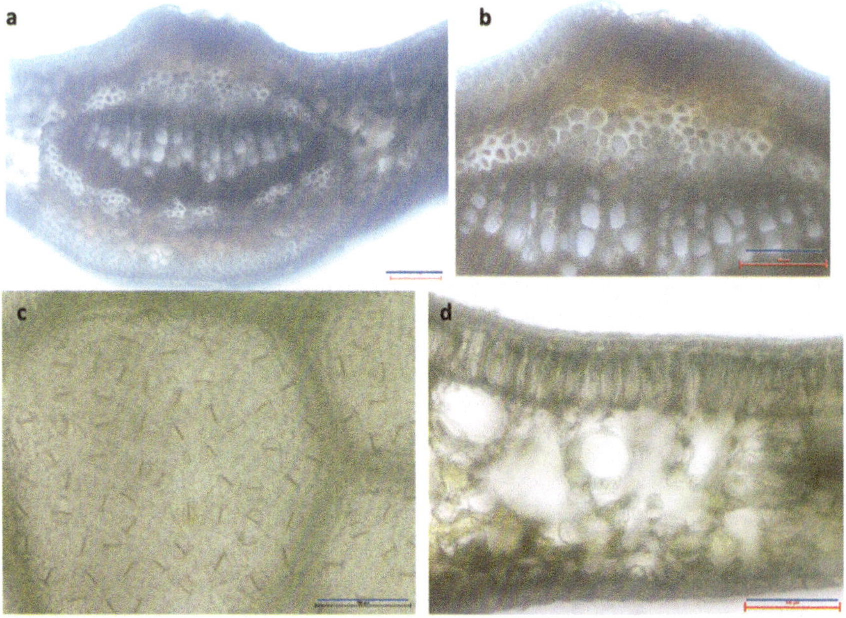

Fig. 4.8 Cross sections of the *C. zeylanicum* leaf and stomata: (**a**) part of a cross section of the leaf blade along the midrib of the leaf showing the leaf anatomical structures; elongated shape vascular system, discontinuous patches of sclerenchymatous cells around the vascular system; (**b**) adaxial surface has a prominent cuticle, which probably contains polyphenols due to bluish-green staining with Toluidine Blue O (Tolonium chloride), arrangement of the vascular tissues; (**c**) stomata on abaxial surface; (**d**) the palisade parenchyma layer is somewhat dense and composed of rectangular, attenuated, and vertical cells, the spongy parenchyma is composed of irregular shape cells, schizogenous secretory cavities are found below the palisade parenchyma. Scale bar 100 μm

cells (Fig. 4.8d). The palisade parenchyma layer is somewhat dense and composed of rectangular, attenuated, and vertical cells (Fig. 4.8d).

4.7.3 Root and Stem Anatomy

Among the wide array of tissues of root, presence of trichomes on the root, cortex layer, pericyclic fibers, mucilage cells, perforation plate structure, inter-vascular pitting, ray composition, axial parenchyma distribution, pit structure of imperforate elements (tracheids, fibers, etc.), oil and mucilage in the cortex and axial parenchyma, sclerenchyma groups are the characters observed in *C. zeylanicum* (Fig. 4.9a). These features are most often employed in assessments of systematic relationships (Patel 1987; Amalesh et al. 2015; Sun et al. 2015). In *C. zeylanicum*, the bark is characterized by secretory cells containing mucilage or essential oil droplets and dark golden brown, dark red or dark-colored pigments, and the presence of islands of sclerenchyma in the pericycle (Fig. 4.9b). These results were

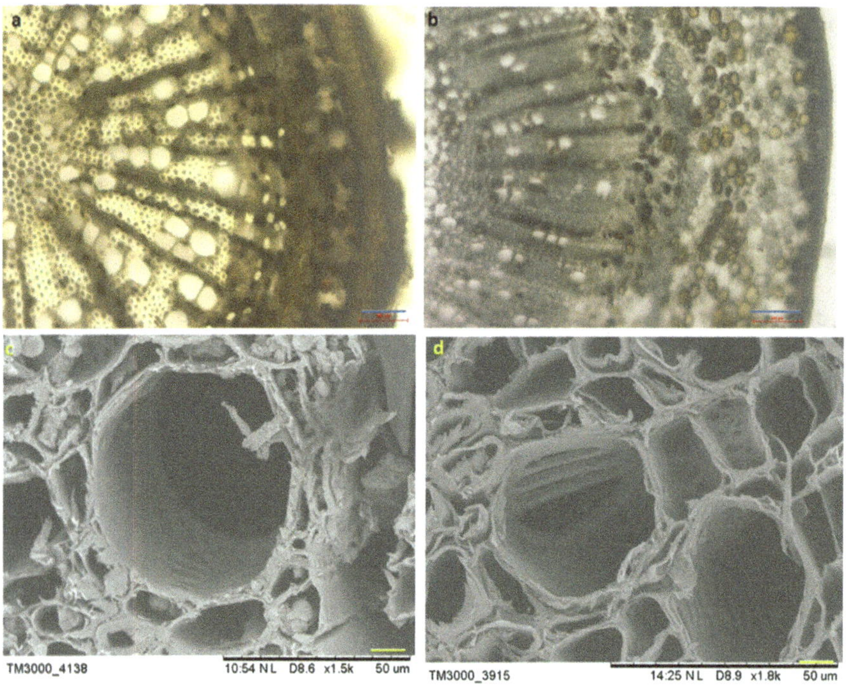

Fig. 4.9 Anatomy of root and stem of *Cinnamomum zeylanicum*. (**a**) *C. zeylanicum* root—3–5 vessel elements transverse to oblique, trichomes on the root, dark-stained bodies in the cortex, mucilage canals and oil in the cortex of *C. zeylanicum* root; (**b**) dark golden brown, dark-colored pigments, islands of sclerenchyma in the pericycle, 3–5 vessel elements transverse to oblique, dark-stained bodies in the cortex of *C. zeylanicum* stem; (**c**) scalariform pits in vessel element, round, aggregated, crystals in fiber tissues in *C. zeylanicum* root; (**d**) scalariform pits in vessel elements of *C. zeylanicum* stem. Scale bar; 50 μm (yellow) and 100 μm (blue)

consistent with those of Shylaja and Manilal (1992), Bakker and Baas (1993), and Geng et al. (2012). Although a taxonomic revision of *Cassytha* (Lauraceae) has been done (Weber 1981) using morphological characters, a detailed description on root anatomy of the members of Lauraceae is lacking. Therefore, a comparative anatomical study of the species of the family Lauraceae cannot be accomplished yet.

4.7.4 Anatomical Features of the Vascular Tissues of C. zeylanicum

4.7.4.1 Petiole and Leaf Vascular Tissue

Vascular bundle of petiole has a simple open arc, which is "C" shaped with slightly curved ends (Fig. 4.6c), discontinuous sclerenchymatous sheath with dark-colored deposits including the cell lumina with ~20–25 collateral vessels (Abeysinghe and

Scharaschkin 2019). The vascular bundle is enclosed by parenchyma cells with distinctive dark-staining secondary (tannin and phenolic) compounds (Fig. 4.6c) and stone cells above the vascular tissue (Fig. 4.6d). Kamel and Loutfy (2001) have observed crescent-shaped vascular strands in worldwide genera of Lauraceae, while Coutinho et al. (2006) have identified V-shaped vascular bundles in some members of Lauraceae. In transverse section along the midrib of the leaf only one central, elongated, and irregular-shaped vascular bundle surrounded by dorsiventrally discontinuous sclerenchymatic sheath can be found (Fig. 4.6a). More than ten collateral vessels that are 1–5 cells thick are arranged vertically, in parallel, and they are spaced from each other by sclerenchyma cells in the vascular bundle. The xylem is positioned toward the upper surface and the phloem is toward the lower (Fig. 4.8a, b).

4.7.4.2 Root and Stem Vascular Tissue

In both root and stem, a radially arranged vascular system is observed. Growth ring boundaries are indistinct. Xylem is diffuse and porous, composed of vessels predominantly solitary in radial or diagonal multiples (usually 2–3 cells) and less frequent are of 4–5 vessel elements transverse to oblique (Fig. 4.9a, b). Shape of the vessels is mostly round (Fig. 4.9c), but oval and irregular shapes are also observed (Fig. 4.9a, b). Large thick-walled vessels are surrounded by thin-walled parenchyma. Paratracheal parenchyma and uniseriate/biseriate rays are present (Fig. 4.9a, b). Inter-cellular patterns of the pittings of *C. zeylanicum* root and stem are opposite (Fig. 4.10a), helical (Fig. 4.10b) and scalariform (Figs. 4.10c, and 4.10d). Vestured pits are found in *C. zeylanicum* (Fig. 4.10a). Different outlines of the perforation plates are found in *C. zeylanicum*; round (Fig. 4.10f) to oval/elongated (Fig. 4.10e). Uni-seriate or bi-seriate rays and marginal axial parenchyma are observed in both stem and root of *C. zeylanicum* (Fig. 4.9a, b).

Wood in Lauraceae is diffuse-porous or semi-ring porous (Patel 1987; Callado and Costa 1997; Loutfy 2009; Andianto et al. 2015; Sun et al. 2015). Growth ring boundaries are distinct or intermediate between distinct and indistinct (Patel 1987; Loutfy 2009; Singh et al. 2015; Sun et al. 2015), solitary vessels in radial or diagonal multiples (usually 2–3 cells) or vessels in radial multiples of 4 or more (Loutfy 2009; Sun et al. 2015). Vessels of most species were evenly distributed (Sun et al. 2015) with simple perforation (Loutfy 2009; Sun et al. 2015) or scalariform perforation (Loutfy 2009; Sun et al. 2015). The intervessel pittings of most species were bordered and alternate (Loutfy 2009; Sun et al. 2015), the outlines of pits were round to oval, sometimes angular, and the pit apertures were round, oval, and lenticular (Sun et al. 2015), vessels were filled with tyloses (Loutfy 2009; Sun et al. 2015), oil and mucilage cells associated with axial and ray parenchyma (Sun et al. 2015), vasicentric axial parenchyma (Loutfy 2009; Sun et al. 2015), fiber tracheids with septate fibers (Loutfy 2009), uniseriate rays and multiseriate rays (Loutfy 2009), and marginal axial parenchyma (Loutfy 2009; Sun et al. 2015) were present.

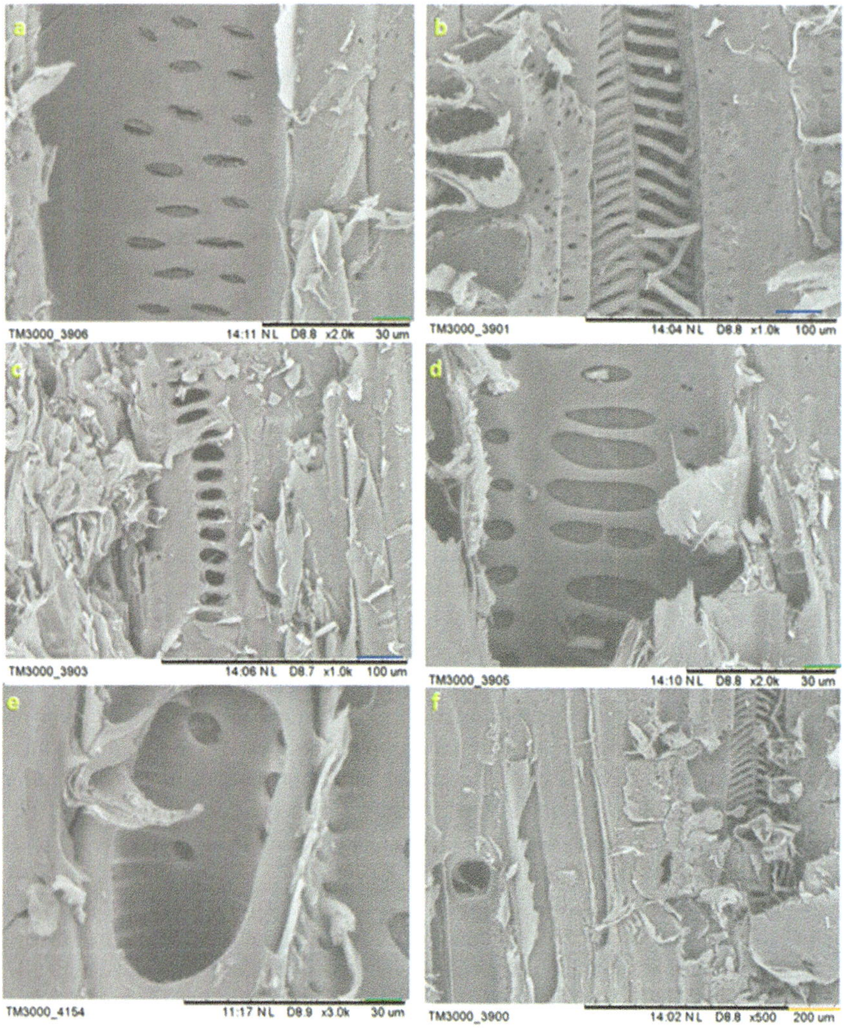

Fig. 4.10 Scanning electron micrographs of *Cinnamomum zeylanicum* root and stem. (**a**) Alternate thickening—*C. zeylanicum* root; (**b**) helical thickenings—*C. zeylanicum* stem; (**c**) simple pitted or bordered; (**d**) scalariform thickenings; (**e**) elongated shape, simple perforation—*C. zeylanicum* root; (**f**) round outline—*C. zeylanicum* stem, perforation plate in fiber cells with its circular perforations

4.7.5 Crystals

Calcium oxalate exists in different crystal forms (simple, aggregates), crystalline inclusions, and shapes (round shape being the predominant form, elongated). In petiole of *C. zeylanicum*, different shapes of crystals (raphides, fine crystalline sand) are found. Calcium oxalate crystals mainly occur in the ground parenchyma

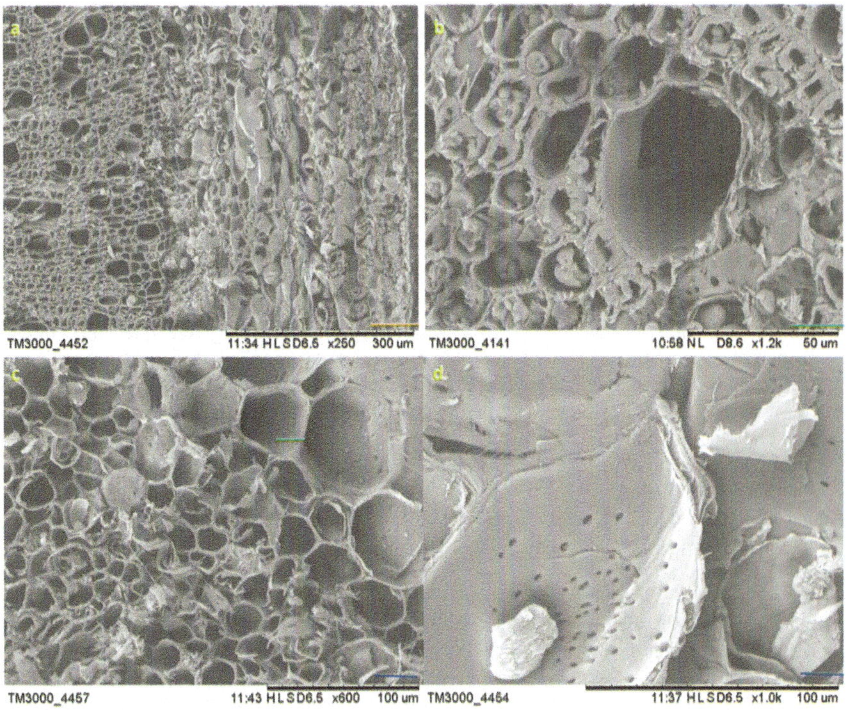

Fig. 4.11 Scanning electron micrographs of *C. zeylanicum* crystals present in different tissues. (**a**) In root cortex mainly globular shape; (**b**) in stem fiber tissues and around the vessel elements; (**c**) in pith parenchyma of root; and (**d**) crystalline inclusions in the stem parenchyma

(Fig. 4.11a) around the vascular sheath (Figs. 4.9c and 4.11b) and pith parenchyma cells (Fig. 4.11c, d). Dark-colored deposits are also common in the mesophyll and are sometimes associated with the phloem of major and minor veins. In SEM, globular-shaped bodies are heavily deposited in cortex parenchyma in both stem and root (Fig. 4.11a). Druses were absent in *C. zeylanicum*.

Acknowledgments P.C.G. Bandaranayake acknowledges the financial assistance provided by the Ministry of Primary Industries and Social Empowerment through the National Science Foundation of Sri Lanka under the special Cinnamon project—Grant No: NSF SP/CIN/2016/01. She thanks the Department of Wildlife Conservation, Sri Lanka, and the Forest Department for permission to collect samples from wild populations in protected areas. P. D. Abeysinghe was financially supported by the Endeavour Research Fellowship (ERF_PDR_2014_4408). Dr Tanya Scharaschkin, Dr. Melodina Fabillo, Dr. Hernán Retamales, and other members of the Plant Structure and Systematics group, EEBS, QUT and technical staff in the Science and Engineering Faculty (SEF) are acknowledged for assistance with logistics and training in plant anatomy techniques. SEM data reported in this chapter were obtained at the Central Analytical Research Facility (CARF) operated by the Institute for Future Environments at QUT, Brisbane, Australia, with assistance from Rachel Hancock and Sanjleena Singh. Ailsa Holland and other staff at the Queensland Herbarium (BRI) are thanked for their advice and logistical support pertaining to sending specimens from Sri Lanka to Australia. Authors are grateful to Director General, Department of Export Agriculture, Kandy

Road, Peradeniya, and Director and staff of National Cinnamon Research and Training Center, Department of Export Agriculture, Thihagoda, Palolpitiya and Sub Research Station, Department of Export Agriculture, Nillambe, for providing cinnamon samples

References

Abeysinghe PD, Scharaschkin T (2019) Taxonomic value of the petiole anatomy in the genus *Cinnamomum* (Lauraceae) found in Sri Lanka. Ruhuna J Sci 10:1–17

Abeysinghe PD, Wijesinghe KGG, Tachida H, Yoshda T (2009) Molecular characterization of cinnamon (*Cinnamomum verum* presl) accessions and evaluation of genetic relatedness of cinnamon species in Sri Lanka based on *Trn*L intron region, intergenic spacers between *trn*T-*trn*L, *trn*L-*trn*F, *trn*H-*psb*A and nuclear ITS. Res J Agric Biol Sci 5:1079–1088

Amalesh N, Nirankush P, Kumar GA, Partha G, Dipankar B, Rahul S, Chandrakant K (2015) Identification of adulterants by pharmacokinetic evaluation: Tvak (*Cinnamomum zeylanicum* Blume.) and Naluka/Cassia [*Cinnamomum cassia* (Nees and T. Nees.) J. Presl]. Int J Res Pharm Biosci 2(2):1–4

Andianto WI, Waluyo TK, Dungani R, Hadiyane A, Hernandi MF (2015) Wood anatomical from Indonesian genus *Cinnamomum* (Lauraceae) and their identification key. Asian J Plant Sci 14(1):11–19

Ariyarathne HBMA, Weerasuriya SN, Senarath WTPSK (2018) Comparison of morphological and chemical characteristics of two selected accessions and six wild species of genus *Cinnamomum* Schaeff. Sri Lankan J Biol 3:11–23

Azad R, Ranawaka R, Senanayake G et al (2016) Morphological variation of cinnamon (C*innamomum verum* Persl) germplasm in Matara District of Sri Lanka. Int J Minor Fruits Med Aromat Plants (IJMFM&AP) 2(1):6–14

Azad R, Kumara KW, Senanayake G, Ranawaka R, Pushpakumara D, Geekiyanage S (2018) Flower morphological diversity of cinnamon (*Cinnamomum verum* Presl) in Matara District, Sri Lanka. Open Agric 3:236–244

Azad R, Jayasekara L, Ranawaka RAAK, Senanayake G, Wasantha Kumara KL, Pushpakumara DKNG, Sudarshanee G (2019) Development of a core collection for Sri Lankan cinnamon germplasm based on morphological characterization using an eco-geographic survey. Aust J Crop Sci 13:1473–1485

Bakker ME, Gerritsen AF, van der Schaaf PJ (1992) Leaf anatomy of *Cinnamomum* Schaeffer (Lauraceae) with special reference to oil and mucilage cells. Blumea 37:1–30

Bakker ME, Baas P (1993) Cell walls in oil and mucilage cells. Acta Botanica Neerlandica 42:133–139

Bandusekara BS, Pushpakumara DKNG, Bandaranayake PCG (2020) Field level identification of *Cinnamomum* species in Sri Lanka using a morphological index. Trop Agric Res 31(4):43–53

Baruah A, Nath SC (1997) Foliar epidermal characters in twelve species of *Cinnamomum* Schaeffer (Lauraceae) from northeastern India. Phytomorphology 47:127–134

Baruah A, Nath SC (2006) Leaf anatomy and essential oil characters of *Cinnamomum pauciflorum* Nees – a potential spice crop from North-East India. J Spices Aromat Crops 15(1):52–56

Baruah A, Nath SC (2007) *Cinnamomum champokianum* sp. Nov.(Lauraceae) from Assam, northeastern India. Nord J Bot 25:281–285

Bhatt SP, Pandya SS (2012) Scanning electron microscopy of leaf of *Litsea chinensis* Lam. Int J Res Pharm Chem 2(2):470–478

Brewbaker JL (1967) The distribution and phylogenetic significance of binucleate and trinucleate pollen grains in the angiosperms. Am J Bot 54:1069–1083

Brewer SW, Stott GL (2017) A new species of *Cinnamomum* (Lauraceae) from the Bladen Nature Reserve, Southern Belize. PhytoKeys 1

CABI (2020) Invasive species compendium; *Cinnamomum verum* (cinnamon). CAB International. Available from: https://www.cabi.org/isc/datasheet/13573#tosummaryOfInvasiveness. Accessed 23 Feb 2020

Callado CH, Costa CG (1997) Wood anatomy of some *Anaueria* and *Beilschmiedia* species (Lauraceae). IAWA J 18(3):247–259

Carpenter RJ, Truswell EM, Harris WK (2010) Lauraceae fossils from a volcanic Palaeocene oceanic island, Ninetyeast Ridge, Indian Ocean: ancient long-distance dispersal? J Biogeogr 37(7):1202–1213

Ceolin GB, Rosito JM, do Canto-Dorow TS (2009) Leaf surface characters applied to Lauraceae taxonomy in a seasonal forest of Southern Brazil. Braz Arch Biol Technol 52(6):1453–1460

Chanderbali AS, van der Werff H, Renner SS (2001) Phylogeny and historical biogeography of Lauraceae: evidence from the chloroplast and nuclear genomes. Ann Mo Bot Gard 88:104–134

Christophel DC, Kerrigan R, Rowett AI (1996) The use of cuticular features in the taxonomy of the Lauraceae. Ann Mo Bot Gard 83(3):419–432

Coutinho DF, Agra MF, Basílio IJLD, Barbosa-Filho JM (2006) Morphoanatomical study of the leaves of *Ocotea duckei* Vattimo (Lauraceae-Lauroideae). Braz J Pharmacogn 16(4):537–544

Dao NK (2004) Chinese cassia. In: Ravindran P, Babu K, Shylaja M (eds) Cinnamon and cassia the genus *Cinnamomum*. CRC Press, Boca Raten, pp 156–184

Dassanayaka MD (1995) A revised handbook to the flora of Ceylon, Revised edn. Oxford & IBH, New Delhi, pp 112–129

Erdelen W (1988) Forest ecosystems and nature conservation in Sri Lanka. Biol Conserv 43:115–135

Erdtman G (1952) Pollen morphology and plant taxonomy. Hafner Publishong Co., New York, pp 6–24

Fleischmann K (1997) Invasion of alien woody plants on the islands of Mahé and Silhouette, Seychelles. J Veg Sci 8:5–12

Forrest DM (1967) A hundred years of Ceylon tea: 1867–1967. Chatto and Windus, London

Geng S, Cui Z, Shu B, ZhaoS YX (2012) Histochemistry and cell wall specialization of oil cells related to the essential oil accumulation in the bark of *Cinnamomum cassia* Presl. (Lauraceae). Plant Prod Sci 15(1):1–9

Gonçalves RDA, Pinheiro AB, Oliveira MA, Do Nascimento RT, Rosalem PF, Garcia VL, Martins AR (2018) Anatomical characters and chemical profile of leaves of three species in Lauraceae family. Rev Bras 28:1–8

Gunatilake HM, Senaratne DMAH, Abeygunawardena P (1993) Role of non-timber forest products in the economy of peripheral communities of knuckles national wilderness area of Sri Lanka: a farming systems approach. Econ Bot 47:275–281

Gunatilleke IAUN, Gunatilleke CVS (1991) Threatened woody endemics of the wet lowlands of Sri Lanka and their conservation. Biol Conserv 55:17–36

Hathurusinghe B, Bandaranayake PCG (2018) Studying unique flowering behavior of *Cinnamomum zeylanicum*: examples from two elite varieties. SLCARP International Research Symposium, 13th and 14th August, Hotel Taj Samudra, Colombo

Helal A, Tagliazucchi D, Verzelloni E, Conte A (2014) Bioaccessibility of polyphenols and cinnamaldehyde in cinnamon beverages subjected to in vitro gastro-pancreatic digestion. J Funct Foods 7:506–516

IUCN (1998) The IUCN red list of threatened species 1998. World Conservation Monitoring Centre. www.iucnredlist.org. Accessed 27 Feb 2020

Jayaprakasha GA, Rao LJ, Sakariah KK (1997) Chemical composition of the volatile oil from the fruits of *Cinnamomum zeylanicum* Blum. Flavour Fragr J 12:331–333

Josy J (1981) Floral biology and variation in cinnamon. Plantation crops symposium IV. Indian Society of Plantation Crops, Mysore

Kamel EA, Loufty MHA (2001) The significance of cuticular features, petiole anatomy and SDS-PAGE in the taxonomy of the Lauraceae. Pak J Biol Sci 4(9):1094–1100

Kostermans AJGH (1957) Lauraceae. Reinwardtia 4:193–256

Kostermans AJGH (1961) The new world species of *Cinnamomum* Trew (Lauraceae). Reinwardia 6:17–24

Kostermans AJGH (1964a) Bibliographia Lauracearum. Bogor Botanic Gardens, Jakarta

Kostermans AJGH (1964b) Family Lauraceae. In: Dassanayake MD, Fosberg FR, Clayton WD (eds) A revised handbook to the flora of Ceylon, vol IX, 1995 edn. Oxford and IBH publishers, New Delhi, pp 105–172

Kostermans AJGU (1973) A forgotten Ceylonese cinnamon tree – *Cinnamomum capparu coronde* Bl. Ceylon J Sci – Biol Sci 10:119–122

Kostermans A (1980) Notes on Ceylonese plants I. In: Arends JC, Boelema G, De Groot CT, Leeuwenberg AJM (eds) Miscellaneous papers 19. Landbouwhogeschool, Wageningen, The Netherlands

Kostermans AJGH (1983) The South Indian species of *Cinnamomum* schaeffer (Lauraceae). Bull Bot Surv India (NELUMBO) 25

Krishnamoorthy B, Sasikumar B, Rema J, Gopalam A, Abraham J (1992) Variability and association studies in cinnamon (*Cinnamomum verum*). J Spices Aromat Crops 1:148–150

Krishnamoorthy B, Rema J, Zachariah TJ, Abraham J, Gopalam A (1996) Navashree and Nithyashree – two new high yielding and high quality Cinnamon – (*Cinnamomum verum* Bercht & Presl.) selections. J Spices Aromat Crops 5:28–33

Kubitzki K, Kurz H (1984) Synchronized dichogamy and dioecy in neotropical Lauraceae. Plant Syst Evol 147:253–266

Kueffer C, Klingler G, Zirfass K, Schumacher E, Edwards PJ, Güsewell S (2008) Invasive trees show only weak potential to impact nutrient dynamics in phosphorus-poor tropical forests in the Seychelles. Funct Ecol 22:359–366

Kumarathilake DMHC (2012) Study on the extinction risks, conservation and domestication of endemic wild *Cinnamomum* species in Sri Lanka. M.Phil, Faculty of Agriculture, University of Ruhuna Sri Lanka

Kumarathilake DMHC, Senanayake SGJN, Wijesekara GAW, Wijesundera DSA, Ranawaka RAAK (2010) Extinction risk assessments at the species level: national red list status of endemic wild cinnamon species in Sri Lanka. Trop Agric Res 21(3):247–257

Leela NK, Vipin TM, Priyanka V, Shafeekh KM, Rema J (2012) *Cinnamomum citriodorum* Thw. – new source of citronellol. J Essent Oil Res 24:507–511

Liyanage ASU (2010) Eco-geographic survey of crop wild relatives. Ministry of Environment and Natural Resources, Peradeniya, Sri Lanka

Liyanage NMN, Bandusekara BS, Kanchanamala RWMK, Hathurusinghe HABM, Rathnayaka AMRWSD, Pushpakumara DKNG, Samita S, Wijesinghe KGG, Jayasinghe GG, Liyanage WK, Bandaranayake PCG (2020) Identification of superior Cinnamomum zeylanicum Blume germplasm for future true cinnamon breeding in the world. Journal of Food Composition and Analysis 96:103747

Liyanage T, Madhujith T, Wijesinghe KGG (2017) Comparative study on major chemical constituents in volatile oil of true cinnamon (*Cinnamomum verum* presl. Syn. *C. zeylanicum* blum.) and five wild cinnamon species grown in Sri Lanka. Trop Agric Res 28:270–280

Loutfy MHA (2001) On the taxonomy of *Apollonias* Nees (Lauraceae). Evidence from general aspects and cuticular features of leaf & fruit. Taeckholmia 21(1):59–80

Loutfy MHA (2009) Wood anatomy and its implications on the taxonomy of *Apollonias* Nees (Lauraceae). Feddes Repert 120:75–90

Mallavarapur GR, Rajeswara Rao BR (2007) Chemical constituents and uses of *Cinnamomum zeylanicum* Blume. In: Jirovetz L, Dung NX, Varshney VK (eds) Aromatic plants from Asia: their chemistry and application in food and therapy. Har Krishan Bhalla & Sons, Dehradun, pp 49–75

Metcalfe CR, Chalk L (1972) Anatomy of the dicotyledons, vol 2, 4th edn. Oxford University Press, London, p 240

MOE (2012) The national red list 2012 of Sri Lanka; Conservation status of the fauna and flora. Ministry of Environment, Colombo, Sri Lanka, viii + 476 pp

Mohanakumar G, Mokashi A, Narayana Swamy P et al (1985) Studies on the floral biology of cinnamon. Indian Cocoa Arecanut Spices J 8:100–102

Nabavi SF, Di Lorenzo A, Izadi M, Sobarzo-Sánchez E, Daglia M, Nabavi SM (2015) Antibacterial effects of cinnamon: from farm to food, cosmetic and pharmaceutical industries. Nutrients 7:7729–7748

Nishida S, van der Werff H (2007) Are cuticular characters useful in solving generic relationships of problematic species of Lauraceae? Taxon 56(4):1229–1237

Nishida S, de Kok R, Yang Y (2016) Cuticular features of *Cryptocarya* (Lauraceae) from Peninsular Malaysia, Thailand and Indo-China and its taxonomic implications. Phytotaxa 244(1):26–44

Pal S (1976) Pollen of Lauraceae. J Palynol 12:55–62

Paranagama PA, Wimalasena S, Jayathilake GS, Jayawardena AL, Senanayake UM, Mubarak AM (2001) A comparison of essential oil constituents of bark, leaf, root and fruit of Cinnamon (*Cinnamomum zeylanicum* (Blum.)) grown in Sri Lanka. J Natl Sci Found Sri Lanka 29:147–153

Patel RN (1987) Wood anatomy of the dicotyledons indigenous to New Zealand 16. Lauraceae. NZ J Bot 25:477–488

Philippe S, Souaïbou F, Jean-Pierre N, Boniface Y, Paulin A, Issaka Y, Chantal M, Dominique S (2012) Chemical composition and antifungal activity of essential oil of fresh leaves of *Cinnamomum zeylanicum* from Benin against six mycotoxigenic fungi isolated from a traditional cheese wagashi. Elixir Appl Biol 49:9787–9791

Pole M (2019) A survey of pliocene to mid-quaternary leaf cuticle from the North Island, New Zealand. Palaeontol Electron 22:1–32

Ranawana KB (2014) Forest dieback in Horton Plains National Park. J Dep Wildl Conserv 2:103–112

Ravindran PN, Shylaja M, Babu KN, Krishnamoorthy B (2004) Botany and crop improvement of cinnamon and cassia. In: Ravindran P, Babu K, Shylaja M (eds) Cinnamon and cassia: the genus *Cinnamomum*. CRC Press, Boca Raton, pp 14–79

Robi AJ, Thomas VP, Sreenivas VK, Udayan PS, Sujanapal P (2018) *Cinnamomum ovalifolium* (Lauraceae) – a new record for India with notes on its lectotypification. Nelumbo 60:110

Rohwer JG (1993) Lauraceae. In: Kubitzki K, Rohwer JG, Bittrich V (eds) Flowering plants dicotyledons: magnoliid, hamamelid and caryophyllid families. Springer, Berlin and Heidelberg, pp 366–391. https://doi.org/10.1007/978-3-662-02899-5_46

Sastri RLN (1963) Studies in the Lauraceae: IV. Comparative embryology and phylogeny. Ann Bot 27:425–433

Sastri RLN (1965) Studies in the Lauraceae: V. Comparative morphology of the flower. Ann Bot 29:39–44

Sedgley M, Griffin AR (2013) Sexual reproduction of tree crops. Academic Press, London

Senanayake UM, Wijesekara ROB (2004) Chemistry of cinnamon and cassia. In: Ravindran PN, Babu KN, Shylaja M (eds) Cinnamon and cassia: the genus *Cinnamomum*. Medicinal and aromatic plants — industrial profiles. CRC Press, London, pp 81–120

Shi G, Xie Z, Li H (2014) High diversity of Lauraceae from the Oligocene of Ningming, South China. Palaeoworld 23:336–356

Shivaprasad D, Prasannakumar C, Somashekar R, Nagaraja B (2015) Reproductive biology of *Cinnamomum sulphuratum* Nees. from wet evergreen forest of Western Ghats in Karnataka. Proc Int Acad Ecol Environ Sci 5:7–15

Shylaja M (1984) Studies on Indian *Cinnamomum*. Ph D thesis, University of Calicut, Calicut, India

Shylaja M, Manilal KS (1992) Bark anatomy of four species of *Cinnamomum* (Lauraceae) from Kerala. J Spices Aromat Crops 1:84–87

Shylaja M, Ravindran PN, Nirmal Babu K (2004) Other useful species of *Cinnamomum*. In: Ravindran PN, Nirmal Babu K, Shylaja M (eds) Cinnamon and cassia: the genus *Cinnamomum*. Medicinal and aromatic plants — industrial profiles. CRC Press, Boca Raton, pp 330–355

Singh MK, Sharma M, Sharma CL (2015) Wood anatomy of some members of family Lauraceae with reference to their identification. J Indian Acad Wood Sci 12(2):137–144

Sritharan R (1984) A study of the genus *Cinnamomum*. M. Phil thesis, Post Graduate Institute of Agriculture, University of Peradeniya, Peradeniya, Sri Lanka
Sritharan R, Jacob VJ, Balasubramaniam S, Bavappa KVA (1993) Palynological and cytological studies of the genus *Cinnamomum*. Acta Hortic 330:107–113
Sun J, Wu J, Wang Z, Gu J, Gao Z (2015) Comparative wood anatomy of 56 species of Lauraceae from Yunnan, China. Braz J Bot. https://doi.org/10.1007/s40415-015-0172-8
Tang B, Xu MHQ, Jin J (2016) Leaf cuticle microstructure of *Machilus maomingensis* sp. nov. (Lauraceae) from the Eocene of the Maoming Basin, South China. Acta Geol Sin 90(5):1561–1571
TURIS (2013) Descriptors for cinnamon (*Cinnamomum verum*). University of Ruhuna, Matara, Sri Lanka
van der Merwe JJM, Van Wyk AE, Kok PDF (1990) Pollen types in the Lauraceae. Grana 29:185–196
Wang Y-H, Avula B, Nanayakkara NPD, Zhao J, Khan IA (2013) Cassia cinnamon as a source of coumarin in cinnamon-flavored food and food supplements in the United States. J Agric Food Chem 61:4470–4476
Weber JZ (1981) A taxonomic revision of *Cassytha* (Lauraceae) in Australia. J Adelaide Bot Gard 3(3):187–262
Weerawardhena SR, Russell AP (2012) Historical land-use patterns in relation to conservation strategies for the Riverstone area, the Knuckles massif, Sri Lanka: insights gained from the recovery of anuran communities. Taprobonica 4:92–102
Weiss EA (2002) Spice crops. CABI Publishing, Oxon
Werner WL, Balasubramaniam S (1992) Structure and dynamics of the upper montane rain forests of Sri Lanka. In: Goldammer JG (ed) Tropical forests in transition. Advances in life sciences. Birkhäuse, Basel. https://doi.org/10.1007/978-3-0348-7256-0_11
Wickramasinghe K, Steele P, Senaratne A (2008) Socio-economic impacts of forest conservation on peripheral communities: case of Knuckles National Wilderness Heritage of Sri Lanka. Paper presented at the governing shared resources: connecting local experience to global challenges, the twelfth biennial conference of the International Association for the Study of Commons, Cheltenham, England
Wijeratne M, Piyasiri KHKL (2015) Conservation of cop wild relatives: a Sri Lankan experience in community participation. Trop Agric Res Ext 18:77–83
Wijesekera ROB, Jayewardene AL (1974) Essential oils II. Chemical constituents of the volatile oil from the bark of a rare variety of cinnamon. J Natl Sci Coun Sri Lanka 2:141–146
Wijesinghe KGG, Pathirana R (2000) Germplasm screening and varietal development in cinnamon. In: Senaratne R, Dayatilake GA (eds) Revitalisation of the cinnamon industry in Sri Lanka: constaints and remedies. Faculty of Agriculture, University of Ruhuna, Matara, pp 15–25
Wuu-Kuang S (2011) Taxonomic revision of *Cinnamomum* (Lauraceae) in Borneo. Blumea Biodivers Evol Biogeogr Plants 56:241–264
Yang Y, Zhang L, Liu B, van der Werff H (2012) Leaf cuticular anatomy and taxonomy of *Syndiclis* (Lauraceae) and its allies. Syst Bot 37(4):861–878
Zeng Z, Liu B, van der Werff H, Ferguson DK, Yang Y (2014) Origin and evolution of the unusual leaf epidermis of *Caryodaphnopsis* (Lauraceae). Perspect Plant Ecol Evol Syst 16:296–309

Chapter 5
Genetics and Molecular Characterization of Genus *Cinnamomum*

Pradeepa C. G. Bandaranayake and D. K. N. G. Pushpakumara

5.1 Introduction

The genus *Cinnamomum* Schaeff. comprises 250 species (Kostermans 1957), found from the Asiatic mainland to Formosa, the Pacific islands, Australia, and tropical America. Genus *Cinnamomum* was first identified from Asia, mainly from the Asia-Pacific region. While earlier taxonomists such as Nees von Esenbeck (1836) recognized some closeness between *Cinnamomum* and neotropical genera such as *Phoebe*, later taxonomists, especially Kostermans (Kostermans 1957, 1961) transferred species from *Phoebe* to *Cinnamomum*. He considered the perianth tube found in the fruit is a good natural character for classification of closely knit genera of Lauraceae (Kostermans 1961). Thus, *Cinnamomum* is no longer an Asiatic genus and it obtained pantropical status occurring in both hemispheres. *Cinnamomum* species are evergreen trees and shrubs characterized by (i) the length of the two basal or sub-basal ascendant veins; (ii) the indumentums (covering of hairs); and (iii) the thalamus cup under the fruit (Kostermans 1980).

Among them, *C. zeylanicum* Blume (synonymous to *C. verum* J. Presl), *C. cassia* J. Presl and *C. camphora* (L.) J. Presl are of economic value and traded in local and world markets. Of them, *C. zeylanicum*, also known as Ceylon cinnamon, or Sri Lankan cinnamon, is considered the "true cinnamon" of commerce. Currently, Sri Lanka is the largest true cinnamon producer in the world. Since 1826, *C. zeylanicum* was considered as a synonym for *C. verum*. In contrast, *C. zeylanicum* is supposed

P. C. G. Bandaranayake (✉)
Agricultural Biotechnology Centre, Faculty of Agriculture, University of Peradeniya, Peradeniya, Sri Lanka
e-mail: pradeepag@agri.pdn.ac.lk; pgunathilake@ucdavis.edu

D. K. N. G. Pushpakumara
Department of Crop Science, Faculty of Agriculture, University of Peradeniya, Peradeniya, Sri Lanka

R. Senaratne, R. Pathirana (eds.), *Cinnamon*,
https://doi.org/10.1007/978-3-030-54426-3_5

to be an endemic species to Sri Lanka and yet to be proved. However, recent molecular taxonomic studies based on whole chloroplast genome sequence data have shown considerable difference among them, possibly leading to taxonomic revision. India is the leading *C. verum* producer in the world. *C. cassia* (L.) J. Presl is commonly known as cassia cinnamon or Chinese cinnamon and much cheaper than Ceylon cinnamon in the world market. *C. camphora* is mainly traded as camphor. Nevertheless, many wild species are also used for different purposes, including medicinal, spice, timber, ornamental and industrial applications. Some are traded together with species having commercial value while some other species are mixed as adulterants. For example, there are eight known cinnamon species in Sri Lanka of which only C. *zeylanicum* is grown and traded as a commercial spice crop. *C. cappara-coronde* Blume, *C. citriodorum* Thw., *C. dubium* Nees, *C. litseaefolium* Thw., *C. ovalifolium* Wight, *C. rivulorum* Kosterm., *C. sinharajaense* Kosterm. are identified as seven endemic wild relatives of *C. zeylanicum* found in Sri Lankan rain forests. Populations of *C. zeylanicum* are also available in natural vegetation of the country. There is also a wide variety of cinnamon plants in farmers' fields and plantations, valuable in terms of quantity and quality of yield and resistance to major pests and diseases, etc. The variation of cinnamon in Sri Lanka has also been identified based on pungency, taste, and texture of the bark such as *pani miris kurundu* (sweet/honey and pungent taste), *pani rasa kurundu* (sweet/honey taste), *thiththa kurundu* (bitter taste), *kahata kurundu* (astringent or kahata flavoured), *weli kurundu* (brittle/sandy nature), *sevela kurundu* (mucilaginous), *naga kurundu,* and *piris kurundu* (Wijesekera et al. 1975). Further, there are two recently released varieties of cinnamon named "Sri Vijaya" and "Sri Gemumu". Hence, as Pushpakumara and Silva (2018) indicated, genetic resources of cinnamon in Sri Lanka include: (i) cinnamon populations in the wild; (ii) diversity maintained in famers fields and plantations; (iii) two recently released new varieties; (iv) germplasm collections at the National Cinnamon Research & Training Center (NCR&TC), Thihagoda, Pallolpitiya; Mid Country Research Station, Delpitiya; Central Research Station, Matale and Bandaranayake Memorial Ayurvedic Research Institute, Nawinna, Maharagama; (v) seven endemic wild relatives of cinnamon; and exotic origin *C. camphora*. Therefore, knowing proper species characteristics, inter- and intra-species diversity is critical for utilization and conservation of cinnamon for industrial and other applications. This chapter discusses genetic and molecular characterization of *Cinnamomum*, both at species level and within species diversity with special emphasis on sustainable utilization of genetic resources of the genus *Cinnamomum*.

5.2 Genetic Characterization of Genus *Cinnamomum*

The literature related to inter- and intra-species genetic diversity is largely on economically important taxa such as *C. zeylanicum*, *C. verum*, *C. camphora*, *C. cassia*, and *C. kanehirae* Hayata. While most of genetic characterization studies are based on morphological traits, some research groups have considered traits such as chemical composition, leaf and bark pungency, and cytology.

5.2.1 Genetic Characterization Using Morphological Traits

Until recent advancement of molecular biology, species identification and description, and inter- and intra-species diversity studies were predominantly based on morphology. While floral morphology is generally important, leaf morphological traits play a critical role in characterization of the members of the genus *Cinnamomum*. Ravindran et al. (2004) reviewed some literature under biosystematics and interrelationship section of their book chapter on botany and crop improvement of cinnamon and cassia.

Previous studies conducted with *C. verum* and related taxa were mainly based on several sets of morphological characters (Ravindran et al. 1991; Manilal and Shylaja 1986; Shylaja 1984), for example, (i) vegetative bud characters, (ii) leaf characters (length, breadth, length/breadth, thickness, epidermal thickness, palisade thickness, thickness of spongy parenchyma, petiole length, stomata type and frequency, length and breadth of guard cells, leaf hairiness, hair frequency, and hair length), and (iii) inflorescence length, nature, and type are important in taxonomic determination of species or taxa.

5.2.1.1 Species Level Diversity

Shylaja (1984) and Ravindran et al. (1996) used principal component analysis (PCA) to recognize the traits responsible for species divergence. They included *C. camphora, C. perrottetii* Meissn., *C. riparium* Gamble, *C. macrocarpum* Hook. f., *C. malabatrum* (Burm. f.) Blume, *C. cassia*, and *C. verum* in their analysis. Interestingly, the first PC—bud type, stomatal type, phyllotaxy, and epidermal thickness—was responsible for the divergence of *C. camphora* from other species. Three of the taxa—*C. perrottetii, C. riparium*, and *C. macrocarpum*—diverged from others due to characters such as hairiness, hair size, and hair frequency, included in the second PC. The inflorescence length, nature, and type, included in the third PC, were responsible for divergence of *C. cassia* from other species. *C. malabatrum* has quite high intra-species variability, and the fourth PC (leaf thickness, palisade, and spongy parenchyma) and the fifth PC (leaf length and breadth) could identify divergence among different collections. These data suggest that the considered morphological traits are sufficient to recognize some of the species included in the study. However, species such as *C. verum* and *C. malabatrum* showed considerable closeness between them.

In another study, Bakker et al. (1992) considered leaf anatomical characters in phenotypic analysis of *Cinnamomum* species. Interestingly, *C. burmannii* Blume and *C. verum* clustered together with over 20 other species. Within the main cluster, *C. burmannii* cluster independently while *C. verum* cluster with 21 other species. *C. cassia* is in cluster 2 and idioblast distribution varies hugely in clusters 1 and 2. Species in cluster 3 have mucilage cells in both mesophyll layers and oil cells in spongy parenchyma. Presence of both idoblastic types in both layers—non-sclerified epidermal cells and weakly or non-sclerified spongy parenchyma—grouped species into cluster 4. While cluster 5 does not have clear discriminating anatomical characters, taxa in cluster 6 do not have mucilage cells in the spongy parenchyma and consist of non-sclerified epidermal cells and spongy parenchyma. Other than that a two-layered sclerified palisade parenchyma and penninerved leaves are also characteristics of cluster 6 members. Cluster 7 members have oil and mucilage cells in both palisade parenchyma and triplinerved leaves. The idioblast distribution pattern discriminates last three clusters where cluster 8 lacks oil cells and clusters 9 and 10 do not have mucilage cells. Interestingly, different *C. camphora* accessions grouped in clusters 4, 9, and 10 suggesting high intraspecific diversity. Most neotropical species included in this study were transferred from other genera such as *Phoebe* and were grouped in clusters 6 and 9.

So far, about 37 *Cinnamomum* species are reported in India (Hooker 1886; Kostermans 1983; Manilal and Shylaja 1986; Baruah and Nath 2000). Baruah and Nath (2007) conducted a systematic census of *Cinnamomum* species in North India including 14 taxa with variable phenotypic characters. They mainly considered morphology of foliar epidermal and venation characters and identified that the studied taxa belong to seven species, namely *C. assamicum* Nath and Baruah, *C. bejolghota* Buch.-Ham., *C. cassia*, *C. iners* Reinw. Ex. Blume, *C. pauciflorum* Nees von Esenbeck, *C. sulphuratum* Nees von Esenbeck and *C. verum*. They have also identified within species variants using the same set of characteristics. For example, *C. bejolghota* has three variants while *C. cassia* has two variants. *C. verum* has three variants of which two of them are also identified as cultivars (Baruah and Nath 2007). One of the variants of *C. bejolghota* was later named as a new species by further characterization of its leaf, panicle, and stem bark essential oils (Baruah and Nath 2007). Similarly, Choudhury et al. (2013) identified six species from Terai and Dooars regions in the foothills of Himalayas using morphological traits. Wuu-Kuang (2011) conducted a taxonomic revision of *Cinnamomum* in Borneo using morphological traits. There are 26 cinnamon species recorded in Borneo and of them 17 species are endemic.

Ariyarathne et al. (2018) studied morphological characters of flowers, leaf, seed and bark, yield of bark and oil, and chemical characters of leaf and bark oil in two *C. zeylanicum* accessions and six wild species. Leaf morphological characters such as leaf weight, shape, venation pattern, leaf apex shape, leaf length and width, petiole length, and leaf area varied among species. Bark characteristics such as bark color, surface texture, odor, and seed color also varied among them (Ariyarathne et al. 2018). Nevertheless, this study consisted of only one replicate from each species except *C. zeylanicum*, which consisted of two cultivars, each having one replicate. The two *C. zeylanicum* cultivars grouped with wild relatives showing they are considerably different from each other.

A recent study included all seven endemic wild species of Sri Lanka collected from their natural habitats and cultivated locations, with each species replicated a minimum of three times. Leaf morphological traits were mainly considered in the analysis and those traits were sufficient to differentiate all the cinnamon species in Sri Lanka; *C. zeylenicum*, *C. citriodorum*, *C. litseaefolium*, *C. dubium*, *C. rivulorum*, *C. capparu-coronde*, *C. sinharajaense*, and *C. ovalifolium*. A leaf morphological index was also developed for field identification of species (Bandusekara et al.

2020). Although Ariyarathne et al. (2018) also studied leaf morphology, they did not develop a morphological index for field level identification.

5.2.1.2 Within Species Diversity

Joy et al. (1998) studied the genetic variability among 234 accessions of *C. verum* established at the Aromatic and Medicinal Plants Research Station in Kerala, India. They mainly considered morphological and yield-related traits, including plant height, canopy spread, flush color, fresh leaf yield, dry leaf yield, leaf oil yield, oil recovery percentage, eugenol yield, and yield index. Plant height of the accessions varied from 108 to 343 cm with an average of 254 cm. While 46% of the leaves were small to medium, 22% had medium to large and 32% had small to large leaves. Flush color varied with 72% medium-colored flushness and 14% light-colored or green flushes. The canopy spread ranged from 78.8 to 311.3 cm with a mean value of 194.5 cm. About 64% of canopies were compact while 24% of them were semi-compact, and loose in 12%. Further, 82% of them had spherical canopy while 13% had semispherical and 6% had linear canopies. While the dry leaf yield varied from 1.28 to 7.88 kg/tree/year, the oil yield varied from 36.45 to 294.69 mL/tree/year. Similarly, the oil recovery percentage and eugenol yield varied from 2.20 to 4.29 and 80.53 to 95.03 mL/tree/year, respectively.

Krishnamoorthy et al. (1988) and Gopalam (1997) studied 239 cinnamon plants. Interestingly, the flush was predominantly green shared with 55% of the trees while the rest had various degrees of purple coloration. They observed a correlation between flush color and quality of yield where purple flushed accessions have about 29% more bark oil than others.

Joy et al. (1998) conducted correlation and path analysis studies in cinnamon. Their analysis showed that yield traits such as fresh leaf yield, leaf oil yield, and eugenol yield are highly correlated among themselves. Further, the morphological traits, plant height, and canopy spread were positively correlated between themselves and yield traits mentioned above. They expanded the analysis to find out multiple regression and path coefficients of highly correlated traits: plant height, canopy spread, fresh leaf yield, leaf oil yield, and eugenol yield. Based on such analysis, the authors suggested that the direct effect of plant height, canopy spread, or fresh leaf yield contributes lesser on the variability of eugenol yield and that would be due to indirect effects. According to the path diagram proposed by Joy et al. (1998), there is a unidirectional relationship among fresh leaf yield, leaf oil yield, and eugenol yield. Though the canopy spread is directly linked to plant height, it is poorly related to the fresh leaf yield. They showed possibility of using leaf yield as an indirect measure of oil or eugenol yield.

Bark oil content also varies among germplasm accessions (Krishnamoorthy et al. 1988). Krishnamoorthy et al. (1991) also studied performance of nine lines and found significant variation in plant height, number of branches per tree, fresh and dry weight of bark, and percentage recovery of bark. Krishnamoorthy et al. (1992) studied a collection of 71 accessions from the largest collection maintained at the Indian Institute of Spice Research (IISR) at Calicut. The leaf length, leaf breadth, leaf size index, fresh weight of bark, recovery of bark (%), bark oleoresin (%), leaf oil (%), and dry weight of bark considerably varied among them. The highest variation was observed in bark dry weight with next observed variation in bark fresh weight, bark oleoresin, and leaf oil. While variation in bark oil content and leaf size index was moderate, there was a high correlation between fresh weight of bark and leaf oil with dry weight of bark. Other morphological traits such as leaf length and breadth also varied considerably. For example, the leaf length varied from 8.75 to 20.69 cm with an average of 13.08 cm. Similarly, the leaf breadth varied from 3.31 to 8.30 cm with an average of 5.06 cm. Paul and Sahoo (1993) studied a field plantation of cinnamon in the Orissa state and found high variability in traits such as plant height with a range of 2.17–3.3 m, stem girth in a range of 7–16.6 cm, leaf oil in the range of 0.38–1.80%, and bark oil in the range of 0.05–2.18%. Interestingly, many groups have observed negative relationship between presence of benzyl benzoate and eugenol content in cinnamon (Paul and Sahoo 1993; Rao et al. 1988; Paul et al. 1996; Sahoo et al. 2000).

Krishnamoorthy and colleagues evaluated the *C. cassia* germplasm maintained at IISR India using morphological and quality parameters (Krishnamoorthy et al. 1999, 2001). In 1999 work, they evaluated four elite lines for bark oil (%), leaf oil (%), bark oleoresin (%), and cinnamaldehyde (%) in bark oil and found a considerable variation among them. The cinnamaldehyde content in leaf oil ranged from 40% to 86% and that of bark oil ranged from 61% to 91%. However, the processing of cassia was done similar to Ceylon cinnamon in the above experiments. They continued working further with more replications and samples and identified three lines to be released as new varieties (Krishnamoorthy et al. 2001).

Wang et al. (2017) studied 179 open-pollinated families of seedlings of *C. camphora* from six provinces of South China using leaf and growth characteristics. Those families had significantly different leaf features such as leaf area, leaf margin circumference, maximum leaf length, maximum leaf width, and leaf length-to-width ratio. Interestingly, all eight traits had high heritability values from 0.623 to 0.974. There was a significant positive correlation between leaf features, growth traits, and both types of traits. PCA showed considerable diversity among the families and over 90% of total variation could be explained with the first three PCs. Cluster analysis grouped 174 families into 16 clusters, each having differences in leaf and growth traits considered (Wang et al. 2017).

Azad et al. (2016) studied morphological variation *of C. zeylanicum* germplasm in Matara district, one of the major cinnamon cultivation areas in Sri Lanka. Forty-seven accessions collected from farmers' fields were assessed using morphological traits; leaf length (LL), leaf width (LW), leaf length-width ratio (LLWR), petiole length (PL), leaf arrangement, leaf shape, leaf apex, leaf base, leaf texture, upper surface color, flush color, bark color, bark surface, and bark fragrance. PCA was conducted with four quantitative characters—LL, LW, PL, and LLWR—and the first two PCs accounted for 88.8% of total variability. The PC-1 explained 56.5% of the total variability was loaded on LL, LW, and PL. The PC-2 accounted for 32.43% of the variation and was loaded on LLWR (Azad et al. 2016). In another study, 269 cinnamon accessions collected from 47 cinnamon-cultivated lands from major cinnamon-growing areas of Matara, Galle, Kalutara, Kurunegala, Ratnapura, and

Humbantota districts were collected and characterized for 15 quantitative and qualitative characters of leaf, stem, and inflorescence (Azad et al. 2019). Leaf characters varied significantly and the leaf length positively correlated with leaf width and petiole length. Hierarchical cluster analysis based on 15 characters grouped the collection into five clusters. Ten percent of accessions were randomly selected from each cluster to develop a core collection with 33 accessions.

Azad et al. (2018) studied morphological diversity of *C. zeylanicum* flowers in Matara district, Sri Lanka. Floral samples were collected from 15 identified locations and assessed for peduncle length (PDL), flower length (FL), flower width (FW), and floral tube length (FTL). Among the characters, variation in tepal shape was distinct, of which the two whorls of tepals of a single flower with two shapes.

As stated earlier, Sri Lankan growers have identified eight different types of Ceylon cinnamon based on pungency, taste, and texture of bark *pani miris kurundu* (sweet/honey and pungent taste), *pani rasa kurundu* (sweet/honey taste), *thiththa kurundu* (bitter taste), *kahata kurundu* (astringent or kahata flavoured), *weli kurundu* (brittle/sandy nature), *sevela kurundu* (mucilaginous), *naga kurundu*, and *piris kurundu* (Wijesekera et al. 1975).

Recently, Liyanage et al. (2020) characterized the largest *C. zeylanicum* collection in the world. A total of 515 germplasm collected from Sri Lanka are established at the Cinnamon Research and Training Centre, Thihagoda, Matara, Sri Lanka. The Department of Export Agriculture, Sri Lanka, maintains this germplasm collection. This collection consists of 288 accessions originated from seeds from the superior parents and 227 accessions clonally propagated from superior parents. Each accession consists of one plant in clonally propagated collection while it is one bush in seed-propagated collection. All the accessions were assessed with 12 morphological and agronomic traits (Table 5.1). All the morphological characters were evaluated according to the cinnamon descriptor defined by TURIS, 2013 Project 2016.

There was a significant variation between seed-propagated germplasm collection and the clonally propagated collection. Vegetatively propagated collection is superior in general. This may either be due to their origin or methods of propagation. This may also be due to cross-pollinated nature of the species which has implication on multiplication of elite material for cultivation, hence needing more research on vegetative propagation of cinnamon. Nevertheless, the authors combined the data and analyzed to identify the overall best-performing accessions and any genetic relationships among them. Overall, the leaf, plant, and stem morphological features varied considerably among accessions (Table 5.2). The yield measured as gram per stem varied from 2.35 kg/stem to 0.01 kg/stem.

5.2.2 Genetic Characterization Using Biochemical Markers

Biochemical properties of cinnamon decide its qualities, and some of the identified chemical constituents such as cinnamaldehyde, eugenol, cinnamic acid, cinnamyl alcohol, coumarin, and cumaric acid and many unidentified chemicals present in

Table 5.1 Morphological traits used for assessing *C. zeylanicum* germplasm

No.	Trait	Description
01	Leaf shape (LS)	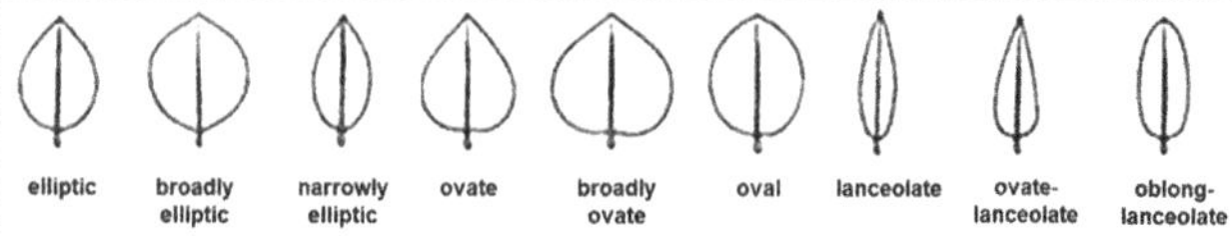
02	Leaf apex (LA)	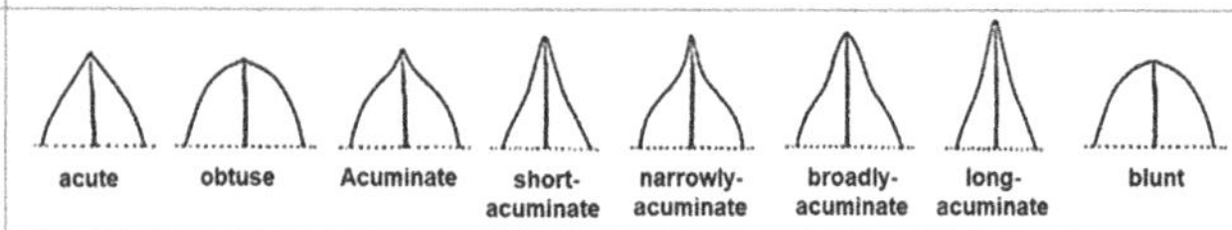
03	Leaf base (LB)	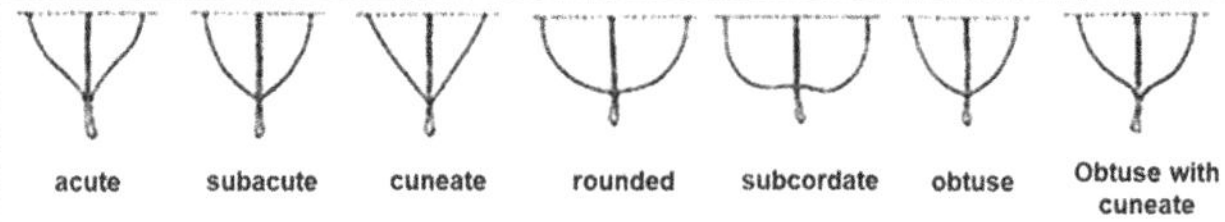
04	Leaf arrangement (LAR)	1. Opposite, 2. sub-opposite, 3. opposite or sub-opposite in different branch but in same plant, 4. opposite to sub-opposite in the same branch in same plant, 5. others
05	Leaf petiole length (LPL)	Average of ten leaves, measured from the detached point of the petiole to starting point of veins (cm)
06	Leaf length (LL)	Average of ten leaves, measured from the leaf apex to leaf base of the leaf blade (cm)
07	Leaf width (LW)	Average of ten leaves, measured at the widest position of the leaf blade (cm)
08	Canopy spread (CS)	Canopy circumference at the height of 1.5 m from the ground level (cm)
09	Tree height (TH)	Maximum height of the tree from ground level (m)
10	Harvestable stems (HS)	Number of harvestable stems available at the time of harvest
11	Erectness (ER)	The harvestable stem was considered from ground level to tip of the branch and score based on available knots and bends. "0"—no bends, "1"—one bends, "2"—two bends, "3"—three bends up to "7"—seven bends
12	Trunk circumference (TC)	Trunk circumference of harvested stem at 1 m height from the ground level (cm)

Note: Followed the descriptor developed by TURIS, 2013 Project 2016. (Liyanage et al. 2020)

small qualities are mainly responsible for overall properties. Biochemical profiles of cinnamon are studied at species level as well as within species.

Variation in flavonoids, terpenoids, and sterols were distinct among *C. verum* and related taxa in Western Ghats of Kerala, India (Shylaja 1984; Ravindran et al. 1991). For example, *C. verum, C. camphora, C. cassia,* and *C. riparium* had different flavonoid, terpenoid, and steroid profiles and could differentiate the species. Further, *C. malabatrum* had high intra-specific variation. Cluster analysis based on flavonoid and triterpenoid data showed independent clustering of *C. verum, C. riparium,*

Table 5.2 Quantitative morphological and agronomic characteristics of *C. zeylanicum* germplasm

	Traits	Average ± SD	Maximum	Minimum
Leaf	Leaf petiole length (cm)	17.9 ± 7.3	3.32	0.62
	Leaf length (cm)	12.3 ± 2.3	22.8	3.0
	Leaf width (cm)	5.8 ± 1.1	9.0	3.0
Plant	Canopy spread (cm)	82.1 ± 55.3	325.0	10.0
	Tree height (m)	2.14 ± 0.78	5.1	0.6
	No. of harvestable stems	2.1 ± 1.2	9	1
Stem	Trunk circumference (cm)	9.4 ± 6.7	25.0	1.0
Yield	Peel/stem proportion	0.16 ± 0.13	1.4	0.001
	Yield (kg/stem)	0.168 ± 0.179	2.358	0.010

Notes: Each accession consisted of one plant, and the values are mean ± SD of total 515 accessions. Both seedling and vegetative-propagated germplasms were considered together. (Liyanage et al 2020).

C. cassia, and *C. camphora*. Analysis showed close clustering of *C. perrottetii* and *C. macrocarpum* as well as *C. malabatrum* and *C. nicholsianum*. Further, the flavonoid complexity is greater in *C. verum* and *C. camphora.*

Intra-specific chemical variability is common in many *Cinnamomum* species. Intra-specific variability of chemicals might be the result of inter-breeding or heavy cross-pollination, segregation, random mutations, and isolation mechanisms leading to speciation. “Polychemism” is a term used by Tétényi (1970) to introduce having more than one chemically distinct form in a species. Ravindran et al. (2004) summarized previously published literature on intra-species chemical diversity and available chemotypes (Table 2.8 of Ravindran et al. 2004). Several species were categorized up to sub-species level based on volatile oil composition (Fujita 1967). According to Tetenyi (1970), the Formosan camphor tree (*C. camphora* spp. formasana) has at least seven distinct chemovarieties, namely Chvar. Borneol, Chvar. Camphor, Chvar. Safrole, Chvar. Sesquiterpene alcohol, Chvar. Cineol, Chvar. Linalool, and Chvar. Sesquiterpene. Interestingly, Chvar. Linalool consists of two chemoforms: chforma 86% linalool and chforma 71% linalool (Fujita 1967). Hirota and Hiroi (1967) and Wan-Yang et al. (1989) identified five chemical races among Chinese camphor trees. Further, Van Khiên et al. (1998) recognized eight Vietnamese camphor chemo-varieties. The same group also recognized chemical segregation of camphor trees into two chemo-varieties. For example, there were four chemoforms among the progenies of a camphor tree rich in camphor, whereas there were six chemoforms among the progenies of a camphor tree rich in linalool.

Grangfu and Yang (1988) studied the chemical composition of cinnamon species found in Hubei province China. Interestingly, there was a good correlation among morphological and chemical traits. Accordingly, the species they studied grouped into three and they proposed evolutionary relationships among them based on the chemical profile and identified three major groups as linalool, eugenol, and cinnamaldehyde. Broad chemical diversity of *C. zeylanicum* oils is also reported by several authors (Thomas et al. 1987; Bernard et al. 1989; Möllenbeck et al. 1997; Variyar and Bandyopadhyay 1989). Thomas et al. (1987) reported influence of the date of harvest on the oil yield with eugenol as the main component of the oil. Five chemotypes of *C. camphora* were identified as camphor, linalool, 1,8-cineole, borneol, and (E)-nerolidol (Shi 1989). The linalool type was present in Malaysia and Vietnam (Jantan and Goh 1992; Dung et al. 1993), while the camphora type was common in Ivory Coast (Pélissier et al. 1995) and the 1,8-cineole type in Madagascar (Möllenbeck et al. 1997). Gu et al. (1990) studied seasonal variation of the chemical composition of variety Linaloolifera.

Chalchat and Valade (2000) studied oils of four cinnamon species grown in Madagascar. They analyzed *C. zeylanicum* collected from wild plants in their natural habitats and the bark oil was (E)-cinnamaldehyde chemotype that also consists of 15.2% of camphor and very small amounts of benzaldehyde (<1%). *C. camphora* was of the 1,8-cineole type, while *C. angustifolium* contained of high portion of hydrocarbons such as pinenes, α-phellandrene, ρ-cymene, β-caryophyllene, and 1,8-cineole.

Hu (1985) analyzed the oil composition of randomly collected leaf samples from 21 provenances in Central, Southern, and Eastern Taiwan and found that cinnamaldehyde was the major constituent of some *C. osmophloeum* provenances while others had eugenol. Based on the leaf essential oil composition they classified *C. osmophloeum* into nine types: cassia type, cinnamaldehyde type, coumarin type, linalool type, eugenol type, camphor type, 4-terpineol type, linalool-terpineol type, and mixed type. Cheng et al. (2004) studied chemical composition of leaf essential oils from eight provenances of *C. osmophloeum*. According to gas chromatography mass spectrometry (GC-MS) and cluster analysis, they classified the leaf essential oil of the eight provenances and their relative content into five chemotypes: cinnamaldehyde type, linalool type, camphor type, cinnamaldehyde/cinnamyl acetate type and mixed type. Interestingly, the mosquito larvicides activity varied among chemotypes, and cinnamaldehyde type and cinnamaldehyde/cinnamyl acetate types had an excellent inhibitory effect against the fourth-instar larvae of *Aedes aegypti*. Cheng et al. (2006) reported the essential oil composition of nine geographical provenances of *C. osmophloeum* using gas chromatography mass spectrometry (GC-MS), and based on cluster analysis and their relative abundance, those were classified into six chemotypes, namely, cinnamaldehyde type, cinnamaldehyde/cinnamyl acetate type, cinnamyl acetate type, linalool type, camphor type, and mixed type. In their antifungal activity assay, cinnamaldehyde type and cinnamaldehyde/cinnamyl acetate type had an excellent inhibitory effect against white-rot fungi *Trametes versicolor* and *Lenzites betulina* and brown-rot fungus *Laetiporus sulphureus*.

Jantan et al. (2008) studied 14 essential oils hydro-distilled from eight cinnamon species, *C. pubescens* Kochummen, *C. impressicostatom* Kosterm., *C. microphyllum* Ridl., *C. scortechinii* Gamb, *C. rhyncophyllum* Miq., *C. cordatum* Kosterm., *C. zeylanicum,* and *C. mollissimum* Hook.f. There were little compositional differences among them though there was a variation in amounts of individual components in some oils. The same group assessed the correlation between chemical composition and antifungal activity of the essential oils of those species and concluded that the strong antifungal activity of the bark and leaf oils of *C. zeylanicum* may be related to the high levels of cinnamaldehyde (44.2%) and eugenol (90.2%) in two types of oils, respectively.

Rana et al. (2012) assessed the chemical variability in the essential oils of 10 *C. tamala* (Ham.) Nees et Eberm. leaf samples collected from different areas of Northeastern states of India. While the oil yield varied from 1.2% to 3.9% (w/w) based on dry weight, the oil composition also varied significantly. The authors suggested utilizing and mass multiplication of accessions having high essential oil and eugenol to use as planting materials.

Chemical polymorphism and composition of leaf essential oils of 26 sources of *C. kanehirae* in Taiwan were studied using gas chromatography mass spectrometry (Cheng et al. 2015). The main constituents in leaf essential oils were linalool, 1,8-cineale, ß-selinene, 1-hexadecyne, and α-cadinol. Based on chemical composition, those were classified into five chemotypes: linalool type, linalool/1,8-cineole type, 1,8-cineole type, linalool/α-cadinol type, and mixed type.

While Hammid et al. (2016) studied essential oils of three cinnamon species found in Sarawak, Ananthakrishnan et al. (2018) studied leaf volatile chemical profiles of eight wild cinnamon species and two chemotypes from the Western Ghats, South India. Hammid et al. (2016) analyzed the essential oils of *C. macrophyllum, C. crassinervium,* and *C. griffithii* collected in Sarawak obtained by hydrodistillation and analyzed by GC-MS. They reported that most of the essential oils were mainly phenylpropanoids and monoterpenes with a small amount of sesquiterpenes. Ananthakrishnan and colleagues identified 112 constituents belonging to monoterpenoid, sesquiterpenoid, and phenyl propanoid classes from by GC-FID and GC-MS and identified monoterpenoids and sesquiterpenoids as the major class of volatile compounds in most of the species studied.

Guo et al. (2017) recently identified five chemotypes: the isoborneol type, camphora type, cineole type, linalool type and borneol type of *C. camphora* at the molecular level based on the multivariate analysis of mass spectral fingerprints recorded from a collection of 150 leaves from each chemotype, using desorption atmospheric pressure chemical ionization mass-spectrometry. A rare chemotype of *C. verum* has also been reported recently (Monteiro et al. 2017). It consists of 65.4% of benzyl benzoate followed by 5.4% linalool, 4% E-cinnamaldehyde, 3.9% α-pinene, 3.4% ß-phellandrene, 3.4% eugenol and 2.7% benzaldehyde. The most common *C. verum* essential oil chemotype reported is eugenol. It has also been reported that chemotypes of *C. verum* leaves contained significant amounts of eugenol, satrol, and benzyl benzoate (Nath et al. 1996; Morsbach et al. 1997). A variety of cinnamon containing benzyl benzoate as the main compound has also been reported by Wijesekera and Chichester (1978).

The largest *C. zeylanicum* germplasm in Sri Lanka consists of 515 accessions and their chemical profiles were analyzed with HPLC as bark and leaf separately (Liyanage et al. 2020). Processed samples were ground into fine powder and 0.25 g of each sample was transferred into clean 50 mL centrifuge tube and extracted with 20.0 mL of 100% methanol, sonicated for 30 min at room temperature, centrifuged at 7800 rpm for 10 min and the supernatant was filtered through a 0.45μm nylon filter. Exactly 10μL of the filtrate was injected onto the HPLC chromatographic system equipped with a reversed-phase ZORBAX Eclipse Plus C18 (250 mm × 4.6 mm, particle size 5μm) column, with a gradient of 0.1 phosphoric acid and acetonitrile (20–100% over 45 min) at a rate of 1 mL/min. The absorbance of *trans*-cinnamaldehyde, eugenol, coumarin, coumaric acid, cinnamyl acetate, and cinnamyl alcohol was measured at 290 nm, 285 nm, 280 nm, 290 nm, 265 nm, respectively, and the concentrations were calculated using standard curves generated with commercial standards. There is a significant variation of considered chemicals among the accessions (Table 5.3; Figs. 5.1 and 5.2). While the average bark cinnamaldehyde content is about 8.9 ± 5 mg/g, some accessions contained over 20 mg/g cinnamaldehyde. Interestingly, the average coumarin content was about 0.034 ± 0.02 mg/g and majority of the accessions did not have detectable levels of coumarin. This is much lower than the average coumarin content in *C. cassia*, about 4.5 mg/g, assayed in the same experimental setup. Liyanage and colleagues have conducted further analysis combining morphological, yield, and biochemical traits and identified overall better performing accessions to be incorporated into future breeding programs. These can also be mass propagated through vegetative methods for release to growers (Table 5.4). The group number 9 and 21 consist of overall superior accessions with high level of favorable chemicals, yield, and very low level of coumarin.

Table 5.3 Evaluation of *C. zeylanicum* germplasm: biochemical traits

	Traits	Average ± SD	Maximum	Minimum
Chemical composition of bark	Cinnamaldehyde (mg/g)	8.912 ± 5.220	24.703	0.076
	Cinnamyl alcohol (mg/g)	0.432 ± 0.554	2.951	0.013
	Cinnamyl acetate (mg/g)	0.430 ± 0.627	3.279	0.009
	Eugenol (mg/g)	0.510 ± 0.580	5.712	0.016
	Coumarin (mg/g)	0.034 ± 0.021	0.137	0.001
	Coumaric acid (mg/g)	0.041 ± 0.000	0.041	0.006
Chemical composition of leaf	Cinnamaldehyde (mg/g)	0.568 ± 1.156	17.726	0.072
	Cinnamyl alcohol (mg/g)	0.206 ± 0.424	3.838	0.011
	Cinnamyl acetate (mg/g)	0.532 ± 1.027	7.541	0.010
	Eugenol (mg/g)	21.706 ± 12.014	75.096	0.112
	Coumarin (mg/g)	0.049 ± 0.033	2.313	0.013
	Coumaric acid (mg/g)	0.074 ± 0.038	0.152	0.004

Notes: Total of 515 accessions was included in the analysis. Bark and leaf chemicals were extracted separately with 100% methanol and analyzed with HPLC system equipped with reverse-phase ZORBAX Eclipse Plus C18 column, with a gradient of 0.1 phosphoric acid and acetonitrile (20–100% over 45 min) at a rate of 1 mL/ min. Average values were calculated considering the samples with higher concentration than limit of detection (LOD) for each chemical. Cinnamyl alcohol (RT14.0 Sig 265), cinnamyl acetate (RT25.6 Sig 265), eugenol (RT20.7 Sig 286), coumarin (RT7.3 Sig 290), coumaric acid (RT12.9 Sig 280), *trans*-cinnamaldehyde (RT17.9 Sig 290). (Liyanage et al. 2020)

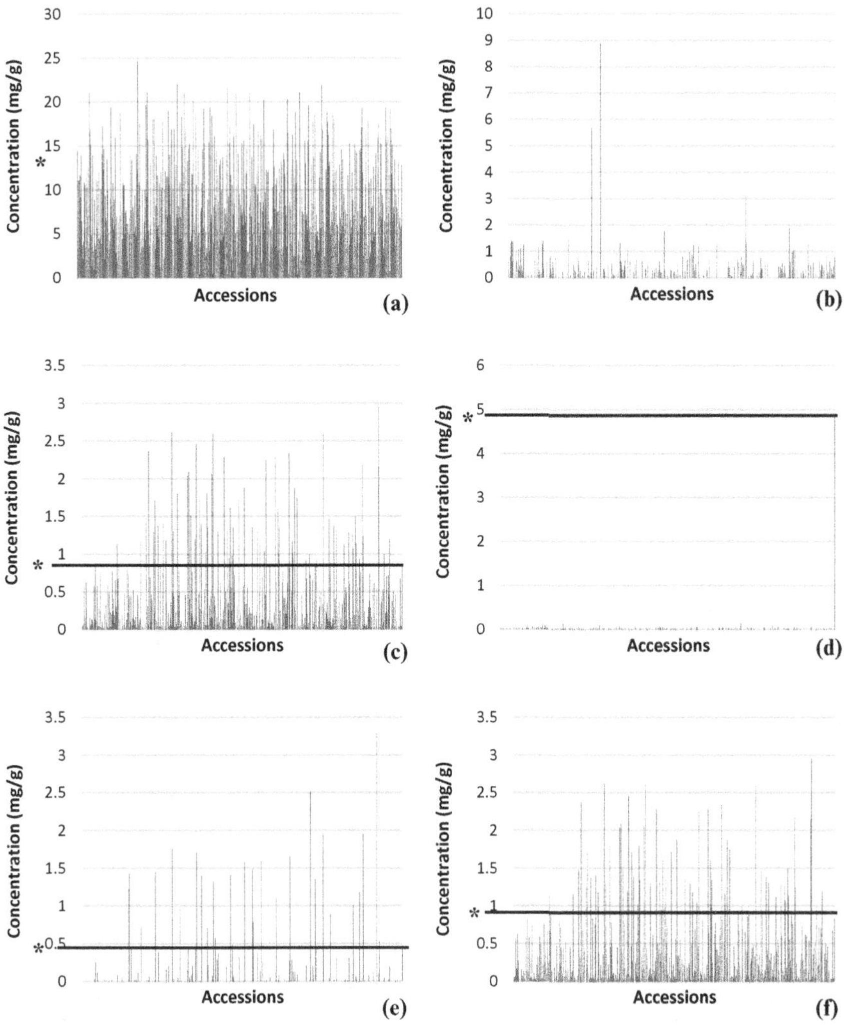

Fig. 5.1 Variation in biochemical composition of *Cinnamomum zeylanicum* bark. (**a**) Cinnamaldehyde, (**b**) eugenol, (**c**) coumaric acid, (**d**) coumarin, (**e**) cinnamyl acetate, and (**f**) cinnamyl alcohol. "*" indicates the average amount present in *C. cassia.* Total of 515 accessions was included in the analysis. Bark and leaf chemicals were extracted separately with 100% methanol and analyzed with HPLC system equipped with reversed-phase ZORBAX Eclipse Plus C18 column, with a gradient of 0.1 phosphoric acid and acetonitrile (20–100% over 45 min) at a rate of 1 mL/ min. Cinnamyl alcohol (RT14.0 Sig 265), cinnamyl acetate (RT25.6 Sig 265), eugenol (RT20.7 Sig 286), coumarin (RT7.3 Sig 290), coumaric acid (RT12.9 Sig 280), *trans*-cinnamaldehyde (RT17.9 Sig 290)

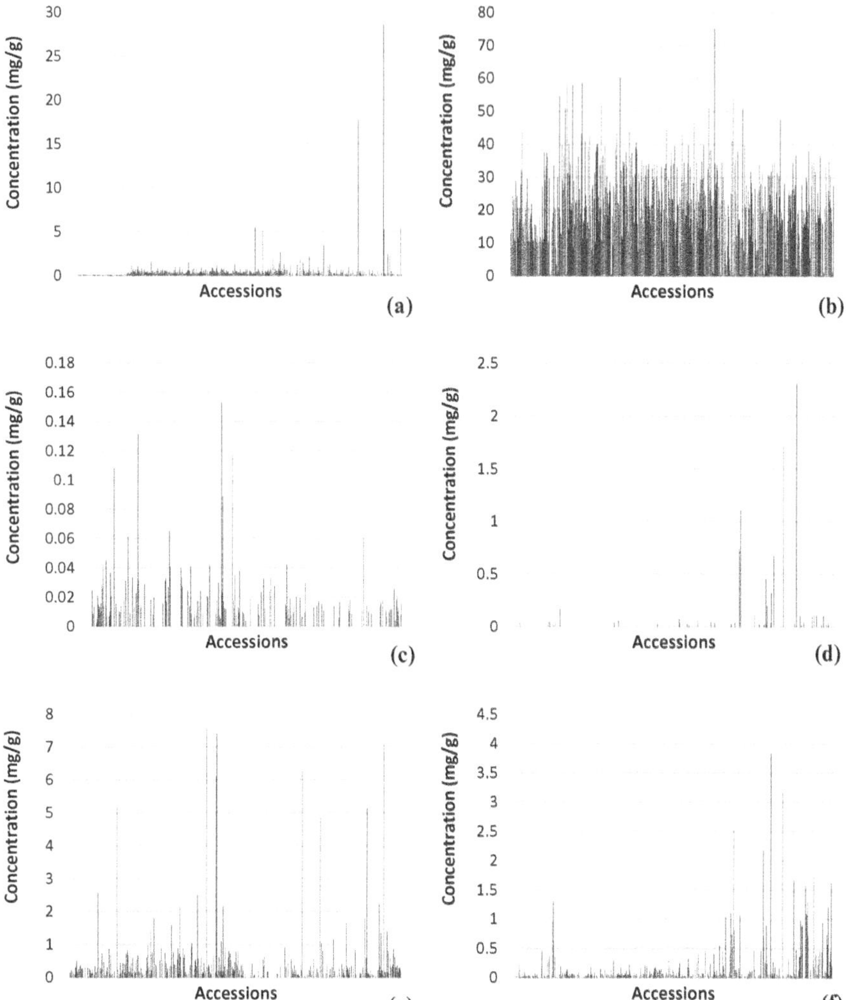

Fig. 5.2 Variation in biochemical composition of *Cinnamomum zeylanicum* leaf. (**a**) Cinnamaldehyde, (**b**) eugenol, (**c**) coumaric acid, (**d**) coumarin, (**e**) cinnamyl acetate, and (**f**) cinnamyl alcohol. Total of 515 accessions was included in the analysis. Bark and leaf chemicals were extracted separately with 100% methanol and analyzed with HPLC system equipped with reversed-phase ZORBAX Eclipse Plus C18 column, with a gradient of 0.1 phosphoric acid and acetonitrile (20–100% over 45 min) at a rate of 1 mL/ min. Cinnamyl alcohol (RT14.0 Sig 265), cinnamyl acetate (RT25.6 Sig 265), eugenol (RT20.7 Sig 286), coumarin (RT7.3 Sig 290), coumaric acid (RT12.9 Sig 280), *trans*-cinnamaldehyde (RT17.9 Sig 290)

Table 5.4 Identified groups based on cluster analysis

Group	Accessions														
1	CRS 001	CRS 015	CRS 032	CRS 063	CRS 079	CRS 093	CRS 118	CRS 123	CRS 133	CRS 136	CRS 142	CRS 163	CRS 203	CRS 209	CRS 243
	CRS 254	CRS 260	CRS 302	CRS 307	CRS 316	CRS 320	CRS 323	CRS 326	CRS 327	CRS 328	CRS 355	CRS 379	CRS 502	CRS 506	CRS 509
	CRS 521	CRS 527	CRS 581	CRS 610	CRS 611	CRS 614	CRS 615	CRS 617	CRS 624	CRS 628	CRS 633	CRS 653	CRS 661	CRS 677	CRS 679
	CRS 690	CRS 707	CRS 734	CRS 740	CRS 757	CRS 779									
2	CRS 562	CRS 575	CRS 592	CRS 635	CRS 637	CRS 668	CRS 674	CRS 700	CRS 702	CRS 715	CRS 716	CRS 737	CRS 741	CRS 742	CRS 748
	CRS 749	CRS 764	CRS 766	CRS 768	CRS 791	CRS 792									
3	CRS 075	CRS 130	CRS 143	CRS 149	CRS 157	CRS 169	CRS 207	CRS 357							
4	CRS 055	CRS 129	CRS 132	CRS 206	CRS 227	CRS 256	CRS 280	CRS 289	CRS 319	CRS 322	CRS 330	CRS 374	CRS 540	CRS 569	CRS 589
	CRS 649	CRS 657	CRS 689	CRS 698	CRS 706	CRS 709	CRS 721	CRS 733	CRS 751	CRS 756	CRS 759	CRS 793			
5	CRS 257	CRS 799													
6	CRS 003	CRS 019	CRS 020	CRS 035	CRS 036	CRS 037	CRS 038	CRS 039	CRS 041	CRS 054	CRS 061	CRS 073	CRS 080	CRS 097	CRS 099
	CRS 101	CRS 104	CRS 116	CRS 135	CRS 148	CRS 166	CRS 172	CRS 183	CRS 188	CRS 194	CRS 196	CRS 211	CRS 229	CRS 230	CRS 255
	CRS 265	CRS 266	CRS 290	CRS 300	CRS 305	CRS 338	CRS 356	CRS 360	CRS 363	CRS 364	CRS 383	CRS 503	CRS 516	CRS 522	CRS 524
	CRS 526	CRS 550	CRS 556	CRS 567	CRS 613	CRS 686									
7	CRS 005	CRS 023	CRS 043	CRS 049	CRS 068	CRS 125	CRS 127	CRS 162	CRS 168	CRS 174	CRS 184	CRS 197	CRS 215	CRS 224	CRS 235
	CRS 253	CRS 270	CRS 275	CRS 285	CRS 306	CRS 329	CRS 336	CRS 339	CRS 365	CRS 370	CRS 371	CRS 376	CRS 510	CRS 512	CRS 513
	CRS 514	CRS 515	CRS 525	CRS 533	CRS 534	CRS 542	CRS 544	CRS 551	CRS 552	CRS 557	CRS 565	CRS 566	CRS 568	CRS 576	CRS 580
	CRS 585	CRS 587	CRS 593	CRS 596	CRS 600	CRS 605	CRS 607	CRS 608	CRS 609	CRS 618	CRS 619	CRS 622	CRS 623	CRS 639	CRS 643
	CRS 647	CRS 650	CRS 651	CRS 652	CRS 654	CRS 659	CRS 660	CRS 663	CRS 665	CRS 671	CRS 672	CRS 678	CRS 680	CRS 681	CRS 691
	CRS 694	CRS 695	CRS 701	CRS 712	CRS 718	CRS 719	CRS 722	CRS 724	CRS 725	CRS 731	CRS 732	CRS 743	CRS 745	CRS 753	CRS 765
	CRS 778	CRS 780	CRS 785	CRS 786	CRS 787	CRS 795									
8	CRS 108	CRS 120	CRS 138	CRS 171	CRS 205	CRS 259	CRS 272	CRS 318	CRS 384	CRS 548	CRS 570	CRS 574	CRS 598		
9	CRS 058	CRS 072	CRS 151	CRS 173	CRS 263	CRS 282	CRS 315								
10	CRS 006	CRS 008	CRS 060	CRS 082	CRS 095	CRS 267	CRS 340	CRS 346	CRS 504	CRS 520	CRS 537	CRS 539	CRS 590	CRS 591	

(continued)

Table 5.4 (continued)

Group	Accessions														
11	CRS 012	CRS 016	CRS 017	CRS 022	CRS 025	CRS 028	CRS 069	CRS 077	CRS 096	CRS 107	CRS 110	CRS 112	CRS 137	CRS 158	CRS 160
	CRS 178	CRS 189	CRS 212	CRS 213	CRS 225	CRS 232	CRS 258	CRS 277	CRS 332	CRS 343	CRS 368	CRS 369	CRS 372	CRS 373	CRS 381
	CRS 535	CRS 545	CRS 547	CRS 606	CRS 669	CRS 670	CRS 692	CRS 747							
12	CRS 541	CRS 687	CRS 723	CRS 752	CRS 798										
13	CRS 010	CRS 011	CRS 018	CRS 026	CRS 042	CRS 105	CRS 139	CRS 185	CRS 223	CRS 252	CRS 291	CRS 292	CRS 295	CRS 344	CRS 351
	CRS 377	CRS 378	CRS 382	CRS 511	CRS 789										
14	CRS 070	CRS 102													
15	CRS 164	CRS 555	CRS 579												
16	CRS 002	CRS 007	CRS 009	CRS 013	CRS 014	CRS 045	CRS 048	CRS 051	CRS 067	CRS 083	CRS 085	CRS 086	CRS 090	CRS 098	CRS 115
	CRS 122	CRS 128	CRS 144	CRS 146	CRS 150	CRS 165	CRS 167	CRS 176	CRS 177	CRS 182	CRS 192	CRS 193	CRS 198	CRS 204	CRS 210
	CRS 216	CRS 218	CRS 221	CRS 240	CRS 245	CRS 251	CRS 261	CRS 264	CRS 293	CRS 303	CRS 304	CRS 317	CRS 333	CRS 341	CRS 345
	CRS 367	CRS 380	CRS 517	CRS 549	CRS 572	CRS 582	CRS 586	CRS 595	CRS 604	CRS 626	CRS 644	CRS 645	CRS 646	CRS 656	CRS 683
	CRS 697	CRS 699	CRS 704	CRS 705	CRS 739	CRS 746	CRS 758	CRS 767	CRS 772	CRS 773	CRS 790				
17	CRS 030	CRS 052	CRS 056	CRS 064	CRS 087	CRS 121	CRS 131	CRS 161	CRS 170	CRS 220	CRS 237	CRS 238	CRS 311	CRS 348	CRS 354
	CRS 577	CRS 584	CRS 620	CRS 627	CRS 675	CRS 761	CRS 796								
18	CRS 029	CRS 200													
19	CRS 021	CRS 024	CRS 034	CRS 040	CRS 053	CRS 078	CRS 126	CRS 140	CRS 141	CRS 152	CRS 175	CRS 202	CRS 214	CRS 246	CRS 284
	CRS 309	CRS 352	CRS 359	CRS 523	CRS 536	CRS 560	CRS 564	CRS 603	CRS 693	CRS 708	CRS 744	CRS 770	CRS 777	CRS 797	
20	CRS 062	CRS 247	CRS 558												
21	CRS 358	CRS 361	CRS 507	CRS 532	CRS 538	CRS 543	CRS 559	CRS 561	CRS 597						
22	CRS 159	CRS 179	CRS 208	CRS 217	CRS 231	CRS 239	CRS 241	CRS 308	CRS 324	CRS 350	CRS 554	CRS 583	CRS 588	CRS 621	CRS 625
	CRS 703	CRS 714	CRS 754												

Note: All the morphological traits described in Table 5.1 and the chemical traits mentioned in Table 5.3 were considered for the cluster analysis. Numbers are the accession numbers introduced at the time of establishment in the field and traceable at the field level (Liyanage et a.l 2020)

5.3 Molecular Characterization

5.3.1 DNA Fingerprinting at Species Level and Within Species

Genetic fingerprinting studies on cinnamon date back to 1980s, starting with allozyme variations. For example, Lin et al. (1997) studied the genetic diversity of *C. kanehirae* Hay using 11 allozyme loci. A total of 164 accessions belonging to four geographic areas of Taiwan were included in the analysis, and seven out of 11 loci were polymorphic on average. The percent heterozygous loci per individual ranged from 13.9% to 21.6% and the number of alleles per loci ranged from 1.7 to 1.9 with effective number of alleles of 1.34–1.54. They partitioned the total genetic diversity into within and among geographic areas and found that the within area variation amounted to 88% of the total variation. According to their data, area 4, located in southeastern Taiwan at lower elevation, showed a higher proportion of polymorphic loci.

Chung et al. (2003) used allozyme loci, wright's fixation index, spatial autocorrelation statistics (Moran's), and co-ancestry measures to examine changes in genetic structure among *C. insularimontanum* Hyata belonging to four age classes in a population in Southern Korea. Their study considered 423 individuals and there was a significant genetic differentiation among the age classes. There was no significant difference among expected heterozygosity among age classes and therefore the authors suggested similar spatial patterns of seed migration from surrounding populations. The average Moran's 1 and co-ancestry estimates indicated random spatial distribution of alleles.

Several research groups have used randomly amplified polymorphic DNA (RAPD) markers to study both inter-species and intra-species genetic diversity of *Cinnamomum*. For example, Joy and Maridass (2008) used 13 RAPD markers to study genetic differences among nine *Cinnamomum* species, including *C. verum, C. citronella, C. camphora,* and *C. glaucense* Nees, and concluded that *C. verum* is closely related to *C. citronella* than the other two. In another study, 15 *C. zeylanicum* accessions collected from different geographical locations of Western Ghats of South India were assessed with 11 RAPD primers (Sandigawad and Patil 2011). Some primers showed higher polymorphism among the accessions and overall analysis suggested higher genetic diversity among the accessions despite their morphological similarity.

Kameyama (2012) isolated and characterized 22 microsatellite loci from *C. camphora* and assessed with 104 adult trees from three populations in Japan. All 22 loci showed clear reproducible single bands with mean number of alleles per locus ranging from 4.1 to 8.0 among populations. Later, Hung et al. (2014) developed 15 novel microsatellite markers for genetic studies of *C. kanehirae*, endemic to Taiwan. Most of those markers were obtained using cross-species amplification procedure and the number of alleles amplified ranged from 2 to 13.

Hung et al. (2017) presented a genetic fingerprinting database for *C. kanehirae* based on 15 microsatellites. These microsatellites could identify the origin of the

C. kanehirae timber sample and its population genetic structure. A total of 817 accessions of *C. camphora* collected from Japan, China, and Taiwan were analyzed with 11 microsatellite markers by Kameyama et al. (2017). There was a strong genetic differentiation between areas: Japan vs. China and Taiwan and less diversity within Japanese population. They also found less consistency between genetic composition of cultivated vs. wild germplasm in a given geographic location suggesting regional transfer or introduction of materials from other regions. There is possible inter-crossing between non-native genotypes and native genotypes.

Genetic variation of *C. tamala* collected from different locations of Uttarkhand and Himalaya regions in India was studied using four RAPD markers and three inter-simple sequence repeat (ISSR) markers (Gwari et al. 2016). With the different approaches used in the study, they could discriminate all the accessions included in the analysis.

Ho and Hung (2011) used leaf morphological characters together with ISSR regions and ITS (rDNA, internal transcribed space) region to study the cladistic relationships among 12 endemic species of *Cinnamomum* in Taiwan. Leaf morphology and ISSR fingerprints could differentiate *Camphora* group and *Cinnamomum* group. The genetic relationship within each group was very close and shared specific 11 bp deletions in all 175 sequences. Nevertheless, in combination with three strategies they were able to differentiate the closely related species from each other. Dong et al. (2016) employed field studies and ISSR fingerprinting techniques to study the genetic diversity and population structure of remnant populations of *C. chago* B.S. Sun & H.L. Zhao in the Yunnan Province, China. Based on some leaf morphological traits and pollen characters, the authors concluded that *C. chago* is a key phylogenetic taxon between the two sections of Asian *Cinnamomum* species, Sect Camphoa (Trew) Meissn and Sect Cinnamomum. The ISSR analysis revealed that it has a moderately high level of intra-species and inter-population genetic diversity.

Cinnamomum species found in Sri Lanka were studied using RAPD and sequence-related amplified polymorphism (SRAP) (Abeysinghe et al. 2014). Fourteen RAPD primers and 20 sets of SRAP primers were used and the amplified products could be categorized into genus-specific, species-specific, and intra-specific markers. A higher genetic diversity is expected in *C. zeylanicum* germplasm in Sri Lanka due to continuous use of seeds as planting material over centuries. To get a general sense of variation generated due to cross-pollination among closely located mother plants, Liyanage et al. (2020) studied morphological features together with biochemical and genetic fingerprints of resulting offspring. Ten seedlings from a mother plant of an elite variety, "Sri Wijaya" released by the Department of Export Agriculture, Sri Lanka, were assessed with six ISSR markers, and the fingerprints were compared with mother plant and possible pollen donors around. A significantly high genetic diversity is created with 2.9 Nei's diversity and 0.44 Shannon's Information Index. Interestingly, the genetic diversity was reflected in leaf morphology and chemical fingerprints of studied individuals.

Bandusekara et al. (unpublished) considered morphological, molecular, and biochemical diversity of endemic wild species of *Cinnamomum* in Sri Lanka.

Samples were collected from different regions representing at least three samples from each wild species, namely, *C. citriodorum*, *C. litseaefolium*, *C. dubium*, *C. rivulorum*, *C. capparu-coronde*, *C. sinharajaense*, and *C. ovalifolium* and the cultivated species, *C. zeylanicum*. There was a significant morphological and biochemical diversity among individuals of the same species collected from different geographical locations. Such diversity was clearly represented by the genetic diversity assessed with six ISSR primers. While the highest diversity was observed among *C. citriodorum* collections, the lowest diversity was reported among *C. sinharajaense*. While Abeysinghe and colleagues studied only four wild species, *C. dubium*, *C. cappara-coronde*, *C. litseaefolium*, and *C. sinharajaense* each represented with two accessions collected from the same germplasm collection, study done by Bandusekera and colleagues consisted of all seven wild species recorded in Sri Lanka.

In some cases, the intron regions of nuclear genes have also been used for studying the population structures and intra-species genetic diversity. For example, Liao et al. (2010) studied 113 individuals of *C. kanehirae* collected from 19 localities of four geographic regions in Taiwan using the Chalcone synthese gene (CHS) intron and leafy (LFY) intron. They could amplify 210 CHS and 170 LFY sequences consisting of 36 and 35 haplotypes respectively. Analysis showed no correlation between genetics and geographic distances. There was a significant among-region genetic difference when comparing eastern to western populations. Nevertheless, among-region diversity was not significant when all four regions were compared together. Further, no genetic structuring was found among the four regions.

5.3.2 DNA Barcoding and Chloroplast Sequencing

Universally accepted DNA barcoding regions are widely being used for species identification purposes. In an early study, in 2001, Chanderbali et al. (2001) studied the phylogenetic relationships among 122 species of Lauraceae representing 44 of 55 recognized genera using the chloroplast and nuclear genomic regions. While the CpDNA regions, trnL-traF, trnT-trnL, psbA-tranH, and rp116, and 5′-end of 26S rDNA resolved major lineages, ITS/5.8S regions of rDNA resolve a terminal clade. Biogeographic history of the family was reconstructed using a combination of molecular, morphological, and temporal dimensions.

Genetic identification of *C. cassia, C. zeylanicum, C. burmannii,* and *C. sieboldii* was done by two chloroplast DNA regions, intergenic spacer region between the trnL 3′-exon and trnF exon (tranL–tranF1GS) and the trnL intron region. Six accessions of *C. cassia*, two accessions of *C. zeylanicum,* one accession each of *C. burmannii* and *C. sieboldii* were included in the study. They identified single nucleotide variation among species in trnL-tranFlGC region and another three sites in the trnL intron. This variation was sufficient to differentiate four considered species. Further, single-strand conformation polymorphism (SSCP) analysis of

PCR products of above two regions could differentiate *C. cassia, C. zeylanicum,* and *C. burmannii* (Kojoma et al. 2002).

Abesinghe et al. (2009) studied genetic variation among *C. verum, C. citriodorum, C. capparu-coronde, C. dubium, C. litseaefolium, C. rivulorum, C. sinharajaense,* and *C. camphora* found in Sri Lanka using cpDNA regions and an ITS region of rDNA. Variation in intergenic spacers between trnL-trnF, trnT-trnL, trnH-ps6A, and trnL intron among species was not sufficient enough to differentiate the species clearly. There were eight different types of ITS variants found among ten *C. verum* accessions studied. While *C. litseaefolium, C. rivulorum,* and *C. sinharajense* had different ITS sequences, there was no within species variation.

Kuo et al. (2010) studied the CpDNA variations in 19 populations of *C. kanehirae* consisting of 94 individuals. Two CpDNA fragments, trnL-trnF and petG-trnP intergenic spacers, were amplified with universal primers and there was a very low nucleotide diversity among 792 bp aligned sequences. Truly identified eight polymorphic sites and six haplotypes and a star-like genealogy were observed suggesting a population expansion.

Barcoding loci rbcL, matK, and psbA–trnH were used for confirming the identification of *C. verum, C. cassia,* and *C. malabatrum* (Swetha et al. 2014). The single nucleotide polymorphisms (SNPs) specific to *C. cassia* were detected in rbcL locus and those could be used for confirming the presence of *C. cassia* as an adulterant in market samples of *C. verum.* Recently Hsu et al. (2019) analyzed 73 geographical strains of *C. osmophloeum* Kanech using partial non-coding internal transcribed spacer 2 (PITS2) of the ribosomal DNA and the trnL-trnF of chloroplast genome. The analysis showed comparatively higher polymorphism in PITS2 region than trnL intron and trnL-trnF Intergenic spacer. The investigators further described the use of PITS2 polymorphisms as a genetic classifier.

Since the universal barcoding regions do not successfully discriminate closer taxa of the genus *Cinnamomum*, chloroplast genome sequencing has become the interest of many research groups. The complete chloroplast genome of *C. kanehirae* Hayata was the first to be sequenced in the family Lauraceae (Wu et al. 2016). Since then several other chloroplast genomes of the genus *Cinnamomum* have been sequenced. A summary of the basic features of the available chloroplast genomes is given in the Table 5.5. All of them share common features of chloroplast genomes in general. Chloroplast genome data have been used to resolve phylogeny of closely related *Cinnamomum* species (Wu et al. 2016, 2017; Chen et al. 2017, 2019; Song et al. 2017; Ren et al. 2019; Li et al. 2019).

Comparative chloroplast genome analysis has also been used in clear identification of morphologically very similar species and authentication of valuable species. For example, *C. kanehirae* is an endemic species in Taiwan and the sole natural host of medicinally important fungus *Antrodia cinnamomea. C. micranthum* is highly similar in morphology to *C. kanehirae* and difficult to distinguish morphologically based on wood structure, and both species grow in similar habitats. Based on chloroplast genome analysis, the authors identified six insertions/deletions (In Dels) and validated experimentally (Wu et al. 2017).

Table 5.5 Main features of published *Cinnamomum* chloroplast genomes

Species	Size (bp)	Special features	References
C. kanehirae	152,700	Inverted repeats (IR alb) of 20,107 bp, Total IR 7 = 40,212, single copy region (LSC) 93,642pb, small single copy (SSC) 18,844 bp GC content = 39.1% Total number of different genes = 112	Wu et al. (2016)
C. micranthum	152,675	LSC—93,662 bp SCC—18,875 IRalb—40,138 GC content = 39.1% Total number of different genes = 112	Wu et al. (2017)
C. camphora	152,570	LSC—93,705 bp SSC—19,093 Two inverted repeat regions Ira/b 19,886 GC content = 39.1% Total number of genes = 123	Chen et al. (2017)
C. japonicum	152,731	Ira/b—20,092 bp for each LSC—93,698 bp SCC—18,849 GC content 39.2% Total number of genes—123	Ren et al. (2019)
C. camphora	152,729	Ira/b—20,074 bp LSC—93,688 bp SSC—18,993 Total number of genes = 127	Li et al. (2019)
C. chago	152,753	Ira/b—20,074 LSC—93,722 bp SSC—18,883 bp GC content = 39.2%	Chen et al. (2019)

Similarly, recent work on comparison of chloroplast genomes of true cinnamon grown in Sri Lanka and India showed considerable differences between them. Further, the previously published *C. verum* chloroplast genome and currently sequenced authentic *C. zeylanicum* chloroplast genomes were similar in length and had only less than 10bp difference in the entire genome, suggesting those two genomes are from the same species (Narampanawa et al. unpublished).

Phylogenetic analysis based on complete chloroplast sequences of wild relatives of Ceylon cinnamon showed interesting relationships among them and possible explanations for morphological and biochemical similarities among them (Narampanawa et al. unpublished). The chloroplast genomes of all endemic *Cinnamomum* species recorded in Sri Lanka *C. zeylanicum*, *C. citriodorum*, *C. litseaefolium*, *C. dubium*, *C. rivulorum*, *C. capparu-coronde*, *C. sinharajaense*, and *C. ovalifolium* were included in the analysis. Results revealed that *C. rivulorum* and *C. dubium* having closer morphological characters and chemical profiles clustered together.

5.3.3 *Genome and Transcriptome Sequencing*

The first *Cinnamomum* genome to be sequenced and published was *C. kanehirae*, commonly known as the stout camphor tree, endemic to Taiwan and under threat of extinction (Chaw et al. 2019). The analysis provided an insight into evolutionary history of the genus. The genome duplication events such as segmental duplications and tandem duplications have contributed to evolutionary history of the genus. This study provided evidence for evolution of diversity of monoterpenes and sesquiterpenes present in the genus *Cinnamomum*.

A draft genome of *C. zeylanicum* has been assembled recently (Narampanawa et al. unpublished). A low coverage of genomic sequence data of wild relatives of Ceylon cinnamon, *C. citriodorum*, *C. litseaefolium*, *C. dubium*, *C. rivulorum*, *C. capparu-coronde*, *C. sinharajaense* and *C. ovalifolium* are also available now. These data would be useful in identifying markers for specific genes and improving breeding efforts.

Several *Cinnamomum* transcriptomes are now available for various applications. For example, Yan et al. (2017) sequenced leaf transcriptome of *C. longepaniculatum* and identified 223 unique genes to be involved in terpenoid metabolism.

The transcriptomics of two chemotypes of *C. camphora* was studied recently (Chen et al. 2018). A total of 67 candidate unigenes were predicted to be involved with terpenoid biosynthesis. Interestingly, 2863 unigenes were differentially expressed in two chemotypes, borneol type and linalool type, of which 1714 were up-regulated and 1149 down-regulated. Li et al. (2018) developed new SSR markers using leaf transcriptomics data of *C. camphora*. Twenty-one polymorphic SSR markers were developed and validated using 45 individuals and examined the cross-species transferability of those to other six related species. The leaf transcriptome of *C. chago* was sequenced, assembled, analyzed, and identified 47 terpenoid biosynthesis genes and 46 fatty acid biosynthesis genes (Zhang et al. 2018). They also have identified 25,654 SSR and 640 SNPs, and based on these 55 EST-SSR primers were developed.

The plant growth environment and maturity stage may decide the quality of oil and bark yield. Therefore, the leaf and bark transcriptome of *C. zeylanicum* was studied at different environmental conditions and maturity stages. The same *C. zeylanicum* variety, "Sri Gemunu," propagated through stem cuttings, grown under three different agro-ecological conditions, was used. While the bark samples were separated into three maturity stages, leaves were separated into two groups. Differentially expressed genes were assessed under considered conditions (Liyanage et al. 2020).

5.4 Future Perspectives

Biotechnological tools, especially the sequencing technologies, are evolving faster than one could imagine. This would provide appropriate data to resolve some issues in phylogenies as well as quality and quantity of the yield. For example, prior to the molecular biology era, scientists used morphological traits for identification of species. Later DNA barcoding technology resolved some of the problems that confronted taxonomists to clearly separate taxa morphologically. Several studies provided evidence that the DNA barcoding could not resolve phylogenies of closely related *Cinnamomum* species. Thanks to the next-generation sequencing (NGS) technologies, chloroplast genome sequencing has now become cheaper and affordable for many labs around the world. Earlier, the chloroplast genome sequencing was done after tedious enrichment procedures carried out in the lab to separate chloroplasts. Nevertheless, currently "skim sequencing" has been the method of choice, where total genomic DNA is sequenced and the chloroplast sequences are separated bioinformatically. This method has reduced the wet lab work and genome sequences are also generated through the same attempt. Therefore, the genomic internal transcribed spacer (ITS) regions are also available for the analysis. This would be an added advantage of identification of closely related species. Nevertheless, only 12 fully annotated chloroplast genomes of *Cinnamomum* are available in the public domain. Given the expenses per GB data, all the fingerprinting and barcoding studies could easily be replaced with skim sequencing of genomic DNA. Bioinformatics capabilities would be the only challenging part for many institutions.

Currently, there is one complete *Cinnamomum* genome in the public domain. More investment on genome sequencing would help in future breeding and conservation efforts. Transcriptome data are also limited compared to other economically important crop species. More investments on transcriptome sequencing would help better understand the relevant biosynthetic pathways and differentially expressed genes in response to different factors. Transcriptomic studies will provide a global overview of gene expression patterns related to terpenoid biosynthesis and other metabolic pathways in *Cinnamomum* species and could help in understating differential accumulation of chemicals of interest in different chemotypes. This would in turn help in improving the quality and quantity of yield. Further, transcriptomics data are used for developing gene-specific SSRs, applicable in future breeding efforts. Such SSRs can also be used for diversity analysis and exploring evolutionary history and genetic differentiation pattern of *Cinnamomum*.

In general, systems biology approach has been employed in recent research work on many crops including rice and wheat. As Walhout and Dekker (2012) suggested this would be the best approach in understanding the biology and resolving field level issues with respect to cinnamon as well. For example, combining morphological, agronomic, biochemical, and gene expression data would provide a better overview of the comprehensive picture and actual contribution of each component and their determination of the system. The transcriptome data would make better sense if relevant biochemical data also show expected pattern or trend. Such systems biology and multidisciplinary research approach in future cinnamon research are encouraged.

Acknowledgments Authors acknowledge the financial assistance provided by the Ministry of Primary Industries and Social Empowerment through the National Science Foundation of Sri Lanka under the special Cinnamon project—Grant No: NSF SP/CIN/2016/01. The authors thank the National Cinnamon Research and Training Center, Department of Export Agriculture, Thihagoda, Palolpitiya and Sub Research Station, Department of Export Agriculture, Nillambe for providing cinnamon samples and the Department of Wildlife Conservation, Sri Lanka and the Forest Department for providing permission to collect samples from wild populations in protected areas. The authors would also like to thank the collaborators of the project and the staff of the Agricultural Biotechnology Centre, Faculty of Agriculture, University of Peradeniya for the continuous support given.

References

Abeysinghe P, Samarajeewa N, Li G, Wijesinghe K (2014) Preliminary investigation for the identification of Sri Lankan Cinnamomum species using randomly amplified polymorphic DNA (RAPD) and sequence related amplified polymorphic (SRAP) markers. J Natl Sci Found 42:175–182

Abeysinghe PD, Wijesinghe KGG, Tachida H, Yoshda T (2009) Molecular characterization of cinnamon (Cinnamomum Verum Presl) accessions and evaluation of genetic Res J Agric Biol Sci 5(6):1079–1088

Ananthakrishnan R, Santhoshkumar ES, Rameshkumar KB (2018) Comparative chemical profiles of essential oil constituents of eight wild Cinnamomum species from the Western Ghats of India. Nat Prod Commun 13:1934578X1801300525

Ariyarathne H, Weerasuriya S, Senarath W (2018) Comparison of morphological and chemical characteristics of two selected accessions and six wild species of genus Cinnamomum Schaeff. Sri Lankan J Biol 3(1):11–23

Azad R, Ranawaka RAAK, Senanayake G, Kumara KW, Wijesinghe KGG, Pushpakumara DKNG, Geekiyanage S (2016) Morphological variation of cinnamon (*Cinnamomum verum* Persl) germplasm in Matara District of Sri Lanka. Int J Minor Fruits Med Aromat Plant 2(1):6–14

Azad R, Kumara KW, Senanayake G, Ranawaka R, Pushpakumara D, Geekiyanage S (2018) Flower morphological diversity of cinnamon (Cinnamomum verum Presl) in Matara District, Sri Lanka. Open Agric 3:236–244

Azad R, Jayasekara L, Ranawaka R et al (2019) Development of a core collection for Sri Lankan cinnamon germplasm based on morphological characterization using an eco-geographic survey. Aust J Crop Sci 13:1473–1485

Bakker M, Gerritsen A, Van Der Schaaf P (1992) Leaf anatomy of Cinnamomum Schaeffer (Lauraceae) with special reference to oil and mucilage cells. Blumea Biodivers Evol Biogeogr Plant 37:1–30

Bandusekara BS, Pushpakumara DKNG, Bandaranayake PCG, Wijesinghe KGG, Jayasinghe GG (2020) Field level identification of Cinnamomum species in Sri Lanka using a morphological index. Trop Agric Res 31(4):43–53

Baruah A, Nath S (2000) Certain chemo-morphological variants of Cinnamomum bejolghota (Buch-Ham) sweet and their relevance to foliar micromorphology. J Swamy Bot Cl 17:19–23

Baruah A, Nath SC (2007) *Cinnamomum champokianum* sp. Nov.(Lauraceae) from Assam, northeastern India. Nord J Bot 25:281–285

Bernard T, Perineau F, Delmas M, Gaset A (1989) Extraction of essential oils by refining of plant materials. II. Processing of products in the dry state: *Illicium verum* Hooker (fruit) and *Cinnamomum zeylanicum* Nees (bark). Flavour Fragr J 4:85–90

Bandusekara BS, Pushpakumara DKNG, Bandaranayake PCG, Wijesinghe KGG, Jayasinghe GG (2020) Field level identification of *Cinnamomum* species in Sri Lanka using a morphological index. Trop Agric Res 31(4):43–53

Chalchat J-C, Valade I (2000) Chemical composition of leaf oils of Cinnamomum from Madagascar: C. zeylanicum Blume, *C. camphora* L., *C. fragrans* Baillon and *C. angustifolium*. J Essent Oil Res 12:537–540

Chanderbali AS, Van Der Werff H, Renner SS (2001) Phylogeny and historical biogeography of Lauraceae: evidence from the chloroplast and nuclear genomes. Ann Mo Bot Gard 88:104–134

Chaw S-M, Liu Y-C, Wu Y-W et al (2019) Stout camphor tree genome fills gaps in understanding of flowering plant genome evolution. Nat Plant 5:63

Chen C, Zheng Y, Liu S et al (2017) The complete chloroplast genome of *Cinnamomum camphora* and its comparison with related Lauraceae species. Peer J 5:e3820

Chen C, Zheng Y, Zhong Y et al (2018) Transcriptome analysis and identification of genes related to terpenoid biosynthesis in *Cinnamomum camphora*. BMC Genomics 19:550

Chen H, Liu C, Han L, Song Y, Tang L (2019) The plastid genome of an oil plants *Cinnamomum chago* (Lauraceae). Mitochondrial DNA Part B4:1733–1734

Cheng S-S, Liu J-Y, Tsai K-H, Chen W-J, Chang S-T (2004) Chemical composition and mosquito larvicidal activity of essential oils from leaves of different *Cinnamomum osmophloeum* provenances. J Agric Food Chem 52:4395–4400

Cheng S-S, Liu J-Y, Hsui Y-R, Chang S-T (2006) Chemical polymorphism and antifungal activity of essential oils from leaves of different provenances of indigenous cinnamon (*Cinnamomum osmophloeum*). Bioresour Technol 97:306–312

Cheng S-S, Lin C-Y, Yang C-K, Chen Y-J, Chung M-J, Chang S-T (2015) Chemical polymorphism and composition of leaf essential oils of *Cinnamomum kanehirae* using gas chromatography/mass spectrometry, cluster analysis, and principal component analysis. J Wood Chem Tech 35:207–219

Choudhury D, Biswas R, Mandal P, Das A (2013) Diversity of *Cinnamomum* Schaeffer (Lauraceae) in Terai and Duars region of West Bengal, India. Pleione 7:441–448

Chung M, Nason J, Epperson B, Chung M (2003) Temporal aspects of the fine-scale genetic structure in a population of *Cinnamomum insularimontanum* (Lauraceae). Heredity 90:98

Dong W, Zhang X, Guansong Y, Yang L, Wang Y, Shen S (2016) Biological characteristics and conservation genetics of the narrowly distributed rare plant *Cinnamomum chago* (Lauraceae). Plant Divers 38:247–252

Dung NX, Van Khien P, Chien HT, Leclercq PA (1993) The essential oil of *Cinnamomum camphora* (L.) Sieb. var. linaloolifera from Vietnam. J Essent Oil Res 5:451–453

Fujita Y (1967) Classification and phylogeny of genus *Cinnamomum* viewed from constituents of essential oils. Bot Mag Tokyo 80:261

Gopalam A (1997) Profile of essential chemical constituents in tree spices. Supplement to cultivation and utilization of aromatic plants, RRL, (CSIR), Jammu, pp 405–23

Grangfu T, Yang Z (1988) Study on numerical chemotaxonomy of *Cinnamomum* in Hubei Province [China]. Acta Phytotaxonomica Sinica (China)

Gu J, Liu L, Zhang Y (1990) Study on the essential oil from the fruits of *Cinnamomum camphora* (L.) Presl var. linaloolifera Fujita. Linchan Huaxue Yu Gongye 10:77–82

Guo X, Cui M, Deng M et al (2017) Molecular differentiation of five *Cinnamomum camphora* chemotypes using desorption atmospheric pressure chemical ionization mass spectrometry of raw leaves. Sci Rep 7:46579

Gwari G, Bhandari U, Naik G, Haider SZ, Chauhan N (2016) Genetic diversity in *Cinnamomum tamala* Nees. Accessions through DNA fingerprinting using molecular markers. Indian J Agric Res 50:446–450

Hammid SA, Assim Z, Ahmad F (2016) Chemical composition of *Cinnamomum* species collected in Sarawak. Sains Malaysiana 45:627–632

Hirota N, Hiroi M (1967) The later studies on the camphor tree, on the leaf oil of each practical form and its utilisation

Ho K-Y, Hung T-Y (2011) Cladistic relationships within the genus *Cinnamomum* (Lauraceae) in Taiwan based on analysis of leaf morphology and inter-simple sequence repeat (ISSR) and internal transcribed spacer (ITS) molecular markers. Afr J Biotechnol 10:4802–4815

Hooker J (1886) Laurineae. Flora Br India 5:116–189
Hsu W-K, Lee S-C, Lu P-L (2019) A useful technical application of the identification of nucleotide sequence polymorphisms and gene resources for *Cinnamomum osmophloeum* Kaneh. (Lauraceae). Forests 10(4):306
Hu T (1985) Natural variation of chemical components of the leaf oil of *Cinnamomum osmophloeum* Kaneh. Bull Taiwan For Res Ind New Ser 78:18
Hung K-H, Lin C-H, Shih H-C, Chiang Y-C, Ju L-P (2014) Development, characterization and cross-species amplification of new microsatellite primers from an endemic species *Cinnamomum kanehirae* (Lauraceae) in Taiwan. Conserv Genet Resour 6:911–913
Hung K-H, Lin C-H, Ju L-P (2017) Tracking the geographical origin of timber by DNA fingerprinting: a study of the endangered species *Cinnamomum kanehirae* in Taiwan. Holzforschung 71:853–862
Jantan IB, Goh SH (1992) Essential oils of *Cinnamomum* species from Peninsular Malaysia. J Essent Oil Res 4:161–171
Jantan IB, Karim Moharam BA, Santhanam J, Jamal JA (2008) Correlation between chemical composition and antifungal activity of the essential oils of eight *Cinnamomum* species. Pharm Biol 46:406–412
Joy P, Maridass M (2008) Inter species relationship of *Cinnamomum* species using RAPD marker analysis. Ethnobot Leafl 2008:58
Joy P, Thomas J, Mathew S, Ibrahim K (1998) Growth, leaf oil yield and quality investigations in cinnamon (*Cinnamomum verum* Presl.). J Med Aromat Plant Sci 20(2):401–406
Kameyama Y (2012) Development of microsatellite markers for *Cinnamomum camphora* (Lauraceae). Am J Bot 99:e1–e3
Kameyama Y, Furumichi J, Li J, Tseng Y-H (2017) Natural genetic differentiation and human-mediated gene flow: the spatiotemporal tendency observed in a long-lived *Cinnamomum camphora* (Lauraceae) tree. Tree Genet Genomes 13:38
Kojoma M, Kurihara K, Yamada K, Sekita S, Satake M, Iida O (2002) Genetic identification of cinnamon (*Cinnamomum* spp.) based on the trnL-trnF chloroplast DNA. Planta Med 68:94–96
Kostermans AJGH (1957) Lauraceae. Reinwardtia 4:193–256
Kostermans A (1961) The new world species of *Cinnamomum* trend (Lauraceae). Reinwardia 6:17–24
Kostermans A (1980) A note on two species of *Cinnamomum* (Lauraceae) described in Hortus Indicus Malabaricus. In: Botany and history of Hortus Malabaricus. Oxford IBH, New Delhi, pp 163–167
Kostermans A (1983) The South Indian species of *Cinnamomum* Schaeffer (Lauraceae). Nelumbo 25:90–133
Krishnamoorthy B, Gopalam A, Abraham J (1988) Quality parameters of cinnamon (*Cinnamomum verum*) in relation to flush colour
Krishnamoorthy B, Rema J, Sasikumar B (1991) Progeny analysis in cinnamon
Krishnamoorthy B, Sasikumar B, Rema J, Gopalam A, Abraham J (1992) Variability and association studies in cinnamon (*Cinnamomum verum*). J Spices Aromat Crops 1:148–150
Krishnamoorthy B, Zachariah TJ, Rema J, Mathew P (1999) Evaluation of selected Chinese cassia (*Cinnamomum cassia* Blume) accessions for chemical quality. J Spices Aromat Crops 8(2):193–195
Krishnamoorthy B, Rema J, Zachariah JT, Mathew P (2001) High quality cassia selections from IISR, Calicut
Kuo D-C, Lin C-C, Ho K-C, Cheng Y-P, Hwang S-Y, Lin T-P (2010) Two genetic divergence centers revealed by chloroplastic DNA variation in populations of *Cinnamomum kanehirae* Hay. Conserv Genet 11:803–812
Li Z, Zhong Y, Yu F, Xu M (2018) Novel SSR marker development and genetic diversity analysis of *Cinnamomum camphora* based on transcriptome sequencing. Plant Genet Resour 16:568–571
Li P, Jia G, Xin G, Cai X (2019) The complete chloroplast genome of *Cinnamomum camphora* (L.) Presl., a unique economic plant to China. Mitochondrial DNA Part B4:2511–2512

Liao P-C, Kuo D-C, Lin C-C, Ho K-C, Lin T-P, Hwang S-Y (2010) Historical spatial range expansion and a very recent bottleneck of *Cinnamomum kanehirae* Hay.(Lauraceae) in Taiwan inferred from nuclear genes. BMC Evol Biol 10:124

Lin T-P, Cheng Y-P, Huang S-G (1997) Allozyme variation in four geographic areas of *Cinnamomum kanehirae*. J Hered 88:433–438

Liyanage NMN, Bandusekara BS, Kanchanamala RWMK, Hathurusinghe HABM, Rathnayaka AMRWSD, Pushpakumara DKNG, Samita S, Wijesinghe KGG, Jayasinghe GG, Liyanage WK, Bandaranayake PCG (2020) Identification of superior *Cinnamomum zeylanicum* Blume germplasm for future true cinnamon breeding in the world. J Food Compos Anal 103747. https://doi.org/10.1016/j.jfca.2020.103747

Manilal K, Shylaja M (1986) A new species of *Cinnamomum* Schaeffer (Lauraceae) from Malabar. Nelumbo 28:111–113

Möllenbeck S, König T, Schreier P, Schwab W, Rajaonarivony J, Ranarivelo L (1997) Chemical composition and analyses of enantiomers of essential oils from Madagascar. Flavour Fragr J 12:63–69

Monteiro IN, Dos Santos MO, Costa-Junior LM et al (2017) Chemical composition and acaricide activity of an essential oil from a rare chemotype of *Cinnamomum verum* Presl on *Rhipicephalus microplus* (Acari: Ixodidae). Vet Parasitol 238:54–57

Morsbach N, Koketsu M, Gonçalves S, Godoy R, Lopes D (1997) Óleos essenciais de cascas e folhas de canela (*Cinnamomum verum* Presl) cultivada no Paraná. Cienc Tecnol Aliment 17:281–285

Nath SC, Pathak MG, Baruah A (1996) Benzyl benzoate, the major component of the leaf and stem bark oil of *Cinnamomum zeylanicum* Blume. J Essent Oil Resear 8:327–328

Nees Von Esenbeck CGD, 1836. Systema laurinarum

Paul S, Sahoo S (1993) Selection of elite cinnamon plants for quality bark production. J Eco Tax Bot 17:353–355

Paul S, Sahoo S, Patra P (1996) Elite *Cinnamomum zeylanicum* genotypes isolated from germplasm of India and Ceylon. Indian J Plant Genet Resour 9:307–308

Pélissier Y, Marion C, Prunac S, Bessière J-M (1995) Volatile components of leaves, stems and bark of *Cinnamonum camphora* Nees et Ebermaier. J Essent Oil Res 7:313–315

Pushpakumara DKNG, Silva GLLP (2018) Chapter 12: Agricultural genetic resources. In: Wijeyaratne MJS, Jayasuriya AHM, Wijayananda NP (eds) Natural resources in Sri Lanka: conditions, trends and prospects. National Science Foundation of Sri Lanka, Colombo 7, pp 249–279

Rana VS, Langoljam RD, Verdeguer M, Blázquez MA (2012) Chemical variability in the essential oil of *Cinnamomum tamala* L. leaves from India. Nat Prod Res 26:1355–1357

Rao Y, Paul S, Dutta P (1988) Major constituents of essential oils of *Cinnamomum zeylanicum*. Indian Perfumer 32:86–89

Ravindran S, Balakrishnan R, Manilal K, Ravindran P (1991) A cluster analysis study on *Cinnamomum* from Kerala, India. Feddes Repertorium 102:167–175

Ravindran S, Manilal K, Ravindran P (1996) Numerical taxonomy of *Cinnamomum* II. Principal component analysis of major species from Kerala, India. In: Taxonomy and plant conservation. IBH Pub. Co, New Delhi, pp 321–332

Ravindran P, Shylaja M, Babu KN, Krishnamoorthy B (2004) Botany and crop improvement of cinnamon and cassia. In: Cinnamon and cassia. CRC Press Boca Raton, pp 30–95

Ren X, Yang C, Xin G et al (2019) Complete chloroplast genome of *Cinnamomum japonicum* (Laurales: Lauraceae), an endangered tree species. Conserv Genet Resour 11:267–269

Sahoo S, Paul S, Patra P (2000) Quality cinnamon production in India. J Med Aromat Plant Sci 22:361–365

Sandigawad A, Patil C (2011) Genetic diversity in *Cinnamomum zeylanicum* Blume.(Lauraceae) using random amplified polymorphic DNA (RAPD) markers. Afr J Biotechnol 10:3682–3688

Shi WY (1989) Study on chemical constituents of essential oils & classification of types from *C. camphora*. Acta Botanica Sinica (China) 31:209–214

Shylaja M (1984) Studies on Indian *Cinnamomum*: Unpublished Ph.D thesis, University of Calicut, India
Song Y, Yu W-B, Tan Y et al (2017) Evolutionary comparisons of the chloroplast genome in Lauraceae and insights into loss events in the Magnoliids. Genome Biol Evol 9:2354–2364
Swetha V, Parvathy V, Sheeja T, Sasikumar B (2014) DNA barcoding for discriminating the economically important *Cinnamomum verum* from its adulterants. Food Biotechnol 28:183–194
Tétényi P (1970) Infraspecific chemical taxa of medicinal plantsE
Thomas J, Greetha K, Shylara K (1987) Studies on leaf oil and quality of Cinnamomum zeylanicum. Indian Perfum 1:249–251
TURIS 2013 Project (2016) Descriptors for cinnamon (*Cinnamomum verum*). University of Ruhuna, Matara
Van Khiên P, Chiên HT, Dung NX, Leclercq AX, Leclercq PA (1998) Chemical segregation of progeny of camphor trees with high camphor cq linalool content. J Essent Oil Res 10:607–612
Variyar P, Bandyopadhyay C (1989) On some chemical aspects of *Cinnamomum zeylanicum*. PAFAI J10:35–38
Walhout M, MVJ D (eds) (2012) Handbook of systems biology: concepts and insights. Academic Press, Oxford
Wang Y, Zeng L, Chen Y et al (2017) Genotypic variations of open-pollinated families of *Cinnamomum camphora* seedlings from South China. J Trop For Sci 29:404–416
Wan-Yang S, Wei H, Guang-yu W, De-xuan G, Guang-yuan L, Yin-gou L (1989) Study on chemical constituents of the essential oil and classification of types from *Cinnamomum camphora*. Acta Bot Sinica 31:209–214
Wijesekera R, Chichester C (1978) The chemistry and technology of cinnamon. Crit Rev Food Sci Nutr 10:1–30
Wijesekera R, Ponnuchamy S, Jayewardene A (1975) Cinnamon. Ceylon Institute of Scientific and Industrial Research, Colombo, p 48
Wu C-C, Ho C-K, Chang S-H (2016) The complete chloroplast genome of *Cinnamomum kanehirae* Hayata (Lauraceae). J Mitochondrial DNA Part A DNA Mapp Seq Anal 27(4):2681–2682
Wu C-C, Chu F-H, Ho C-K, Sung C-H, Chang S-H (2017) Comparative analysis of the complete chloroplast genomic sequence and chemical components of *Cinnamomum micranthum* and *Cinnamomum kanehirae*. Holzforschung 71:189–197
Wuu-Kuang S (2011) Taxonomic revision of *Cinnamomum* (Lauraceae) in Borneo. Blumea Biodivers Evol Biogeogr Plant 56:241–264
Yan K, Wei Q, Feng R, Zhou W, Chen F (2017) Transcriptome analysis of *Cinnamomum longepaniculatum* by high-throughput sequencing. Electron J Biotechnol 28:58–66
Zhang X, Zhang Y, Wang Y-H, Shen S-K (2018) Transcriptome analysis of *Cinnamomum chago*: a revelation of candidate genes for abiotic stress response and terpenoid and fatty acid biosyntheses. Front Genet 9:505

Chapter 6
Germplasm Resources, Breeding Technologies and the Release of Cinnamon Cultivars 'Sri Wijaya' and 'Sri Gemunu' in Sri Lanka

Sheron N. Weerasuriya and Ranjith Pathirana

6.1 Introduction

6.1.1 Distribution of the Genus Cinnamomum

The plant family Lauraceae to which cinnamon belongs, encompasses about 50 genera and 2500–3000 species of trees and shrubs widely represented in Asian and American tropics, also with large numbers in Australia and Madagascar (van der Werff and Richter 1996). The genetic resources of a cultivated species can be discussed in terms of variability of that species alone or in a broader sense by considering the diversity of related species. It is worthwhile to discuss both these aspects in the case of cinnamon because there are many other species within the genus that are used as cheap substitutes for true cinnamon and also may well be used in future for transfer of valuable genetic traits.

The number of species described in the genus *Cinnamomum* varies from 100 to 341 depending on the taxonomic separators used. These species are spread across Asia, Australia, the Pacific islands and Fiji (Kostermans 1957; Purseglove 1974). The transfer of 68 American neotropical species of *Phoebe* Nees. in the 1960s (Kostermans 1961) substantially increased the number of species in the genus and gave it a pantropical status. Asiatic *Cinnamomum* species are characterised by a swollen pedicel and a fruit cup with remnants of tepals. The main identifying features of *Cinnamomum* species, according to Kostermans (1980), are the length of

S. N. Weerasuriya
National Cinnamon Research and Training Center, Department of Export Agriculture, Thihagoda, Sri Lanka

R. Pathirana (✉)
Plant & Food Research Australia Pty Ltd., Waite Campus Research Precinct, Urrbrae, SA, Australia
e-mail: Ranjith.pathirana@plantandfood.co.nz

R. Senaratne, R. Pathirana (eds.), *Cinnamon*,
https://doi.org/10.1007/978-3-030-54426-3_6

the basal and sub-basal ascendant veins, the covering of hairs (indumentum) and the thalamus cup under the fruit.

Cinnamomum cassia Presl. is the major producer of cassia (also called cassia cinnamon and Chinese cinnamon). Indonesian cassia, *Cinnamomum burmannii* Blume., is endemic to Java and Sumatra, and is the other major cultivated species of cinnamon. It is exported mainly to the United States (Krishnamoorthy et al. 1997). *Cinnamomum culitlawan* (L.) Kosterm. is a native of Malucca, Ambon and adjacent islands. Its bark has a clove smell and the oil is rich in eugenol. The cultivation of this species was attempted in India and Malaysia in the nineteenth century without success (Purseglove 1974; Weiss 2002). Northeastern parts of India are home to *Cinnamomum tamala* (Buch.-Ham.) Nees. & Eberm. (Krishnamoorthy et al. 1997), *Cinnamomum bejolghota* (Buch.-Ham.) and *Cinnamomum impressinervium* Meissner (Ravindran and Babu 2004). Commercial Indian cassia is mainly *C. tamala* (Buch.-Ham.) Nees. & Eberm., its leaves are widely used as a spice and its oil has many medicinal uses (Sharma and Nautiyal 2011). *Cinnamomum loureiroi* Nees., native to Vietnam was earlier considered to be exported in large quantities from Vietnam and was coined the names Vietnamese or Saigon cassia. According to Dao (2004), *C. loureiroi* Nees. is a rare species, and what is exported from Vietnam is now considered as *C. cassia*. Because of different harvesting and processing practices in Vietnam to that in China, the product is distinctly different and hence was considered to be derived from a different species (Dao 2004; Ravindran et al. 2004).

Kostermans (1983) described 13 species of *Cinnamomum* that occur in different parts of South India, and Krishnamoorthy et al. (1997) report eight of these to be endemic to the Western Ghats. These include *C. malabatrum* (Burm.f.) Bl., *C. perrottetii* Meissn., *C. riparium* Gamble., *C. keralanese* Kostem., *C. travancoricum* Gamble., *C. sulphuratum* Nees., *C. wightii* Meissn. and *C. heyneanum* Nees. *C. macrocarpum* is found only in the Nilgiris range along with some of the endemic species mentioned for the Western Ghats. Another five species exist in the Western Ghats although they are not endemic to that region (Krishnamoorthy et al. 1997). *C. glaucescens*, occurring in Nepal and *C. deschampsia* Gamble. in Malaysia have also been described as substitutes and adulterants of cinnamon (Weiss 2002).

Cinnamomum camphora has been heavily exploited as a source of camphor in Japan and Taiwan until the Second World War, then introduced to India during the 1950s. Owing to the availability of cheap synthetic camphor, the international demand for the natural form has declined in recent years. The use of *C. camphora* as a source of Ho leaf oil, on the other hand, has expanded in recent years and Chinese Ho wood oil has largely displaced the use of rosewood as a source of natural linalool (Frizzo et al. 2000).

A comprehensive review of morphological and anatomical features as well as biosystematics and nomenclature has been published in a review by Ravindran et al. (2004). Fujita (1967) looked at the classification and phylogeny of 25 *Cinnamomum* species in terms of the chemical constituents of their essential oils and proposed a cubic system of the genus. He also found that *C. tamala* Nees. has constituents of essential oils that are very similar to *C. zeylanicum*.

6.1.2 Cinnamomum *Species Endemic to Sri Lanka*

In addition to being the centre of origin of true cinnamon, *C. zeylanicum* Blume., Sri Lanka is home to another seven Asian cinnamons: *C. citriodorum*, *C. litseaefolium*, *C. ovalifolium*, *C. sinharajaense*, *C. capparu-coronde*, *C. rivulorum* and *C. dubium*. However, none of these are exploited commercially (Sritharan et al. 1993; Pathirana 2000).

6.1.3 Cytology

Chromosome studies undertaken by various workers have revealed that all *Cinnamomum* species contain a somatic number of 24 chromosomes, and hence, the genus has a basic chromosome number of 12 (Kostermans 1957; Sharma and Bhattacharya 1959; Okada and Tanaka 1975; Sritharan et al. 1993). Sritharan et al. (1993) found that fixing the root tips in the morning between 11 and 11.30 a.m. gave a significantly higher number of cells undergoing mitosis and suitable for cytological studies.

6.2 Distribution and Genetic Resources of *Cinnamomum zeylanicum* Blume.

Among the *Cinnamomum* species, a special place is occupied by *C. zeylanicum* Blume. for its delicate flavour and aroma. As a result, European importers have separate standards for Ceylon cinnamon (*C. zeylanicum*) and cassia (*C. cassia*) (CBI 2019).

6.2.1 Origin and Distribution of C. zeylanicum

True cinnamon or commercial Ceylon cinnamon is endemic to Sri Lanka (Kostermans 1983). South India and the Tennesirim hills of Myanmar have also been noted for natural populations of *C. zeylanicum* (De Guzman and Siemonsma 1999) and it is very likely that its introduction to these areas in the past centuries has resulted in natural populations. This is confirmed by the high degree of invasiveness of *C. zeylanicum* in these regions as shown by Fleischmann (1997) and CAB International (2019). Among 107 alien woody species in two of the main islands of the Seychelles, *C. zeylanicum* was confirmed as one of the most prominent invaders, accounting for 18.9% of the total invasions recorded. Endemic bulbul bird (*Hypsipetes crassirostris*) and introduced mynah bird (*Acridotheres tristis*) were the main dispersal agents. Moreover, some of the descriptions of *C. zeylanicum* (*C. verum*) collected from India have been attributed to taxonomic errors (Kostermans 1983). According to CAB International (2019), *C. zeylanicum* is present in 10 Asian countries, including eight states in India, 12 African countries, 9 countries in the Central American and Caribbean Region, 5 South American countries and almost all main islands of the Pacific. It has become an invasive species in several island nations (CAB International 2019).

C. zeylanicum is a tree of the wet tropics and grows naturally from sea level to 1500 m, but most of the productive plantations occur from sea level to 500 m in Sri Lanka and somewhat higher in India. It is in these areas that the natural diversity of *C. zeylanicum* can also be observed. In Sri Lanka, the areas of diversity coincide with humid, wet areas where forests have been cleared for plantation agriculture, so the natural populations of most of the *Cinnamomum* species are threatened (Pathirana 2000), although establishment of cinnamon plantations in the low and mid-country wet zone of Sri Lanka in the period 1767–1770 by the Dutch (Weiss 2002) helped to prevent further erosion of genetic resources through forest harvest.

An average rainfall of 2000–2500 mm is optimum in the major production areas: Sri Lanka, India and Seychelles (Purseglove 1974; Weiss 2002). Ranatunga et al. (2004) considered 1250–2500 mm rainfall with temperatures of 20–30 °C as optimal for production of quality cinnamon in Sri Lanka.

6.2.2 *Germplasm Resources* of C. zeylanicum *Blume.*

Despite the importance of cinnamon as a spice from ancient times, only a limited effort has been made to scientifically describe the large genetic variation that has been generated in this species through natural evolution and through selection in the past few centuries of cultivation. Compared with other tree spices such as clove and nutmeg, cinnamon is more genetically diverse (Krishnamoorthy et al. 1997). Nevertheless, with less than three centuries in cultivation since the Dutch established the first plantations in Sri Lanka and the long periods between replanting, cinnamon can be considered to be in its early stages of domestication. Considering cinnamon's value as a spice crop, it is logical that the intraspecific classification in Sri Lanka has been based more on organoleptic characteristics than on morphological characters. The vernacular names in Sinhalese identify different types as *Peni Miris Kurundu* (sweet & hot cinnamon), *Thitta Kurundu* (bitter cinnamon), *Peni Kurundu* (sweet cinnamon), *Naga Kurundu* (snake cinnamon), *Veli Kurundu* (sandy cinnamon implicating the granular nature), *Sevel kurundu* (slimy cinnamon) and *Kahata Kurundu* (acrid cinnamon) (Anonymous 1996). However, none of these types can be clearly identified in cultivated cinnamon at present (Wijesinghe and Pathirana 2000).

Research and development of cinnamon began in post-independence Sri Lanka by the Department of Minor Export Crops (the current Department of Export Agriculture—DEA), but systematic study of the germplasm was made possible only after the National Cinnamon Research and Training Center (NCRTC) (Formerly Cinnamon Research Station) opened at Thihagoda in the Matara District in southern Sri Lanka where more than 30% of the national cinnamon crop is produced. Cinnamon production in the Matara District (6035 Mt) comes a close second only to the adjacent Galle District (6683 Mt) (Kumari Fonseka et al. 2018).

A systematic evaluation of 11 elite cinnamon progenies was conducted by Perera (1984), which led to the identification of W-37/3, W-16-1 and W-17/9 as lines with high bark quality. As a result of a field survey for elite material conducted by the NCRTC in the 1980s, 700 superior accessions were identified from the growing areas. The selections were based on morphological characters and pungency of the leaf petiole (Wijesinghe and Pathirana 2000). During the second phase of the germplasm enhancement programme, 400 superior selections were established from green-wood cuttings at the NCTRC. Screening of the collection for characters based on a selection index began in 1996. The characters considered for indexing included the total number of branches, number of harvestable branches, erectness of stem, susceptibility to major pests and diseases, peeling ability, bark pungency and dry weight/fresh weight ratio (Wijesinghe and Pathirana 2000—Table 6.1).

One hundred superior genotypes selected from this study were then screened for bark oil content. Eight lines recorded more than 3% bark oil on a dry weight basis, compared with a mean of 1.68% in the 100 lines tested, with a range of 0.49–3.70% (Wijesinghe and Pathirana 2000—Table 6.2). Ten superior lines from this study were established in multi-location yield trials in the final stage of the screening process. As the outcome of long-term evaluation of these 10 lines in multi-location trials, two lines were released as superior cinnamon cultivars for the first time in 2009 in Sri Lanka, called 'Sri Gemunu' and 'Sri Wijaya'. The variety 'Sri Gemunu' has a distinctive sweet and pungent flavour, and an aroma characteristic of Ceylon cinnamon, produces average bark yield of 1200–1300 kg/ha/year, and has high bark oil content (3.25–3.6%), leaf oil (3.6–3.9%) and cinnamaldehyde content (80–83%) of bark oil. Variety 'Sri Wijaya' produces high bark yield of 1600–1800 kg/ha/year with a higher content of eugenol (88–92%) in leaf oil. These varieties are clonally

Table 6.1 Selection indices developed by the NCTRC for selecting superior cinnamon genotypes

Character	Total score
Total stems	10
Number of harvestable stems	5
Erectness of stem	15
Major pests	10
Major diseases	10
Peeling ability	10
Pungency of the bark	10
Yield ratio: dry weight/(fresh weight × 100)	30
Total	100

Table 6.2 Bark oil percentage of 20 superior selections after screening 100 clones in the Matara District, Sri Lanka

Clone ID	Bark oil (%)	Clone ID	Bark oil (%)
VP-241	3.7	VP-110	2.6
VP-126	3.6	VP-218	2.56
VP-203	3.5	VP-219	2.43
VP-307	3.19	VP-098	2.41
GACG 1-1	3.1	VP-340	2.34
VP-179	3.09	VP-028	2.3
VP-308	3.07	VP-049	2.24
VP-334	3.05	VP-161	2.24
VP-116	2.7	VP-085	2.17
VP-05	2.65	VP-186	2.12

Mean of 100 clones tested—1.68%; range—0.49–3.7%

Fig. 6.1 Cultivars 'Sri Gemunu' (left) and 'Sri Wijaya' (right) released for cultivation in Sri Lanka by the Department of Export Agriculture as a result of clonal selection

propagated and distributed to growers. The morphological features of two released varieties in Sri Lanka are shown in Fig. 6.1.

Wijesinghe and Pathirana (2000) have shown the danger of genetic erosion of this indigenous crop because of the development and spread of vegetatively propagated elite clonal material. To ensure the conservation of diverse genetic resources, they proposed the establishment of isolated seed gardens of cinnamon in the Intermediate and Wet Zones of Sri Lanka. The seed-derived plantations established over the past few centuries in Sri Lanka, initially from elite selections by the growers themselves and subsequently by the Department of Export Agriculture, are a valuable genetic resource for the further improvement of cinnamon. Further sampling from these populations for the genetic collections at the NCTRC will add value to this resource.

The other major collection of true cinnamon has been established at the Indian Institute of Spices Research (IISR) at its experimental farm in Peruvannamuzhi in Kerala State where most of India's cinnamon is grown (Krishnamoorthy et al. 1996). This collection consists of about 300 accessions of *C. zeylanicum* and another 64 wild *Cinnamomum* species (Krishnamoorthy et al. 1988, 1997). Accessions from the commercially important *C. zeylanicum* are derived mainly from selected plants of cinnamon estates of Ancharakandi in Kannar District and Mangalamcarp Estate in Wynad, both in Kerala. Plants in these oldest cinnamon estates were established from seeds brought from Sri Lanka by the British in the nineteenth century (Ravindran et al. 2004). Another 105 Sri Lankan progenies from 13 accessions were added to this collection in 1970 and 1979 (Krishnamoorthy et al. 1997). Thus, the two major collections in India and Sri Lanka have a common ancestry. Nevertheless, selection within the original heterogeneous collections under local conditions over several generations of cultivation has resulted in genetic variation for agronomic characters, including tolerance to various environmental factors. Thus, for example, there are distinct populations of cinnamon adapted to the drier conditions of the Intermediate Zone of Sri Lanka and for the higher rainfall areas in the Wet Zone.

In India, testing of 12 Sri Lankan progenies introduced in the 1970s along with another 291 superior lines selected at IISR resulted in the release of Sri Lankan line SL 53 and Indian line IN 189 for cultivation with the names 'Navashree' and 'Nithyashree', respectively (Krishnamoorthy et al. 1996). According to Peter and Nirmal Babu (2009), five high yielding varieties of cinnamon have been recommended for release.

6.3 Genetic Studies on Agronomic and Quality Traits of *C. zeylanicum* Blume.

Most of the genetic studies in cinnamon have focused on the variability for characters that contribute to the yield or quality of the bark or oil. Even morphological and anatomical studies have been directed mainly towards the identification of structures related to oil biosynthesis. For instance, a study conducted at NCRTC to investigate different morphological features of *C. zeylanicum* in relation to yield and quality of bark and leaf revealed seven different leaf types on the basis of leaf morphology and a significant positive correlation between bark yield and leaf characters. The trees with large round leaves showed high bark yield while plants with inwardly curved leaves showed higher cinnamaldehyde percentage in bark oil. High leaf oil content was found in trees with small round leaves (Wijesinghe and Gunarathna 2001). However, the mode of inheritance of such characters has not been published, owing to limited or no crosses, the heterozygous nature of the crop and, being a tree crop, the long time required for each generation. Nevertheless, few studies of the genetic variation in agronomic characters have been conducted.

6.3.1 Genetic Variation Within C. zeylanicum

Krishnamoorthy et al. (1992) analysed nine characters for variation and correlations in a collection of 71 cinnamon accessions from India and Sri Lanka maintained at Peruvannamuzhi, India. A high level of variability was observed for the dry and fresh weight of the bark (8–305 g with a mean of 64.7 g and 30–840 g with a mean of 207 g, respectively), the oleoresin content of the bark (1.3–20.0% with a mean of 8.5%) and the oil content (0.72–4.8% with a mean of 2.0% for leaf oil and 0.5–3.85% with a mean of 1.81% for bark oil). Recovery of bark was also highly variable, ranging from 10.7% to 80% with an average of 32.1%, and the data were skewed towards lower recovery rates. Leaf length, breadth and leaf size index (derived from length and breadth) were not as variable as the yield parameters, probably because selection for yield may have indirectly selected for an optimum leaf size. Correlations were performed on these data, and as expected, the highest correlation was between dry and fresh weights of bark. There were significant positive correlations between dry weight of the bark and leaf oil percentages, and a negative correlation between bark oil and leaf oil. The authors used data from the collection of varieties, hence these correlations can be categorised as genotypic and will be useful in planning selection. However, the results are based on the measurements from a single plant of each of the accessions and need to be interpreted cautiously. Similar results were shown while evaluating ten clonally propagated cinnamon accessions with elite characteristics selected from germplasm collection at NCRTC, Sri Lanka. Bark yield, oil percentage of bark and leaf, cinnamaldehyde percentage in bark oil and eugenol percentage in leaf oil, showed highly significant differences among the selections.

Krishnamoorthy et al. (1991) studied the progeny performance of nine lines and found a significant variation in plant height, number of branches per tree, fresh and dry bark weight and bark recovery. Considering the heterozygous nature of cinnamon and its pollination behaviour, high variability can be expected among progenies even from the same plant after controlled hybridisation, and should be taken into consideration in breeding programmes.

Joy et al. (1998, as quoted by Ravindran et al. 2004) studied leaf oil yield in a collection of cinnamon accessions in Odakkali, India and found a variation of 36.5–294.7 mL/tree/year. However, oil content is highly dependent also on the weather conditions and therefore multi-year, multilocation studies would reveal the genotype × environment interaction useful for selecting high oil genotypes. In a collection of 239 cinnamon accessions Krishnamoorthy et al. (1988) found 55.6% plants to have green flush colour and the others to be purple. A significantly higher bark oil percentage (1.84%) was recorded in varieties with purple flush colour than in varieties with green flush (1.43%). There were no significant differences in bark oleoresin content or leaf oil percentage between the two groups.

Analysis of morphological characters from 47 cinnamon accessions collected within Matara district, the second largest cinnamon cultivation district in Sri Lanka, resulted in nine groups indicating variation for leaf and bark characters (Azad et al.

2016). Although the authors used 14 characters in their study, 11 of those were related to leaf morphology. The same authors have studied the morphology of a segregating population from two plants selected in a plantation and reported a high variability in leaf characters (Azad et al. 2015), as expected in a highly heterozygous species.

6.4 Interspecific Variation in *Cinnamomum*

Variation has been extensively studied in relation to the taxonomy of different *Cinnamomum* species. Balasubramanian et al. (1993) used petiole anatomy to characterise four species. They found differences in shape of the petiole and the vascular strands in *C. camphora*, *C. zeylanicum*, *C. macrocarpum* and *C. tamala.* Using 20 morphological characters in a cluster analysis involving eight *Cinnamomum* species, Ravindran et al. (1991) showed that *C. malabatrum* is closely related to *C. zeylanicum.* The species *C. camphora*, *C. perrottetii*, *C. riparium* and *C. cassia* were very distinct types and formed independent clusters.

More recently, Ariyarathne et al. (2018) conducted a study on morphological and chemical characteristics of two selected true cinnamon accessions ('Sri Wijaya' and 'Sri Gemunu') and six wild species of the genus *Cinnamomum* (Schaeff.) endemic to Sri Lanka, that is, *C. citriodorum*, *C. capparu-coronde*, *C. litseaefolium*, *C. revulorum*, *C. dubium* and *C. sinharajaense*. They examined morphological characters of flower, leaf, seed and bark and bark yield and leaf oil content, as well as the chemical characters of leaf and bark oil. A dendrogram (Fig. 6.2) was constructed and a diagnostic key (Fig. 6.3) was developed based on oil yield, and morphological and chemical characters. The analysis revealed that the studied species could be

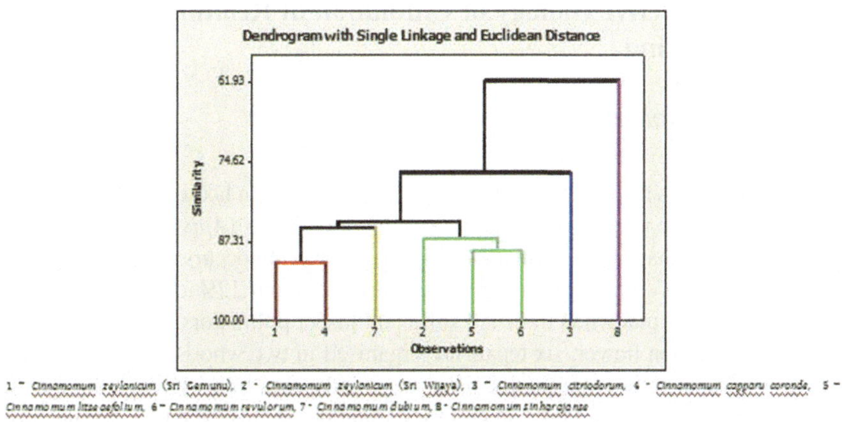

Fig. 6.2 Dendrogram developed by Ariyarathne et al. (2018) illustrating the relationship between six *Cinnamomum* species endemic to Sri Lanka and two *C. zeylanicum* cultivars released by the Department of Export Agriculture, Sri Lanka

1. Pinnate venation	*C. citriodorum*
1. Three veined venation with prominent midrib and two lateral veins	
2. Eugenol present in leaf oil	
3. Cinnamaldehyde absent in leaf oil	*C. sinharajaense*
3. Cinnamaldehyde present in leaf oil	
4. Leaf lamina shape - oval	
5. Leaf apex acute	*C. liseaefolium*
5. Leaf apex ovate to elliptic	*C. zeylanicum* ('Sri Gemunu')
4. Leaf lamina shape - lanceolate	
5. Leaf apex acute	*C. zeylanicum* ('Sri Wijaya')
5. Leaf apex acuminate	*C. capparu-coronde*
2. Eugenol absent in leaf oil	
3. Contains benzyl benzoate in bark oil	*C. revulorum*
3. Benzyl benzoate absent in bark oil	*C. dubium*

Fig. 6.3 Key to *Cinnamomum* species endemic to Sri Lanka including the two cultivars of *C. zeylanicum* released by the Department of Export Agriculture. (Modified from Ariyarathne et al. 2018)

grouped into five clusters. *C. sinharajanse* was entirely different from the rest of the tested species and formed a separate cluster. 'Sri Gemunu' and 'Sri Wijaya' being improved accessions of *C. zeylanicum* exhibited less variability in leaf size, colour, fragrance etc. but showed high variations in shape and the apex of the leaf (Ariyarathne et al. 2018).

A comparison of leaf morphology of the two released cinnamon cultivars and the cinnamon species endemic to Sri Lanka is presented in the Fig. 6.4.

6.5 Reproductive Biology of Cinnamon in Relation to Breeding

6.5.1 *Morphology of the Floral Organs*

Cinnamon flowers are produced on long green peduncles in lax terminal or axillary panicles. They are insect pollinated, mostly by honey bees (*Apis cerana*, *A. corea*, *A. dorsata*) in India (Mohankumar et al. 1985). Flowering occurs during November to March in most cinnamon-growing countries and about 12% of open-pollinated flowers produce fruits when there are sufficient insect pollinators.

In the cinnamon flower, six tepals are organised in two whorls with three tepals per whorl. It has a trimerous androecium, essentially with four whorls, although the outer whorl is reduced and represented by staminodes. Fertile stamens in cinnamon show valvular dehiscence (Ravindran et al. 2004). The ovary is superior, unilocular with a single pendulous anatropous ovule and a long style. The stigma is dry and papillate (Heslop-Harrison and Shivanna 1977).

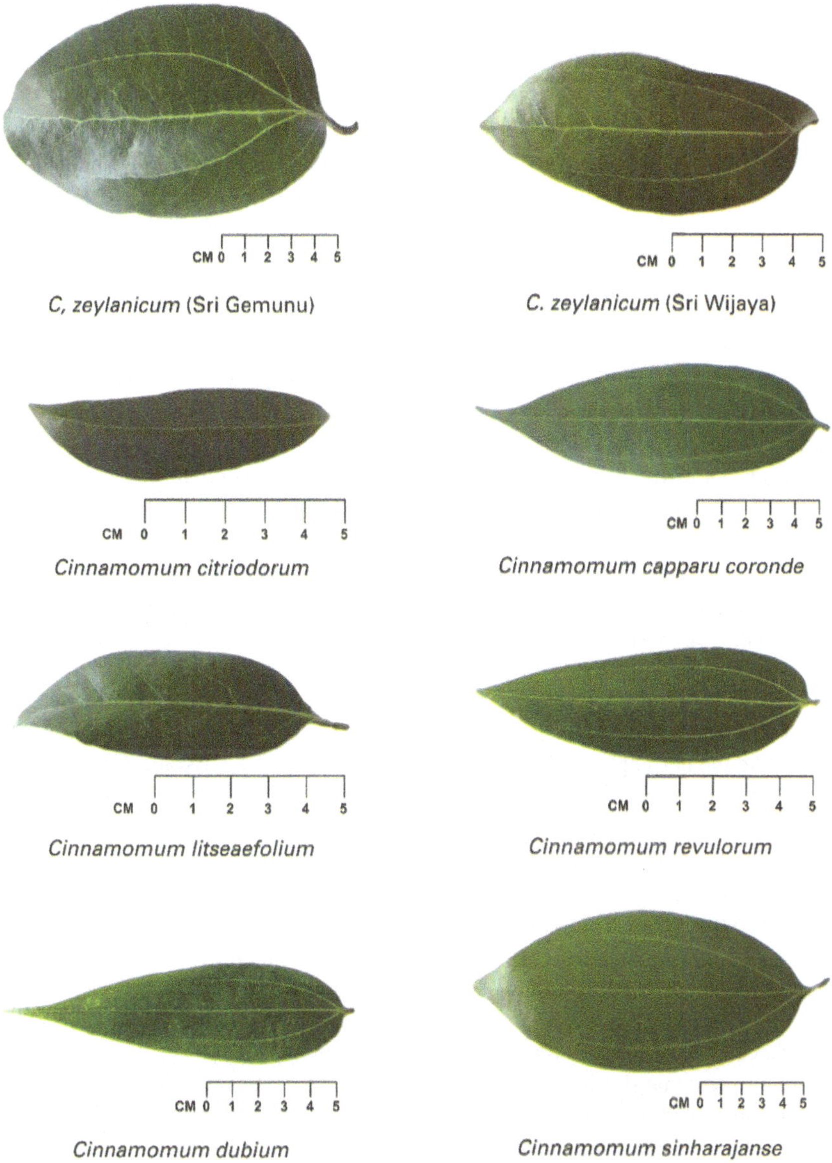

Fig. 6.4 Leaf morphology of cinnamon species native to Sri Lanka in comparison with two released cultivars of *C. zeylanicum*

6.5.2 Synchronised (Protogynous) Dichogamy as a Mechanism Ensuring Cross-Pollination

Many species of the Lauraceae show dichogamy, both in old and new world taxa (Kubitzki and Kurz 1984). A conspicuous trait of the floral biology of Lauraceae is a very pronounced protogyny, which Kubitzki and Kurz (1984) described as a prerequisite to the phenomenon of synchronous dichogamy, in which each flower opens twice, either in the morning or in the afternoon of the first day, and the next day either in the afternoon or in the morning, respectively. This phenomenon has also been confirmed in *C. zeylanicum* in Sri Lanka (Wijesinghe and Pathirana 2000). In Sri Lanka, flowering season starts in November and lasts till early March. In cinnamon, it is now known that the maturing of female phase precedes the male phase by about 24 h ensuring outcrossing, thus this phenomenon is identified as protogynous dichogamy.

Synchronous dichogamy also occurs in avocado (*Persea americana*), the other economically important species of Lauraceae (Sedgley and Griffin 1989). Synchronous dichogamy in *P. americana* requires the presence of two morphs, whose flowering has to be complimentarily synchronised. Pollen transfer is possible only between individuals belonging to the two morphs, providing that rain or other disturbing factors do not lead to an asynchrony of flower opening. Thus, synchronised dichogamy represents a kind of temporal dioecism. Periods of light and dark have been implicated in the control of synchronous dichogamy of avocado (Sedgley 1985).

Unlike the hermaphroditic Amazonian taxa of Lauraceae, Asiatic *Cinnamomum* species produce ample nectar from the glands flanking the inner stamens and the staminodes corresponding to the fourth staminal whorl. In the female phase of flower opening, the outer stamens lie pressed against the petals, and the stigma appears brilliantly white and receptive. In the intermediate phase, the petals close and the stigma starts wilting. First, dark spots appear on the stigma until it becomes discoloured and is no longer receptive. In the second phase of flower opening, stamens stand up prominently in the centre of the flower and shed their pollen. Pollen remains attached to the inner surface of the valves, which are by now bent upwards. More nectar is produced during the second opening. All flowers of a given tree maintain the same diurnal rhythm of development, in that flower opening occurs either in the evening and the following morning, or in the morning and the evening of the following day. This excludes any possibility of pollination between flowers of the same tree. The phenomenon can be considered as the forerunner of dioecy present in some Amazonian taxa of Lauraceae described by Kubitzky and Kurz (1984). Both morphs appear to occur randomly in any given population of cinnamon, even within the same seed-propagated variety.

6.5.3 Controlled Pollination

Although the synchronised dichogamy described above ensures cross-pollination, it is customary among breeders to emasculate flowers the day before controlled pollinations are made. For success in controlled pollination, it is essential to select plants from two morphs in which androgenesis coincides with gynogenesis. Although pollinations can be made with preserved pollen or by artificially creating conditions for asynchrony of flowering so that pollen from flowers of the same morph can be used, it is possible that other incompatibility systems may prevent fertilisation success. Furthermore, during high humidity, plants that shed pollen in the afternoon may still be shedding viable pollen the following morning. During rainy days, asynchrony may even lead to the simultaneous availability of pollen and receptive stigmas on one plant. Attempts by Kubitzky and Kurz (1984) and Wijesinghe and Pathirana (2000) to achieve such geitonogamous pollinations failed, thus demonstrating the existence of other fertility barriers between morphs. Weerasuriya et al. (2016) were successful in hybridising 'Sri Gemunu' (Type A variety) with 'Sri Wijaya' (Type B Variety). After artificially pollinating 250 flowers taking into consideration the active period of the male and female stages, they produced 46 F_1 seedlings. The evaluation of this hybrid population is ongoing (Weerasuriya et al. 2016).

For emasculation, the tepals can be opened up with a pair of forceps while holding the base of the flower with the other hand; the stamens are then carefully detached. After emasculation of selected flowers from an inflorescence, the other flowers should be removed and the inflorescence covered by a bag to prevent any uncontrolled pollination. On the next day, pollen from the male parent can be directly transferred onto the stigma of the female parent, either by using detached male flowers or a soft brush to collect the pollen. Cinnamon pollen clump on the opened valve of the pollen sac, so it is relatively easy to pollinate selected plants using detached flowers of the male parent. Alternatively, anthers can be separated from the rest of floral parts before dehiscence and pollen can be extracted from freshly opened flowers into containers.

Pollen viability should be assessed by staining tests or by in vitro germination, especially if stored pollen is being used for pollination. Popular stains for this purpose include acetocarmine or iodine. For in vitro germination tests an agar solidified 28% sucrose solution containing 20 ppm boric acid and about 2.5 mL yeast extract may be used. Selective crossings may also be achieved by growing the plants of separate morphs of two selected parent varieties in isolation, but there is no scientific information on how to select these before flowering. Therefore, this may be achieved by growing a number of plants from the two varieties selected for crossing in an isolated plot, observe the flowering behaviour when they reach the reproductive stage and remove all plants of one morph from the first variety and all plants of the other morph of the second variety to ensure coincidence of gynogenesis of one variety with the androgenesis of the second.

6.6 Handling Hybrid Seeds and Populations

Once the fruit matures, the pulp should be removed from freshly collected berries and the seeds extracted, washed and air-dried. Uneven maturity has to be considered when harvesting berries. As with many tropical tree species, cinnamon is recalcitrant (Pathirana 2000) and therefore seeds have to be sown for germination without delay. Samaraweera (2000) reports that without special treatment, germination of seeds can drop to 40% within a period of 3 weeks after harvest, and a complete loss of viability can occur by the seventh week. However, in our recent experiments, we found that air-dried seeds stored in airtight bags, treated with 0.05% Folicure® solution can maintain a viability of above 80% for 4–6 months (Weerasuriya et al. unpublished). Seed size also seems to affect seed germination and seedling growth, with larger seeds recording higher germination percentages and improved seedling vigour (Samaraweera 2000).

A moist seedbed in a screen house results in better germination than in the open field. Seedlings 8–10 months old can be transplanted into unprotected fields during the rainy season. Field establishment is 80–90% when seedlings are raised in polythene bags compared with 60–70% establishment for seeds germinating in a seedbed prior to transplanting in the field (Abeykoon 2000). Therefore, seed from cross-pollinations should be raised in bags before transplanting into the field.

Seed progeny of each cross should be planted in separate plots and evaluated for germination and vigour of growth. The crosses can be handled in different ways in tree species such as cinnamon, depending on the breeding plan. They can be individually assessed for the characters of interest, and promising plants can serve as parents for establishing clonal progenies for further evaluation for release as varieties. This approach allows quicker release of varieties. Maintaining records of yield and other agronomic characters in parents and progeny of different combinations of crosses allows the identification of varieties with good combining ability for use in the future. On the other hand, if the objective of the breeding programme is to combine superior characters of two selected varieties, then it is more efficient if the progenies are grown in isolation, the inferior plants culled, the superior plants allowed to flower and seeds collected for the second and subsequent generations. In each generation, only the plants with superior characteristics should be allowed to produce seed.

6.7 Implications of the Juvenile Period and Vegetative Propagation in Cinnamon Breeding

6.7.1 Juvenile Period in Relation to Breeding of Cinnamon

Most long-term tree breeding programmes are based upon several generations derived from controlled crosses. The selected superior plants can serve as candidates for variety release while serving at the same time as parents for producing the next seedling generation. The genetic progress of a cinnamon-breeding programme would depend on the length of the juvenile period of each generation, and approaches to shorten the juvenile period would considerably increase the genetic gain.

Zimmerman (1972) showed a correlation between plant height and flowering in fruit trees, which may be related to a minimum number of nodes being required to overcome the juvenile period. Hence, as in fruit crops, heavy fertilisation enhancing vigorous growth may help in achieving early flowering in cinnamon too. Another approach is to train the tree to a single stem in dense plantations as is used for chestnut (Zimmerman 1971). Agronomic and physiological approaches to shorten the juvenile period, specifically focused on cinnamon, would certainly help in long-term breeding.

6.7.2 Vegetative Propagation in Relation to Breeding of Cinnamon

Almost all existing cinnamon plantations have been established from seedlings, mainly from selected parent plants but not from seedling progenies of controlled pollinations. In this regard, work by Haldankar et al. (1994) is unique. The researchers initially used 300 selected seedlings from Sri Lankan cinnamon collection at IISR in Calicut for screening under coconut and then vegetatively propagated the best four plants to establish a yield trial in Maharashtra State, India. Breeding line B-IV gave the highest bark yields for both fresh weight (230 g) and dry weight (84.5 g). Their approach was to screen the seedlings on the basis of organoleptic properties and use selected seedlings for further screening for yield and chemical composition of the oil. As a result, line B-IV isolated has a high percentage of cinnamaldehyde (70.2%) and eugenol (6.9%) in the bark oil. This line was later released as Konkan in the Konkan region of Maharashtra State (Ravindran et al. 2004).

Seedling progenies of the elite lines already developed in Sri Lanka (Wijesinghe and Pathirana 2000) and India (Krishnamoorthy et al. 1996) may serve as useful resources for further screening for elite material in their progenies. Parental lines for the development of elite populations can then be identified using data on yield and chemical composition of bark and leaf oil of individual plants from heterogeneous populations.

The recent successes in vegetative propagation of cinnamon both in Sri Lanka (Dayatilake 2000; Samaraweera 2000) and described in Chap. 7 of this book) and in India (Benerjee et al. 1982; Hegde et al. 1989; Ranawere et al. 1994, 1995; Thirunavoukkarasu 1996) have resulted in the possible production of clonal progeny from selected elite plants. Unfortunately, there are no reports so far published on the field performance of vegetatively propagated material in cinnamon, except for the study by Haldankar et al. (1994) on four selected clones. Unlike other plantation crops such as tea, cinnamon is grown by smallholders with minimal inputs

(Kularatne 2000). The plants have a long productive period in the field and because of a prolific root system are able to survive and yield under adverse weather with minimum inputs. Moreover, the populations have good tolerance to biotic stresses as no outbreaks of diseases or pests have been reported. In other crops where production is based on clonal material, for example, coffee, significant disease outbreaks have already occurred in Sri Lanka. Therefore, it is essential that the diversity of cinnamon be maintained, even under clonal propagation, by the release of several clones of diverse origin. Disease and pest build-up need to be monitored carefully and the performance of clonal material should be compared with improved seedling populations before converting already established seedling plantations into clonal ones.

6.8 Breeding Strategy and Objectives

Cinnamon growers in the past have consciously or unconsciously attempted to develop favourable populations in their fields by selecting superior phenotypes from natural populations. This grower-driven mass selection has resulted in the current diversely adapted populations of cinnamon in different parts of Sri Lanka and other cinnamon growing countries today. Faster gains in desirable characteristics can be expected if more sophisticated and targeted breeding approaches are adopted. Experiences from the more advanced perennial crop breeding programmes can be used in developing the strategies for cinnamon breeding.

6.8.1 The Need for Long-Term Planning, Funding and Collaboration

Perennial crop breeding is particularly costly in terms of land, labour, time and infrastructure facilities. It is therefore essential that short- and long-term breeding plans are well thought-out and executed, and supported by assured long-term funding. In this regard, partnership arrangement between central government, provincial agencies and private grower associations should be an essential feature. Obviously, the nature and scope of the breeding programme should depend on the needs and demands of the cinnamon industry. Encouraging the growers to contribute financially to the breeding effort will give them ownership of the programme and at the same time the breeders will have a source of long-term funding. However, private funding should not be allowed to hijack the long-term breeding objectives by overemphasising frequent variety releases as experienced in private breeding programmes.

The relatively high cost per plant due to larger size and longer generation time of cinnamon will be compensated to a great extent by resorting to clonal propagation for the mass production of elite material in the future. Nevertheless, a faulty strategy can nullify all breeding efforts and therefore proper scientific planning of the breeding programme is of utmost importance. Control of pollination is crucial and should be planned from the very outset; without such controls an entire 5–8-year breeding cycle can be lost. The information generated by a programme will be of no use for future breeders unless the characteristics of elite parental plant types can be described and documented. With more sophisticated molecular tools becoming available for breeders, an attempt must be made to enter into long-term collaboration with a laboratory that has capabilities in molecular markers.

6.8.2 Cultivated and Wild Gene Pools

Another important aspect of any breeding plan is the maintenance and use of a diverse gene pool that is known to contain desirable phenotypes. In this respect, Sri Lankan breeders have the luxury of the cultivated and wild gene pools of *C. zeylanicum*. The cultivated crop has a higher gene frequency of desirable traits compared with natural populations and therefore a very strong justification will be needed if extensive use of the wild gene pool is contemplated. If traits such as disease and insect pest resistance are to be emphasised, then the parents carrying these traits should first be screened for the overridingly important characteristics, such as bark and oil yield and quality to justify their inclusion in the breeding programme.

Moreover, clonal propagation represents monoculture in its most extreme form, with a single genotype spread over a wide area. With the two major cinnamon producing countries, Sri Lanka and India, both at the brink of making extensive use of clonal propagation for plantation establishment, the specific resistance of the crop will be negated by the natural selection of the pathogen or the pest unless the resistance was selected for durability. When deciding on conversion of seedling plantations to clonal, advantage of monocropping for the product quality will have to be evaluated against maximising the vulnerability of the crop to biotic and abiotic stress factors. The experience of the tea and rubber industries will be useful in this regard. Unlike rubber and tea, cinnamon is indigenous to Sri Lanka and plantations exist side by side with natural populations. Hence, there is a constant influx of pathogens and pests specific to cinnamon from the natural populations. It is therefore important to embark on a multiclonal strategy from the very outset, even within the same plantation.

With oil quality being an important aspect of breeding, there needs to be good analytical facilities with trained staff to support the breeding programme. In this regard, identifying the genotypes with a high frequency of genes for particular chemical traits and using them with high bark and oil yielding parents will help in combining genes for quality and yield.

6.9 Unconventional Breeding Methods

6.9.1 Interspecific Hybridisation and Embryo Culture

There are no reports on attempts to produce interspecific hybrids of *Cinnamomum* species. *C. zeylanicum* has the best quality among the different species and apparently does not need the infusion of genes from other related species for improvement. A similar argument applies to disease and pest resistance. Close relationships that seem to exist between *C. zeylanicum*, *C. tamala* and also some species indigenous to Sri Lanka (see Chap. 4 of this book for details), should allow successful interspecific hybridisation but this would be more useful in understanding taxonomic relationships than in developing new cultivars.

6.9.2 Induced Mutations

Induced mutations have been used widely to improve perennial fruit species, and the availability of vegetative propagation techniques should allow fixing of any useful mutants in a mutation breeding programme. There have been no serious attempts to improve cinnamon through mutation induction techniques and there are no reports on the effects of radiation on seeds or vegetative organs. Therefore, establishing growth reduction curves is a prerequisite (Pathirana 2011) before initiating mutation breeding programmes in cinnamon.

With the development of techniques for clonal propagation, induced mutations will become increasingly more valuable for improvement of cinnamon, because this technique allows the improvement of a single trait without affecting the other characters, as no hybridisation steps are involved (Pathirana 2011). Therefore, as a first step, the development of protocols including dose response for different mutagens will help future breeding using this technique.

6.9.3 In Vitro Techniques for Genetic Improvement

There are several reports on successes in tissue culture regeneration in cinnamon (Rai and Jagdishchandra 1987; Babu et al. 1998; Dayatilake 2000). One possible use of this technology would be crop improvement through in vitro mutagenesis as well as somaclonal variation. Tissue and cell culture techniques may find other applications in cinnamon breeding, such as in distant hybridisation through embryo rescue, induction of polyploidy and in-plant genetic transformation.

6.9.4 Molecular Genetic Approaches

Molecular genetic approaches have the potential to increase the efficiency of cinnamon breeding, understanding evolution of the genus and the genetic relationships among species. First, mapping populations involving different variety groups need to be developed. Thereafter, the construction of high density (HD) maps by developing microsatellite (SSR) and/or single nucleotide polymorphic (SNP) markers, building a multi-parent consensus map of cinnamon by comparative quantitative trait loci (QTL) analyses, and developing user-friendly molecular marker sets for future practical applications, can follow. These individual populations and the HD maps can be used to perform QTL analyses on agronomic characters, including growth, flowering, yield, oil content and quality. If several identical characters are evaluated in different genetic backgrounds, it will be possible to compare QTL locations in several genetic backgrounds. However, traditional methods for developing molecular markers are costly, time consuming and laborious. Considering the increasing affordability of the use of next generation sequencing (NGS) platforms, marker development based on high throughput sequencing for crops such as cinnamon where molecular data are limited, will be more efficient. The advent of NGS has allowed the development of SSR markers in species such as cranberry for which molecular genetics data were lacking (Zalapa et al. 2012). NGS technologies have also facilitated the development of novel SNP platforms for genotyping by sequencing (Bhat et al. 2016). For the implementation of such highly desirable DNA marker-based breeding strategies for cinnamon, it is important for the two cinnamon improvement groups in India and Sri Lanka to collaborate and to find partners in laboratories with facilities for molecular biology research. Continuous funding of such research is essential for any practical outcome to be anticipated.

6.10 Conclusions

Despite being the oldest known spice, cinnamon has been in cultivation for less than three centuries since the Dutch established the first plantations in Ceylon (Sri Lanka) in 1765. The crop is still in the early stages of domestication. There is appreciable genetic variation in agronomic and quality traits that could be used in breeding programmes. Plant breeding programmes with limited genetic collections have been established in Southern Sri Lanka and Kerala State, India. However, the crop genetic diversity still exists largely in the plantations established using seedlings, in growers' lands, and also in closely related indigenous species in the wild. Therefore, further collection and establishment of *ex situ* germplasm collections is important, especially with the release of new cultivars which are being clonally propagated and established, replacing seed-derived plantations. Establishment of isolated seed gardens from selected elite materials could also help to conserve the genetic diversity of cinnamon. Limited success in controlled pollination of elite material has been reported and the progenies are under evaluation at the National Cinnamon Research and Training Center, Sri Lanka. Synchronous (protogynous) dichogamy exhibited by the genus makes controlled pollinations difficult and therefore most of the elite material developed to date is based on progenies derived from open pollination of selected plants. This method commenced in the 1980s at the National Cinnamon Research and Training Center, Sri Lanka, resulting in the release of two cultivars, 'Sri Gemunu' and 'Sri Wijaya'. Agronomic and physiological studies leading to shortening of the juvenile phase and the availability of quick and reliable analytical facilities for oil quality are important cornerstones in improving the cinnamon crop. Advanced genetic techniques, such as molecular marker-based selection, need to be introduced into the breeding programmes to identify useful progeny and reject the undesirable ones early. High throughput next generation sequencing platforms should be considered as molecular data are limited. Considering the long-term nature of breeding programmes with this perennial crop, careful planning, availability of funds in the long term and collaboration among scientists of relevant disciplines and grower associations are important.

References

Abeykoon AMD (2000) Improved agronomic practices of cinnamon. In: Senaratne R, Dayatilake GA (eds) Revitalization of the cinnamon industry in Sri Lanka: constraints and remedies. Faculty of Agriculture, University of Ruhuna, Kamburupitiya, Sri Lanka. pp 47–54

Anonymous (1996) Cinnamon: cultivation and processing. Technical bulletin 5. Department of Export Agriculture, Peradeniya, Sri Lanka

Ariyarathne HBMA, Weerasuriya SN, Senarath WTPSK (2018) Comparison of morphological and chemical characteristics of two selected accessions and six wild species of genus *Cinnamomum* Schaeff. Sri Lankan J Biol 3:11–23

Azad R, Senanayake G, Kumara KLW, Ranawaka RAAK, Pushpa-Kumara DKNG, Wijesinghe KGG, Geekiyanage S (2015) Morphological variation within progeny and deviations from mother plant reveal the allele richness in *Cinnamomum verum* germ-plasm: a case study from Deiyandara, Matara collection at the early vegetative stage. Trop Agric Res Ext 18(4):163–167

Azad R, Ranawake RAAK, Senanayaka G, Wasantha Kumara KL, Pushpakumara DKNG, Wijesinghe KGG, Geekiyanage S (2016) Morphological variation of cinnamon (*Cinnamomum verum* Persl) germplasm in Matara District of Sri Lanka. Int J Minor Fruits Med Aromat Plants 2:1

Babu KN, Minoo D, Geetha SP, Samsudeen K, Rema J, Ravindran PN, Peter KV (1998) Plant biotechnology – its role in improvement of spices. Indian J Agric Sci 68:533–547

Balasubramanian A, Jacob T, Savanna S (1993) Petiolar anatomy as an aid to the identification of *Cinnamomum* species (Lauraceae). Indian Forester 119:583–586

Benerjee DP, Chatterjee BK, Sen S (1982) Air layering of cinnamon (*Cinnamomum zeylanicum* Nees.). South Indian Hort 30:272–273

Bhat JA, Ali S, Salgotra RK, Mir ZA, Dutta S, Jadon V, Tyagi A, Mushtaq M, Jain N, Singh PK, Singh GP, Prabhu KV (2016) Genomic selection in the era of next generation sequencing for complex traits in plant breeding. Front Genet 7:221

CAB International (2019) Invasive species compendium; *Cinnamomum verum* (cinnamon). https://www.cabi.org/isc/datasheet/13573#tosummaryOfInvasiveness. Accessed 22 Sept 2019

CBI (2019) Exporting cinnamon to Europe. Centre for Promotion of Imports from Developing Countries, Ministry of Foreign Affairs, The Netherlands. https://www.cbi.eu/market-information/spices-herbs/cinnamon/. Accessed 1 Aug 2019

Dao NK (2004) Chinese cassia. In: Ravindran P, Babu K, Shylaja M (eds) Cinnamon and cassia the genus *Cinnamomum*. CRC Press, Boca Raton, pp 156–184

Dayatilake GA (2000) Possibilities of cloning cinnamon *in-vivo* and *in-vitro*. In: Senaratne R, Dayatilake GA (eds) Revitalization of the cinnamon industry in Sri Lanka: constraints and remedies. Faculty of Agriculture, University of Ruhuna, Kamburupitiya, Sri Lanka. pp 26–37

De Guzman SS, Siemonsma JS (1999) Plant resources of South East Asia, No. 13, Spices. Backhuys Publ., Leiden

Fleischmann K (1997) Invasion of alien woody plants on the islands of Mahe and Silhouette. Seychelles J Vegetation Sci 8:5–12

Frizzo CD, Santos AC, Paroul N, Serafini LA, Dellacassa E, Lorenzo D, Moyna P (2000) Essential oils of camphor tree (*Cinnamomum camphora* Nees & Eberm) cultivated in Southern Brazil. Braz Arch Biol Technol 43:313–316

Fujita Y (1967) Classification and phylogeny of the genus *Cinnamomum* viewed from the constituents of essential oils. Bot Mag Tokyo 80:261–271

Haldankar PM, Nagwekar DD, Patil JL, Gunjate RT (1994) Varietal screening for yield and quality in cinnamon. Indian Cocoa Arecanut and Spices J. 18:79–81

Hegde KR, Sulikeri GS, Hulamani NC (1989) Effect of growth regulators and pre-girdling treatments on rooting of cinnamon (*Cinnamomum verum* Presl.) air layers. South Indian Hort 37:329–332

Heslop-Harrison Y, Shivanna KR (1977) The receptive surface of the angiosperm stigma. Ann Bot 41:1233–1258

Joy PP, Thomas J, Mathew S, Ibrahim KK (1998) Growth, leaf oil yield and quality investigations in Cinnamon (*Cinnamomum verum*). J Med Aromatic Plants 28:401–406

Kostermans AJGH (1957) Lauraceae. Reinwardtia 4:193–256

Kostermans AJGH (1961) The new world species of *Cinnamomum* Trew (Lauraceae). Reinwardtia 6:17–24

Kostermans AJGH (1980) A note on two species of *Cinnamomum* (Lauraceae) described in Hortus Indicus Malabaricus. In: Manilal KS (ed) Botany and history of Hortus Malabaricus. Oxford IBH, New Delhi, pp 163–167. (Cited from Ravindran et al. 2004)

Kostermans AJGH (1983) The South Indian species of *Cinnamomum* Schaeffer (Lauraceae). Bull Bot Surv India 25:90–133

Krishnamoorthy B, Gopalam A, Abraham J (1988) Quality parameters of cinnamon (*Cinnamomum verum*) in relation to flush colour. Indian Cocoa Arecanut Spices J 12:38–39

Krishnamoorthy B, Rema J, Sasikumar B (1991) Progeny analysis in cinnamon. Indian Cocoa Arecanut Spices J 14:124–125

Krishnamoorthy B, Sasikumar B, Rema J, Gopalam A, Abraham J (1992) Variability and association studies in cinnamon (*Cinnamomum verum*). J Spices Aromat Crops 1:148–150

Krishnamoorthy B, Rema J, Zachariah TJ, Abraham J, Gopalam A (1996) Navashree amd Nithyashree – two new high yielding and high quality cinnamon (*Cinnamomum verum* Bercht & Presl.) selections. J Spices Aromat Crops 5:28–33

Krishnamoorthy B, Sasikumar B, Rema J, George JK, Peter KV (1997) Genetic resources of tree spices and their conservation in India. Plant Genet Res Newsl 111:53–58

Kubitzki K, Kurz H (1984) Synchronised dichogamy and dioecy in neotropical Lauraceae. Plant Syst Evol 147:253–266

Kularatne MAR (2000) Present status of cinnamon industry in Sri Lanka. In: Senaratne R, Dayatilake GA (eds) Revitalization of the cinnamon industry in Sri Lanka: constraints and remedies. Faculty of Agriculture, University of Ruhuna, Kamburupitiya, pp 8–13

Kumari Fonseka DLC, Aluthgamage HN, Wickramaarachchi WWUI (2018) Present situation of cinnamon industry in Southern Sri Lanka. Int J Curr Res Biosci Plant Biol 5(8):63–70. https://doi.org/10.20546/ijcrbp.2018.508.009

Mohankumar GN, Mokashi AN, Narayanaya Swamy P, Prabhakar N, Devar KV, Reddy DNR, Nalawadi UG (1985) Studies on the floral biology of cinnamon. Indian Cocoa, Arecanut & Spice J 8(4):100–102

Okada H, Tanaka R (1975) Karyological studies in some species of Lauraceae. Taxon 24:271–280

Pathirana R (2000) A phytosociological study of angiosperms in lowland wet tropical forests of Sri Lanka. Final report, project no. 7584 of Biodiversity Support Program, World Wildlife Fund. Faculty of Agriculture, University of Ruhuna, Sri Lanka

Pathirana R (2011) Plant mutation breeding in agriculture. CAB Rev Perspect Agric Vet Sci Nutr Nat Resour 6:1–20

Perera KP (1984) Genetic evaluation of some cinnamon (*Cinnamommum verum* Presl.) progenies. M.Phil. Thesis, Postgraduate Institute of Agriculture, University of Peradeniya, Peradeniya, Sri Lanka

Peter KV, Nirmal Babu K (2009) Advances in spice breeding. Paper presented at "Plant Breeding in the Post Genomics Era", Proc Second National Plant Breeding Congress, 1–3 March 2006, Coimbatore, India

Purseglove JW (1974) Tropical crops, dicotyledons. Longman, London

Rai R, Jagdishchandra KS (1987) Clonal propagation of Cinnamomum zeylanicum Breyn. by tissue culture. Plant Cell Tiss Org Cult 9:81–88

Ranatunga J, Senanayake UM, Wijesekera ROB (2004) Cultivation and management of cinnamon. In: Ravindran PN, Babu KN, Shylaja M (eds) Cinnamon and cassia the genus cinnamon. CRC Press, Boca Raton, pp 121–129

Ranawere VS, Nawale RN, Khandekar RG (1994) Effect of rooting media on air layering of cinnamon (*Cinnamomum zeylanicum* Blume). Spice India 7:19–21

Ranawere VS, Nawale RN, Khandekar RG, Magdun MB (1995) Effect of season on air layering of cinnamon (*Cinnamomum zeylanicum* Blume). Indian Cocoa Spices Arecanut J 19:81–84

Ravindran PN, Babu KN (2004) Botany and crop improvement of cinnamon and cassia. In: Ravindran PN, Babu KN, Shylaja M (eds) Cinnamon and cassia the genus cinnamon. CRC Press, Boca Raton, pp 14–79

Ravindran S, Balakrishnan R, Manilal KS, Ravindran PM (1991) A cluster analysis of study on *Cinnamomum* from Kerala. Feddes Repert 102:167–175

Ravindran PN, Shylaja M, Babu KN, Krishnamoorthy B (2004) Botany and crop improvement of cinnamon and cassia, in: Ravindran P, Babu K, Shylaja M (Eds.), Cinnamon and Cassia The Genus *Cinnamomum* CRC Press, Boca Raton, pp 14-79

Samaraweera DN (2000) Propagation of cinnamon. In: Senaratne R, Dayatilake GA (eds) Revitalization of the cinnamon industry in Sri Lanka: constraints and remedies. Faculty of Agriculture, University of Ruhuna, Kamburupitiya, pp 38–45

Sedgley M (1985) Some effects of day length and flower manipulation on the floral cycle of two cultivars of avocado (*Persea Americana* Mill. Lauraceae) a species showing protogynous dichogamy. J Exp Bot 36:823–832

Sedgley M, Griffin AR (1989) Sexual reproduction in tree crops. Academic Press, New York

Sharma AK, Bhattacharya NK (1959) Chromosome studies on four different species of *Cinnamomum*. Jap J Bot 17:43–54

Sharma G, Nautiyal AR (2011) *Cinnamomum tamala*: a valuable tree from Himalayas. Int J Med Aromat Plants 1(1):1–4

Sritharan R, Jacob VJ, Balasubramaniam S, Bavappa KVA (1993) Palynological and cytological studies of the genus *Cinnamomum*. Acta Hort 330:107–112

Thirunavoukkarasu M (1996) Induction of rooting in leafy shoot cuttings of *Cinnamomum zeylanicum* Breyn. by IBA. Indian Forester 122:258–262

van der Werff H, Richter HG (1996) Toward an improved classification of Lauraceae. Ann Mo Bot Gard 83(3):409–418. https://doi.org/10.2307/2399870

Weerasuriya SN, Munasinghe RS, Wijesinghe KGG (2016) Hybridisation among selected true cinnamon (*Cinnamomum zeylanicum* Blume.) accessions for varietal improvement. Proc

Annual Symposium, National Science Foundation of Sri Lanka, 30 June-1st July 2016, BMICH, Colombo

Weiss EA (2002) Spice crops. CABI Publ, Oxon

Wijesinghe KGG, Pathirana R (2000) Germplasm screening and varietal development in cinnamon. In: Senaratne R, Dayatilake GA (eds) Revitalization of the cinnamon industry in Sri Lanka: constraints and remedies. Faculty of Agriculture, University of Ruhuna, Kamburupitiya, pp 15–25

Wijesinghe KGG, Gunarathna WDL (2001) Characterization of true cinnamon (*Cinnamomum zeylanicum* Blume.) based on leaf morphology and their relationship with yield and quality. Sri Lanka Association of the Advancement of Science. Proc. 57th Annual Session. Part 1, p 42

Zalapa JE, Cuevas H, Zhu H, Steffan S, Senalik D, Zeldin E, McCown B, Harbut R, Simon P (2012) Using next-generation sequencing approaches to isolate simple sequence repeat (SSR) loci in the plant sciences. Am J Bot 99:193–208

Zimmerman RH (1971) Flowering in crabapple seedlings. Methods of shortening the juvenile phase. J Am Soc Hort Sci 96:404–411

Zimmerman RH (1972) Juvenility and flowering in woody plants: a review. Hort Sci 7:447–455

Chapter 7
Ecology, Agronomy and Management of Cinnamon (*Cinnamomum zeylanicum* Blume)

D. N. Samaraweera, Sheron N. Weerasuriya, A. S. Karunaratne, S. Subasinghe, and Ranjith Senaratne

7.1 Introduction

Sri Lanka had been a prominent cradle of ancient spice trade, and from time immemorial Arabs and Chinese had visited Sri Lanka (then Ceylon) via the silk route to purchase cinnamon and other spices, thus Ceylon was synonymous with cinnamon. Cinnamon (*Cinnamomum zeylanicum* Blume), which is indigenous to Sri Lanka, occupies an exalted position among the spices grown in the country. Given its exquisite aroma and culinary values, and many other useful properties, cinnamon has found wide applications in many industries, including food and beverage, liqueur, pharmaceutical, nutraceutical, cosmeceutical and perfumery. Sri Lanka holds a virtual monopoly of cinnamon in the world, accounting for around 90% of the global trade of true cinnamon with exports to over 40 countries, while the rest being supplied by Madagascar and Seychelles. Sri Lanka annually exports around 15,000 MT principally in the form of quills of different grades with little value addition.

Currently, there is about 31,000 ha of land under cinnamon cultivation in Sri Lanka, which is mainly cultivated in the Galle (35%), Matara (27%), Ratnapura (12%), Kalutara (6%) and Hambantota (6%) districts. The Southern province thus accounts for nearly 70% of the national cinnamon production, which has increased only marginally over the past two decades. It currently hovers around 17,000 kg year^{-1}. Cinnamon is mainly grown in small holdings of about 0.5 ha with

D. N. Samaraweera · S. N. Weerasuriya
National Cinnamon Research and Training Center, Department of Export Agriculture, Palolpitiya, Thihagoda, Sri Lanka

A. S. Karunaratne
Faculty of Agricultural Sciences, Sabaragamuwa University of Sri Lanka, Sabaragamuwa, Sri Lanka

S. Subasinghe · R. Senaratne (✉)
Faculty of Agriculture, University of Ruhuna, Mapalana, Kamburipititya, Sri Lanka

R. Senaratne, R. Pathirana (eds.), *Cinnamon*,
https://doi.org/10.1007/978-3-030-54426-3_7

over 100,000 growers. Its average bark yield is in the range of 400–600 kg ha^{-1} year^{-1}, whereas yields exceeding 1000 kg ha^{-1} year^{-1} have been obtained under good management. This industry provides livelihood to over 350,000 families, including around 30,000 peelers and 16,000 processors. However, a cheap substitute of cinnamon, namely, cassia (*C. cassia* and *C. burmannii*) produced in countries such as Vietnam, China, Indonesia and Malaysia is also traded as cinnamon in the global market; it contains a high content of coumarin in the bark, that is, about 5 mg/kg—a compound detrimental to health, including carcinogenic and hepatotoxic properties (Abraham et al. 2010; Blahova and Svobodova 2012). In contrast, only traces of coumarin occur in *C. zeylanicum* (BfR 2006; WHO 1999). Yet because of the much lower price of cassia, its share in the global market has been steadily increasing over the years; consequently, the ratio of cinnamon to cassia that stood at 53:47 in 1975 changed to 13:87 in 1990s (Lankage 2015). Despite an appreciable content of coumarin in cassia, the Food and Drug Administration of the USA still does not make a distinction between true cinnamon and cassia, and the former being 3–4 times more expensive than the latter, cassia is predominantly available in most of the supermarkets, whereas cinnamon is available only from specialty suppliers.

In view of the wide industrial applications of true cinnamon and burgeoning green economy, there is growing demand for cinnamon in the global market. Present export revenue from cinnamon is about US$200 million per annum. Nevertheless, due to its wide industrial applications and growing demand for natural food additives with the advent of green economy, the local cinnamon industry shows promise as a multi-billion-dollar industry. However, low productivity and low volume of production of this crop are serious constraints to meeting the growing global demand. In the circumstances, it becomes imperative to augment cinnamon production in Sri Lanka through development of high-yielding promising varieties, improved agro-technology and management and expansion of its cultivation.

7.2 Cultivated Varieties

In Sri Lanka, cinnamon, (*C. zeylanicum* Blume) appears to have originated in the central hills, such as Kandy, Matale, Belihuloya, Haputale, Horton plains and Sinharaja forest range, where seven other endemic wild species of cinnamon also occur, that is, *C. capparu-coronde* Blume, *C. citriodorum* Thw., *C. dubium* Nees, *C. litseaefolium* Thw., *C. ovalifolium* Wight, *C. rivulorum* Kosterm and *C. sinharajaense* Kosterm. However, only *C. zeylanicum* is commercially valuable as a spice.

Cinnamon is a cross-pollinating species, which has made cinnamon plantations in the country highly heterogeneous, resulting in a considerable intra-specific variation in yield and quality. There had not been any long-term crop improvement programmes in the past and seeds obtained from healthy plants had been used in establishing cinnamon plantations. Therefore, the National Cinnamon Research and Training Center (NCR&TC) of the Department of Export Agriculture at Thihagoda, Matara in Sri Lanka collected seeds and cuttings of promising cinnamon genotypes

Fig. 7.1 Morphological differences in leaves of 'Sri Wijaya' (**a**) and 'Sri Gemunu' (**b**) varieties

Table 7.1 Special characteristics of 'Sri Vijaya' and 'Sri Gemunu' selections of *C. zeylanicum*

Attribute	'Sri Wijaya'	'Sri Gemunu'
Bark yield (kg ha^{-1})	1600–1800	1200–1300
Bark oil (%)	1.4–1.6	3.2–3.6
Leaf oil (%)	2.9–3.1	3.60–3.9
Cinnamaldehyde content in bark oil (%)	50.0–55.0	80.0–83.0
Eugenol content in leaf oil (%)	90.0–94.0	88.0–92.7

[a]Technical Bulletin (2015)

from the major cinnamon growing areas in the country and established a cinnamon germplasm collection comprising over 550 accessions. Of these, 10 selected accessions were subjected to rigorous screening for yield and quality and the two most promising lines named 'Sri Gemunu' and 'Sri Wijaya' were released for cultivation in 2009. The two varieties can be distinguished from leaf shape (Fig. 7.1), and morphological differences between the two varieties are discussed under Sect. 6.2 of Chap. 6 of this book. As evident from Table 7.1, 'Sri Wijaya' produces a higher bark yield than 'Sri Gemunu', but the latter has a higher content of bark oil and cinnamaldehyde and superior organoleptic properties (Technical Bulletin 2015).

In view of the growing demand for the newly released superior varieties, NCR&TC embarked upon a rapid multiplication programme using shoot cuttings and issued over 12,000 vegetatively propagated planting material (i.e. over 3000 of 'Sri Wijaya' and 9000 of 'Sri Gemunu' varieties) up to 2018. Augmentation of planting material production has been constrained due to limited mother plants of those varieties, which has affected establishment of new plantations using these superior genotypes. Therefore, establishment of adequate plots of mother plants of the said varieties for year-round plant material production is critically important to improve productivity and strengthen the cinnamon industry.

7.3 Ecological Requirements

7.3.1 Climatic Conditions

Rainfall, insolation, temperature and wind are the major climatic factors affecting the growth and yield of cinnamon.

7.3.1.1 Rainfall

Being a tropical plant with origin in the central hills of Sri Lanka, cinnamon is better adapted to the humid areas (>2500 mm of annual rainfall); however, it can be grown in the low (<300 m amsl) and mid (>300 and <900 m amsl) altitudes receiving an annual rainfall between 1750 and 3500 mm. During the Dutch rule, cinnamon had been mainly established in the coastal belt of the Southern and Western provinces of the country, which has later gradually expanded inland, and presently it is successfully cultivated even in the Badulla district (500–800 m amsl). However, it is not suitable for areas with prolonged dry periods, but with supplementary irrigation, it could be grown in the agro-ecological zone termed the dry zone of Sri Lanka receiving an annual rainfall less than 1750 mm.

7.3.1.2 Temperature

The most suitable temperature regime for cinnamon cultivation lies between 25 and 32 °C. The temperature regime in wet and intermediate zones of the country is favourable for plant growth. However, incidence of certain diseases and poor growth performance could assume economic proportions under overcast sky conditions, which are common at higher altitudes. This is discussed in detail under 'Diseases' in Chap. 8.

7.3.1.3 Day Length and Insolation

Day length has a profound effect on the growth and yield of any plant and cinnamon is no exception. However, cinnamon being a tropical crop grown mainly close to the equator with no drastic variation in day length, hardly any information on the effect of day length on its growth and development is available. However, cinnamon generally comes into flowering from late December to end of February. Therefore, short-day conditions during November and December may induce flowering of cinnamon. Reduced growth of cinnamon has been observed at higher altitudes due to reduced insolation resulting from frequent cloud cover. Cinnamon, being heliophytic (sun loving), demands bright sun for optimum growth and yield and shade could affect bark thickness and stem erectness. However, exposure of seedlings or root cuttings to direct sunlight in the early phase of nursery period could cause leaf scorching. Therefore, adequate shade should be provided in the nursery to prevent sun scorching of tender leaves, particularly when they are purple in colour. Higher growth and yield of cinnamon have been observed in monoculture than when intercropped, due partly to shading under intercropping.

7.3.1.4 Wind

Wind is generally not a problem in the growth and yield of cinnamon. However, when grown in high altitudes, strong winds could damage branches, affect stem elongation and increase water loss due to evapotranspiration. Therefore, establishment of wind breaks along the boundary of cinnamon plantations has proven useful to minimize wind damage at high altitudes.

7.3.1.5 Effect of Climate on Development of Organoleptic Properties

Organoleptic properties such as flavour, taste, aroma and colour are due to secondary metabolites occurring in plants whose biosynthesis is affected by climatic factors, such as light intensity, day and night temperature and water stress (Morison and Lawlor 1999; Mishra 2016). Beverages and food products with superior organoleptic properties fetch a premium price. Therefore, many studies have been conducted to understand the factors affecting flavour development. American ginseng plants exposed to longer periods of sunlight had a higher content of root ginsenoside than those exposed to shorter periods (Li et al. 1996). Drought caused oxidative stress and increased flavonoids and phenolic acids in willow leaves (Larson 1988). Salick et al. (2009) reported that changing climate, particularly the temperature stress, could affect the chemical composition of secondary metabolites and other compounds that plants produce. It has been reported that the complex balance of about 50 chemicals contributes to flavour development in tea (Nowogrodzki 2019).

Flavour in tea is attributable to the development of terpenoids, which are climate dependent. Characteristic superior aroma in tea is developed in the Dimbulla region during January and February and in the Uva region during July and August in Sri Lanka during which desiccating winds and low temperatures ranging from 20 °C during day to 1–6 °C at night exist (Wickremasinghe 1974). Similarly, production of highly prized flavour tea in Yunnan Province in South-western China depends upon climatic conditions (Ahmed et al. 2010, 2014). In addition, climatic conditions affect the quality of coffee (Bertrand et al. 2012) and wine (Jones et al. 2005).

Given the wide applications of cinnamon in the food and beverage industry, production of cinnamon with rich organoleptic properties assumes considerable importance. However, hardly any studies have been carried out to ascertain the effect of climatic and edaphic factors on organoleptic properties of cinnamon. Such studies will be important to identify if any regions, seasons or genotypes exist for production of cinnamon with specific quality characteristics and, if so, to develop geographical indications and to make appropriate agronomic interventions to minimize the impact of unfavourable climatic conditions on the development of flavour and organoleptic properties.

7.3.2 Edaphic Requirements

Cinnamon can be grown in a wide range of soils varying from white silicatious sandy soils ('silver sands') in Negombo to loamy and lateritic gravelly soils in the Southern costal belt and interior of Sri Lanka. The most suitable soil for growing cinnamon is Red Yellow Podzolic (RYP) (Typic Paleudults), which is the predominant soil occurring in the wet (>2500 mm of annual rainfall) and intermediate (<2500 and >1750 mm of annual rainfall) zones of Sri Lanka. Cinnamon thrives in soils belonging to *Boralu*, *Dodangoda*, *Malaboda* and *Weddagala* series. In addition, Reddish Brown Latosols (RBL) in the low country and Immature Brown Loam (IBL) soils in the mid- and intermediate-country wet zones are suitable for cinnamon cultivation. In addition to climate, the quality of bark, economically the most important component, is affected by soil factors as well, and the best quality cinnamon is produced in white sandy soils in the Negombo district. However, recently introduced Alba grade—thin and smooth with a quill diameter of 6–7 mm comparable to that of a pencil fetching the highest price—is produced mainly from cinnamon grown in Red Yellow Podzolic soils in the Southern province of Sri Lanka.

Cinnamon established in shallow soils does not grow well and the yield is poor due to restricted root growth. It being a perennial, a soil depth of about 1 m is desirable. Its roots can penetrate deeper soil layers even through cracks in soil parent material and there are no restrictions with respect to soil physical properties. However, stony and steep lands are not suitable for commercial cultivation. Cinnamon does not tolerate extreme pH regimes and a pH in the range of 5.5–6.5 ensures good growth and yield. When soils in the wet and intermediate zones become acidic due to leaching following heavy precipitation, the application of dolomite is necessary to keep the soil within the said range (Fig. 7.2). Ill-drained conditions are not suitable for cinnamon as submerged conditions affect root growth.

7.3.3 Soil and Water Conservation

It is important to establish appropriate soil and water conservation measures to make the most of the water from precipitation, minimize loss of soil and fertility due to erosion and to mitigate the impact of climate change. In this connection, structural (i.e. construction of terraces, contour bunds, contour drains, ridges, lock and spill drains and retention ditches) and agronomic (establishment of windbreaks, grass barriers and cover crops) interventions prove important. The former should be completed before the onset of the rainy season to minimize soil erosion and protect

Fig. 7.2 A cinnamon plantation at Eheliyagoda, Ratnapura in Sri Lanka, affected due to soil acidity (left) and improved growth after the application of dolomite (right)

inherent soil fertility. Use of heavy earth-moving equipment, such as bulldozers and excavators, is not recommended and natural drains should be used as main drains as far as possible.

Research conducted at NCR&TC showed that the use of cover crop *Arachis pintoi* protects the soil while improving the chemical, physical and biological soil properties (Samaraweera unpublished). Planting of cinnamon should be done along the contours on steep lands following the recommended spacing to reduce soil erosion and improve water conservation. Drains should be cleaned annually to minimize surface run-off and soil erosion. Proper construction of contour drains, contour bunds, stone bunds and lock and spill should be done before crop establishment.

7.4 Identification of Suitable Areas for Expansion of Cinnamon Cultivation

In many developing countries, selection of land for crop production is often done in an ad hoc manner without matching the crop requirements with the climatic and edaphic potential of the area. Consequently, crops are grown under suboptimal conditions, resulting in low yield. Stated differently, there is a considerable gap between the potential yield and the actual yield, which besides causing inefficient use of resources, natural and otherwise, poses a serious issue for increasing food production to meet the needs of growing population in a context of dwindling resources. Therefore, crop establishment and management in a matched crop–climate–land resource scenario through land suitability analysis is an urgent and integral need to sustain optimum agricultural productivity.

As per the Food and Agriculture Organization of the United Nations (FAO) EcoCrop model for suitability assessments (FAO 2012), there are two ecological ranges for a given crop, each one defined by a pair of parameters (absolute range and optimum range) for each variable (i.e. temperature and rainfall). When the conditions over the growing season (i.e. temperature and rainfall) at a particular location are beyond the absolute threshold, the conditions are not suitable for the crop;

Table 7.2 Suitable areas for cinnamon cultivation in four major cinnamon growing districts of Sri Lanka categorized according to FAO suitability criteria (FAO 2012)

	Galle		Matara		Kalutara		Ratnapura	
Suitability class	km^2	%	km^2	%	km^2	%	km^2	%
Highly unsuitable	5.01	0.32	0.85	0.07	16.9	1.04	0	0
Unsuitable	73.44	4.67	110.14	8.77	260.39	15.96	146.97	4.58
Moderately suitable	725.44	46.09	334.87	26.67	919.81	56.37	2441.00	75.99
Suitable	743.01	47.21	621.67	49.51	330.52	20.26	624.21	19.43
Highly suitable	26.90	1.71	188.09	14.98	104.12	6.38	0.00	0.00

when they are between absolute and optimum thresholds, suitability ranges from 1 to 99%, and when they are within the optimum conditions, the suitability score is 100%. Accordingly, the FAO recommended a land suitability assessment approach for crops in terms of suitability rating, that is, highly suitable, suitable, moderately suitable, unsuitable and highly unsuitable based on climatic, topographic and edaphic characteristics (FAO 2012). Such suitability assessment proves important in identifying potential areas, formulating appropriate management decisions, recommending ameliorative measures and adopting best practices for increasing productivity. This paradigm shift involving a holistic and mechanistic approach for optimizing the climate, soil and water resources for enhanced productivity and profitability is of great value and relevance to developing countries, including Sri Lanka.

By deploying this approach, land suitability for cinnamon cultivation was assessed in four major cinnamon growing districts of Sri Lanka, namely, Galle, Matara, Rathnapura and Kalutara, and its results are given in Table 7.2 and Fig. 7.3. As evident therefrom, the Matara district has the highest extent of 'highly suitable' land, while the Galle district has the highest extent of 'suitable land' for cinnamon.

7.5 Propagation of Cinnamon

Cinnamon can be propagated by both seeds and cuttings. Protocol and techniques for producing planting material of cinnamon through micropropagation have also been developed.

7.5.1 Propagation by Seeds

Propagation by seeds is the easiest and most widely adopted method to produce planting material of cinnamon. When grown under favourable conditions, cinnamon comes into flowering within 30–36 months after planting of seedlings in the field. Flowering of cinnamon generally occurs from late December to the end of February in Sri Lanka with occasional sporadic flowering at other times. Only about 20–22%

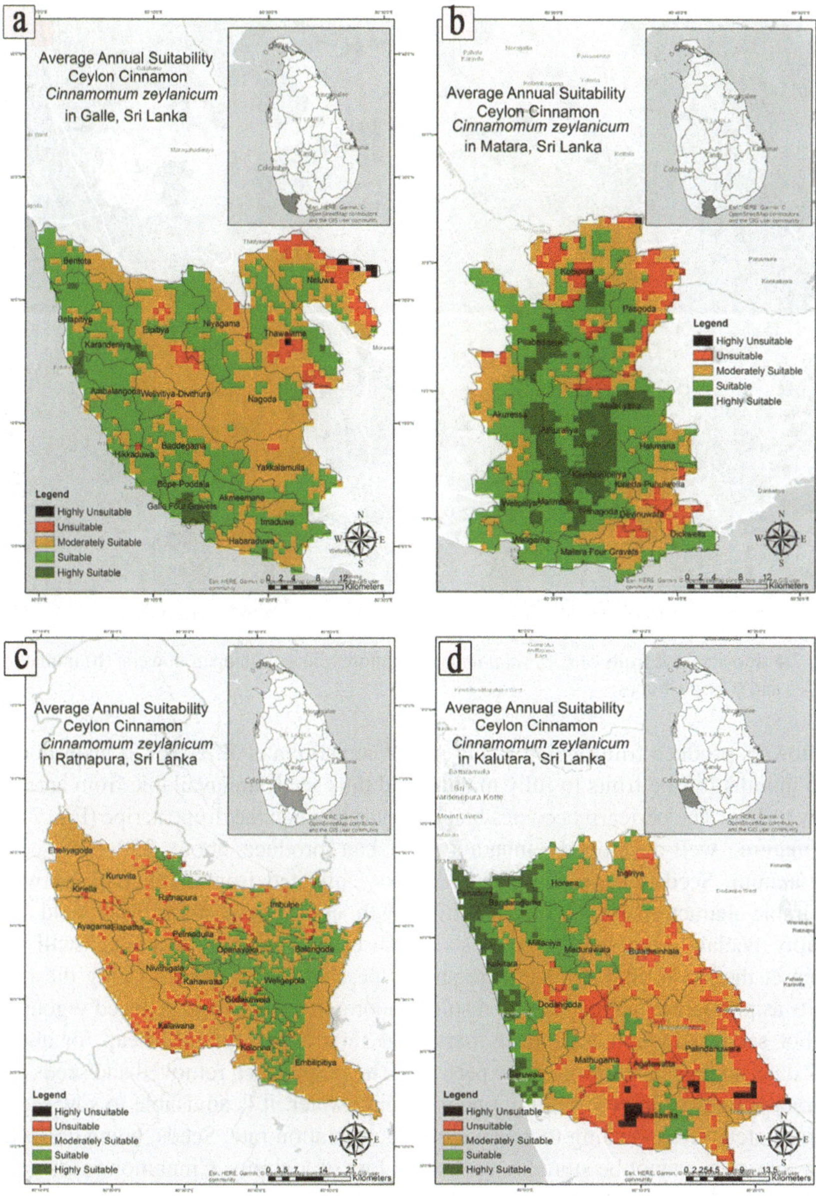

Fig. 7.3 Suitability mapping for cinnamon cultivation in Galle, Matara, Kalutara and Ratnapura districts of Sri Lanka according to FAO (2016)

Fig. 7.4 Flowers and fruits of *C. zeyalanicum*: (**a**) inflorescence with open flowers, (**b**) immature drupes and (**c**) ripe drupes

of flowers produce fruit, which is a drupe (Weerasuriya et al. 2016). It takes about 4–5 months for the fruits to fully mature, and thus seeds are available from around May to July. The pericarp becomes blackish purple when the drupe is ripe (Fig. 7.4). A mature, well-developed cinnamon tree can produce about 250–500 seeds per annum. Seeds for propagation should be collected from mother plants with desirable characteristics, that is, healthy growth, erectness of stems, high yield and quality, peelability and free from pests and diseases. All seeds in a bunch will not ripen at the same time. Hence, care should be exercised to harvest only the ripe fruits as premature harvesting can result in poor germination and reduced vigour.

For seed extraction, ripe fruits harvested should be kept in a heap for about 1–2 days under shade to soften the pericarp. The pulp is then removed and seeds are extracted by washing thoroughly under running water. It is advisable to sow seeds immediately after washing to achieve high germination rate. Seeds, being desiccation sensitive, cannot be stored long without loss of viability. Cinnamon seeds generally lose 40–60% of viability within 4 weeks and completely by 6–8 weeks (Technical Bulletin 2015).

Botanically, cinnamon flower is complete and its flowering follows protogynous dichogamy. Though both androecium and gynoecium occur within the same flower, they become functional at different times, minimizing chances for self-pollination. Stigma becomes receptive on the first day, followed by the dehiscence of anthers on the second day; hence, cinnamon is protogynous. This is an evolutionary development to enhance genetic diversity. As a result, planting material produced from seeds is heterogeneous, showing considerable plant-to-plant variation in growth, bark yield and oil quality (Ponnuswami et al. 1982; Krishnamoorthy et al. 1988, 1991, 1992; Paul and Sahoo 1993), peelability and morphological characters (Krishnamoorthy et al. 1992; Vadivel et al. 1981; Wijesinghe and Pathirana 2000).

7.5.1.1 Methods of Prolonging Seed Viability

Krishnamoorthy and Rema (1987) reported complete loss of viability in cinnamon seeds within 40 days of harvest. Seasonal nature of seed production confined to about 2 months in a year combined with short longevity of cinnamon seeds affords hardly any flexibility for the timing of seed planting. Therefore, studies have been conducted to extend the longevity of cinnamon seeds. Loss of viability of depericarped seeds could be checked up to 4 weeks when stored under shade with good ventilation (Wijesinghe 1997). Samaraweera (1998) found that when cinnamon seeds are stored in dried saw dust or air-dried river sand, seed viability exceeding 90% could be maintained up to 6 weeks. Here, the dryness of the storage medium is crucially important as moisture could lead to fungal infection and seed germination.

Weerasuriya et al. (2017) have shown that seed desiccation beyond a certain limit will drastically affect seed viability. Fully mature, freshly harvested seeds have a moisture content of about 40%, which progressively decreases over time due to desiccation. Drop of water content below 20% of the fresh weight would render seeds inviable (Weerasuriya et al. 2017). Therefore, cinnamon seed is desiccation sensitive and shows a critical moisture content of around 20% below which germination ability is completely lost. Weerasuriya et al. (unpublished) further found that when freshly harvested ripe fruits are depericarped, thoroughly washed, treated with 0.05% Folicur® solution (fungicide containing Tebuconazole) for 45 min and air-dried for 5–7 days and when stored under room temperature in sealed cellophane bags (<5 kg each), high seed viability (>80%) could be maintained up to 4–6 months.

7.5.2 Nursery Management

Selection of an appropriate site for nursery is a prerequisite for production of healthy and vigorous planting material. Cinnamon being light demanding, there should be adequate sunlight with provision for shading during the early stages of the nursery period. In addition, factors such as site hygiene, good drainage, availability of clean water and easy access for after-care operations and logistics prove important.

Given the rapid loss of seed viability, it is necessary to prepare nurseries before the fruit harvesting season begins. Seeds could be planted in rows about 12 cm apart at 3 cm intervals in raised seedbeds. Seeds germinate within 2–3 weeks and when seedlings reach a height of 10 cm, they should be transferred to polythene bags

Fig. 7.5 Planting material raised from seeds of cinnamon: (**a**) seedling nursery and (**b**) seedling pot ready for field establishment

(20 cm × 12.5 cm) filled with a potting mixture containing equal parts of top soil, river sand, coir dust and dried cow dung.

Given possible damage to seedlings when transferring from seedbed to polybags, resulting in uneven growth of seedlings, and labour-intensive nature of the operation, direct seeding in polythene bags is preferable, where 8–10 seeds are directly planted in polythene bags (20 cm × 12.5 cm) filled with the potting mixture discussed above (Fig. 7.5). To produce planting material for 1 ha, 30–35 kg of seeds is required. As birds relish the pericarp of ripe drupes, use of bird netting is necessary for collecting adequate seed material for establishing large-scale plantations. As fruits in a bunch do not ripen at the same time, picking has to be done daily for a few weeks for the harvest of ripe drupes, which is time-consuming and tedious; yet it is important so as to obtain seeds of high viability.

During the first 2 months of the nursery period, temporary shade must be provided to protect seedlings from sunburn. Watering should be done as and when required. Management of pests, diseases and weeds is important to produce high quality planting material. This has been dealt with in detail in Chap. 8 under ‘Pest and disease management in nurseries’. Under good management, seedlings will be ready for field establishment within 6 months.

7.5.3 Vegetative Propagation (VP)

As described above, seasonality of seed production and rapid loss of seed viability combined with cross-pollination in cinnamon pose a challenge to producing planting material from seeds. In addition, limited seed availability is a constraint to stepping-up planting material production to meet the increasing demand. In the circumstances, it is imperative to produce planting material through vegetative means, such as cuttings, air-layering, grafting and micropropagation, producing 'true-to-type' plant populations.

7.5.3.1 Propagation by Cuttings and Air-Layering

Cinnamon is a hard-to-root woody perennial. Therefore, plant growth substances, auxins, have been used to induce rooting in green and semi-hard stem cuttings and air-layers of cinnamon. Dayatilake (2000) observed that the application of indole butyric acid (IBA) and naphthaleneacetic acid (NAA) induced adventitious roots in semi-hard and green wood and in air-layers of cinnamon. Rema and Krishnamoorthy (1993) reported that the application of IBA (2000 ppm) and IAA (2000 ppm) resulted in over 65% rooting in terminal shoots, while there was 22.5% rooting with NAA (500 ppm) in softwood cuttings and 45% rooting with IBA (2500 ppm) in hardwood cuttings of cinnamon.

Semi-hardwood cuttings were found to be ideal for air-layering in cinnamon (Ranaware et al. 1994; Rema and Krishnamoorthy 1993). Air-layering of cinnamon using gallic acid (100 ppm)—a phenolic compound—resulted in 80% rooting (Banerjee et al. 1982). Rooting can also be induced in non-girdled shoots treated with NAA (2500 ppm) either alone or in combination with IBA (100 ppm) (Bhat et al. 1989). Dayatilake (2000) reported that rooting is easier with air-layering than with cuttings and the propagules produced by the former are larger and more vigorous, and thus they can be established in the field earlier than those produced by cuttings. However, air-layering is suitable only for small-scale planting. Dayatilake (2000) recommended single nodal semi-hardwood cuttings for large-scale propagation of cinnamon.

It is important to select promising mother plants for obtaining cuttings, and the criteria followed for selecting mother plants for seed collection could be used in this regard. However, only upright shoots should be collected so that the resulting trees will have a more erect architecture. It is important to manage mother plants through proper training and pruning to facilitate production of more erect shoots for which separate plots of mother plants should be maintained.

Withering of shoots and single nodal cuttings can affect rooting and vigour of resulting plants. Therefore, shoots detached from mother plants and single nodal cuttings taken should be immediately kept in water under shade to prevent desiccation. In taking single nodal cuttings, nodes with active buds should be selected and about one-third of the leaf blade must be removed for improved rooting and a higher

Fig. 7.6 Stages in the vegetative propagation of cinnamon: (**a**) single nodal cuttings, (**b**) planting of cuttings in nursery pots, (**c**) vegetative propagation nursery and (**d**) vegetatively propagated nursery pot ready for acclimatization

success rate. It is important to excise due care to prevent withering of cuttings until planted in polythene bags (20 cm × 12.5 cm) filled with potting mixture described before. Soil should be well moistened, firmed and compacted prior to planting. A leaf cutting may be about 3 cm long with about 2.5 cm below the bud and about 0.5 cm above the bud, with both lower and upper cuts made at a slant as shown in Fig. 7.6a; 2–3 cuttings are planted per pot (Fig. 7.6b). After watering well and spraying with a copper-based fungicide, the bags should be covered with sealed thick transparent polythene (500 gauge) to form a temporary propagator (Fig. 7.6c). Propagators should be set up in a shade net house with a 60–70% shade level to prevent exposure to direct sunlight. In addition to vigour of cuttings and climatic conditions, factors such as genetic background, soil properties, nitrogen level in the rooting medium, physiological state of the mother plant at the time of taking cuttings, type of wood, carbohydrate reserves in the cutting and so on affect rooting (Rema and Krishnamoorthy 1993; Wijesinghe and Gunarathna 2003; Purushotham et al. 1986)

Propagators will be first opened 4 weeks after planting, and thereafter at fortnightly intervals for after-care operations, such as watering, weeding and general management practices. By 10 weeks, plants are ready for hardening (Fig. 7.6d) and can be planted in the field when 6–8 months old.

7.5.4 *Micropropagation of Cinnamon*

Micropropagation has the potential for rapid multiplication of selected plant types with desired characteristics. Cinnamon being a woody perennial, is much more difficult to clone in vitro than herbaceous plants, which present no difficulties in sterilization (Torres 1988). Nevertheless, Subasinghe et al. (2016) were successful in developing a protocol for micropropagation of cinnamon.

7.5.4.1 Surface Sterilization and Culture Establishment

Subasinghe et al. (2016) observed minimum browning of explants (33.1%) in 15% NaOCl with 20-min exposure and successful establishment of cultures of isolated embryos in 7 days in half strength of MS (Murashige and Skoog 1962) basal medium supplemented with 1.5 mg l^{-1} BAP + 0.2 mg l^{-1} IAA. They further observed that in the same culture medium, embryonic axis with half of cotyledon intact as explant was better for establishment of in vitro embryo cultures of cinnamon than embryonic axis without cotyledon. However, axillary buds taken from greenhouse-grown plants were not successful when NaOCl was used for surface sterilization. Since cinnamon contains high levels of polyphenolic compounds, 6-min exposure to 0.1% $HgCl_2$ was used to surface sterilize the cinnamon axillary buds and it was effective for establishment of axillary bud cultures. Higher percentage of clean cultures (69.23%) and lower browning percentage (61.5% with no browning) were achieved as a result (Hettiarachchi 2007).

7.5.4.2 Shoot Proliferation

Hettiarachchi (2007) reported that the lowest number of shoots (1.22) was initiated in the plantlets with 1.0 g l^{-1} activated charcoal, 0.1 g l^{-1} citric acid and the control treatment (without antioxidants), while the highest number of shoots (1.29) was initiated in plantlets with the treatments of 1.0 g l^{-1} polyvinylpyrrolidone (PVP) and 0.1 g l^{-1} ascorbic acid in the medium after 14 days of culture. However, the number of shoots initiated in the plantlets did not vary significantly with or without addition of antioxidants or adsorbents in the culture medium. After 14 days of culturing of axillary buds taken from greenhouse-grown plants, the effect of individual presence of 1.0 g l^{-1} PVP, 1.0 g l^{-1} activated charcoal and 0.1 g l^{-1} citric acid in the culture medium on non-browning of plantlets was significantly higher than in the control. When considering the cumulative effect (mean of 7 and 14 days data) on non-browning of tested plantlets, the highest mean rank value was reported in plantlets with 1.0 g l^{-1} activated charcoal. The highest shoot initiation was observed in Anderson's Rhododendron medium (Anderson 1984) (2.2 shoots per bud), while the lowest was (0.707 shoots per bud) in Vitis (V) and MS media (Hettiarachchi 2007).

Subasinghe et al. (2016) have shown that the presence of activated charcoal in the culture medium leads to a higher rate of stem elongation of in vitro cultured cinnamon plantlets. Therefore, it is beneficial to add activated charcoal to the culture medium to achieve a higher growth rate. The application of appropriate concentration of GA_3 (i.e. up to 2.0 mg l^{-1}) to the culture media is advantageous to induce growth parameters, such as stem elongation and the number of shoots and leaves of in vitro plantlets of cinnamon. However, the relationship between GA_3 concentration and root length is inverse (Subasinghe et al. 2016). In 1998, Kumar and Kumar revealed that if there is no rooting in proliferated shoots, the possible cause of it is the effect of GA_3 presence in the medium. Therefore, shoots have to be subcultured in GA_3-free medium for some time before shifting to rooting medium.

Different concentrations of BAP (1.0, 2.0, 3.0 mg l^{-1}) with NAA (0.1, 0.2, 0.5 mg l^{-1}) were tested for the proliferation of axillary buds originating from explants collected from greenhouse-grown plants. The effect was tested by selected parameters, that is, the number of shoots initiated per bud, time taken for multiple bud formation and so on. Results revealed that the combination of 3.0 mg l^{-1} BAP and 0.1 mg l^{-1} NAA was optimum for the higher multiplication rate with a mean value of 5.6 shoots per bud. Further, when considering the individual effects of BAP concentrations on shoot proliferation of axillary buds, 3 mg l^{-1} BAP resulted in the highest multiplication rate (3.1 shoots/bud). However, there was an inverse relationship between the levels of NAA in the medium and the number of shoots initiated. The lowest number of shoots initiated/bud (1.86) was observed in 0.5 mg l^{-1} NAA, while the highest (3.86) was in 0.1 mg l^{-1} NAA in the medium (Hettiarachchi 2007).

7.5.4.3 Adventitious Root Induction in In Vitro Propagation of Cinnamon

Subasinghe et al. (2016) reported that 0.1 mg l^{-1} NAA is effective irrespective of different BAP concentrations over the other tested NAA concentrations during 8 weeks in media with regard to root length. Furthermore, it is clear that lower concentrations of NAA promote elongation of adventitious roots in cinnamon. They also reported that combination of the highest level of BAP (4 mg l^{-1}) with 0.1 mg l^{-1} NAA in the medium resulted in highest root length, but Davies (1995) reported that BAP, like most frequently used cytokinins, inhibits root formation and therefore is left out from rooting media. In the research by Subasinghe et al. (2016), reason for the effectiveness of the highest concentration of BAP on root elongation may not be considered as a direct effect, but as a combined effect with NAA.

For the induction of adventitious roots, axillary buds taken from greenhouse-grown plants were cultured in full strength MS medium incorporated with 3 mg l^{-1} BAP and 0.5 mg l^{-1} NAA with different concentrations of activated charcoal (0.1, 0.2, 0.5, 1.0 g l^{-1}). In the presence of activated charcoal, multiple bud formation was inhibited, but the highest root length (9.5 cm) was observed when 1.0 g l^{-1} of activated charcoal was incorporated in the culture medium. However, the effect of other treatments on adventitious root formation of axillary bud cultures of cinnamon was not beneficial (Hettiarachchi 2007). Different stages of cinnamon micropropagation using axillary bud and embryo explants are shown in Fig. 7.7.

7.5.4.4 Establishment of Rooted Plantlets of Cinnamon

Acclimatization is the most crucial stage of tissue culture technology on which the ultimate success depends. Subasinghe et al. (2016) reported that among the different potting media used for acclimatization of in vitro rooted plantlets taken from shoots formed in embryo culture of cinnamon, medium with only coir dust resulted in the highest survival (90%) after 5 weeks of growth in the greenhouse and ready for field planting, while the coir dust and sand (1:1) potting medium gave comparatively better growth performance than the medium containing only top soil, where all the plants died after 5 weeks. Similar to embryo-derived in vitro plantlets, in vitro rooted axillary buds (initial explants taken from greenhouse-grown plants) planted in four different potting media also showed that coir dust medium was the best, with highest survival (85%) after 6 weeks in the greenhouse and ready for field planting, while all the rooted plantlets died in the media with top soil alone after the first week of acclimatization (Hettiarachchi 2007).

7.5.5 Canopy Characteristics and Other Attributes of Plants Propagated by Seeds and Single Nodal Cuttings

Cinnamon naturally grows up to a height of about 10–12 m and has a lifespan of over 100 years. However, when commercially grown, bushes are maintained at a height of around 2.5–3.0 m. Seed propagated plants have a robust deep tap root with a well ramified root system and have an economic lifespan of over 50 years, and productive plantations older than 100 years are not uncommon under good management and climatic conditions. On the other hand, VP plants are sustained with an adventitious root system, which makes them more vulnerable to abiotic stresses such as drought and deluges that are becoming increasingly frequent with the advent of climate change. However, the adventitious root system in VP cinnamon may become more ramified and vigorous under adverse climatic conditions, partly compensating for the lack of a tap root and a robust root system as in seedling cinnamon. As VP stands in Sri Lanka are still young, that is, the oldest VP plantations are only about 20 years old, it is premature to understand their economic lifespan, but it is likely to be less than that of the seed propagated plantations.

As evident from Fig. 7.8, VP plants have a more spreading growth habit, whereas seed propagated ones have a more compact canopy. Therefore, more attention has to be paid to VP cinnamon in training and pruning the bush until the first few harvests are done, that is, up to about 5 years, which is rather labour intensive. Given the high cost and scarcity of labour, this will add to the cost of maintenance,

(a) (b) (c) (d)

Fig. 7.7 In vitro initiation and different phases of tissue culture of cinnamon using axillary bud and embryo explants: (**a**) In vitro establishment of axillary bud cultures in half-strength MS (Murashige and Skoog 1962) medium without plant growth regulators, (**b**) enhanced stem elongation and leaf initiation in in vitro embryo cultures on half strength MS basal medium supplemented with 1.0 g l^{-1} activated charcoal, (**c**) multiple shoot formation in axillary bud cultures of cinnamon in full strength MS medium supplemented with 3.0 mg l^{-1} 6-benzylaminopurine (BAP) and 0.1 mg l^{-1} 1-naphthaleneacetic acid (NAA) and (**d**) adventitious root induction in axillary bud culture of cinnamon in full strength MS medium supplemented with 3 g l^{-1} BAP, 0.5 g l^{-1} NAA and 1.0 g l^{-1} activated charcoal

particularly in the early stages of establishment during which period the farmer receives no income from cinnamon. Therefore, it is not uncommon that many farmers still prefer seed propagated material over VP planting material. However, because of cross-pollination in cinnamon, plantations established by seedlings show considerable variation in yield and quality. If new value-added products, including

Fig. 7.8 Comparison of canopy characteristics of seed and vegetatively propagated plants at different growth stages: (**a**) young seed propagated plants, (**b**) young vegetatively propagated plants, (**c**) seed propagated bush and (**d**) vegetatively propagated bush

drugs, nutraceuticals and cosmeceuticals, are to be developed for the global market, it is of paramount importance to ensure consistency, reproducibility and traceability of such products to achieve market penetration. Therefore, production of true-to-type planting material is of prime importance to enhance the cinnamon industry, which can be achieved only through VP. However, homogeneous plantations with no diversity can suffer more than genetically heterogeneous plantations in the event of any outbreak of pests or diseases or unfavourable climatic conditions.

7.6 Land Preparation and Field Establishment

Land for cinnamon cultivation should be selected based on suitability for the crop. Land suitability maps have been developed for four major cinnamon growing districts (Fig. 7.3), where five groups have been identified; highly suitable, suitable, moderately suitable, unsuitable and highly unsuitable. Such maps should be consulted where available in identifying land suitable for cinnamon cultivation.

Cinnamon being light demanding, its growth and yield is affected by shading. Hence, it is mainly cultivated as a monocrop. If it is to be established in an uncleared area, trees and shrubs that could shade or compete with cinnamon should be removed with minimal disturbance to soil, leaving any trees along the boundary that can serve as a windbreak and a source of green manure. After clearing the site, soil and water conservation measures should be established, particularly when the slope of land is greater than 5%, as described earlier in this chapter. Use of heavy earth-moving equipment should be avoided in land development and preparation to minimize disturbance to soil and biodiversity and reduce soil erosion. Land preparation should be confined to opening planting holes and establishment of soil and water conservation measures impacting minimally on the ground vegetal cover. This will minimize soil erosion, fertility decline and loss of biodiversity while maintaining a desirable micro-climate.

If the selected land had been previously cropped to rubber, all rubber roots have to be removed to reduce the risk of white root disease affecting cinnamon. Cinnamon can be planted in straight lines on flat lands and on contours on sloping land in planting holes of 30 cm × 30 cm × 30 cm. Top soil can be used to fill planting holes. It is recommended to incorporate 25 g of rock phosphate per planting hole to ensure development of a vigorous root system for early crop establishment. Spacing recommended for cinnamon by the Department of Export Agriculture in Sri Lanka is 120 cm × 90 cm for flat lands and 120 cm × 60 cm for sloping lands. It is good to establish rows as far as practicable in the east–west direction to intercept more light, reduce mutual shading and facilitate intercropping with suitable short-term crops during the early stages of establishment.

Samaraweera et al. (2001) investigated the effect of plant spacing and the number of plants per planting point on bark yield of cinnamon under varying spatial arrangements while maintaining the plant density constant. It clearly showed that closer intra-row spacing, that is, 120 cm × 45 cm (three plants per hole or 0.54 m^2 per planting point) produced substantially higher bark yield than the recommended wider spacing of 120 cm × 90 cm (six plants per hole or 1.08 m^2 per planting point). This may have been due to reduced inter-plant competition as a result of less overcrowding; thus, the former may have had a greater number of more vigorous harvestable stems per unit area with less lateral canopy spread. However, it is important to have an inter-row spacing of about 120 cm to facilitate cultural practices such as weeding, fertilization and harvesting. When established at a spacing of 120 cm × 90 cm with six plants per planting hole, a hectare carries around 9000–9250 plants.

Though cinnamon is mainly grown as a monocrop, it can also be grown as an intercrop. Dias and Sumanasena (1998) investigated the feasibility of growing cinnamon under coconut with different spacing and special arrangements. Highest growth and yield were obtained when cinnamon was planted at a spacing of 120 cm × 60 cm in three rows in between coconut rows spaced 8.5 m apart. Pathirathne and Edirisinghe (2003) the studied agronomic and economic viability of rubber/cinnamon intercropping under varying spacing of rubber ranging from 8.4 to 18.0 m. They found that inter-row spacing of 14.4, 15.6 and 16.8 m in rubber was suitable for viable rubber/cinnamon intercropping systems.

7.7 After-Care Operations

7.7.1 Gap Filling

Plant mortality could occur due to various reasons and the resulting low density of plants is an important contributory factor to low productivity of cinnamon. Therefore, timely infilling becomes important to ensure optimal plant density, efficient utilization of land and effective weed management. Vacancies occurring in the first 2 years after establishment should be filled periodically as it is difficult to be done in mature stands. The first gap filling could be done 3 months after planting with normal seedlings in polybags. In later stages, 18-month-old seedlings with a height of about 45 cm raised in large polybags (30 cm × 25 cm) should be used. Therefore, it is important to raise special planting material in the nursery for the purpose of infilling.

7.7.2 Irrigation

Cinnamon is grown in wet and intermediate agro-climatic zones of Sri Lanka where the annual rainfall is greater than 1750 mm (https://esdac.jrc.ec.europa.eu/content/agro-ecological-regions-sri-lanka). Water scarcity is the major constraint to grow cinnamon in the dry zone. However, it is possible to grow cinnamon in the dry zone where irrigation facilities are available. Microclimatic conditions in a cinnamon cultivation can affect not only the yield and quality, but also the peelability of bark. Therefore, maintenance of favourable microclimate through windbreaks, mulch and green vegetal cover by eliminating only competitive weeds is important. If prolonged dry weather conditions exist during harvesting time, supplementary irrigation is important to facilitate peeling. Establishment of a water harvesting facility can provide drip irrigation during dry spells in the dry zone.

7.7.3 Weed Management

Weeds when not properly managed, particularly during the early stages of establishment, can affect the growth and yield of cinnamon. It is recommended to keep a circle with a radius of about 45 cm from the plant free from competitive weeds to prevent suppression of crop growth. Clean weeding is not recommended and growth of weeds in other areas can be checked through periodic slashing. Use of chemical weedicides should be avoided as far as possible. If circumstances demand, chemical weedicides may be used strictly adhering to the instructions given so that cinnamon products will not carry any residues. This is of paramount importance as importing countries have laid down stringent requirements in terms of food quality and safety, that is, maximum residue level (MRL) and some even 'zero tolerance'. When planted at recommended spacing, cinnamon assumes a complete ground cover within a period of about two and a half years, after which hardly any weeding is required. However, harvesting stems at maturity will open up space for weed growth; therefore, it is important to manage weeds after each harvest to facilitate regeneration growth.

7.7.4 Nutrition and Fertilization

Soil fertility management is one of the major factors affecting sustainable production of cinnamon, which poses a great challenge to cinnamon farmers in Sri Lanka (Samaraweera 2011). Unlike in other crops, all harvestable stems, including leaves and twigs, of cinnamon are removed at harvest, which will result in loss of considerable amounts of nutrients from the field (Table 7.2) (Samaraweera unpublished data); this will invariably lead to decline in soil fertility. Therefore, it is important to replenish the nutrients so lost to maintain productivity. In cinnamon, bark constitutes economically the most important part. Hence, this factor has to be taken into account in formulating fertilizer mixtures that promote vegetative growth (Table 7.3)..

Studies carried out in Sri Lanka have shown that of the three major nutrients—N, P and K—nitrogen has the greatest effect on bark yield, followed by phosphorous. Major cinnamon growing areas in the country receive an annual rainfall exceeding 2500 mm, and thus the soils tend to become acidic, resulting in phosphorous

Table 7.3 Loss of nutrients from the field at harvest of cinnamon (kg ha^{-1} year^{-1}) (Samaraweera unpublished)

	Plant part				
Nutrient	Bark	Leaves	Scrapings	Wood	Total
Nitrogen (N)	4.1	81.3	2.6	19.4	107.4
Phosphorous (P)	0.6	10.1	0.3	3.8	14.9
Potassium (K)	5.9	61.7	4.3	18.1	90

Fig. 7.9 (**a**) New shoots emerging after harvest of cinnamon stems and (**b**) the suitable stage for fertilization after harvest

fixation and increased availability of Al^{3+} and Fe^{3+}, which could prove toxic under extreme conditions. Application of up to 1000 kg of dolomite depending on the acidity and soil properties can remedy such situations (Fig. 7.2).

Recommended fertilizer mixture for cinnamon in Sri Lanka includes 210 kg of N, 60 kg of P_2O_5 and 135 kg of K_2O ha^{-1} $year^{-1}$ (Technical Bulletin 2015). It contains urea, rock phosphate and muriate of potash in the ratio of 2:1:1 by weight. Rate of fertilizer application depends on the age of the stand. During the first year (3 months after planting) 300 kg ha^{-1}, second year 600 kg ha^{-1} and thereafter 900 kg ha^{-1} $year^{-1}$ should be applied in two instalments. After harvesting, fertilizer should be applied when the foliage of new emerging shoots turns from reddish to light greenish in colour (Fig. 7.9). Rock phosphate which is imported, can be replaced by Eppawala rock phosphate (natural phosphate fertilizer in Sri Lanka) without affecting yield (Heenkende et al. 2003; Samaraweera et al. 2012). Ranasinghe et al. (2013) observed improved availability of phosphorous in the soil following the application of mycorrhizae.

7.7.5 Training and Pruning

In its natural habitat, cinnamon plant grows to a height of about 10 m. Therefore, it has to be trained and pruned periodically to maintain it healthy with multiple stems at a convenient height of about 2.5–3.0 m to derive economic benefits. When the plant is about 1.5–2 years old, its side branches and leaves in the lower half of the bush, which are mainly 'parasitic', should be removed. This will promote plant vigour, straight growth and firmness of the bark, facilitating peeling after harvest. Greenish bark of the stem turns brown when it matures, with which cinnamaldehyde content, the most valuable compound in bark oil, increases and eugenol content decreases. Therefore, premature harvest will result in poor quality bark oil. When plants reach the age of about 3 years, main matured stems (2–3) should be coppiced at a height about 5–6 cm from the collar with the cut facing inward at an angle of around 45°. This will induce emergence of new shoots outwardly, preventing crowding of shoots. When harvesting, it is important to leave 2–3 healthy stems for the next harvesting cycle, which will ensure vigorous growth of the bush as well as rapid regeneration. Three months after harvest, unhealthy and weak shoots have to be removed leaving maximum of three shoots per stem cut. It is recommended to remove any unhealthy and weak branches 3 months before harvest to ensure efficient distribution of assimilates and facilitate peeling.

7.7.6 Pests and Diseases

Pests and diseases affecting cinnamon in juvenile, mature and post-harvest stages are dealt in detail in Chap. 8 of this book.

7.8 Production of Organic Cinnamon

Organic agriculture is a holistic production management system that promotes and enhances agro-ecosystem health, including biodiversity, biogeochemical cycles and soil biological activity. The concept of organic production is not new to Sri Lankan cinnamon farmers, who traditionally utilized indigenous knowledge and followed the principles and practices of sustainable and environment-friendly farming.

In view of serious environmental and health hazards associated with conventional agriculture and increasing socio-economic standard of people, there is a growing demand for organic products across the world. Global organic food and beverages market is expected to reach US$211.44 billion by 2020, growing at a compound annual growth rate of 15.7% from 2014 to 2020 (https://orgprints.org/28077/7/28077.pdf). The number of organic farmers in 2017 in the world stood at 2.9 million with a total of 69.8 million hectares, representing a 20% growth in extent over 2016 (https://www.ifoam.bio/en/news/2019/02/13/world-organic-agriculture-2019). Australia has the largest organic agricultural area (35.6 million hectares), followed by Argentina (3.4 million hectares) and China (three million hectares).

Cinnamon leaves which are discarded at pruning and harvest can be used to make compost. In addition, loppings from windbreaks and green belts can be used as green manure. Long-term experiments conducted on applying cinnamon compost at the rate of 20 MT ha^{-1} showed a significant increase in organic matter, total nitrogen, available phosphorus and sulphur and soil electrical conductivity and lowering of soil pH (Samaraweera, unpublished). It was also observed that replacement of inorganic fertilizer with organic fertilizer by 50% resulted in increased bark yield by around 10%. However, organic cinnamon production is faced with challenges due to low yield with limited soil nutrients associated with organic inputs and prohibitive cost incurred in certification and accreditation of organic products in keeping with the requirements of the major importing countries, particularly in Europe and North America.

7.9 Harvesting

When plant reaches the age of about 3 years, brown, mature stems of 3–5 cm in diameter should be coppiced at a height of about 5–6 cm from the collar, with the cut facing inward at an angle of about 45° (Fig. 7.10), and two harvests can be taken per year. When the age of stand is 5 years and above, three or more stems can be harvested from each bush at 6-month intervals under good management. Peak yield is generally achieved by 8–10 years, which will continue for about 40 years in seedling propagated stands, after which it tends to decline. However, under good management and climatic conditions, high yield is sustained up to about 50 years with seedling-derived plantations. Over 30,000 ha of land is under cinnamon cultivation in Sri Lanka. However, due to dearth of peelers, only around 20% is harvested and peeled twice a year, while about 70% once a year and 10% once in 2 years, incurring a heavy potential loss of yield and export revenue to the country.

Prior to harvesting of cinnamon, peelers randomly select a few bushes and open a slit in the bark to see if the bark is peelable. If not, harvesting will be differed until conditions, intrinsic and extrinsic, become favourable for peeling. Peeling the bark—economically the most important component of the tree—involves a time-consuming, labour-intensive process, which accounts for around 60% of the cost of production of cinnamon. This aspect is discussed in detail in Chap. 9 titled 'Harvesting and Processing'. Sri Lanka exports around 15,000 MT of cinnamon bark annually, of which over 90% is accounted for by quills; they are made up of thin slivers of cinnamon bark rolled into a cigar-shaped cinnamon stick. To make quills, the bark has to be separated from the cinnamon stem through a process called peeling. Several factors, including genetic, physiological, environmental, management, post-harvest conditions and skill of the peeler, affect peelability of cinnamon.

Considerable variation exists among genotypes in terms of peelability, and promising varieties with high peelability are recommended for new planting and replanting of cinnamon. Physiological state of the plant has a profound effect on the tenacity with which the bark is attached to the stem. During physiologically active stages of growth and development such as flushing, flowering and seed bearing, bark is more tenaciously held to the stem due probably to active translocation taking place in the phloem. Therefore, it is better to harvest when the tree is not physiologically very active. Shading due to intercrops and water stress can also affect peelability. Therefore, weeding prior to harvest is not recommended as it could expose the soil to direct sun, causing water stress. In addition, peelability is affected by post-harvest practices and skill of the peeler.

Fig. 7.10 Well-developed cinnamon stems ready for harvest

Average bark yield of cinnamon in Sri Lanka is around 500 hg ha^{-1}, whereas yields exceeding 1000 kg ha^{-1} can be obtained under good management. However, mean productivity of Chinese cinnamon (*C. cassia*) and Indonesian cinnamon (*C. burmannii*) is about 1600 and 1000 kg ha^{-1}, respectively. Farmers growing those species harvest them when stems are 6–20 years old, and thus their bark is thicker and heavier, resulting in a higher yield compared to that of Ceylon cinnamon.

Sri Lankan cinnamon exporters should look for foreign buyers who do not require cinnamon in the traditional bundle form. For instance, more than 85% of cassia is exported as powder or as pieces (chips). Much of Sri Lankan cinnamon is exported to Mexico and South American countries in the form of quills and a fair percentage goes to the European Union. Sri Lanka should look at new and emerging markets in other parts of the world, including Asia and non-traditional markets in Russia, Uzbekistan, and Kazakhstan, which are seeking spices from Sri Lanka. Sri Lanka can then supply such foreign buyers cinnamon in powdered form, or as chips, at a much lower price as peeling accounts for 60% of the cost of production. Mass markets are not an option and the industry needs to look for niche markets for novel specialized products.

7.10 Concluding Remarks

Given the wide range of industrial applications of cinnamon, including food and beverage, liqueur, perfumery, nutraceutical, cosmeceutical and oral care industries and burgeoning green economy, there is growing global demand for Ceylon cinnamon. Because of its exquisite organoleptic properties, including colour, flavour, taste and aroma coupled with insignificant presence of coumarin, a carcinogen, in its bark, Ceylon cinnamon, if strategically and vigorously marketed, could command a clear competitive edge in the global market over its competitor cassia, which is of inferior quality and carries appreciable amounts of coumarin. Therefore, cinnamon, which is indigenous to Sri Lanka, shows great promise as an agro-industrial crop with great export potential and if that is properly harnessed, it could potentially surpass tea, an introduced crop to Sri Lanka, as a foreign exchange earner.

However, low productivity of cinnamon, limited extent cultivated, low export volume, dearth of peelers and lack of value addition are key constraints to enhancing the cinnamon industry in Sri Lanka. Agronomic interventions such as development of high-yielding varieties with desirable canopy characteristics, improved agro-technology and crop management, optimizing agro-climatic and land potential for enhanced yield and quality, expansion of cultivation into climatically and edaphically suitable areas and development of efficient, novel and less labour-intensive peeling technology will contribute in no small measure to transform cinnamon into a multi-billion dollar industry in Sri Lanka. Compliance of cinnamon products with the stringent quality and safety requirements of the importing countries, such as good agricultural practices (GAP), good manufacturing practices (GMP), Hazard Analysis and Critical Control Points (HACCP) and ISO 22000, are of prime importance. Most of the farmers presently use smartphones. Therefore, information and communication technology can be deployed as a powerful tool on a mobile platform in improving the yield, quality and profitability of cinnamon through networking with the key stakeholders such as input suppliers and service providers. In addition, it can provide information services on weather and markets, and technical know-how and advice on production, processing and marketing for growers.

Acknowledgement The authors wish to express their appreciation to Mr. Gamini Wijesinghe, former Director of the National Cinnamon Research and Training Centre of the Department of Export Agriculture for sharing his knowledge and experience with them.

References

Abraham K, Wöhrlin F, Lindtner O et al (2010) Toxicology and risk assessment of coumarin: focus on human data. Mol Nutr Food Res 54:228–239

Ahmed S, Stepp JR, Orians C, Griffin T, Matyas C et al (2014) Effects of extreme climate events on tea (*Camellia sinensis*) functional quality validate indigenous farmer knowledge and sen-

sory preferences in tropical China. PLoS One 9(10):e109126. https://doi.org/10.1371/journal.pone.0109126
Ahmed S, Unachukwu U, Stepp JR, Peters CM, Long C, et al (2010) Pu-erh tea tasting in Yunnan, China: correlation of drinkers' perceptions to phytochemistry. J Ethnopharmacol 132:176–185
Anderson WC (1984) A revised tissue culture medium for shoot multiplication of Rhododendron. J Amer Soc Hort Sci 109:343–347
Banerjee DP, Chatterjee BK, Sen S (1982) Air layering of cinnamon (*Cinnamomum zeylanicum* Nees). South Indian Hortic 30:272–273
Bertrand B, Boulanger R, Dussert S, Ribeyre F, Berthiot L, Descroix F, Joët T (2012) Climatic factors directly impact the volatile organic compound fingerprint in green Arabica coffee bean as well as coffee beverage quality. Food Chem 135(4):2575–2583
BfR (2006) High daily intakes of cinnamon: health risk cannot be ruled out. BfR Health Assessment No.044/2006, 18 Aug 2006
Blahova J, Svobodova Z (2012) Assessment of coumarin levels in ground cinnamon available in Czech retail market. Sci World J 2012:263851
Bhat V, Hegde D, Sulikeri GS (1989) Effect of growth regulators on rooting of air layers of cinnamon (*Cinnamomum zeylanicum*). J Agric Sci 2:328–329
Davies PJ (1995) Guava. In: Parthasarathy VA, Bose TK, Das P (eds) Biotechnology of horticultural crops. Naya Prokash, India, pp 277–282
Dayatilake GA (2000) Possibilities of cloning cinnamon *in-vitro* and *in-vivo*. In: Senaratne R, Dayatilake GA (eds) Revitalization of cinnamon industry in Sri Lanka: constraints and remedies. Proc symposium. Faculty of Agriculture, University of Ruhuna, Kamburupitiya, pp 26–35
Dias KA, Sumanasena HA (1998) Intercropping cinnamon with coconut in the low country intermediate zone of Sri Lanka. In: Senaratne R, Dayatilake GA (eds) Revitalizaation of cinnamon industry in Sri Lanka: constraints and remedies. Faculty of Agriculture, University of Ruhuna, Kamburupitiya, Sri Lanka. p 55
FAO (2012) Assessing suitable land: the methodological framework and its implementation. Food and Agriculture Organization of United Nations, Rome
FAO (2016) The state of food and agriculture: Climate change, agriculture and food security, Food and Agriculture Organization of the United Nations, Rome
Heenkende AP, Samaraweera DN, Gunarathne WDL (2003) Role of Rock Phosphate on performance of immature cinnamon. In: Proceedings part 1– abstracts of the 59th Annual session, Sri Lanka Association for the Advancement of Science, Colombo, Sri Lanka, p 40
Hettiarachchi CS (2007) In Vitro Propagation of Cinnamon (*Cinnamomum verum* Presl) Using Embryos and Axillary Buds, Unpublished MPhil thesis, Department of Crop Science, Faculty of Agriculture, University of Ruhuna, Sri Lanka
Jones GV, White MA, Cooper OR et al (2005) Climate change and global wine quality. Clim Chang 73:319–343. https://doi.org/10.1007/s10584-005-4704-2
Krishnamoorthy B, Rema J (1987) Nursery techniques in tree spices. Indian Cocoa Arecanut Spices J 11:83–84
Krishnamoorthy B, Gopalam A, Abraham J (1988) Quality parameters of cinnamon (*Cinnamomum verum*) in relation to flush colour. Indian Cocoa Arecanut Spices J 12:38
Krishnamoorthy B, Rema J, Sasikumar B (1991) Progeny analysis in cinnamon. Indian Cocoa Arecanut Spices J 14:124–125
Krishnamoorthy B, Sasikumar B, Rema J, Gopalam A, Abraham J (1992) Variability and association studies in cinnamon (*Cinnamomum verum*). J Spices Aromat Crops 1:148–150
Lankage J (2015) Cinnamon stakeholders hail budget proposal, establishment of Cinnamon Development Authority in particular. Daily Mirror, Colombo, Sri Lanka. 8 Dec 2015
Larson RA (1988) The antioxidants of higher plants. Phytochemistry 27:969–978
Li TSC, Mazza G, Cottrell AC, Gao L (1996) Ginsenosides in roots and leaves of American ginseng. J Agric Food Chem 1996:712–720
Mishra T (2016) Climate change and production of secondary metabolites in medicinal plants: a review. Int J Herb Med 4(4):27–30

Morison JIL, Lawlor DW (1999) Interactions between increasing CO2 concentration and temperature on plant growth. Plant Cell Environ 22:659–682

Murashige T, Skoog F (1962) A revised medium for rapid growth and bio assays with tobacco tissue cultures. Physiologia Plantarum 15:473–497

Nowogrodzki A (2019) How climate change might affect tea. Nat Outlook 566:S10–S11

Pathirathne LSS, Edirisinghe JC (2003) Agronomic and economic viability of rubber (*Hevea brasiliensis* Muell.Arg)/(*Cinnamomum verum* Presl) intercropping systems involving wider inter row spacing in rubber plantations. J Rubber Res Inst Sri Lanka 86:46–57

Paul SC, Sahoo S (1993) Selection of elite cinnamon plants for quality bark production. J Econ Tax Bot 17:353–355

Ponnuswami V, Irulappan I, Annadurai S, Vadivel E (1982) Variability studies in cinnamon (*Cinnamomum zeylanicum*, Blume.). South Indian Hortic 30:159–160

Purushotham K, Sulladmath UV, Vishveshwara S (1986) Biochemical status of coffee suckers and its relation to rooting. Indian Agric 30:67–74

Ranaware VS, Nawale RN, Khandekar RG (1994) Effect of rooting media and wood maturity on rooting in air layering of cinnamon (*Cinnamomum zeylanicum* Blume). Spice India 7(6):19–21

Rema J, Krishnamoorthy B (1993) Rooting response of elite cinnamon (*Cinnamomum verum* Bercht & Presl.) lines. J Spices Aromat Crops 2:21–25

Ranasinghe RKRS, Nanayakkara CM, Sumanasena HA, Samaraweera DN (2013) Nutrient management through bio-fertilizers for export agricultural crops. Proc Symp Minor Export Crops (Ed. B. Marambe). Peradeniya, Sri Lanka. 243–258

Salick J, Fangb Z, Byg A (2009) Eastern Himalayan alpine plant ecology, Tibetan ethnobotany, and climate change. Global Environ Chang 19(2):147–155

Samaraweera DN (1998) Effect of storage conditions and length of storage period on germination of cinnamon seeds. In: Proc. part 1 abstracts, 54th Annual session of the Sri Lanka Association for the advancement of science, pp 137–138

Samaraweera DN (2011) Improved soil fertility management practices for optimization of cinnamon yield. In: Proceedings of the international symposium technology to reach future aspirations of Ceylon cinnamon. Faculty of Agriculture, University of Ruhuna, Sri Lanka, 18–19 July, p 11

Samaraweera DN, Gunarathne WDL, Wickramasinghe PJ (2001) Effect of plant spacing and number of plants per planting point on bark yield of *Cinnamomum verum* Presl. (Cinnamon). In: Proc. part 1 abstracts, 57th annual session of the Sri Lanka association for the advancement of Science, p 28

Samaraweera DN, Heenkende AP, Ekanayake JS (2012) Evaluation of Eppawala rock phosphate and imported rock phosphate for direct application in cinnamon (*Cinnamomum zeylanicum* Blume). In: Marambe B (ed) Proceedings of the symposium of minor export crops 16–17 August 2012, Peradeniya, Sri Lanka, p 17

Subasinghe S, Hettiarachchi CS, Iddagoda N (2016) In-vitro propagation of cinnamon (*Cinnamomum verum* Presl.) using embryos and in vitro axillary buds. J Adv Agric Technol 3(3):164–169

Technical Bulletin (2015) Department of Export Agriculture, Sri Lanka

Torres KC (1988) Tissue culture techniques for horticultural crops. Van Nostrand Reienhold, New York, pp 87–152

Vadivel E, Pannuswami V, Irulappan I, Dharmaraj G (1981) Vegetative propagation in cinnamon (*Cinnamomum zeylanicum*). South Indian Hortic 29(4):231–232

Weerasuriya SN, Munasinghe RS, Wijesinghe KGG (2016) Hybridization among selected true cinnamon (*Cinnamomum zeylanicum* Blume.) accessions for varietal improvement. In: Abstracts of National Science Foundation Research Summit empowered by research and innovation, 7th–8th July. BMICH, Colombo, p 118

Weerasuriya SN, Munasinghe RS, Wijesinghe KGG (2017) Effect of seed moisture content and storage condition on viability of true cinnamon (*Cinnamomum zeylanicum* Blume.) seeds.

In: Annual symposium of minor export crops. Plant Genetic Resource Center, Gannoruwa, pp 31–38

Wickremasinghe RL (1974) The mechanism of operation of climatic factors in the biogenesis of tea flavor. Phytochemistry 13:2057–2063

Wijesinghe KGG (1997) Effect of storage period and removal of pericarp on cinnamon seed germination. In: Proceedings part 1 abstracts, 53rd annual session of the Sri Lanka Association for the advancement of science, p 100

Wijesinghe KGG, Gunarathna WDL (2003) Response of elite cinnamon (*Cinnamomum verum* Presl.) lines for vegetative propagation. In: Proceedings part 1 abstracts, 59th annual session of the Sri Lanka Association for the Advancement of Science, p 47

Wijesinghe KGG, Pathirana R (2000) Germplasm screening and varietal development of cinnamon. In: Senaratne R, Dayatilake GA (eds) Revitalization of cinnamon industry in Sri Lanka: constraints and remedies. Proc symposium. Faculty of Agriculture, University of Ruhuna, Kamburupitiya, pp 15–25

WHO (1999) World Health Organisation Monograph on Selected Medicinal Plants. World Health Organization, Geneva, Switzerland. Vol -1, pp 99-100

Chapter 8
Pests and Diseases of Cinnamon (*Cinnamomum zeylanicum* Blume)

G. G. Jayasinghe, R. H. S. Rajapakse, K. L. Wasantha Kumara, D. Ratnasekera, and N. K. B. Adikaram

8.1 Introduction

8.1.1 Importance

Cinnamon (*Cinnamomum zeylanicum* Bloom), belonging to the Family Lauraceae, is indigenous to Sri Lanka and believed to have originated in the central hilly areas of the island. The tree is medium sized and evergreen and the flush is a bright light-red color. The bark of cinnamon, growing up to 10 mm thick, is light brown and possesses a strong, spicy, and aromatic but pleasant smell and a burning taste. The bark of the tree is used worldwide as a valuable spice. From the bark and the leaf, an essential oil is distilled (Kostermans 1996). Cinnamon is cultivated as low bush, about 2–3 m tall, to make harvesting easier. The extent of cinnamon cultivation in Sri Lanka is around 33,000 ha (Anonymous 2018). The country supplies almost 90% of the total true-cinnamon requirement to the world (Anonymous 2017).

G. G. Jayasinghe (✉)
National Cinnamon Research & Training Center, Department of Export Agriculture, Palolpitiya, Sri Lanka

R. H. S. Rajapakse · K. L. W. Kumara · D. Ratnasekera
Department of Agricultural Biology, University of Ruhuna, Faculty of Agriculture, Mapalana, Kamburupitiya, Sri Lanka

N. K. B. Adikaram
National Institute of Fundamental Studies, Kandy, Sri Lanka
e-mail: nimal.ad@nifs.ac.lk

R. Senaratne, R. Pathirana (eds.), *Cinnamon*,
https://doi.org/10.1007/978-3-030-54426-3_8

8.1.2 Insect Pests and Diseases of Cinnamon

In spite of the presence of a mixture of essential oils, with antimicrobial, antifungal, or insecticidal properties, in almost every part of the cinnamon (*C. zeylanicum*) (Singh and Maurya 2005), many arthropod pests and fungal pathogens are known to attack the plant. Over 70 species of insect pests have been reported from the cinnamon crop in Sri Lanka (Rajapakse and Kulasekara 1982) and India (Premkumar et al. 1994; Anandaraj and Devasahayam 2004). The majority of them are, however, not economically important. Understanding of the local pest complex is crucial for their management and to keep their population below a damaging economic threshold. The incidence and severity of pests and diseases are highly dependent on the growth stage of cinnamon. In the nursery stage, the young leaves are extremely susceptible to leaf blight and cinnamon thrips leading to collapse of plants and a higher rate of motility, especially during the rainy season. In addition, leaf miner, gall makers, and leaf eating caterpillars also attack young leaves at the nursery stage, the management of which is somewhat easier than that of thrips and leaf blight. The cinnamon crop, from nursery to the first harvesting stage, is susceptible to various pests, root grubs, thrips, leaf galls makers, shot hole borers (Scolytidae), and several diseases including white root disease. Leaf blight, leaf miner, leaf webber, and other minor pests can damage young plants, but the damage is not economically important.

Rough bark disease (RBD), caused by *Phomopsis* sp., is the most common and destructive fungal disease that lowers the yield and the quality of the cinnamon crop. The wood boring moth (*I. cinnamomumi*) is most devastating to mature cinnamon plantations lowering the productivity significantly. White root disease, brown root rot, thrips, stem and stripe cankers, shoot boring weevil, and vertebrate pests are identified as occasional pests that affect the cinnamon crop under conducive weather conditions. Leaf blight, leaf miner, leaf galls, leaf webbers, and leaf eating caterpillars, though present in mature cinnamon plantations, are also of minor importance (Jayasinghe et al. 2016).

Cinnamon thrips (*Dicromothrips* spp.), gall forming mites (*E. boisi)*, gall forming louse (*Trioza cinnamomi*), leaf miner (*Acrocercops* sp.), and root grubs are also occasional pests that damage cinnamon at different stages of cultivation. Red bark beetle (*Zeuzera coffeae*), shoot boring weevil (*Alcipes clauses*), leaf webber (*Sorolopa* sp.), cinnamon butterfly (*Chilasa clytia*), atlas moth (*Attacus atlas)*, seedling borer (*Xyleborus arquatus*), and leaf eating caterpillar, including the lappet moth (*Phyllodesma* sp.), are considered as minor pests.

Leaf, stem, and root diseases can inflict significant yield losses in cinnamon. Leaf blight (*Colletotrichum gloeosporioides* (Penz.) Penz. & Sacc.) is a common disease at all stages of the crop when rainy and humid weather conditions prevail. White root disease (*Rigidoporus microporus* (Fr.) Overeem) and brown root rot (*Phellinus noxius* (Corner) G. Cunn) are very common during the dry seasons when the plants are under stress. The incidence of stem canker and stripe cankers (*Phytophthora cinnamomi* Rands) is occasional. Black sooty mold (*Stenella* sp.), blisters on fruits (*Exobasidium cinnamomi* Petch), horse hair blight (*Marasmius crinis-equi* F. Muell. ex Kalchbr.), and algal leaf spot (*Cephaleuros virescens* Kunze ex E.M.Fries) are common but of minor importance.

Several vertebrate pests, namely, barking deer, rat deer, stag, peacock, squirrels, and hares are also likely to cause significant damage when the cinnamon crop is in close vicinity to forest covers.

8.2 Insect Pests of Cinnamon

8.2.1 Major Insect Pests

8.2.1.1 Cinnamon Wood Borer (*I. cinnamomumi* Tosevskiis)

Level of Importance The larvae of several species of clearwing moth (Sesiidae) are important wood-boring pests (Duckworth and Eichlin 1978). Cinnamon wood boring moth (*Ichneumenoptera cinnamomumi*) Tosevskiis, also known as cinnamon clearwing moth, is native to Sri Lanka and is the most destructive to the cinnamon crop in Sri Lanka (Dharmadasa and Jayasinghe 2000; Jayasinghe et al. 2006).

Symptoms The larvae of the cinnamon clearwing moth feed and tunnel on the stem at ground level resulting in depletion of food reserves, weakening and breakage of stems, die-back of shoots, and rotting of the pruned stems without producing new shoots. Development of numerous adventitious roots, above the damaged point, is a common and characteristic symptom of cinnamon clearwing moth infestation (Jayasinghe 2013).

Life Cycle Females of CWBM deposit eggs in crevices on the base of cinnamon bushes. Eggs are reddish to dark brown and 1 mm long. The incubation period of eggs is about 1 week. Five larval stages span a period of 40–55 days. The larvae bore into the bark and later tunnel into the heartwood. Larvae have a light pink body with a dark brown head; hence, the pest is called pink stem borer. Pupation takes place beneath the bark which lasts about 35–40 days (Fig. 8.1). Emerged adult moths survive for about 3–7 days and male to female population ratio is 2:1. With the increase of air temperature during February to May in Sri Lanka, the population of male moths is reduced significantly (Dharshanee et al. 2008). The moth completes two to three life cycles per year.

Management Biological characteristics and the behavior of the pest make it rather difficult to adopt management practices. Three different management strategies have been introduced in Sri Lanka to manage the CWBM and each practice has its own strengths and weaknesses.

Cultural Methods Although heaping soil at the base of the cinnamon bush, referred to as earthing-up, minimizes laying of eggs on preferred sites and has been found to

Fig. 8.1 The damage caused by cinnamon wood boring moth (*Ichneumenoptera cinnamomumi* Tosevskiis) in cinnamon plants: (**a**) initial damage, (**b**) subsequent severe damage, (**c**) damaged bark, (**d**) development of adventitious roots above the damage point, (**e**) the adult moth, (f) larvae, (**g**) pupae, and (**h**) earthing-up prevents moth damage. (Photographed by G. G. Jayasinghe)

be highly (90%) effective in managing the pest (Jayasinghe and Wickramasinghe 2001; Jayasinghe 2013), the method is difficult to practice in sloping land and plantations with jutted out stem bases, owing to soil erosion.

Chemical Control Insecticides, effective in the control of the pest, are available. However, due to hazardous effect of synthetic chemicals to the environment and possible residual effects in the cinnamon bark to be exported, the use of chemical insecticides is not preferred (Jayasinghe and Wickramasinghe 2001).

Biological Control A novel alternative control method for cinnamon clearwing moth is the use of a pheromone, [(*E*,*Z*)-3,13-octadecadien-1-ol and (*E*,*Z*)-3,13-octadecadienyl acetate] (Grassi et al. 2002; Jayasinghe et al. 2006; Dharshanee et al. 2008), that disrupts the mating pattern of the moth. The slow rate of control and the length of the period taken to manage the pest are drawbacks in pheromone-based management.

Integrated Approaches It is often difficult to rely upon individual strategies to manage the CWBM. Selection of a suitable combination of strategies, depending upon the prevailing climatic conditions, the land size, topography, etc., is advantageous. The most profitable strategy for managing the cinnamon clearwing moth is combining field application of insecticides, immediately after harvest, with "earthing-up" (heaping soil at the base of the bush) and selective pruning, about 3 months after harvest. In well-managed fields with flat terrain, the most suitable and sustainable singular practice is "earthing-up," twice a year (Jayasinghe 2013).

CWBM infestation at certain levels can bring about 50% yield loss through reduced dry-bark yield and a reduced number of harvestable stems per bush or of bushes lost due to pest damage. Two "economic injury levels" (EIL), based on the percentage of damaged bushes (8%) and the number of pests in 100 bushes (10), were established using Benefit Cost Ratio. This can be used in decision-making concerning the management of wood boring moth in cinnamon fields (Jayasinghe 2015).

8.2.1.2 Cinnamon Leaf Gall Makers

Level of Importance Gall forming pests generally cause little damage to plants, as such they do not adversely affect the bark yield since the affected plant parts continue to photosynthesize with near normal efficiency. The yield and the quality of leaf oil might be changed significantly due to gall formation. However, leaf gall infestation represents a prominent pest damage in young cinnamon cultivations and nurseries.

Taxonomy Forty-eight species of psyllids are reported (Hollis and Martin 1997) from lauraceous host plants and a significant majority (72%) of them belongs to the family Triozidae. Mani (1973) reported an unknown psyllid that also induces gall formation in cinnamon in India. Jumping plant louse (*T. cinnamomi*; Homoptera: Triozidae) is principally associated with the foliage of cinnamon causing leaf galls.

Symptoms In cinnamon, two conspicuous leaf gall types are observed:

(i) Upper surface leaf galls, caused by jumping plant louse (*T. cinnamomi*), a homopteran (Rajapakse and Kulasekara 1982)
(ii) Lower surface leaf galls, induced by *Eriophyes boisi*, a mite belonging to the family *Eriophyidae* (Perera et al. 1985)

Both pests are plant suckers and form galls on the leaf blade as their habitat. Feeding by *Eriophyes boisi* or *T. cinnamomi* causes abnormal cell division and formation of galls. Each gall type is identical in shape, but their dimensions are variable. The galls caused by the psyllid (plant louse) and the mite can be distinguished easily by their morphology.

Oil distillation requires cinnamon leaves without galls or with lower levels of galls, in order to maximize a quality product, as leaf gall infestation not only reduces the quantity but also the eugenol content of oil which is required to be maintained above 85% for higher returns at the international market (Dalandawatta et al. 2015). The use of infested cinnamon leaves with galls as raw materials in the cinnamon oil industry will only increase the cost of production per unit weight of oil, and will not render maximum yields in distillation terms.

Management

Cultural Methods The use of healthy plant material and resistant varieties and improving sanitary conditions in nurseries would reduce gall formation. In mature plantations selective pruning helps to reduce pest population.

Chemical Control In situations of intense infestation, a systemic insecticide may be used (Sahabandu et al. 1998)

8.2.1.3 Upper Surface Leaf Galls: The Jumping Plant Louse (*T. cinnamomi*) or Psyllids

Homoptera: Triozidae

Symptoms The insect galls are solitary and widely spread on the upper surface of the leaf (epiphyllous) but not on the veins. They are conical shaped, unilocular, hard, and yellowish-green. The galls measure 2–3 mm in height and 1–2 mm in thickness at the base (Mani 1973). Only one insect lives within each gall until it reaches maturity and the mature insects leave the gall, making a hole on the lower surface of the leaf which turns the gall brown and dry (Fig. 8.2).

Taxonomy and Life Cycle Upper surface leaf galls are caused by the jumping plant louse (*T. cinnamomi*) belonging to Homoptera: Triozidae). Numerous generations are found within a year, each generation lasting for 35–40 days.

T. cinnamomi prefers younger, growing, and incompletely expanded leaves to mature or over matured leaves. Therefore, the gall initiation in cinnamon appears to be largely restricted to young and tender leaves. *T. cinnamomi* also exhibits a strong preference for the top crown level of the tree (Rajapakse and Ratnasekera 1997).

In a study, conducted to determine the nutrient composition of various categories of leaves, it was found that there was no significant relationship between crude protein level and the abundance of galls (Rajapakse and Ratnasekera 1997). However, significant differences were observed in crude fats in young leaves. Although a relationship between a preference for young leaves and crude proteins which is well known for other Homoptera (Warren and Moran 1978) this study has not supported the same.

Fig. 8.2 Insect galls in cinnamon leaf: (**a**) galls formed on the upper surface of cinnamon leaves, (**b**) the upper leaf gall maker (*Trioza cinnamomi*) (×400), (**c**) dried galls after the insects had left the galls. (Photographed by G. G. Jayasinghe)

8.2.1.4 Lower Leaf Galls (*Eriophyes boisi*; Acarina; Eriophydae)

Symptoms Mite-infested galls, mostly formed on the lower leaf surface, that is, hypophyllous, are comparatively larger, more irregular, and softer than the insect galls that are present on the veins and the apical bud. Sometimes the entire apical bud becomes a mite gall and does not develop into a leaf. The mite-infested galls, initiated on the lower surface of young leaves, are ovoid or irregularly conical with a ridged surface, pinkish in color initially and becoming green on maturity. The gall cavity has long hairs and a rugose surface, the color is greenish or yellowish and somewhat ridged. The lower surface is covered by a thin layer of cells which ruptures to permit emergence of the adult (Mani 1973).

Life Cycle Large numbers of four-legged, worm-like Eriophyes mites live in a single gall (Fig. 8.3). Mature mites are spread out of split-dried galls by wind during the dry season. There may be numerous generations per year whose length varies according to the season. After emergence of the adult, the gall dries up and turns dark brown in color. The gall cavity is lined by a small but closely packed mass of cells that lack chlorophyll, surrounded by sclerenchyma and parenchyma cells which also lack chlorophyll. The gall formation results in retardation of growth of plants at young and nursery stages.

Infestation by either the mite or gall insect significantly reduces the leaf oil content by 18 to 43% and lowers the oil quality (Perera et al. 1985). The decline of cinnamon oil content is greater with the increase in the severity of infestation, that is, 10.5% reduction at 25.6% severity which increases to 74.26% at 97.26% severity following mite infestation; 25.9% reduction at 22.7% severity increases to 66.5% at 62.9% severity, after insect gall infestation. Leaf infestation reduces the eugenol content of cinnamon oil that needs to be maintained above 85% for higher returns in the international market (Dalandawatta et al. 2015).

Management The severity of the incidence of leaf galls in cinnamon and the extent of damage vary with the degree of plant resistance and the season. The possibility

Fig. 8.3 Mite galls in cinnamon: (**a**) initial stage of damage, (**b**) galls formed on lower surface of cinnamon leaves, (**c**) lower leaf gall maker (*Eriophyes boisi*) in cinnamon (×400). (Photographed by G. G. Jayasinghe)

of initiation of both types of galls is greater with the development of new flush with the rain. Infestation becomes critical mostly in nurseries and also in young plantations as the affected plants show retardation of growth. Similar infestations will not cause much economic damage in mature plantations.

Once the initial injury is evident, control measures will have to be implemented before the mites and psyllids become established within the plant, since the damage, once caused, cannot be reversed.

Cultural Methods Periodical selective pruning is the best management practice to remove *Trioza* and *Eriophyes*-infested, gall-bearing shoots, in addition to application of chemicals.

Biological Control Biological control of eriophyid mites and psyllid is difficult because they live inside and feed on galls, which protects the mites from predators. In addition, effective natural enemies for eriophyid mites are not known.

Chemical Control Insecticides/miticides must be applied to prevent further infestation, when initial symptoms become apparent in a few newly emerged flushes (Sahabandu et al. 1998). However, only a limited number of effective miticides are available for controlling eriophyid mites.

8.2.1.5 Cinnamon Thrips

Level of Importance Infestation by cinnamon thrips may result in significant yield losses in mature cinnamon plantations while severe growth retardation can occur in nurseries and plants at young stage. The thrips that infest cinnamon crops have not been identified and their taxonomy is not known.

Symptoms Thrips damage results from the piercing and rasping action of their cone-shaped mouth parts. Thrips damage can be easily identified by the symptoms. Initially minute, dark green spots appear on the leaf and with time these turn into white or silver. If widespread, these can give a silvery and streaked appearance (Fig. 8.4).

Life Cycle The life cycle of thrips from egg to adult may be completed within 2 weeks. Thrips damage is greatest after periods of hot, dry weather. Cool, rainy weather reduces thrips populations and damage.

Management

Cultural Methods Good crop management and field sanitation generally keep thrips damage to a minimum. Thrips have a wide range of hosts including numerous weed species. Weed management in and out of cinnamon plantations reduces the level of thrips.

Fig. 8.4 Thrips damage in cinnamon: (**a**) thrips infestation of new shoots emerged after a harvest, (**b**) damage to young shoots, (**c**) die-back of the plant after thrips damage, (**d**) leaf shedding after thrips damage, (**e**) damage to nursery plants, (**f**) nymph (×10) sucking sap underside the leaves, (**g**) adult thrips (×100). (Photographed by G. G. Jayasinghe)

Biological Control Effective biological control methods using natural predators and parasitoids need to be developed.

Chemical Control Management practices involving treatment with an effective insecticide have been developed for nurseries and younger plantations. Use of synthetic insecticides in mature plantations is not, however, desirable because of possible environmental effects and residual issues concerning the final product.

8.2.1.6 Cinnamon Red Borer (*Z. coffeae* Nietn. Thysanoptera: Cossidae)

Cinnamon red borer (*Z. coffeae* Nietn.) is a generalist pest, that is, one found in a vast range of crop plants. This is one of the earliest known pests, first described in 1861 by John Nietner and was first recoded in Sri Lanka by Rutherford (1913).

Symptoms The damage to the host plant is caused by the caterpillars of the cinnamon red borer. The caterpillars bore into the bark and twigs of different host plants and make tunnels inside the stem and the root. When tunneling is in progress, withering of leaves is common in the cinnamon plant and at severe stage, the stems or whole bushes die off. Dark brown excreta, coming out of the holes made by the pest, is characteristic for identification of the pest. The feces are globular in shape and 3–5 mm in diameter, bigger than the excreta of the cinnamon wood boring moth.

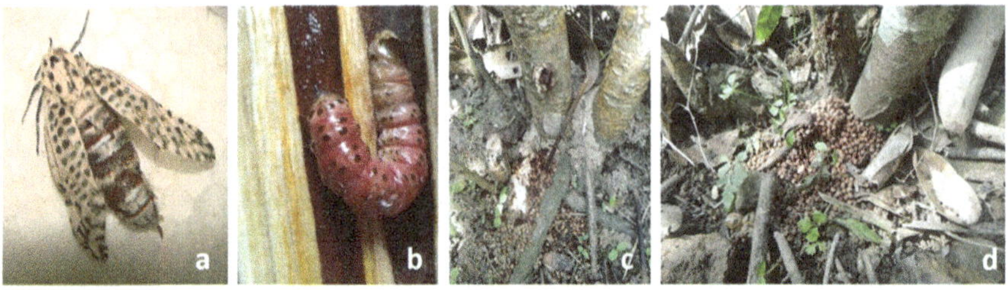

Fig. 8.5 Cinnamon red borer: (**a**) adult borer (*Zeuzera coffeae*), (**b**) larvae living inside the cinnamon stem, and (**c**, **d**) nature of the damage. (Photographed by G. G. Jayasinghe)

Life Cycle Adult moth is white in color and 30–40 mm in size when the wings are expanded. Males are comparatively smaller in size than females. Small black dots are present in the forewings and the outer margin of hind wings (Fig. 8.5). The abdomen is long and posteriorly tapering. Bipectinate antennae are present and proboscis absent in both sexes. Emerging young larvae are slender, soft, and red in color. The eggs that are deposited by copulated female moths in rows on host plants are hatched after 10 days. The larvae bore into the stems or twigs start making tunnels after feeding on the pith of plants. Larvae become fully matured in 4–5 months and are dark red in color and 40 mm long. After 3–4 weeks of pupation, the emerging moth comes out from a hole that the larvae make before pupation. It takes 5–6 months to complete the life cycle (Chang 1984).

Management

Cultural Methods Removing and burning of severely infested branches or twigs is effective in reducing the pest population.

Chemical Control Since larvae live safely inside the stem or twigs in host plants, chemical spraying will not be very effective. Instead a cotton bud, soaked with an insecticide, may be inserted into the tunnel through the hole that the larvae have made in cinnamon stems.

Integrated Approach An integrated management system, using larval parasitoids with chemical, mechanical and cultural methods, has been developed (Ahmad 2017).

8.2.1.7 Cinnamon Root Grubs (*Anomala* sp. Scarabaeidae)

White grubs are among the most destructive pests in certain cinnamon growing areas, especially when the cinnamon plants are at young stage.

Life Cycle Some beetle grubs, including *Anomala* species, feed on the roots of grasses, and younger plants. Population of root grubs might be very high in soils rich in organic matter. Though mature cinnamon plants show resistance to root grub,

Fig. 8.6 Root grubs (*Anomala* sp.) damage in cinnamon, (**a–c**) nature of the damage and (**d**) grubs living in soil. (Photographed by G. G. Jayasinghe)

newly established young cinnamon plants are most vulnerable to the pest (Fig. 8.6). Damage can be more severe with the onset of the rainy season, after a long dry period, because starving grubs start feeding on roots voraciously after spending an inactive period during the dry spell.

Management

Cultural Methods Clean weeding is not recommended when grubs damage is severe.

Chemical Control Soil drenching with a suitable chemical where new cinnamon seedlings are established after removing all damaged plants is recommended. Alternatively, the chemical may be added to the planting sites, a day before planting. These are effective in managing root grubs, when the cinnamon plants are still at young stage. Repeated application of the chemical after 2 weeks kills newly hatched larvae since eggs tend to resist the insecticide to a certain extent.

8.2.1.8 Cinnamon Shot-Hole Borer (Scolytidae)

Shot-hole borer damage is very common in cinnamon when the seedlings are over-matured more than 2 years at nurseries. Though the exact species of the shot-hole borer in cinnamon is not known, the morphology of the shot-hole borer, the symptoms of infested plants and the damage caused are almost same as those of *Xylosandrus compactus* (Eichhoff).

Biology Tiny Scolytidae beetles deposit eggs only in sufficiently matured stems, that is, when the stems reach the girth of more than 5 mm which is needed for establishment of beetle galleries. Galleries are infected with a symbiotic fungus that the beetle carries in its mouth parts. Females lay eggs in galleries. Eggs hatch and larvae feed on the fungus. Off-white tiny eggs, typical C-shaped legless larvae of weevil,

Fig. 8.7 Shot-hole borer damage in cinnamon: (**a–c**) nature of damage, and (**d**) adult beetle. (Photographs courtesy Janaka Chandana)

and white colored pupae can be seen in the galleries. Newly emerged adults, with fungus inoculums, come out from galleries and spread to new host plants.

Symptoms Infested seedlings show leaf and stem necrosis extending from the entrance hole. Five to seven days after gallery formation, the twigs, branches, or the entire plants start wilting (Fig. 8.7). Cinnamon seedlings collapse due to the breakdown of the stem at the point of pest damage.

Management

Cultural Practices Removal and destruction of infested plants is the most effective and recommended cultural practice.

Chemical Control Chemical control is essential when the infestation is severe.

Biological Control Biological control and breeding for resistance would be economically feasible and ecofriendly, but only a few efforts on these lines have been reported (Walgama 2012).

8.2.2 Minor Insect Pests

8.2.2.1 Cinnamon Butterfly, the Common Mime (Lepidoptera: Papilionidae) (*Papilio clytia* (Synonym: *C. clytia*)

P. clytia is a species belonging to the swallow tail butterfly family, Papilionidae, which is common in Sri Lanka, India, and some South East Asian countries. The larvae feed on young leaves of cinnamon. There are two different color forms for each sex, "clytia" and "dissimilis" (van der Poorten and van der Poorten 2011).

Level of Importance Cinnamon butterfly is not an economically important pest in Sri Lanka. However, *C. clytia*, is one of the major pests in cinnamon in India, responsible for severe damage (Singh et al. 1978).

Life Cycle The butterfly lays eggs singly on the upper and lower surfaces of young leaves. The eggs are spherical, waxy looking, and orange-yellow in color. The larvae feed on tender and slightly mature leaves. In cases of infestation, only the midribs of leaves with portions of veins are left behind or the entire plant is defoliated. The larval period (five instars) lasts for 11–17 days. The fully-grown larvae are pale yellow with dark stripes on the sides and measure about 2.5 cm in length. The caterpillars resemble bird droppings (defensive mimicry) at initial stages. The pupae are cylindrical and remarkably well camouflaged by fixing on to a branch so that they resemble the end of a broken twig. The pupal period lasts for 11–13 days and the total life cycle takes 24–36 days (Rajapakse and Wasantha Kumara 2007). Bell (1912) reported that the moth is seen in large numbers during monsoon months in India and spends the dry months as pupae (Fig. 8.8). Eggs are heavily parasitized by the egg parasitoid *Telenomus remus* (Rajapakse and Wasantha Kumara 2007).

Management

Cultural Practices Hand removal of larvae and pupae in lighter infestations is effective (Butani 1983).

Chemical Control Spraying tender and partly matured leaves with an effective insecticide may control the pest in severe infestations.

8.2.2.2 Atlas Moth, *A. atlas* (Linnaeus 1758)

Level of Importance The Atlas moth is said to be the world's largest moth and was first described by Linnaeus (1758) as a pest in cinnamon. Giant larvae voraciously feed on younger and semi-mature leaves of cinnamon, causing heavy damage to

Fig. 8.8 Cinnamon butterfly (*Papilio clytia*): (**a**) initial damage to cinnamon, (**b**) mature larvae, (**c**) pupae, and (**d**) the adult. ((**b**, **d**) Photographed by G. G. Jayasinghe; (**a**) (http://krishimala.com/catalogue/cinnamon-butterfly); and (**c**) (https://commons.Wikimedia.org/wiki/File:Chilasa_clytia_pupa.jpg))

cinnamon plants. Since the number of larvae is limited, the damage is not economically significant. The Atlas moth is present in Asian countries, including India, Sri Lanka, and Philippines, and depends on a wide variety of host plant range, including cinnamon (Ahmed 2013).

Life Cycle The lack of mouth parts in adult moths makes them incapable of feeding on leaves, but they survive for a few days until fertilization takes place and eggs are laid, using food reserves built up during the larval stage. A powerful pheromone-mediated mating system helps fertilization and production of eggs. Oval-shaped eggs are deposited on the leaves of host plants and caterpillars are hatched from eggs after a few days. Larvae are grown gradually up to a length of 40 mm which, after five moltings within 60 days, are moved to a suitable place to pupate (Fig. 8.9).

Pupae start making strong silken cocoons around a suitable leaf and about 12 days after the pupae are formed the final molting takes place inside the cocoon. Eclosion occurs 25–30 days from pupating. Emerging adults spend 2 hours hanging out their wings, by pumping air through them, until their wings become fully functioning. The male moth remains within the pupal case until a night before flying off to seek a virgin female to mate with.

8.2.2.3 The Leaf and Shoot Webber (*Orthaga vitialis*): (Lepidoptera: Pyralidae)

Level of Importance The leaf and shoot webber (*Orthaga vitialis*) infests cinnamon crop but the damage caused is not significant or economically important in Sri Lanka. In India, the damage by the shoot and leaf webber, *Sorolopha archimedias*, has been reported as a major pest (Singh et al. 1978).

Life Cycle The adult leaf and shoot webber (*Orthaga vitialis*) is a yellow brown to dark brown moth and the larvae are very active, brown in color, and web the leaves and the terminal shoots into clusters (Fig. 8.10).

A webbed cluster of leaves harbors several larvae. The larvae are gregarious at the beginning and feed by scraping the leaf surface. Pupation takes place within the webbed-up cluster for 11–14 days. The larval period extends to about 28–30 days (Rajapakse and Wasantha Kumara 2007).

Fig. 8.9 Atlas moth (*Attacus atlas*): (**a**) adult moth, (**b**) larva, and (**c**) with pupal case. ((**a**, **c**) Photographed by G. G. Jayasinghe and (**b**) Photograph caurtesy Chinthaka Vidhanapathirana

Fig. 8.10 Cinnamon leaf and shoot webber (*Orthaga vitialis*): (**a**) nature of damage, (**b**) adult moth, and (**c**) larvae. (Photographed by G. G. Jayasinghe)

Management

Cultural Practices Regular inspection of the plantation and pruning infested clusters followed by destruction is a good cultural control method. Since the leaf and shoot webber is not an economically important pest in cinnamon in Sri Lanka, removing and destruction of nests are sufficient to manage the pest.

8.2.2.4 Common Blue Bottle (*Graphium sarpendum*); Lepidoptera; Papilionidae)

The blue bottles are common butterflies and seen year-round in cinnamon growing areas.

Life Cycle In flight, the beautifully contrasting fluorescent blue and black are unmistakable and the sexes are similar (Fig. 8.11). The common blue bottle and Tailed Jay, *Graphium agememnon*, are among the fastest nectar feeders of butterflies found in Sri Lanka (van der Poorten and van der Poorten 2011).

Eggs are laid on the lower surface of tender leaves and are completely round, smooth and yellow when they are first laid. The eggs last for 5–6 days. The first instar caterpillar is very spiny and of a smoky color. They are very sluggish and at first keep to the undersurface of leaves, feeding on them, and later they seem to favor the midrib on the upper surface of the cinnamon leaf. As the caterpillar grows older, it becomes green and quite well concealed by its green camouflage. The larval stage (five instars) lasts 29–31 days. They pupate on the underside of the leaves, stalks or small branches and the pupal period lasts for 19–20 days. The life cycle is about 59–60 days. The larvae of the Common Jay *G. dosonis* are also reported to be feeding on the tender leaves of cinnamon (Bell 1912).

Fig. 8.11 Common blue bottle (*Graphium sarpendum*): (**a**) young larvae (http://butterflycircle.blogspot.com/2009/01/life-history-of-common-bluebottle.html), (**b**) mature larvae Photograph caurtesy Chinthaka Vidhanapathirana, (**c**, **d**) adults (https://en.wikipedia.org/wiki/Graphium_sarpedon#/media/File:Common_bluebottle,_Keitakuen,_Osaka.jpg)

8.2.2.5 The Leaf Miner, *Acrocercops* spp. (Lepidoptera: Gracillaridae)

Level of Importance Leaf miner (*Acrocercops* spp.) is listed in Sri Lanka as a minor pest of cinnamon. However, *Conopomorpha civica* and *Phyllocnistis chrysophthalma* Meyer are considered as major leaf mining pests in India (Devashayam and Koya 1997; Butani 1983) where *C. civica* infestation of cinnamon seedlings has caused over 20% loss.

Life Cycle The adult is a tiny silvery moth. The females lay minute flat eggs singly on the lower surface of th.3e leaves, close to the midrib. The eggs are hatched in about 2–6 days. The pale gray larvae enter the leaf tissue by mining (Fig. 8.12). They feed on tissues between the upper and lower epidermis of tender leaves resulting in linear mines that end in blister like patches.

Symptoms The infested leaves become crinkled and the mined areas dry up, leading to formation of large holes on the leaves. The mined leaves turn pale and curl up and the development of young leaves is retarded.

Management Appropriate insecticide sprays, at nursery and young stages of plants and during the emergence of new flush, are generally effective in preventing leaf miner infestations.

8.2.2.6 Shoot Boring Weevil: *A. clauses*: Curculionidae

This is a localized weevil, found in hilly areas like Ratnapura District (Sabaragamuwa Province) in Sri Lanka, where relatively cooler weather conditions prevail.

Life Cycle The shoot boring weevil (*A. clauses*) feeds on young shoots, emerging after a harvest and also pruning. They lay eggs inside young, developing shoots. Larvae emerging after 5–6 days also start feeding on young shoots and the larval period ends after four moltings. Larvae pupate within 20–30 days and pupae take another 15–20 days to become adults. The male weevil is slenderer than the female. Each female lays a single egg in a day and 10–20 eggs in its life time of 30–40 days. Multiple mating can occur during the life time of a weevil. Weevils complete –three

Fig. 8.12 Cinnamon leaf miner (*Acrocercops* spp.): (**a**) mild damage, (**b**) severe damage. (Photographed by G. G. Jayasinghe)

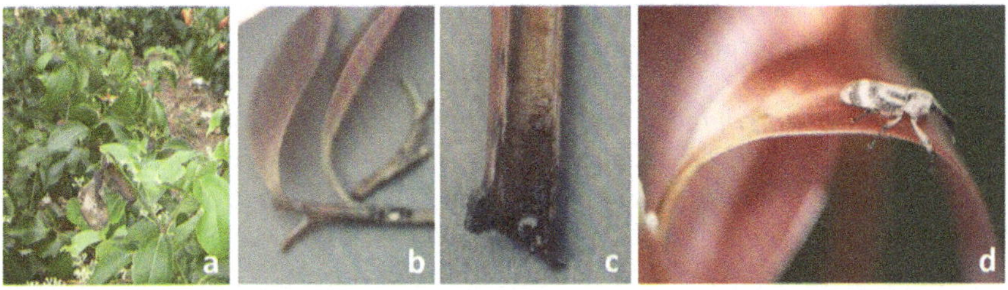

Fig. 8.13 Cinnamon shoot boring weevil (*Alcipes clauses*): (**a**) symptoms of damage, (**b**) eggs inside the stem, (**c**) the larvae, and (**d**) adults living freely. (Photographed by G. G. Jayasinghe)

to four life cycles within a year when young shoots are available (Fig. 8.13) (unpublished data).

Symptoms Damaged plants take a rosette-like appearance where multiple stems and branches do not become straight enough for processing into cinnamon sticks of good visual quality.

Management

Cultural Practices Pruning or removal of damaged shoots followed by their destruction will lower the pest population and infestation of plants. Depending on the population size, hand picking of adults, and their destruction may be carried out.

Chemical Control When the damage is likely to be severe, application of an effective insecticide would be needed.

8.2.2.7 Fruit Borer (*Alcides morio* Heller) (Coleoptera: Curculionidae)

Level of Importance Grubs (*A. morio* Heller) bore tunnels leading to the cinnamon seed and feed on the inner contents of the seed. The damage is economically significant since cinnamon is propagated through seeds.

Life Cycle The mature grub has a brownish head with a whitish body which attains 8–10 mm in length. The females are larger than the males. Pupation takes place inside the seed and lasts for 7–9 days. The weevil makes a circular hole on the seed coat to emerge. They are dirty black in color and not active. The longevity of the beetle is 5–7 days (Premkumar 1988).

8.2.2.8 Leaf Eating Caterpillar *Euproctis fraterna* (Lepidoptera: Lymantriidae)

The larvae of *E. fraterna* feed voraciously on leaves. Initially the larvae scrape the green matter leading to skeletonization of leaves. Later the larvae move into other parts of the plant and finally defoliation occurs. The larval stage of this hairy caterpillar lasts for 13–29 days and the pupal period is 9–20 days, making the total life cycle lasting for 6–7 weeks (Rajapakse and Kulasekara 1982).

8.2.2.9 Leaf Eating Caterpillar *Dasychira mendosa* (Lepidoptera: Lymatriidae)

Due to the broader nature of their host range, the larvae of the moth *D. mendosa* feed on the foliage of a range of plant species, including cinnamon. Breeding continues throughout the year during which there are probably five or six generations. The moth, with pale yellow hind wings and forewings, is irregularly patterned with various brown shades. The female lays large masses of eggs. Feeding by larvae results in defoliation and the larval period lasts for 21–28 days. A fully grown hairy larva has a reddish head and a grayish or yellowish body, striped with red and long, dense dorsal tufts of whitish hairs. They pupate in loose cocoons, made of silk and hairs and the pupal period lasts for 11–12 days (Rajapakse and Kulasekara 1982). Application of insecticides would keep most of the caterpillars under control.

8.2.2.10 Looper Caterpillar (*Thalassodes* spp.) Lepidoptera: Gracillaridae

Thalassodes species are looper caterpillars, found feeding on developing cinnamon leaves. The larval period lasts for 17–18 days. Larvae possess a color the same as that of newly emerging shoots, and assume a characteristic pose on the twig, which is often mistaken for a leaf petiole. The pupal period lasts for 7–8 days (Rajapakse and Kulasekara 1982).

8.2.2.11 Other Minor Pests

Agroploce aprobola (Lepidoptera: Encosmidae), the scale *Ceroplastes rubens* Maskell (Homoptera: Coccidae), the sucking bug *Coptosoma pygmaeum* Mont (Heteroptera: Plataspidae), and *Leptocentrus obliquus* Walker (Homoptera: Membracidae) are considered as minor pests of cinnamon in Sri Lanka (Rajapakse and Kulasekera 1982). The chrysomelid beetles, *Coenobium lateralis* Weise, *Cryptocephalus snillus* Suffr, *Cryptocephalus virgule* Suffr, and *Podagric abadia* Harold were also found feeding on developing cinnamon leaves and shoots in Sri Lanka (Rajapakse and Kulasekara 1982). Damage by the larvae of *Leucopho lispinguis* Burmeister (Coleoptera: Scarabeidae) in nurseries of cinnamon has been reported.

Bhumannavar (1991) reported the presence of the tortricid *Sorolopha archimedias* (syn. *Eudemiopsis archimedias*) in cinnamon in South Andaman, India. Damage by the chaffer beetle *Popillia complanata* and a leaf beetle *Singhala helleri* has been reported in India (Singh et al. 1978). *Oecophylla smaragdina* F. (Hymenoptera: Formicidae) are a nuisance as the adult ants form nests from the leaves (Rajapakse and Wasantha Kumara 2007).

8.3 Diseases of Cinnamon

Despite the higher number of diseases reported in *C. zeylanicum*, only a few are significantly affecting the yield or the quality of cinnamon. Among them, rough bark disease (*Phomopsis* sp.), stripe canker (*Phytophthora cinnamomi* Rands), foot rot (*Fusarium oxysporum* Schltdl.), stem canker, leaf blight (*C. gloeosporioides* (Penz.) Penz. & Sacc.), white root disease (*R. microporus* (Fr.) Overeem.), and brown root rot (*P. noxius* (Corner) G. Cunn.) are common in cinnamon causing significant yield losses.

Diseases like blisters on fruit (*E. cinnamomi* Petch), gray leaf blight (*Pestalotia* sp.), thread blight (*Marasmius equicrinis* F. Muell. ex Berk.) (Dassanayake et al. 2009), witches' broom (Phytoplasma), and algal leaf spot (*C. virescens* Kunze ex E.M.Fries) are considered minor, occurring mostly in badly managed cultivations.

8.3.1 Major Diseases in Cinnamon

8.3.1.1 Rough Bark Disease (*Phomopsis* sp.)

Level of Importance Rough Bark Disease (RBD) has become a major threat to the cinnamon industry in Sri Lanka, causing severe losses of yield and quality of the product.

Pathogen The causal agent has been identified as *Phomopsis* sp. (Jayasinghe and Ratnasoma 2013). In another study, the involvement of several fungal species has been shown (D. M. de Costa, Personal communication). A scab-like condition, similar to RBD and affecting the quilling efficiency, was also reported (Kumara 1999a).

Symptoms Disease development initiates with the appearance of tiny black dots in 4–6 months old, greenish, semi-hardwood stems. With the advancement of disease, the black dots enlarge and turn into brown color lesions with a prominent black color margin which is characteristic of RBD (Fig. 8.14c). Within 2–3 months, the disease reaches a critical or irrecoverable stage where the fungus invades the xylem tissues of the stem causing blockage of water uptake, and also minerals, and interveinal chlorosis of foliage (Fig. 8.14e) (Jayasinghe et al. 2017).

The infected plants at a critical stage may exhibit branch die-back, leaf shedding, and/or poor stem growth above the diseased tissues. Symptoms of deficiency of nutrients, N, P, K, Mg, Ca, or Fe, appear when the disease progresses, although the soil nutrient levels remain unchanged at different stages of disease progression, suggesting that RBD infection blocks the upward movement of mineral nutrients in the plant. RBD reduces the quality, color, thickness, and the moisture content of bark and the quantity of bark oil (Jayasinghe et al. 2018). The efficiency of peeling is also adversely affected as diseased barks are difficult to be peeled off.

Though the pattern of progression of rough bark disease resembles a typical growth curve, it shows a slower disease progress within the first 10 weeks after initiation (Jayasinghe et al. 2018). This lengthy initial period of slower disease progress appears to be the easiest stage for management of the disease with lesser effort and cost.

Fig. 8.14 Rough bark disease (*Phomopsis* sp.) in cinnamon: (**a**) Initial stage of the damage, (**b**) and (c) moderate level of damage, (**c**) severe damage with bark splitting, (**d**) final stage of the disease, the leaves show interveinal chlorosis with leaf necrosis. (Photographed by G. G. Jayasinghe)

Disease Management

Cultural Practices The fungus survives on diseased crop residues. The spread of RBD can be controlled by removing and destroying diseased plants and by adoption of appropriate cultural practices such as harvesting at correct intervals and removing infected branches periodically.

Chemical Control Protectant fungicides, 1% Bordeaux mixture, copper-based fungicides, copper hydroxide or copper oxychloride and systemic fungicides have been tested with considerable success (Jayasinghe and Ratnasoma 2013).

8.3.1.2 Leaf Blight (*C. gloeosporioides*)

Level of Importance Leaf blight, caused by *C. gloeosporioides* (Anandaraj and Devasahayam 2004; Kumara 1999a), can be observed in almost every cinnamon growing area in Sri Lanka. The disease directly affects the foliage.

A moderate level of foliar damage (18%) had been reported due to this disease in Matara District (Southern Province, Sri Lanka); however, any significant correlation between the cinnamon yield and the disease severity has not been established (Kumara 1999a).

Critical Factors Affecting the Disease The fungus initiates infections in younger leaves during rainy weather. Shady and humid conditions can predispose the plants to infection and increase the severity of disease. Although the disease appears in cinnamon crop at all stages of growth, the plants bearing new flush and those at the nursery or young stages can be heavily affected, resulting in the collapse of seedlings and retardation of growth. A considerably higher disease incidence was observed in poorly weeded lands with greater shade and planting densities, and improper pruning practices (Kumara 1999a, b).

Symptoms The symptoms in young leaves include the development of small, brown specks on the leaf lamina which later coalesce to form larger, irregular necrotic lesions. These may be spread further giving to the entire leaf a scorched appearance (Fig. 8.15). The lesions remain as large, brownish areas on older leaves. The entire diseased areas may later become papery with dark brown margins. Sometimes the central portion of the spot is shed, giving the lesion a shot-hole appearance. Lesions may extend from either the tip of the leaf or from the leaf margins. With severe infections, lesions may become larger than a half of the leaf. In some seedlings, the infection spreads to the stem causing the leaves to be shed and finally leading to die-back (Karunakaran and Nair 1980).

Fu and Chang (1999) have reported the occurrence of brown to black spots on *Colletotrichum*-infected leaves of *Cinnamomum verum* in Taiwan. These spots later coalesce into larger areas and the infected leaves are shed. The pathogen was identified as *C. gloeosporioides* or its sexual form, *Glomerella cingulata*. Partial drying

Fig. 8.15 Leaf blight disease (*Colletotrichum gloeosporioides*) in cinnamon, (**a**) initial symptom, (**b**) damage in young plantation, (**c**) severe damage, (**d**) nursery plants severely affected in the rainy season. (Photographs (**a**, **b**, **d**) by G. G. Jayasinghe and (**c**) by K. L. Wasantha Kumara)

of the seedlings due to *C. gloeosporioides* (*G. cingulata*) was also demonstrated, 7 days after artificial inoculation of healthy cinnamon seedlings with the pathogen (Bhat et al. 1988). This can be another stage of the disease or a symptom under the prevailing conditions.

Pathogen The genus *Colletotrichum* comprises a highly diverse group of pathogens infecting a wide range of plant hosts. Recent molecular studies have enabled researchers to identify many new *Colletotrichum* species and understand the taxonomical position of species that were known for decades and identified using morphological characteristics.

The entry and infection of host tissues by *Colletotrichum* spp. generally starts with the germination of conidia and the formation of infection structures, appressoria, that facilitate entry through the host cuticle, and epidermal cell walls. Infected leaves, twigs, etc., function as sources of inoculum which is generally dispersed by rain splash.

Disease Management

Cultural Practices Removal and destruction of infected parts improve sanitary conditions and lower fresh *Colletotrichum* infections.

Chemical Control Many effective, systemic fungicides are available to combat *Colletotrichum* infections in the field. Copper-based fungicides are also recommended to control the disease. Application of water-soluble sulfur on the emerging new flush, prior to initiation of leaf blight symptoms at nursery stage as a preventive measure, was found to be very effective in preventing *C. gloeosporioides* infection (Dharshanee et al. 2010).

8.3.1.3 White Root Disease (*R. microporus;* syn. *R. lignosus*)

Level of Importance White root disease, caused by *R. microporus* (Fr.) Overeem (Syn. *Rigidoporus lignosus*), is regarded as one of the most destructive root diseases in rubber (*Hevea brasiliensis*) plantations. The fungus, causing white root rot of tropical crops, was first described as a pathogen of rubber.

Critical Factors That Affect the Disease The disease is commonly found in cinnamon when it is planted in lands that had previously been used for rubber cultivation. The cinnamon tree is somewhat resistant to the fungus. However, under stress conditions during dry periods or when the cinnamon crop is cultivated in close vicinity to rubber plantations or in lands where cinnamon was planted after uprooting rubber, the plant is vulnerable to white root disease. *R. microporus* persists in dead or live root debris for a long time and causes new infections in healthy plants.

Symptoms Externally, yellowing and subsequent shedding of leaves, wilting, and die-back of branches or the whole tree are the main above ground symptoms of white root disease. The external white rhizomorphs are firmly attached to the root and the collar which become yellowish and later reddish. The leading edge of mycelium that is advancing appears as a continuous sheath, like a fan on the bark surface.

Pathogen The fungus belongs to the Phylum Basidiomycota and is classified under Agaricomycotina, Agaricomycetes, Polyporales, and the Family Meripilaceae. Fruit bodies or basidiocarps are formed at advanced stage of the disease or after the death of the infected trees (Fig. 8.16). The basidiocarp is brownish-orange with a bright yellow margin when fresh, while the lower surface appears reddish-brown and shows characteristic concentric zones (Omorusi 2012). The fungus has a wider host range, extending to over 100 woody plant species (Jayasuriya and Thennakoon 2007).

Disease Management Integrated approaches are vital for long-term management of white root disease in cinnamon.

Cultural Practices Application of sulfur dust to planting holes is recommended at the time when cinnamon cultivations are established, especially when in close vicin-

Fig. 8.17 Brown root rot disease (*Phellinus noxius*) in cinnamon, (**a**) above ground symptoms, (**b**) root symptoms, (**c**) leaf symptoms, and (**d**) basidiocarp development in dead trees or stumps. (Photographed by G. G. Jayasinghe)

ity to rubber plantations or lands where rubber was cultivated previously. Infected, dead plants and their roots must be uprooted and burned as a preventive measure to stop spreading the disease.

Chemical Root bases should be cleaned and an effective fungicide must be applied to the base of infected plants.

Biological Possible use of *Trichoderma harzianum* isolates, antagonistic to *R. microporus*, in the control of white root disease has been investigated and some *T. harzianum* strains collected from rubber established soils have shown significant control of pathogen in vitro (Jayasuriya and Thennakoon 2007).

8.3.1.4 Brown Root Rot (*P. noxius* (Corner) G. Cunn.)

Brown root disease is caused by the fungus, *P. noxius* (Chang and Yang 1998), especially when *C. zeylanicum* plants are grown under unsatisfactory drainage conditions or in shade. The fungus is known to have a wide host range (Ann et al. 2002). Brown root disease caused by *P. noxius* is also reported in *Cinnamomum camphora* (Chang 1992).

The fungus belongs to Basidiomycota and classified under the Phylum Basidiomycota, Agaricomycotina, Agaricomycetes, Hymenochaetales, and the Family Hymenochaetaceae.

Symptoms Leaves in infected trees turn brown and show wilting symptoms, and some bushes immediately die. Leaf wilting in brown root rot affected trees takes place faster than those in plants with white root disease. The fungus produces thin, hard and uneven basidiocarps which are initially yellowish-brown with a white margin and later on become brown and then dark gray when the host plant dies (Fig. 8.17).

Disease Management

Cultural Lands which were previously used for rubber cultivation or lands in close vicinity to established rubber plantations should be avoided for cinnamon cultivation. Provision of sufficient sunlight to the cinnamon crop by removing nearby trees, branches, etc., that cut off the sun and improving drainage would help reduce the incidence of disease. Infected and dead plants must be uprooted with the root system and burnt to destroy the pathogen.

Chemical Root bases of plants, suspected to be infected, or plants with initial symptoms, should be cleaned and treated with an effective fungicide. Alternatively, sulfur may be applied to the bases of infected plants.

Fig. 8.16 White root disease (*Rigidoporus microporus*) in cinnamon: (**a**) initial symptom, (**b**) a plant that died due to the disease, (**c**) basidiocarps develop at the final stage of the disease, and (**d**) clearly visible fungal mycelium in infected roots. (Photographed by G. G. Jayasinghe)

8.3.1.5 Stripe Canker (*Phytophthora cinnamomi*)

Stripe canker of cinnamon is caused by an Oomycete, *Phytophthora cinnamomi* (Rands 1922). *P. cinnamomi* attacks shoots and young stems of cinnamon. Rands (1922) first reported *P. cinnamomi* causing severe losses to forest trees and avocado (*Persea americana*) and as the causal agent of stripe canker of *C. verum*. The pathogen also infects other cinnamon species, *C. camphora, C. culitlawan* and *C. sintok* (Rands 1922; Djafaruddin and Hanafiah 1975).

P. cinnamomi, isolated from pineapple, was also reported pathogenic on cinnamon, but with reduced virulence (Anandaraj and Devasahayam 2004). This may indicate the possibility of cross infection by *P. cinnamomi.*

Symptoms Stripe canker is found on the trunks and branches of *C. verum* and *C. burmannii*, particularly of young trees in Indonesia (Mehrlich 1934). Vertical stripes are seen on the stems with amber color exudates at the advancing margins which harden later. Vertical stripes of dead bark are most numerous near ground level (Fig. 8.18). The disease is prevalent mostly on ill-drained soils.

Pathogen *P. cinnamomi*, belonging to the Phylum Oomycota, is no longer considered a true fungus to be grouped within the Kingdom Fungi. Oomycota members that produce motile spores and grow as hyphae with cellulose containing walls, are included within the Kingdom Chromista together with brown algae and diatoms. *P. cinnamomi,* is largely a pathogen of woody plants.

Disease Management Improving soil drainage may keep the disease incidence at a lower level. In India, field sanitation such as removal and destruction of affected parts and wound dressing with tar have been recommended as control measures (Anandaraj and Devasahayam 2004).

Fig. 8.18 Cinnamon Stripe Canker (*Phytophthora cinnamomi*): (**a**) initial stage of infection, (**b**) rupturing of the bark, (**c**) longitudinal split of bark, (**d**) final stage of the disease and the start of the die-back. (Photographed by G. G. Jayasinghe)

8.3.2 Minor Diseases

8.3.2.1 Gray Leaf Spots/Blight (*Pestalotia* sp. Petch)

Causal Organism Gray blight, caused by *Pestalotia* sp. was reported as one of the commonest diseases of cinnamon in Sri Lanka and it has been reported in India as well (Narendra and Rao 1972). A disease with similar symptoms on *C. verum*, causing foliar damage up to 90% was reported from India and the causal agent was identified as *Pestalotia palmarum* (Karunakaran et al. 1993). Similarly, in the Dominican Republic and Pakistan, the disease with similar leaf spot symptoms was reported to be caused by *Pestalotia furierea* in bay leaves (*cinnamomum tamala*) (Ciferri and Fragosa 1927; Wadud et al. 2017).

Symptoms Small, yellowish brown spots appear as initial symptoms on the cinnamon leaves and later on the spots turn gray with a sharp border and spread into the leaf lamina. In older lesions, dark acervuli are produced which appear as black dots in the center of the lesion. The disease causes severe damage and defoliation in cinnamon.

8.3.2.2 Black Sooty Mold (*Stenella* spp.)

Sooty molds are saprophytic fungi, living epiphytically on leaves and stems, forming black mycelial mats. They belong to several different families of Dothideales (Ascomycota) and are particularly abundant in the tropics. *Stenella* spp. appear to be the sooty molds in cinnamon and do not show any host preference. Colonies may consist of mixed populations of eight or more species. The disease is not economically important and therefore, fungicidal control is not recommended in Sri Lanka (Bavappa et al. 1996).

Since the sooty molds do not cause any damage to plants, except for reducing sunlight falling on leaves and reducing photosynthesis, their presence in cinnamon

Fig. 8.19 Sooty molds and algal leaf spots that appear under unsatisfactory crop management conditions, (**a**) cinnamon growing under shade, (**b**, **c**) sooty mold (*Stenella* spp.) on cinnamon leaves, (**d**, **e**) algal leaf spots (C. virescens). (Photographs (**a**, **b**, **d**, **e**) by G. G. Jayasinghe, and (**c**) by K. L. Wasantha Kumara)

is not considered as an adverse situation. Sooty molds are very common in cinnamon, especially under shady condition (Fig. 8.19a–c).

8.3.2.3 Red Rust or Algal Leaf Spots (*C. virescens*)

Growth of algal colonies on leaves is common, especially when proper field sanitation practices have not been adopted in the cultivation. Algal growth is not considered important in terms of any damage to the host plant. The condition is referred to as red rust caused by the brown alga, *C. virescens*. Small, orange or brownish spots with a velvet appearance are seen on the leaf surface and their enlargement is not common or is very slow (Fig. 8.19d–e). Algal growth on leaf surface is generally encouraged by shady conditions (Bavappa et al. 1996).

Red rust can be avoided by adopting proper sanitation practices and by clearing nearby large trees and reducing shade. There are no other control measures practiced as the condition is observed occasionally in cinnamon lands.

8.3.2.4 Pink Disease (*Corticium salmonicolor* Berk. & Broome)

This disease has been reported from cinnamon in Sri Lanka, India, and Indonesia and the causal organism is identified as *C. salmonicolor* (Weiss 2002).

Symptoms The diseased areas appear first as pale pinkish white encrusts on stems and branches which later spread into larger areas in the bark. Pink disease can lead to death of smaller shoots at advanced stages of the disease.

Pathogen Pathogen has a wide host range and infects large trees, mango, jackfruit, custard apple, etc. The fungus is classified under the Phylum Basidiomycota, Agaricomycotina, Agaricomycetes, Corticiales, and the Family Corticiaceae. The spread of the disease can be slowed down by removing and burning of affected plants.

Fig. 8.20 Swollen fruit disease (*Exobasidium cinnamomi*) in cinnamon: (**a**) initial stage and (**b**, **c**) advanced stage of the disease. (Photographs by G. G. Jayasinghe)

8.3.2.5 Other Diseases

E. cinnamomi infects leaves and fruits cinnamon in Sri Lanka (Weiss 2002). Initially the fungus infects the leaves producing small, yellowish concave spots. Grayish-white spore bodies are produced on the lower surface of the leaf. Infected fruits show large, swollen, multiple blisters which make the fruits shrink and wrinkle (Bavappa et al. 1996) (Fig. 8.20). The disease is described as swollen fruit.

In another report, *Diplidia* spp. were reported as causing small light brown patches on cinnamon stems leading to stem blight and subsequent death of young seedlings in the nurseries (Da Camera Edes 1933).

Aecidium cinnamomi, Leptosphaeria spp., and *Colletotrichum capsici* cause leaf spots in cinnamon (Weiss 2002; Prakasam 1991). Further, Hosagoudar (1984) has described a fungus, *Caeoma keralense* (Syn. *Caeoma keralensis*), causing hypertrophy and witches' broom on young shoots of cinnamon trees in India.

8.4 Vertebrate Pests and Their Management in Cinnamon

Vertebrate pests also often pose problems to cinnamon cultivations. Different strategies need to be employed against vertebrate pests as most of them are protected animals. The strategies need to be economically viable, environmentally sound, and socially accepted.

Most vertebrate pests damage the newly generated shoots after a harvest (Fig. 8.21a). Damage to shoots can significantly slow down plant growth and reduce the yield. After damage to young shoots, the coppiced cinnamon plant does not have the erect stems that are essential for the preparation of cinnamon quills.

Peacock (Fig. 8.21f), mouse deer (Fig. 8.21g), barking deer (Fig. 8.21h), squirrels, hares, and stag are the most common vertebrate pests in cinnamon cultivations that are in close proximity to forest reserves. Most vertebrates are nocturnal, making identification difficult. The nature of the damage is often used to identify the pest. Ultrasonic sound devices with different frequencies (Fig. 8.21e) that were tested in

Fig. 8.21 Vertebrate pests in cinnamon and their management: (**a**) nature of damage (most vertebrate pests eat newly emerged shoots that develop after a regular harvest), (**b**) hiding new shoots behind the remaining twigs, (**c**) temporary fencing using locally available material, (**d**) sound devices used for repelling animals, (**e**) ultrasonic sound devices used to repel animals, (**f**) peacock (*Pavo cristatus*), (**g**) mouse dear (*Moschiola kathygre*), and (**h**) barking dear (Muntiacus muntjak malabaricus)

repelling peacock from cinnamon cultivations provided a significant reduction of damage to cinnamon.

References

Ahmad I (2017) Integrated pest management of *Zeuzera coffeae* Nietner: an efficient approach to reduce the infestation of walnut trees. Pak J Zool 49(2):693–698

Ahmed SA (2013) *Attacus atlas* L. (Lepidoptera: Saturniidae): a new record on *Ailanthus excels* Roxb from Assam, India. World J Zool 8(2):127–130

Anandaraj M, Devasahayam S (2004) Pest and diseases of cinnamon and cassia. In: Ravindran PN, Babu KN, Shylaja M (eds) The genus *Cinnamomum*: medicinal and aromatic plant – industrial profiles. CRC Press, New York, pp 239–259

Ann PJ, Chang TT, Ko WH et al (2002) *Phellinus noxious*, Brown root rot of fruit and ornamental trees in Taiwan. Plant Dis 86(8):20–26

Anonymous (2017) Analysis of cinnamon, pepper and cardamom value chains in Sri Lanka. Institute of policy studies of Sri Lanka, p 19

Anonymous (2018) Export agricultural crops, stat book. Department of Export Agriculture, Peradeniya, pp 13–19

Bavappa KVA, Ruettiman RA, Gunarane WDL, Abeykoon AMD et al (1996) Cinnamon: cultivation and processing. Technical Bulletin 5, Department of Export Agriculture, Ministry of Agric., land and forestry, Sri Lanka, p 8

Bell TR (1912) The common butterflies of the plains of India. J Bombay Nat Hist Soc 21:517–544

Bhat MN, Hegde RK, Hiremath PC, Naid KS et al (1988) A note on the occurrence of a dieback on cinnamon in Karnatake. Curr Res Univ Agric Sci Bangalore 17(11):153

Bhumannavar BS (1991) New records of *Sorolopha archimedias* on cinnamon in South Andaman. J Andaman Sci Assoc 7(2):82–83

Butani DK (1983) Spices and pest problems – 2. Cinnamon Pestic 17(9):32–33

Chang CP (1984) The morphology and life history of *Zeuzera coffeae* Nietner on grapevine in Taiwan. Plant Protect Bull Taiwan 26:145–153

Chang TT (1992) Decline of some forest trees associated with brown root rot caused by *Phellinus noxius*. Plant Pathol Bull 1(2):90–95

Chang TT, Yang WW (1998) *Phellinus noxius* in Taiwan: distribution, host plants and the pH and texture of the rhizosphere soils of infected hosts. Mycol Res 102:1085–1088

Ciferri R, Fragosa GR (1927) Parasitic and saprophytic fungi of the Dominican Republic. Bot R Soc Expanola Hist Nat 27(6):267–280

Dalandawatta NP, Jayasinghe GG, Chandrasena G, Weerawansha ANR et al (2015) Assessment of oil yield and quality in cinnamon (*Cinnamomum zeylanicum* Blume) leaves under different severity levels of two types of leaf galls (*Trioza cinnammomi, Erioplytes boisi*). In: Marambe B (ed) Annual symposium of minor export crops, vol 3, 13–14 August. Peradeniya, Sri Lanka, pp 91–96

Dassanayake NC, Wanigasundara WADP, Balasuriya A, Amarathunga MKSLD et al (2009) A field assessment of the factors affecting Horse Hair Blight (*Marsmius equicrinis*) in tea in Ratnapura districts. J Agric Sci 4(2):59–66

Devashayam S, Koya KMA (1997) IPM in spices- challenges for the future. In: Kumar NKK, Verghese A (eds) Proc. First Naiona Symposium on pest management in horticultural crops. Indian Inst. Spices Res, Bangalore, pp 157–164

Dharmadasa M, Jayasinghe GG (2000) A clearwing moth (*Synanthedon* spp.), a new pest damage in cinnamon cultivations and its damage severity in Sri Lanka. Proc Sri Lanka Assoc Adv Sci Colombo:91 (Abs)

Dharshanee HLC, Dharmadasa M, Nugaliyadda L, Wijesinghe KGG et al (2008) Development of a pheromone based management method for cinnamon wood boring moth (*Ichneumoniptera cinnamomumi*) (Lepidoptera: Sesiidae). Proc Natl Symp Fac Agric Univ Ruhuna Sri Lanka:31

Dharshanee HLC, Jayasinghe GG, Wijesinghe KGG et al (2010) Evaluation of the efficacy of water soluble Sulfur to control leaf blight diseases (*Colletotrichum gloeosporioides*) and leaf gall mites (*Eriophyes boisi*) of cinnamon (*Cinnamomum zeylanicum* Blume) nursery. Proc Sri Lanka Assoc Adv Sci Part 1:36(Abs)

Djafaruddin M, Hanafiah A (1975) Research on cinnamon diseases especially canker (*Phytophthora cinnamomi*). West Sumatra, Ris Mc, Bogor, Indonesia, p 76

Duckworth WD, Eichlin TD (1978) The type material of Central and South American clearwin moth (Lepidoptera: Sesiidae). Smithsonian Contribution in Zoology 261:1–28

Edes DC (1933) Contribution to mycoflora of Portugal century X. Rev Agron 20:5–63

Fu CH, Chang TT (1999) Anthracnose of cinnamon. Taiwan J Forest Sci 14:513–515

Grassi A, Zini M, Forno F et al (2002) Mating disruption field trials to control the current clear-wing moth, *Synanthedon tipuliformis*: a three year study. Bulletin – OILB/SRO 25(9):69–76

Hollis D, Martin JH (1997) Jumping plant lice attacking avocado pear trees *Persea americana* in the New World, with a review of Lauraceae feeding among psylloids. Bull Entomol Res 87:471–480

Hosagoudar VB (1984) Two interesting fungi on *Cinnamomum malabatrum* from Idukki, Kerala, India. J Econ Taxon Bot 5(1):209–211

Jayasinghe GG (2013) Integrated management of cinnamon clear wing (*Ichneumoniptera cinnamomumi* Tosevski) using earthing-up, insecticides and sex pheromones. In: Marambe B (ed) Proc. ASMEC, pp 73–79

Jayasinghe GG (2015) Determination of economic injury levels of wood boring moth (*Ichneumoniptera cinnamomumi* Tosevski) with reference to the age of cinnamon (*Cinnamomum zeylanicum* Blume) plantation. In: Marambe B (ed) Proc. ASMEC, vol 3, 13–14 August. Peradeniya, Sri Lanka, pp 62–69

Jayasinghe GG, Ratnasoma HA (2013) Efficacy of fungicides for managing rough bark disease (*Phomopsis* sp.) in cinnamon (*Cinnamomum zeylanicum* Blume). Proc Sri Lanka Assoc Adv Sci:35 (Abs)

Jayasinghe GG, Wickramasinghe PJ (2001) Control measures of wood boring moth (*Synanthidon* sp.) in cinnamon (*C. verum* Presl.). Proc 57th Ann Sessions Proc Sri Lanka Assoc Adv Sci Part 1:59 (Abs)

Jayasinghe GG, Gunaratne WDL, Dharshanee HLC, Griepink FC, Louwaars NP, Stol W et al (2006) Environmentally-sound insect control in cinnamon: a feasibility study on the use of insect pheromones to replace large-scale use of insecticides, report 119. Plant Research International B.V, Wageningen

Jayasinghe GG, Ketakumbura KHMPK, Wijesinghe KGG, Hemachandra KS, Weligamage S et al (2016) Relationship between pest and disease incidences and agronomic operations implemented by farmers in cinnamon (*Cinnamomum zeylanicum* Blume) fields in Southern Sri Lanka. Sri Lankan J Food Agric 2(1):33–38

Jayasinghe GG, Kumara KLW, Wijayawardhana MWGC, Wijesinghe KGG, Priyangika KMM et al (2017) Rough bark disease of cinnamon (*C. verum*): disease symptoms development and the causal agent with special reference to its morphology, histopathology and nutritional states of affected plants. In: Marambe B (ed) Proc. Symposium Minor Export Crops, vol 4, 16–17 March 2017. Peradeniya, Sri Lanka, pp 63–72

Jayasinghe GG, Kumara KLW, Sandamali SAAW, Maddegoda C, Tharanga KHGM et al (2018) Studying nutritional and physicochemical changes with the disease progression of rough bark disease (*Phomopsis* spp.) in cinnamon (*Cinnamomum zeylanicum* Blume). Proc. 2nd Internat. Res. Symp. Sri Lanka Council for Agric. Res. Policy, Taj Samudra Hotel, Colombo, 9th–10th Aug 2018 (Abs)

Jayasuriya KE, Thennakoon BI (2007) Biological control of *Rigidoporus microporus*, the cause of white root disease in rubber. Ceylon J Sci (Bio Sci) 36(1):9–16

Karunakaran P, Nair MC, Das L et al (1993) Grey blight disease of cinnamon (*Cinnamomum verum* Bercht. & Presl.) leaves. J Spices Aromat Crops 2(1–2):66–67

Karunakaran P, Nair MC (1980) Leaf spot and die back disease of *Cinnamomum zeylanicum* caused by *Collectrotricum gloesporioides*. Plant disease 64:220–221

Kostermans AJGH (1996) Lauraceae. In: Dassanayake MD, Fosberg FR, Clayton WD (eds) A revised handbook to the flora of Ceylon, vol 9. Amerind Publishing, New Delhi, pp 105–172

Kumara KLW (1999a) Effect of some agronomic practices on the incidence of leaf spot disease (*Colletotrichum gloeosporioides)* in cinnamon (*Cinnamomum zeylanicum* Blume) in southern region of Sri Lanka. SLAAS-University of Ruhuna Seminar on "Pest control in the next millennium". Proc Sri Lanka Assoc Adv Sci, 29th Nov.–3rd Dec 1999, Colombo, Sri Lanka (Abs)

Kumara KLW (1999b) Study of leaf galls, leaf spots and scab-like condition of cinnamon (*Cinnamomum zeylanicum* Blume) in Matara District. Proc Sri Lanka Assoc Adv Sci, 29th Nov.–3rd Dec 1999, Colombo, Sri Lanka, p 65

Linnaeus C (1758) Systema naturae, 10th edn. https://en.wikipedia.org/wiki/Attacus_atlas

Mani MS (1973) Plant galls of India. Macmillan Co., India Ltd, New Delhi

Mehrlich FP (1934) Physiological specialization in *Phytophthora* species. Phytopathology 24:1139–1150

Narendra DV, Rao VH (1972) Some addition to the fungi of India. Sydowia 26:282–284. https://www.zobodat.at/web4beta/pdf/Sydowia_26_0282-0287.pdf

Omorusi VI (2012) Effect of white root rot disease on *Hevea brasiliensis* (Muell. Arg.): challenges and control approach. https://doi.org/10.5772/54024, http://creativecommons.org/licenses/by/3.0. Retrieved on 1 Sept 2018

Perera HAS, Sritharan R, Perera KP et al (1985) Some studies on cinnamon galls in Sri Lanka. Sri Lankan J Agric Sci 22:23–27

Premkumar T (1988) *Alcides morio* Heller (Curculionidae: Coleoptera) cinnamon fruit borer. Entomon 13:187

Premkumar T, Devasahayam S, Koya KMA et al (1994) Pests of spice crops. In: Chadha KL, Rethinam P (eds) Advances in horticulture, Plantation and spice crops part-2, vol 10. Malhotra Publishing House, New Delhi, pp 787–823

Prakasam V (1991) Red leaf spot of Cinnamon in lower Pulney hills of Tamil Nadu. Indian Cocoa, Aricanut and Spices Journal 14(3):123

Rajapakse RHS, Kulasekara VK (1982) Some observations on the insect pests of cinnamon in Sri Lanka. Entomon 7:221–223

Rajapakse RHS, Ratnasekera D (1997) Studies on the distribution and control of leaf galls in cinnamon caused by *Trioza cinnamoni* Boselli in Sri Lanka. Int J Trop Agric 55(1–4):53

Rajapakse RHS, Wasantha Kumara KL (2007) A review of identification and management of pests and diseases of cinnamon (*Cinnamomum zeylanicum* Blume). Trop Agric Res Ext 1:1–10

Rands RD (1922) Stripe canker of cinnamon caused by *Phytophthora cinnamomi* n. sp. *Meded.* Inst Voor Plantenziekten 54:53

Rutherford A (1913) *Zeuzera coffeae* (Red borer; Coffee borer). Trop Agric 41:486–488

Sahabandu TKH, Rajapakse R, Ratnasekera D et al (1998) The effect of three pesticides and pesticide interval of *Cinnamomum veram* galls in Southern Sri Lanka. Madras Agric J India 85(7):9

Singh G, Maurya S (2005) Antimicrobial, antifungal and insecticidal investigations on essential oils: an overview. Nat Prod Res 4(3):179–192

Singh V, Dubey OP, Nair CPR, Pillai GB et al (1978) Biology and bionomics of insect pests of cinnamon. J Plant Crops 6:24–27

van der Poorten G, van der Poorten N (2011) New and revised description of the immature stages of the some butterflies in Sri Lanka and their larval food plants (Lepidoptera: Papillionidae) The Journal of Research on the Lepidoptera 44:111–127

Wadud MA, Fahim AHF, Naher MS, Sarker MB, Uddin MJ, Humauan MR et al (2017) Management of grey leaf spot/blight disease of bay leaf (*Cinnamomum tamala*). Eco-friendly Agric J 10(07):90–95

Walgama RS (2012) Ecology and integrated pest management of *Xyleborus fornicatus* (Coleoptera: Scolytidae) in Sri Lanka. J Integr Pest Manag 3:A1–A8

Warren JW, Moran VC (1978) The influence of the host plant on the population dynamics of *Acissa rusettae* (Homoptera: Psyllidae). Ecol Entomol 3:313–321

Weiss EA (2002) Essential oil crops. CAB international, UK, p 191

Chapter 9
Cinnamon Process Technology

K. D. N. Weerasinghe and N. P. G. Pushpitha

9.1 Introduction

In the wild, cinnamon grows to a height of 10–15 m and stem girth of mature cinnamon reaches 30–50 cm. However, in plantations the plants are pruned and trained, maintaining a bushy architecture, without harming the entire bush, to facilitate the cinnamon quill making. Only one or two peelable stems are harvested from the bush at a time, while keeping few immature stems for the next harvest. This process induces the emergence of new shoots at the base of the plant for future harvesting (Wijesekara et al. 1975).

In a new plantation, when the bushes attain the age of about 24 years, a height of about 2 m, and stem girth of around 3–6 cm, they are ready for harvesting. Cinnamon processors (peelers) collect the stems sufficient for the day's work. The harvested stems are bundled up and transported to peeling sheds where the stems are cleaned, brushed, and washed in a tub of water to remove extraneous material and to keep the stems fresh for peeling (Technical Publication 1996; Weerasinghe 2011).

9.2 Traditional Cinnamon Quill Making Process

The technological process traditionally used in Sri Lanka for cinnamon quill processing is unique to the Ceylon cinnamon industry. From field harvesting to producing the final product, that is, cinnamon quills, there are a number of steps involved: transporting to peeling sheds, cleaning and washing, removing knots, scraping,

K. D. N. Weerasinghe (✉)
Faculty of Agriculture, University of Ruhuna, Kamburupitiya, Sri Lanka

N. P. G. Pushpitha
Ocean University of Sri Lanka, Crow Island, Mattakkuliya, Colombo, Sri Lanka

R. Senaratne, R. Pathirana (eds.), *Cinnamon*,
https://doi.org/10.1007/978-3-030-54426-3_9

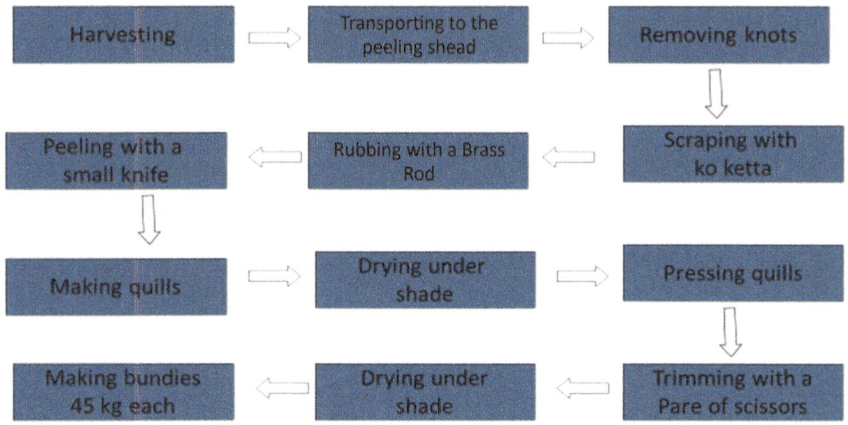

Fig. 9.1 Flowchart of traditional processing of Ceylon cinnamon (Weerasinghe and Pushpitha 2010)

rubbing, peeling, drying, and quill making (Basnayake and De Silva 2019). The different stages of this process are shown in Fig. 9.1 and explained in more detail in the next sections.

9.2.1 Field Harvesting

The ideal period for harvesting cinnamon for producing good quality quills is during the rainy season, and in Sri Lanka during May–August, the period coinciding with the South-West monsoon. Due to better sap flow during this period, the bark peels off easily. Harvesting begins in the early morning, 5.30–6.00 am. Skilled peelers visually identify mature shoots in a bush for harvesting, leaving the immature shoots for the next harvest. Before harvesting, the peeler makes a cut in the mature stem and lifts the bark to test the peelable nature of the bark of each stem, which he identifies as mature (Fig. 9.2, left). If there is any difficulty in detaching the bark, the peeler will reject the shoot (Pushpitha 2006).

All the field harvesting operations are done using a simple curved sharp knife (Fig. 9.2) with a long handle. The blade is made of quality steel and local blacksmiths usually fabricate it from a disposed leaf spring of motor vehicles.

The first action of the peeling is to remove knots on the stem, for which stem is held in one hand and the curved harvesting knife in the other. After removing the knots, the patch appears to be of button shape (Fig. 9.3a).

9.2.1.1 Scraping

Scraping is the removing of the epidermal tissue layer from the stems (Fig. 9.3b). A traditional hand tool named "*Koketta*" is used for this purpose. There are two types

Fig. 9.2 Stages of harvesting cinnamon for processing. Left, making a test cut; middle, harvesting mature shoots; right, bundled stems being transferred to peeling shed

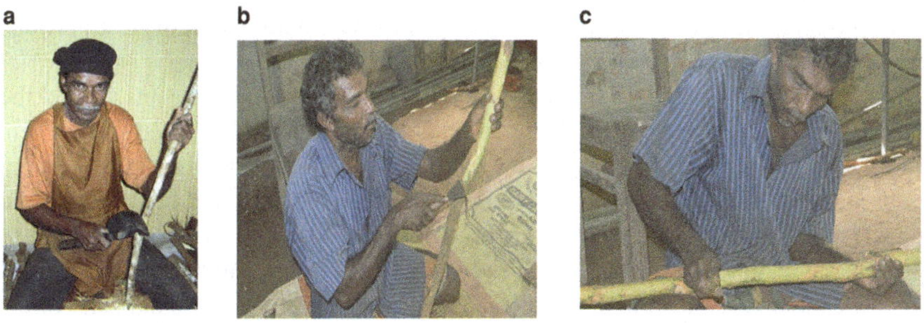

Fig. 9.3 (**a**) Removing knots with "*Keththa*". (**b**) Scraping with "*Koketta*". (**c**) Rubbing with a brass rod

of tools: one with a curved sharp blade and the other with a blade and small handle. The latter is called "*Sawthtuwa*" in Sinhalese.

The curvature of the blade is selected to match the diameter of the stems. Sometimes, peelers may use two or three scraping tools with different curvature for efficient scraping. Understanding the physical quality of the stems (roughness and maturity) is important for proper scraping. Therefore, the mechanization of this step is complicated. Stillness and physical properties of the stems (stick diameter, number of knots, straightness, etc.) have an effect on the scraping time.

9.2.1.2 Rubbing

Historically, bark was detached without rubbing. Subsequently, a piece of the wooden rod was used for rubbing. Later, it was replaced by a copper rod, which was subsequently replaced by the brass rod. The average diameter, length, and weight of the brass rod are 15 mm, 203 mm, and 1.1 kg, respectively. Rubbing is the most laborious step in cinnamon processing, which loosens the bark to detach from the core of the stem. Due to the exhausting nature of the rubbing process, only male peelers are employed for this (Pushpitha 2006). Time taken for rubbing varies with the diameter of the stem, its evenness, number of knots, season, cultivar, etc. For the stems harvested during suboptimal periods, extra effort has to be applied for rubbing. After 4–6 hours of rubbing, the productivity of the peeler diminishes. This can lead to bark damage and poor quality of quills. About 40–60 strokes at 9–30 N of vertical force have to be applied around a selected length of the stem for proper rubbing. During the rubbing process, bark sap oozes indicating proper rubbing. However, extreme rubbing can damage the bark (Gunasena et al. 1997; Pushpitha 2006).

9.2.1.3 Peeling

Peeling (bark detachment from the stem) is a skilled and time-consuming step. Just after rubbing, the stem is examined to decide the maximum length of bark portions that can be peeled off to make the outer cover of the quills.

Once the appropriate bark removing style has been decided, two cuts around the stems are made with a maximum length of intervals using a small pointed knife. Then, a longitudinal slit is made from one end to the other of the selected portion. Subsequently, the knife is inserted carefully between the bark and hardwood while raising and detaching the bark to peel it off. During this process, an attempt is made to prevent any damage to the bark (Fig. 9.4b). Finally, another longitudinal slit is made opposite the first slit and the bark is detached to two halves. When the diameter of the stem is high enough, the bark can be divided into three or even four strips along the stem. The unpeeled bark left on the inner side of bends or around the knots and other bark abnormalities are removed separately as small strips to fill the interior of the quills.

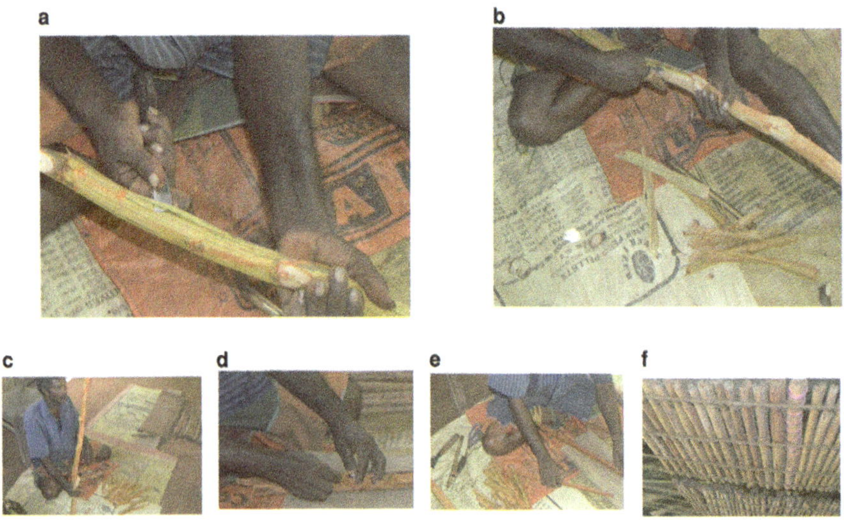

Fig. 9.4 (**a**) Peeling process using a small knife. (**b**) Removing rough particles. (**c**) Stick without bark. (**d**) Quill packing. (**e**) Trimming. (**f**) Quill drying

Fig. 9.5 Cinnamon bales made up of standard quills ready for transport

Although the ideal time for cinnamon harvesting is during the monsoon period, cinnamon can be harvested throughout the year except during the dry periods; in cinnamon growing areas in Sri Lanka, this is about 2 months. Peeling is quite difficult during the dry periods when soil moisture depletes. Mulching practice using cinnamon leaves is good to conserve soil moisture (Wijesekara et al. 1975; Gunarathne 2011)

After peeling the bark, long and intact peels that are used to make the outer cover of the quills are dried in the shade for 2–3 hours. During this period, bark curls inverted. In the rainy season, this period is extended up to 5–8 hours. At present, racks made out of coconut rope or steel are used to shorten the drying period (Fig. 9.4f).

Cinnamon quills are prepared by experienced peelers to maintain uniform thickness from end to end. Bark halves are packed one inside the other until cigar-like tubes are formed (Fig. 9.4d). The hollow inside of the quill is packed with pieces of thin bark, which are unsuitable to make the outer cover of the quill. A pair of scissors and a measuring stick of 107 cm, with a wooden lifter ("*Pethi Kotuwa*" in Singhalese), are used for quill making. When it reaches the required length, the end is trimmed with scissors and it is gently lifted and kept on a mat for further drying. They are trimmed with scissors when it is necessary.

Quills are air dried and pressed by hand to stack properly. Quills are covered by gunny bags or cadjan leaves to protect from the sun. The processed quills are stacked together to make a bundle of 45 kg, which are named as bales for transporting and marketing purposes (Fig. 9.5) (Administration Reports 2009–2018).

A skilled cinnamon peeler can produce 4–5 kg of dried processed cinnamon in a day. To achieve this target, the peeler has to peel about 50 harvested stems, working for 10–15 hours. Shortcomings of the existing method are low efficiency and high labor cost, which is about 50% of the farm gate price (cost of production).

9.3 Cinnamon Products

As discussed earlier, cinnamon quills represent the most valuable product from cinnamon. However, there are several other useful by-products generated during the processing of quills, which are classified into three major commercial groups: quills,

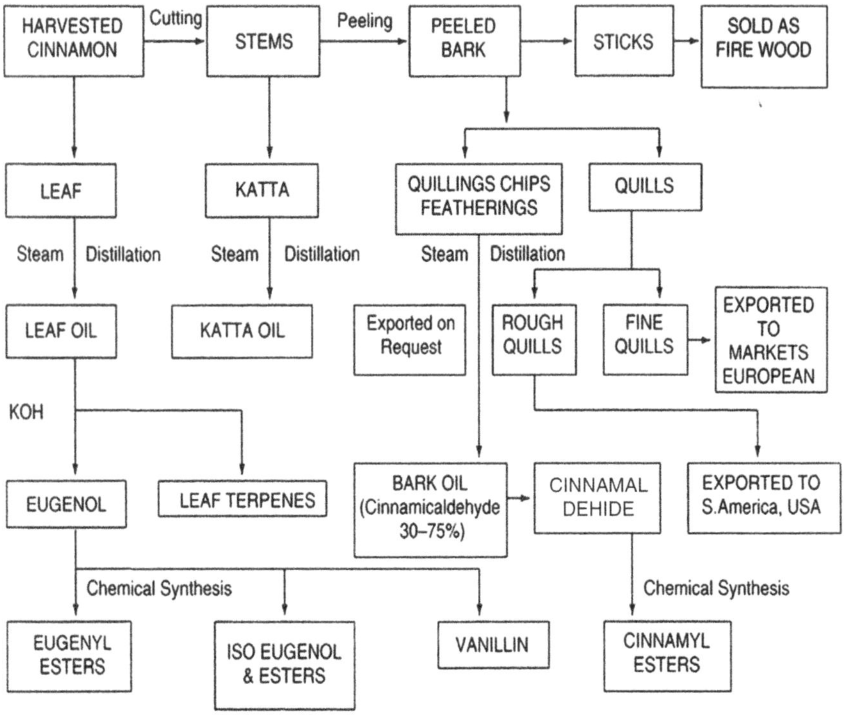

Fig. 9.6 Flowchart of products manufactured from cinnamon. (Adapted from Wijesekara et al. (1975))

quillings, and scrapes (*Katta*) that go for oil distillation (Dayananda 2011). The marketable products coming out from the cinnamon process industry are demonstrated in Fig. 9.6.

9.3.1 Cinnamon Quills

Cinnamon quills are the major product of the cinnamon industry. Quills are graded to nine different classes based on the diameter and number of quills contained in 1 kg of the market product (Table 9.1) (Fig. 9.7).

Maximum diameter of quilling representative of each grade and the number of quills constituting a kilogram of quills are given in Table 9.1. In lower grades, the diameter of quills increases up to 38 mm, constituting only seven sticks to form 1 kg. Price variation at the farm gate can be from LKR 2800 per kg (equivalent of US$ 16) for Alba grade to LKR 1600 per kg (US$ 9) for lower grades (Information on Export Agriculture Crops 2005-2009).

Table 9.1 Cinnamon quill grades

S. No. (Standard index)	Grade	Maximum diameter of quillings (mm)	No. of quills per kg
I	Alba	6	45
II	C5 sp	8	33
III	C5	12	27
IV	C4	16	22
V	M5	16	22
VI	M4	19	18
VII	H1	23	11
VIII	H2	25	9
IX	H3	38	7

Source: Specification for Cinnamon (1972, 1978, 1992)

Fig. 9.7 Traditional quill making technique

The most superior grade is the Alba grade constituting very thin diameter (<6 mm) quills with 45 quills/kg.

9.3.1.1 Quillings

Quillings are the broken pieces and splits of all grades of cinnamon quills. The main characteristics of quillings are their shape and size. The aroma and taste of quillings are the same as the quills, even though they are marketed as medium quality cinnamon.

Cinnamon featherings are feather-like pieces of inner bark consisting of shavings and small pieces of bark leftover in the process of making quills. Scraping of the bark or small twigs and stalks of cinnamon shoots including a minimum quantity of chips are also considered as featherings. The product is marketed as medium quality cinnamon.

Chips are not peeled out from the stem. Instead, they are scraped off greenish-brown mature thick hand-picked pieces of the bark, which are inferior quality cinnamon. Outer bark obtained by beating or scraping the shoots is also included in chips.

Chips are graded into two categories:

Grade 1: Those containing small featherings obtained by scraping very small immature twigs. They may contain a small amount of other bark materials as well.
Grade 2: Those containing inner and outer bark and pieces of wood.

Depending on the extent of being free from extraneous matter such as refuse and dust, the chips are cleaned by washing or bleaching and are further divided into four types (type 1, 3, 0, and 00).

Hygienic conditions, free from dust, mold, and other pathogenic organisms are important aspects of the production of cinnamon quills, quillings, featherings, and chips. Consumers directly use them in most of the food formulation and preparation (Specification for Cinnamon 1978, 1992).

9.4 Technology Shift from Traditional Processing to Machine Processing

From the foregoing, it should be evident that the cinnamon processing is still not modernized and is a labor-intensive cottage industry. It is a laborious and tedious process requiring a high level of hand-skill and knowledge. The technical know-how is passed down from generation to generation, within the same groups of families. Lack of mechanization of the processing industry is the critical factor hindering cinnamon expansion programs to meet the present-day demand.

Mechanization and automation to transform the technological process into a flow line operation is the solution to achieve greater productivity and reduce labor cost. The productivity improvement is difficult as long as the traditional cottage industrial processing is maintained. Even though up to 35–50% of sales revenue per kilogram of the finished product is paid as wages to the peelers, the younger generation does not come into the industry due to social attitudes linked to the prevailing caste system in the cinnamon sector to date.

A peeler can process about 2.0–3.5 kg of quills per day. As per the present market price of about LKR 2000 per kg (US$ 12), the peeler's income exceeds LKR 2500 (US$ 13.8) a day. As such, the most crucial issue faced by the industry today is the availability and accessibility of the peelers (Gunarathne 2011). Due to the dearth of peelers, there exists high labor demand, and peeler dominated pricing and sales arrangements. Growers are reluctant to invest in upgrading the plantations due to the high labor cost for processing. Educating and training the peelers is one of the promising ways to make them understand the importance of making quills of finer grades. Realizing the importance of peeler training, the Sri Lankan government introduced a certification system for peelers jointly with the Vocational Training

Institute of Sri Lanka (VTI), Spice Council of Sri Lanka, and the University of Ruhuna to transfer the traditional cottage industry system to a market-oriented process. The training of new peelers is important for a good quality product. Alternatively, diversifying products without depending on the quill form of cinnamon and the mechanization of peeling have been identified as possible avenues to revitalize and safeguard the cinnamon industry. In this endeavor, new market opportunities are to be explored and strategies have to be developed for product diversification (National Postharvest Research and Developmental Action Plan 2011).

As described earlier, the current quill peeling and post-harvest operations are carried out as a traditional cottage industry. However, to meet the different standards linked to Eurogap, HACCP, food safety, health and safety, and phytosanitary standards imposed on food products, the cinnamon industry is moving toward the machine industry to maintain flow line operations (Weerasinghe and Pushpitha 2010; Basnayake and De Silva 2019). The new trends in the cinnamon industry are to produce certified cinnamon to cater to the emerging world demand. In 2006–2008, a GTZ program assisted a group of nine producers and exporters of premium cinnamon to form a Union of U10 to produce quality cinnamon. U10 is intended to be an ISO and HAACP certified consortium, which is committed to push the boundaries of the cinnamon industry, focusing on the development of plantations, factories, and businesses in Sri Lanka (Weerasinghe and Pushpitha 2010). Another program initiated by GTZ has upgraded peeling centers to ISO 22000 certified factory level. Dahanayake Walauwa model factory, at Kosgoda in Galle district, Carlston Estate factory in Matara District are two premier modernized cinnamon processing centers with well-organized structures and plans (Fig. 9.8).

Majority of the labor employed in modern factories are females (Fig. 9.9), and they work in 8-hour working shifts. In the factory processing line, only one trained laborer handles the rubbing process, using the RUWEEKA rubbing machine developed by the University of Ruhuna (Weerasinghe et al. 1998a, b, 2006) to provide

Fig. 9.8 External view of a cinnamon peeling factory (Carlston Estate)

Fig. 9.9 Women employees in new cinnamon factories

peeling materials to about 18-member workforce of female peelers. On average, peelers process 3 kg of cinnamon quills per day per person who receives a daily wage of LKR 800 (US$ 4.40). An additional LKR.100 per kg is added to the laborers for additional production of quills. This new technology adoption is still to be improved in factory production lines for which selected factory managers, supervisors, and peelers are to be trained to change the traditional, labor-intensive method.

9.5 Constraints to Increase of Production and Value-Addition Through Processing

The main reasons hindering the cinnamon expansion are linked to small-scale farm units, adverse weather conditions during harvesting and post-harvest processing, poor storage facilities, lack of awareness about quality requirements among growers, lack of premium prices for high-quality products, and improper market information flow. Formation of cluster groups and associations is seen as a remedy for these shortcomings.

A detailed analysis and description of the traditional cinnamon process adopted in Sri Lanka is described in a report submitted to UNIDO (Weerasinghe 2011). The recommendations made to UNIDO by the above study are condensed as follows:

1. Assist to develop database and knowledge based on market potential and consumers for cinnamon and cassia.
2. Research and development to process technology innovation and adoption and to introduce new and added value products.
3. Assist in the active promotion of marketing and public relations improvement in Latin American countries.

4. Promote research program to select productive lines, breeding programs, biotechnology, and tissue culture to introduce high quality and high yielding planting materials faster.
5. Assist in the development of appropriate machinery, tools, and technology innovations that reduce labor-intensive processes; making a flow line process.
6. Develop centers of excellence and training centers for technology transfer, awareness building, and quality assurance. Establishment of quality testing centers at village levels.
7. Introduce the cluster approach for the organization, management and production of cinnamon, and its processed products.

The recommendations for the Department of Export Agriculture (DEA) include the following:

1. Strengthen the extension arm by recruiting an adequate number of extension personnel.
2. Strengthen the cinnamon research sector by recruiting more research staff and expanding the laboratory space and facilities.
3. Promotion program to replant and adopt recommended agronomical and agricultural practices to increase the production.

In 1997, the University of Ruhuna commenced a research program to study the whole processing steps to develop appropriate machinery, tools, and technology innovations to reduce labor and deskill tasks to make a flow line process (Pushpitha 2006), focusing on the following:

- Replacing the laborious steps by new strategies (transport, rubbing, etc.)
- Shorten the time-consuming steps (scraping, rubbing, drying, etc.)
- Attention on peelers health and comfort
- Process modernization to attract younger people to the industry
- Promote phytohygenic production to adhere to GMP
- Product diversification

9.6 Machinery Designs and Adoption

The technology shift from floor operations to the table processing that has taken place as a result of the research and development programs is demonstrated in Fig. 9.10. During the past 10 years, devices have been developed and adopted to ease the most difficult act of the rubbing process in cinnamon quill processing. Benches and devices have been introduced for table processing to incorporate flow line operations. Cinnamon dryers have been successfully introduced. A brief description of these equipment is given below.

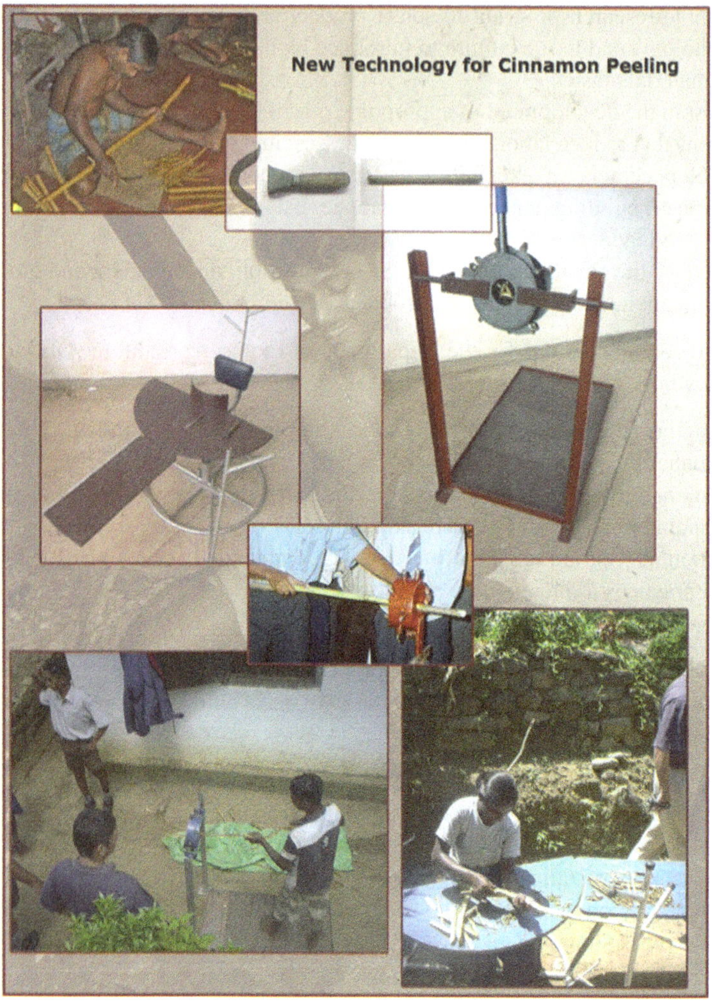

Fig. 9.10 Demonstration of technology shift in cinnamon peeling industry from floor operations to table operation (Weerasinghe 2011).

9.6.1 Instruments Available for Mechanization of Cinnamon Peeling Industry

9.6.1.1 Cinnamon Rubbing Machine RUWEEKA_CG (Patent No. 11869)

This device replaces the heavy labor involvement in rubbing by increasing the efficiency of up to 59.7%. Operation could be done by inserting the scraped sticks through the spring-loaded spindles of the machine by moving cinnamon sticks several time; here, a stick could be rubbed in 25 seconds (Fig. 9.11) (Gunasena et al. 1997; Weerasinghe et al. 1999, 2006).

Fig. 9.11 RUWEKA-CG rubbing device (Ruhuna-Weerasinghe-Karunarathne- Chandima Gunasena)

9.6.1.2 RUWEEKA-PG (Ruhuna-Weerasinghe-Karunarathne-Palliaguru)—International Patent Classification A23N 7/10, 11/00 B27J 3/00 Go5G 17/00;patent No. 17342

RUWEEKA-PG is the improved design by scaling up and incorporating the suitable mechanism to insert the sticks. This device works in the same principle as RUWEEKA–CG, and stems up to 5.6 cm diameter that can be inserted to the device (Weerasinghe and Pushpitha 2005a).

A simple mechanism is incorporated into the device to facilitate the inserting of sticks to the rubbing device. Pressing or pulling the handle of the mechanism spring-loaded spindles can expand at once to insert the stick and then shrink it. It is a machine mounted in a stand and could be rotated around its horizontal axis (Weerasinghe and Pushpitha 2005a). This helps to change the rubbing ends of the sticks easily (Fig. 9.12).

During 2004–2010, further modifications have been incorporated to the RUWEEKA–PG machine to overcome the difficulties associated with the rubbing of stems harvested during dry periods, for rough peeling. The tension adjustment mechanism is incorporated into the spindles with an adjustment screw to regulate the tension for efficient rubbing. The applicability of this machine was tested with success in August 2011, and the machine was developed in collaboration with the industry (Fig. 9.13).

Fig. 9.12 RUWEKA-PG with the stand and cam mechanism to insert sticks

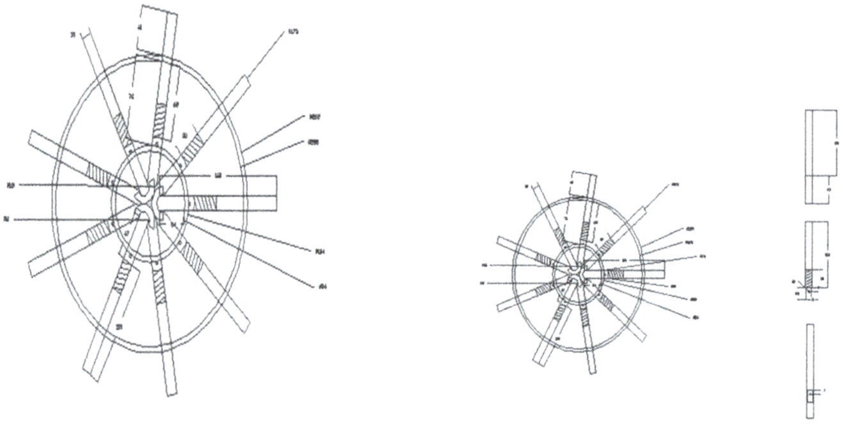

Fig. 9.13 Modified RUWEKA machine for cinnamon rubbing

9.6.1.3 Cinnamon Processing Bench

This device appeared to be the most appropriate cinnamon processing bench so far developed, which can be adopted to be used for the entire peeling process (scraping, hand rubbing, peeling, and quills making) for the production of quality cinnamon efficiently, adhering to peelers ergonomics comfort and health (Weerasinghe and Pushpitha 2005b).

Different types of benches and tables are designed by different agencies to convert the floor processing process into table processing (Annual Report 2010; Basnayake and De Silva 2019). However, the cinnamon peeling bench shown

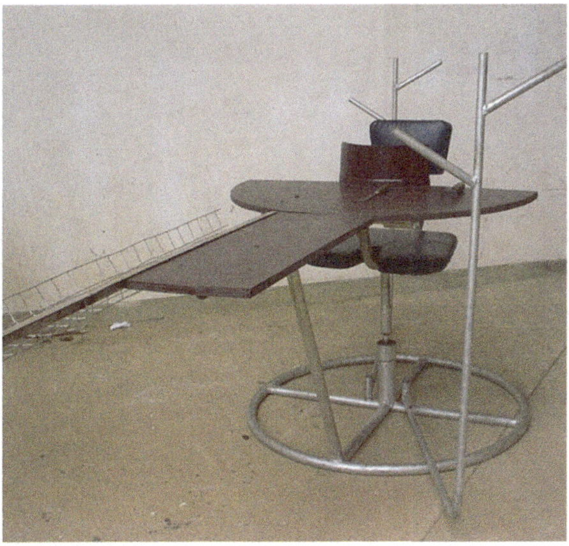

Fig. 9.14 Cinnamon processing bench introduced by Department of Agricultural Engineering, University of Ruhuna. Industrial Design 7628. (Weerasinghe and Pushpitha 2005b)

in Fig. 9.14 is designed to economize the space and facilitate all the operations (peeling, rubbing, and quills making) in a single table, with rotating stick handling supporters to ease the operations adhering to correct ergonomics. Benches that are used in modified factories (e.g., Palolpitiya Process Center) are demonstrated in Fig. 9.15.

Different types of dryers are introduced to the industry for efficient drying (Fig. 9.16), which are widely used in all the processing centers.

9.7 Cinnamon Process Centers

The machinery developed for the cinnamon peeling industry during the last 10 years has made a distinctive change to organize cinnamon peeling from cottage level to industrial scale while adhering to the GAP, HACCAP, and GMP requirements.

The DEA has designed a medium-sized process center of 4.6 × 9.5 m at an estimated cost of LKR 200,000 (US$ 1112). The DEA provides a subsidy of LKR 40000/= to construct such units. A model process center is being constructed at Cinnamon Research Center at Palolpitiya in Matara District, incorporating the existing machinery for teaching and demonstration purposes. The plan and the view of a model process center designed by the DEA are depicted in Fig. 9.17 (Adminis tration Reports 2009-2018).

Under the GTZ program, nine central process units with an area of about 800–1000 m^2 have been constructed in Galle and Matara districts. In these

Fig. 9.15 Cinnamon benches used in the Palolpitiya cinnamon processing factory in Matara district, Sri Lanka

Fig. 9.16 Cinnamon drying sheds

complexes, central processing hall with identified sections for scraping, rubbing, quill making, and separate drying cabinets and storage facilities are established. By the end of 2010–2011, such process centers introduced units to process >1000 kg/day in one center. These factories are designed for a workforce of about 50 peelers who could process about 150–200 kg/day. The raw materials for the centers are received from the surrounding plantations.

Some of the recommendations for sector development based on the findings of different programs are as follows:

1. Strengthen central cinnamon peeling factories by introducing new technologies and hygienic conditions to receive ISO 22000 through clustering and branding.
2. Undertake human resource development that is an essential component to achieve the quality production (training for the managers, technicians, cinnamon

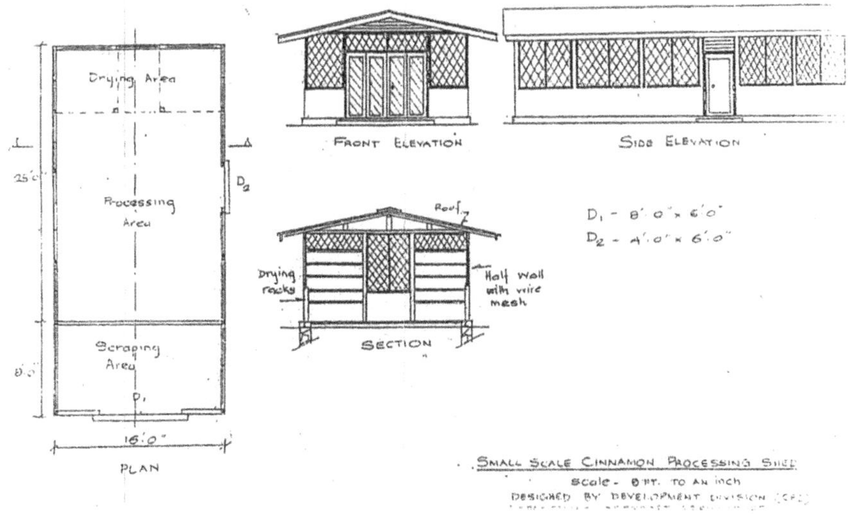

Fig. 9.17 Plan and view of the cinnamon process center designed by the Department of Export Agriculture, Sri Lanka

peelers, planters, and technologists) for which curricular development and training programs are to be initiated.

3. Mechanization of cinnamon processing and the introduction of the new inventions to the newly established processing units.
4. Formation of cinnamon clusters, a survey on existing small holder associations, strengthening of the associations, formation of export villages, and linking with international trade.
5. Strengthen University–Industry collaboration with Spice Association for machinery and equipment development and adoption.

References

Administration Reports (2009–2018) Department of Export Agriculture, Gatambe, Peradeniya, Sri Lanka

Annual Report (2010) Central Bank of Sri Lanka. Table 76, Volume and value of exports of other agricultural products

Basnayake BMRL, De Silva A (2019) Economics of ergonomics of Ceylon cinnamon industry. Ceylon cinnamon value chain development project (Book No. 3). Sabaragamuwa University, Sri Lanka

Dayananda KR (2011) Improved cinnamon distillation technology. Abstract of the paper presented to international symposium, "Technology to reach future aspirations of Ceylon cinnamon", 18–19th July 2011. Faculty of Agriculture, University of Ruhuna, Kamburupitiya

Gunarathne WDL (2011) Current situation and future trends of cinnamon industry. Abstract of the paper presented to international symposium, "Technology to reach future aspirations of Ceylon cinnamon", 18–19th July. Faculty of Agriculture, University of Ruhuna, Kamburupitiya

Gunasena CP, Weerasinghe KDN, Karunarathne G (1997) Technological aspects of cinnamon processing for future mechanization. Proceedings of the Sri Lanka Association for the advancement of science. 53rd annual session, p 108

Information on Export Agriculture Crops (2005–2009) Publications by the Statistics Division, Economic Research Unit, Department of Export Agriculture, Peradeniya, Sri Lanka

National Postharvest Research and Developmental Action Plan (2011) Identification of research priorities and policy issues 2011–2015. The National Committee on Post-Harvest Technology, Sri Lanka Council for Agricultural Research Policy, Colombo, Sri Lanka

Pushpitha NPG (2006) Design and construction of a cinnamon processing device to intensify the cinnamon peeling, M.Phil thesis. Faculty of Agriculture, University of Ruhuna, Kamburupitiya, Sri Lanka

Specification for Cinnamon (1978) Specifications for Cinnamon Products Part 2 – quillings, featherings and chips (First revision). Sri Lanka Standards Institute, Colombo, Sri Lanka

Specification for Cinnamon (1992) Specifications for Cinnamon Products Part 1 – quills (Second revision), Sri Lanka Standards Institute, Colombo, Sri Lanka

Specification for Cinnamon (1972) Specifications for Oil of Cinnamon Bark. Sri Lanka Standards Institute, Colombo, Sri Lanka

Technical Publication (1996) Cinnamon Cultivation and Processing. Department of Export Agriculture, Getambe, Peradeniya, Sri Lanka

UNIDO (2004) Reinvigorating Sri Lanka's Ceylon cinnamon exports. Available at: https://www.unido.org/news/reinvigorating-sri-lankas-ceylon-cinnamon-exports Accessed 24 October 2020

Weerasinghe KDN (2011) Harvest and the postharvest processing operations of the cinnamon sector in the southern region of Sri Lanka. UNIDO, Unpublished report TF/SRL/06/001/17-52

Weerasinghe KDN, Pushpitha NPG (2005a) Patent No. 17342, "A poly cam mechanism to improve the cinnamon mechanization process", International patent classification A23N 7/10, 11/00 B27J 3/00 Go5G 17/00, 30th Nov 2005

Weerasinghe KDN, Pushpitha NPG (2005b) Registration of industrial design no;7628 cinnamon processing bench, 24th Nov 2005

Weerasinghe KDN, Pushpitha NPG (2010) Sri Lanka's cinnamon industry. Geogr Rev 23:20–22

Weerasinghe KDN, Gunasena CP, Karunarathne G (1998a) Improvement of cinnamon processing technology through appropriate mechanization. Proceedings of 54th session of the Sri Lanka Association for the Advancement of Scence, Annual sessions part 1, p 122

Weerasinghe KDN, Gunasena CP, Karunarathne G (1998b) Technological aspects of cinnamon peeling and it's appropriate mechanization. Proceedings of the workshop on revitalization of cinnamon industry in Sri Lanka: constraints and remedies. Faculty of Agriculture, University of Ruhuna, Sri Lanka

Weerasinghe KDN, Gunasena CP, Karunarathne G (1999) Patent No 11869 "Cinnamon peeling device", International patent certification, B,27J,-3/00.A23N-7/10,11/00, 3rd Dec 1999

Weerasinghe KDN, Liyanage MDS, Silva MATD (2006) Present situation and future trends of cinnamon industry in Sri Lanka. Report of the cinnamon conference

Wijesekara ROB, Ponnuchamy S, Jayewardene AL (1975) Cinnamon: natural products monograph. Ceylon Institute of Scientific and Industrial Research, Colombo, pp 1–7

Chapter 10
Chemistry and Bioactive Compounds of *Cinnamomum zeylanicum* Blume

P. A. Paranagama, K. R. Dayananda, and Jinasena W. Hewage

10.1 Introduction

Genus *Cinnamomum* comprises about 250 species, of which 8 occur in Sri Lanka, namely, *Cinnamomum zeylanicum* (Blume Linn.) or Ceylon cinnamon and wild species, i.e., *Cinnamomum dubium* Nees (Sinhala: Sewel Kurundu or Wal Kurundu), *Cinnamomum ovalifolium* Wight, *Cinnamomum litseaefolium* Thwaites (Sinhala: Kudu Kurundu), *Cinnamomum citriodorum* (Sinhala: Pangiri Kurundu), *Cinnamomum rivulorum* Kostermans, *Cinnamomum sinharajaense* Kostermans, and *Cinnamomum capparu-coronde* Blume (Sinhala: Kapuru Kurundu) (Laurentius 1972); Sritharan et al. 1994).

Ceylon cinnamon (*C. zeylanicum* Blume (Linn.)) (*Kurundu* in Sinhala and cinnamon in English) is an economically important common species in Sri Lanka. It is an evergreen and endemic plant to Sri Lanka. Bark and leaves of Ceylon cinnamon are strongly aromatic, and essential oils are commercially extracted from cinnamon plants (Senanayake 1990; Paranagama 1990).

Chinese cassia cinnamon (*Cinnamomum cassia*) is the other most widely available species mainly in China. An important difference between *C. zeylanicum* and *C. cassia* is presence of higher amount of coumarin (1, 2-benzopyrone) in the latter, which is a secondary metabolite with strong carcinogenic and hepatotoxic properties (Abraham and Preiss-Weigert 2010). Therefore, Sri Lankan cinnamon has a higher economic value in the world market. However, since *C. cassia* is cheap and

P. A. Paranagama (✉)
Department of Chemistry, University of Kelaniya, Kelaniya, Sri Lanka
e-mail: priyani@kln.ac.lk

K. R. Dayananda
Retired Senior Research Officer, Industrial Technology Institute (successor to CISIR), Colombo, Sri Lanka

J. W. Hewage
Department of Chemistry, Faculty of Science, University of Ruhuna, Matara, Sri Lanka

R. Senaratne, R. Pathirana (eds.), *Cinnamon*,
https://doi.org/10.1007/978-3-030-54426-3_10

abundantly available, commercial Ceylon cinnamon available in the market is adulterated with essential oil of *C. cassia*. Research conducted on chemical constituents of essential oils of wild cinnamon varieties is fragmentary. However, it is reported that they have high medicinal and ethno-botanical values (Kumarathilaka et al. 2010).

10.2 Extraction of the Essential Oil from Cinnamon

The essential oils of cinnamon can be extracted using different extraction techniques used to isolate volatile constituents. The technology used in the production of essential oil is an important element to improve the yield and quality. The traditional distillation methods of extraction of essential oils are of great significance and are still being operated in Sri Lanka. Hydrodistillation and steam distillation are the most commonly used methods. Hydrodistillation is the most favored method of production of essential oils from cinnamon. However, supercritical fluid extraction method is used to produce essential oils and oleoresins from cinnamon in industrial scale as this method helps to obtain a high-quality and high-yielding product (Paranagama 1990; Senanayake et al. 1978).

The methods given below can be employed to extract the essential oils from cinnamon in small scale. Cinnamon stem bark, leaf, root, and fruits are the common parts used for extraction of essential oils, and chemical constituents in essential oils of the four parts are different from each other (Paranagama et al. 2001).

10.2.1 Steam Distillation Methods

This is a common method used in the laboratory to extract essential oils from cinnamon. Plant materials obtained from cinnamon are shade-dried for 2 days and subjected to steam distillation for 3 h at a rate of 150 ml/h. The aqueous layer is saturated with NaCl and extracted with dichloromethane (100 × 3 ml). The organic layer is evaporated on a rotary evaporator at 35 °C, and the residue is dried on anhydrous Na_2SO_4. Further, the remaining solvent is evaporated under a stream of nitrogen. The laboratory steam-distilled essential oil of cinnamon gives volatile oil with pungent odor.

10.2.2 Modified Likens and Nickerson Apparatus

Some of the low volatile chemical constituents in cinnamon cannot be distilled without decomposition, and they are usually obtained by using a Likens and Nickerson apparatus. This method is different from steam distillation method as volatiles are trapped into an organic solvent. A shade-dried sample of cinnamon

(100 g) is chopped, mixed with water (500 ml), and extracted for 4 h in a modified Likens and Nickerson apparatus using isopentane (20 ml) (Fig. 10.1). Cooling water in the condenser is connected to the apparatus and maintained at 10 °C using a water-cooling unit. Dry ice/acetone condenser is employed beyond the main condenser area to prevent any loss of volatiles (Paranagama 1990).

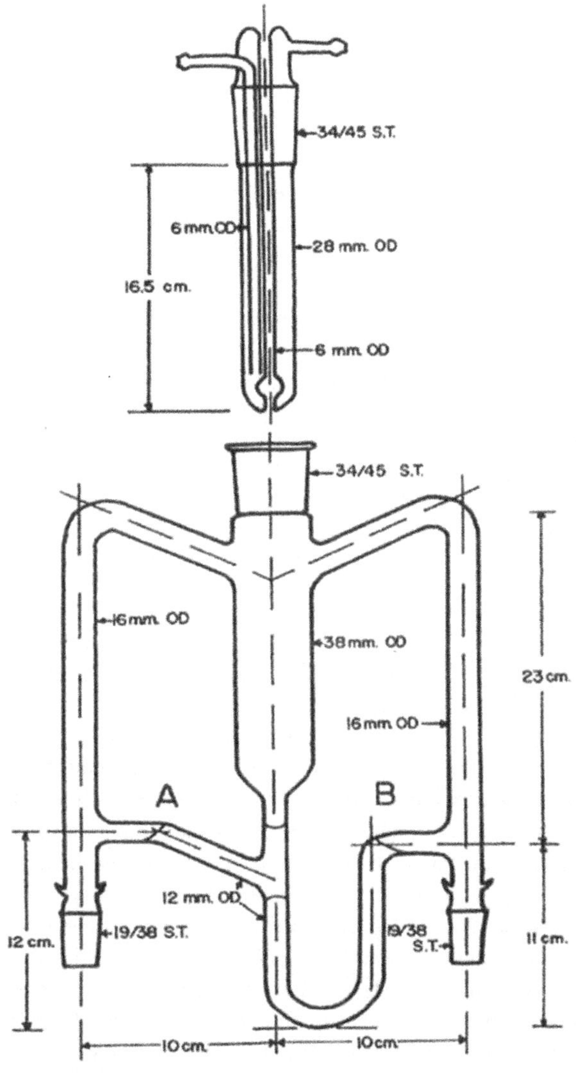

Fig. 10.1 Likens and Nickerson apparatus

10.2.3 Chemical Constituents of Cinnamomum zeylanicum

Gas chromatography (GC) is used as a powerful technique to analyze chemical constituents in the essential oil of cinnamon, and combined gas chromatography and mass spectroscopy (GC-MS) is used to identify the chemical constituents in the essential oils. Characteristic GC patterns given by the essential oil under the same conditions and retention time (RT) can be used as criteria for identification of the constituents. However, any deviation from the normal GC pattern of the essential oil could be due to the adulterants added to the test sample (Paranagama 1990). Adulterants are the common problem associated with the essential oil of cinnamon. This is done deliberately either by adding cassia oil, petroleum hydrocarbons (kerosene, diesel), coconut oil, and synthetic chemicals or mixing low-grade oils with quality oils.

10.2.3.1 Chemical Constituents in Essential Oils from Bark, Leaf, Root, and Fruits of Cinnamon (*C. zeylanicum* Blume) Grown in Sri Lanka

Cinnamon bark is used as a spice to improve flavor of local food preparations, and cinnamon bark and leaf oils are widely used in food flavors, cosmetics, pharmaceuticals, and indigenous medicine. The bark, leaf, root, and fruit of cinnamon separately produce pale yellow essential oils with a pleasant, spicy odor, and the yields are 1.2%, 0.75%, 2.0%, and 0.5% respectively.

Paranagama et al. (2001) analyzed chemical constituents of the essential oils of air-dried samples of fruits, bark, leaves, and roots of *C. zeylanicum* by modified Likens and Nickerson method. The essential oils were analyzed using gas chromatography with FID detector with Supelcowax 10 capillary columns (30 m × 0.32 mm; 0.25 pm film thickness). The GC program used for this analysis is 50 °C (2 min), 50–210 °C at 2 °C/min, 210 °C (30 min) with helium as the carrier gas. The injector and detector temperatures are 250 °C, and 0.1 μL of the oil is injected in split-less mode with a sampling time of 0.5 min; direct interface at 250 °C; ionization voltage of 70 eV; and ion source temperature of 250 °C.

Individual constituents are identified by comparison of their mass spectra in the database and retention indices with literature data. Investigations on the composition of cinnamon bark, leaf, root, and seed oils revealed that they are rich in monoterpenoids and phenylpropanoids and differed significantly in chemical composition. Cinnamaldehyde, eugenol, camphor, and cadinene are the major constituents present in cinnamon bark, leaf, root, and fruit, respectively (Table 10.1). This unique characteristic of cinnamon has created much interest in the biosynthetic pathways of its chemical constituents. Thus cinnamon offers a variety of oils with varied aroma and flavor to the food and beverage industry. On the other hand, camphor is the main constituent in the root bark, but it does not seem to have commercial value

Table 10.1 Chemical constituents of cinnamon fruit, bark, leaf, and root oils from Sri Lanka

Compound	Relative abundance			
	Fruit	Bark	Leaf	Root
α-Pinene	2.19	3.34	0.73	5.70
Unknown	–	1.10	0.08	0.57
Camphene	0.29	0.63	0.29	2.77
β-Pinene	1.61	0.61	0.26	3.45
Sabinene	–	0.26	–	1.51
α-Phellandrene	0.43	0.14	0.65	4.92
Myrcene	–	2.70	0.77	0.43
α-Terpenene	0.08	1.30	1.10	1.05
Limonene	1.0	1.2	0.3	6.2
β-Phellandrene	0.07	–	–	2.09
(Z)β-Ocimene	0.03	0.14	–	0.28
Terpenene	0.05	0.16	–	0.57
(E)β-Ocimene	0.02	0.13	–	0.94
P-Cymene	0.01	1.91	0.92	1.38
Terpinolene	0.30	0.21	0.61	0.47
Linalool	0.08	3.70	2.77	0.13
Terpinen-4-ol	0.27	0.40	0.11	1.90
α-Terpineol	0.64	0.70	0.28	3.94
α-Fenchyl alcohol	0.41	–	–	–
Isoborneol	0.70	0.08	–	0.68
Sabinol	–	–	0.20	–
1,8-Cineole	0.05	4.60	0.51	6.39
Methyl chavicol	–	–	–	0.19
P-Cymen-7-01	–	–	–	0.06
Cinnamyl alcohol	–	0.16	0.09	0.12
2-Phenylethyl alcohol	–	0.47	–	–
Coumarin	–	0.36	–	–
Benzaldehyde	0.50	0.61	0.14	–
Hydrocinnamaldehyde	–	0.80	0.12	0.09
Camphor	–	–	–	47.42
Piperitone	–	–	–	0.24
2-Phenylethyl acetate	–	0.18	–	0.05
3-Phenylpropyl acetate	–	0.38	–	0.03
Cinnamaldehyde	0.3	60.5	2.7	0.1
Methyl cinnamate	–	0.27	0.09	0.10
(Z)-Cinnamyl acetate	0.10	8.78	1.00	0.12
Benzyl benzoate	–	1.10	4.01	0.16
Eugenyl acetate	1.00	0.40	0.64	–
Linalyl acetate	–	–	–	0.10
Eugenol	0.45	4.15	76.74	0.60
Methyl isoeugenol	0.22	–	–	0.21

(continued)

Table 10.1 (continued)

Compound	Relative abundance			
	Fruit	Bark	Leaf	Root
Isoeugenol	0.32	0.08	0.07	–
Safrole	1.79	0.08	0.08	0.04
Methyl eugenol	–	0.15	–	1.32
Methoxy eugenol	0.20	–	–	0.12
α-Cubebene	2.71	–	–	0.17
α-Ylangene	5.63	0.70	0.14	0.68
β-Caryophyllene	1.41	8.00	3.47	0.03
α-Humulene	1.08	1.30	0.57	0.62
β-Farnesene	1.58	–	–	0.12
α-Gurjunene	2.32	–	–	–
β-Cadinene	2.78	–	–	–
Gurjunene	0.70	–	–	–
α-Muurolen	4.40	–	–	–
γ-Cadinene & δ-Cadinene	36.00	–	–	–
Cadina-1,4-diene	1.59	–	–	–
δ-Cadinene	5.64	–	–	–
Calamenene	0.17	–	–	–
Bisabolol	0.35	–	–	–
Nerolidol	0.84	–	–	–
Cadinol	1.12	–	–	–
Elemol	0.13	–	–	–
Sequiphellandrol	0.30	–	–	–
T-Cadinol	7.70	–	–	–
T-Muurolol	1.05	–	–	–

Table 10.2 Relative abundance of monoterpene, sesquiterpene, and phenylpropanoid fractions in cinnamon bark, leaf, root, and fruit oils

Fraction	Bark	Leaf	Root	Fruit
Monoterpenes	25.3	6.7	95.2	6.7
Phenylpropanoids	64.8	85.4	2.2	0.9
Sesquiterpenes	8.7	4.7	0.7	83.6

unlike the leaf and bark oils. The fruits of the cinnamon plant are also aromatic and possess a sweet spicy aroma.

The components identified by GC/MS and relative retention times in the essential oils of cinnamon bark, leaf, root, and fruit from Sri Lanka are listed in Table 10.1. The constituents of the essential oils are monoterpenes, sesquiterpenes, and phenylpropanoids. The oil of cinnamon fruit is rich in sesquiterpenoids.

Amounts of monoterpenes, sesquiterpenes, and phenylpropanoids fractions in bark, leaf, root, and fruit oils from *C. zeylanicum* in Sri Lanka are given in Table 10.2. The phenylpropanoid fraction is prominent in leaf and bark oils (Table 10.2), amounting to 85.4% and 64.8%, respectively, but present in very low levels in root and fruit oils (2.2% and 0.9%, respectively). On the other hand, cinnamon fruit oil is found to be rich in sesquiterpenes (83.6%), while cinnamon root oil contained more than 95% of monoterpenes.

The composition of cinnamon fruit oil is widely different from cinnamon leaf, bark, and root oils. The fruit oil does not contain high concentrations of components such as eugenol, cinnamaldehyde, and camphor that are major constituents of the leaf, bark, and root oils of cinnamon, respectively. The intense, characteristic aroma of cinnamon fruit oil is therefore most probably due to its sesquiterpene fractions, particularly the major components δ and γ- cadinene, cadinol, and β-caryophyllene. *Cinnamomum zeylanicum* bark and leaf oils have been analyzed by several investigators from Sri Lanka using GC and GC-MS. All of them reported that essential oils of cinnamon bark and leaf contain monoterpenes, oxygenated monoterpenes, sesquiterpenes, and phenylpropanoids with cinnamaldehyde (60–70%) and eugenol (70–75%) as major constituents in *C. zeylanicum* bark and leaf oils (Paranagama et al. 2001; Abeywickrama 2009; Liyanage et al. 2017).

The fruit oil of cinnamon grown in India had been investigated by Jayaprakasha et al. (1997) and Mallavarapu and Ramesh (2000). The major constituents of the oil obtained from Kerala are reported to be β-caryophyllene and (E)-cinnamyl acetate, whereas the oil from Bangalore constituted α-pinene, β-pinene, β-caryophyllene, α-muurolene, δ- and γ-cadinene, and muulol.

10.3 Chemical Constituents in Wild *Cinnamomum* Species in Sri Lanka

Identification of chemical constituents of essential oils in wild cinnamon species is very important as these findings are very helpful to develop new cinnamon varieties, identify economically important wild cinnamon species, guide conservation programs of wild relatives, produce value-added products and sustainably utilize wild cinnamon species. The limited studies have been reported on chemical constituents of the volatiles in wild species of cinnamon grown in Sri Lanka. *C. zeylanicum* has the highest economic importance as it has very high global demand, and *C. zeylanicum (Sri Gamunu)* and *C. zeylanicum* (*Sri Vijaya*) have been introduced as the two varieties with high percentage of essential oils. *Cinnamomum capparu-coronde* (Sinhala: Kapuru Kurundu), *C. sinharajaense* Kostermans, *C. rivulorum* Kostermans, *C. dubium* Nees (Sinhala: Sewel Kurundu or Wal Kurundu), *C. ovalifolium* Weight, *C. litseaefolium* Thwaites (Sinhala: Kudu Kurundu), and *C. citriodorum* (Sinhala: Pangiri Kurundu) have been reported as the wild cinnamon species available in Sri Lanka (Ariyarathne et al. 2018).

10.3.1 Yield of Essential Oils of Leaf and Bark of True and Wild Cinnamon Species in Sri Lanka

It is reported that the essential oils obtained after the hydrodistillation of the bark and leaf of *Cinnamomum* species provided different color and odor (Table 10.3). *C. zeylanicum* leaf and bark oils are golden in color, while the color of other species is pale yellow. *C. zeylanicum* and *C. citriodorum* bark oils show a pleasant odor. Odor of *C. citriodorum* leaf oil is similar to that of citronella grass oil, and pleasant geranium-like odor is reported in *C. dubium* leaf oil (Liyanage et al. 2017).

The yields of essential oils obtained from leaf and bark of *C. zeylanicum* and wild cinnamon species have been estimated on dry weight basis and presented in Table 10.4. It was reported that the oil content in cinnamon species varied significantly. The highest cinnamon leaf oil yield (3.26%) is observed with *C. zeylanicum*, whereas the essential oil obtained from *C. rivulorum* leaf shows the lowest yield (0.43%). Among the wild varieties of cinnamon available in Sri Lanka, the highest essential oil content was reported from leaf and bark of *C. sinharajaense* (Liyanage et al. 2017).

The investigations on chemical constituents of essential oils in bark and leaf of wild cinnamon species are considerably different from true cinnamon. Most important and abundant chemical constituents in cinnamon bark and leaf oils were cinnamaldehyde and eugenol. Depending on the part of the plant and species, cinnamaldehyde and eugenol quantities are varied (Table 10.5). In the bark oils, the highest cinnamaldehyde and eugenol contents were found in 'Sri Gamunu' (68.50%) and 'Sri Vijaya' (13.20%), respectively. Among the wild cinnamon species, *C. sinharajaense* has the highest cinnamaldehyde content (65.51%) and lowest eugenol content (0.51%). *C. riviulorum* bark oil contains the highest percentage of 1,8-cineole (30.36%) (Table 10.5).

The highest eugenol content (Table 10.6) was observed in leaf oil of 'Sri Gamunu' (92.89%), and cinnamaldehyde percentage in leaf oil is comparatively less than 3%. *C. sinharajaense* has the highest eugenol content among wild species. The highest

Table 10.3 Odor characteristics of stem bark and leaf oils of different *Cinnamomum* species (Liyanage et al. 2017)

Species name	Leaf oil	Bark oil
Cinnamomum verum	Pleasant clove-like odor	Pleasant cinnamon-like odor
Cinnamomum dubium	Pleasant geranium-like odor	Pleasant
Cinnamomum rivulorum	Slightly unpleasant clove-like	Pleasant clove-like odor
Cinnamomum sinharajaense	Slightly unpleasant	Pleasant
Cinnamomum citriodorum	Pleasant strong citronella grass oil-like odor	Pleasant cinnamon-like odor
Cinnamomum capparu-coronde	Pleasant	–

Table 10.4 Essential oil content in leaf and bark of true and wild cinnamon species in Sri Lanka (Liyanage et al. 2017)

Species	% leaf oil content on dry weight basis	% bark oil content on dry weight basis
Cinnamomum verum	3.26	3.06
Cinnamomum dubium	0.86	0.51
Cinnamomum rivulorum	0.43	3.53
Cinnamomum sinharajaense	2.41	3.53
Cinnamomum citriodorum	0.92	0.82
Cinnamomum capparu-coronde	1.49	n.d.

benzyl benzoate content (39.10%) is found in *C. dubium* leaf oil. *C. riviulorum* leaf oil contains the highest linalool content (9.46%). *C. zeylanicum* leaf oil has the highest β-caryophyllene content (10.96%) (Fig. 10.2).

10.4 Nonvolatile Constituents of *C. zeylanicum*

Ceylon cinnamon is mainly used in flavor, fragrance, and pharmaceutical industries mainly due to the presence of essential oils in the bark and leaf. However, it is now documented that many interesting chemical substances, other than those present in essential oils, have been reported from Ceylon cinnamon and cassia. It is noted that some of these compounds are responsible for the pharmacological and toxicological properties of cinnamon. Identification of diterpenes cinnazeylanine and cinnacssiols was reported (Senanayake and Wijesekera 2005). Aqueous fraction of cinnamon has been analyzed to isolate nonvolatile constituents such as alkaloids, saponins, flavonoids, and proanthocyanidins which are responsible for various biological activities observed with cinnamon. There are many reports on the coumarin content in essential oils of Ceylon cinnamon and cassia as commercial Ceylon cinnamon available in the world market is adulterated using bark oil of cassia. Therefore, this chapter describes coumarin content in Ceylon cinnamon and cassia, analysis of coumarin content in cinnamon bark, and toxicity of coumarin.

10.4.1 Proanthocyanidins

Bioactivity of aqueous extract of cinnamon bark obtained from *C. zeylanicum* has been studied, and proanthocyanidins were isolated as the bioactive compounds in the aqueous extract (Taher et al. 2006). Only the bark of *C. zeylanicum* contained a major phenolic metabolite of doubly linked proanthocyanidins (A-Type). In the same study, it was reported that proanthocyanidins isolated from *C. zeylanicum* are

Table 10.5 Chemical constituents of essential oils obtained from bark of true and wild *Cinnamomum* species in Sri Lanka

Chemical compounds	C. zeylanicum	Sri Vijaya	Sri Gamunu	*C. sinharajaense*	*C.dubium*	*C. riviulorum*	*C. capparu*
α-Pinene	0.16	Nd	nd	nd	nd	0.77	nd
β-Pinene	0.13	Nd	nd	nd	nd	0.76	nd
Sabinene	3.55	Nd	0.31	nd	2.76	3.51	nd
Myrcene	0.14	Nd	nd	nd	nd	1.33	0.41
α-Terpinene	1.24	0.20	0.76	0.36	0.98	8.84	0.90
β-Phellandrene	nd	0.69	2.18	1.49	nd	nd	3.40
y-Terpinene	0.30	0.58	0.82	2.90	0.79	8.84	1.65
(E)-β-Ocimene	nd	Nd	nd	nd	5.87	nd	nd
p-Cymene	nd	Nd	nd	0.72	nd	nd	nd
Terpinolene	0.12	Nd	nd	nd	nd	1.44	0.86
Linalool	nd	Nd	nd	nd	nd	10.14	nd
Terpinen-4-0l	nd	Nd	nd	0.81	nd	nd	9.64
α-Terpineol	0.82	Nd	0.56	nd	nd	14.86	nd
1,8-Cineole		Nd	nd	nd	nd	30.36	nd
Benzaldehyde	0.13	Nd	0.16	0.20	0.22	nd	nd
Hydrocinnamaldehyde		Nd	nd	1.13	nd	nd	nd
Cinnamaldehyde	52.41	51.10	68.50	65.51	59.17	0.81	63.29
(Z)-Cinnamyl acetate	19.59	23.07	18.11	14.55	21.83	5.09	nd
Benzyl benzoate	7.74	7.88	1.19	7.85	8.19	3.76	4.83
Eugenol	8.89	13.20	4.74	0.51	2.13	1.09	5.37
Methyl eugenol	nd	Nd	nd	nd	nd	nd	2.09
β-Caryophyllene	2.98	2.58	1.77	2.94	2.91	nd	2.47
α-Humulene	0.67	0.59	0.39	0.65	0.61	0.91	nd
o-Cymol	0.49	Nd	0.24	nd	nd	3.69	2.50
l-Phellandrene	0.62	0.11	0.26	0.39	0.39	3.79	1.96

Table 10.6 Chemical constituents of essential oils obtained from leaf of true and wild *Cinnamomum* species in Sri Lanka

Chemical compound	*C. zeylanicum*	Sri Vijaya	Sri Gamunu	*C. sinharajaense*	*C. dubium*	*C. rivulorum*	*C. capparu-coronde*
α-Pinene	0.64	1.24	nd	nd	5.98	4.74	3.66
Camphene	0.22	0.54	nd	nd	1.15	1.63	1.05
β-Pinene	0.34	0.54	nd	nd	2.88	4.74	2.11
Sabinene	nd	nd	nd	nd	0.16	nd	
α-Phellandrene	nd	1.62	nd	nd	nd	0.26	0.12
Myrcene	nd	0.20	nd	nd	2.42	1.06	0.87
α-Terpinene	0.44	0.37	nd	nd	0.63	0.29	
Limonene	nd	nd	nd	nd	3.26	nd	
β-Phellandrene	1.16	1.26	nd	nd	nd	nd	
(Z)-β-Ocimene	nd	nd	nd	0.21	nd	0.09	
y-Terpinene	nd	0.37	nd	nd	nd	0.24	0.19
(E)-β-Ocimene	nd	nd	nd	nd	1.30	0.35	0.31
p-Cymene	nd	1.21	nd	0.63	nd	nd	
Terpinolene	0.43	0.31	nd	nd	5.38	1.11	1.09
Linalool	3.07	7.61	nd	2.06	7.70	9.46	9.11
Terpinen-4-01	0.39	0.30	nd		0.66	0.29	0.28
α-Terpineol	nd	nd	nd	0.35	nd	1.08	0.98
α-Fenchyl alcohol	nd	nd	0.27	nd	nd	nd	
1,8-Cineole	nd	nd	nd	nd	3.11	1.31	
Benzaldehyde	nd	0.22	nd	nd	nd	0.12	
2-Phenylethyl acetate	nd	nd	nd	nd	0.06	nd	
Cinnamaldehyde	0.31	0.88	2.12	0.52	0.32	1.38	1.26
(Z)-Cinnamyl acetate	5.01	nd	nd	nd	0.13	0.17	0.16
Benzyl benzoate	2.34	1.42		18.73	2.16	37.34	39.10
Eugenol	69.2	73.07	92.89	76.29	30.64	30.93	31.78
Methyl eugenol	0.15	nd	nd	nd	0.35	nd	0.36
α-Cubebene	nd	nd	nd	nd	1.85	nd	
β-Caryophyllene	10.96	8.83	2.75	nd	3.16	3.07	2.90
α-Humulene	2.55	nd	0.56	0.55	2.18	nd	
o-Cymol	1.38	nd	nd	0.63	nd	nd	4.66
1-Phellandrene	0.64	nd	nd	nd	26.10	0.34	

different from proanthocyanidins present in other cinnamon species. In addition, presence of proanthocyanidins had been reported in the bark of *C. burmannii* and *C. cassia* and the root of *C. camphora*, as the major phenolic metabolites. The acetone extract of cinnamon bark contains both essential oil and proanthocyanidins, and cinnamtannin B1 (Fig. 10.3) had been identified as the proanthocyanidin. Presence of flavan-3-ol moieties in the structure of cinnamtannin B1 was confirmed

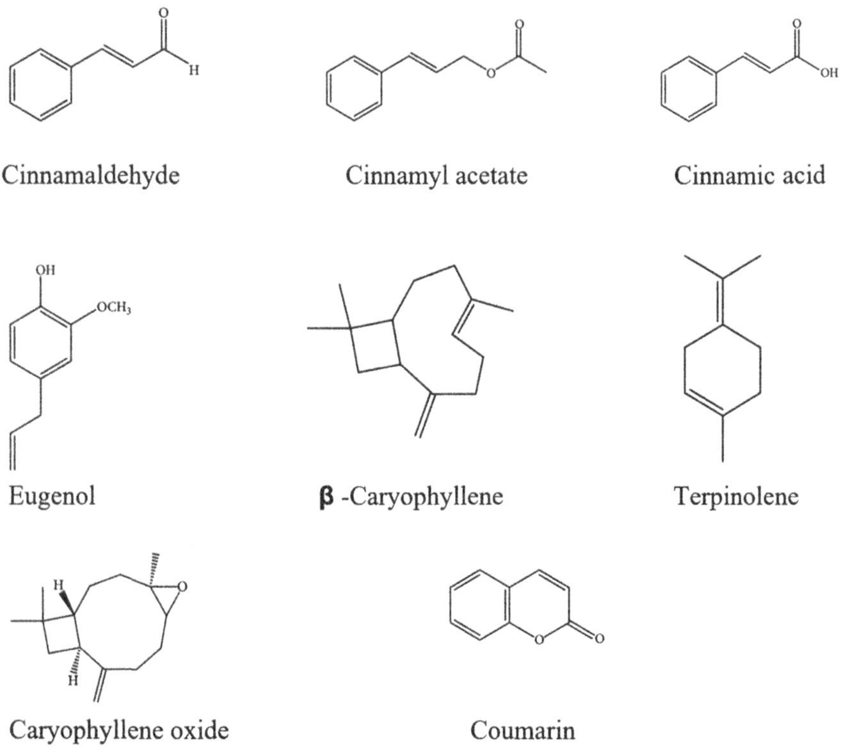

Fig. 10.2 The chemical structures of some important constituents of cinnamon species

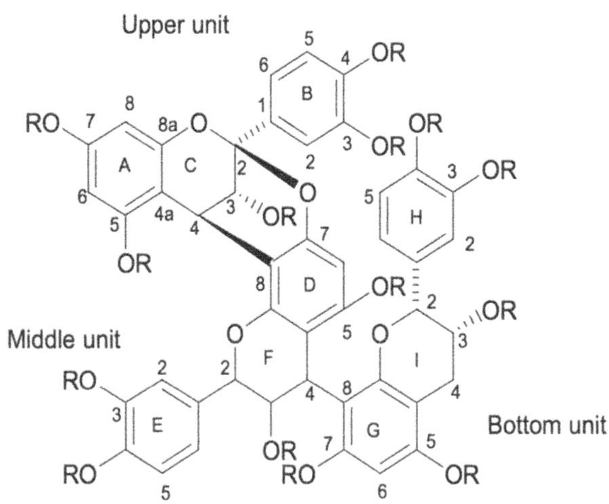

Fig. 10.3 Cinnamtannin B1

using the UV ($\lambda_{max\,278}$ nm) and the IR spectra (3400 cm^{-1} for the hydroxyl group and 1615 cm^{-1} for aromatic ring) of cinnamtannin B1. White amorphous crystals of cinnamtannin B1 have been isolated from the acetone extract, and the purity was confirmed showing a single peak in the HPLC analysis.

10.4.2 Alkaloids, Saponins, and Flavonoids in Ceylon Cinnamon

Investigation of phytochemicals present in bark and leaf of Ceylon cinnamon has been reported by several researchers (Senanayake and Wijesekera 2005; Paliwal et al. 2018; Dhanalaxmi and Vastrad 2014; Weerasekra et al. unpublished data) and indicated presence of glycosides, tannins, alkaloids, saponins, and flavonoids in Ceylon cinnamon using the phytochemical screening methods with different chemical reagents. Phytochemical screening of bark and leaf of Ceylon cinnamon had been reported by Paliwal et al. (2018) and Dhanalaxmi and Vastrad (2014). In these studies, aqueous or alcoholic extracts have been extracted to screen for the presence of phytochemicals, and the total phenols of the extracts have been estimated using Folin-Ciocalteu reagent with gallic acid as the reference standard. It was also stated that these phytochemicals may be responsible for pharmacological properties of cinnamon. The results revealed the presence of alkaloids, phenolics, flavonoids, saponins, tannin, lignin, and steroids in the cinnamon bark and leaf (Table 10.7).

A book chapter written by Senanayake and Wijesekera (2005) reported the isolation of alkaloids from *Cinnamomum* spp. and *C. camphora*. Isolation of benzylisoquinoline alkaloids norcinnamolaurine, (−) cinnamolaurine, (+) reticulin, aporphine, and (+) corydine has been reported from the bark of *Cinnamomum laubatii* collected from Australia, and it is stated that norcinnamolaurine is unique to the bark of this cinnamon species, whereas isolation of reticulin had also been reported from both *C. laubatii* and *C. camphora* (Gellett and Summons 1970). The chemistry of

Table 10.7 Qualitative phytochemical screening of extracts of *C. zeylanicum* (Paliwal et al. 2018; Dhanalaxmi and Vastrad 2014)

		Nature of the extract from			
		Bark		Leaf	
Plant chemicals	Test performed	MeOH	Aqueous	MeOH	Aqueous
Alkaloids	Mayer's test/Wagner's test/ Dragendoff's test	+ +	++	++	++
Saponins	Foam test	+	+	−	+
Steroids	Lieberman test	+	+		
Phenolic compounds and tannins	Ferric chloride test/gelatin test/ lead acetate test	++	+ +	++	++
Flavonoids	Ammonia test Sodium hydroxide	+ +	++	++	++
Terpenoids	Salkowski test	+	+	+	+

norcinnamolaurine and cinnamolaurine has been studied and reported; both compounds have optical rotations of opposite signs at the sodium D-line, and they both possess the same absolute configuration (Senanayake and Wijesekera 2005).

Recent studies on Ceylon cinnamon by Wijeweera et al. (2019a, b) have reported the genotype, maturity, agroecological, and part of the plant material dependence of quantity and quality of bioactive compounds, especially alkaloids, saponins, flavonoids, and polyphenols. Wijeweera et al. (unpublished data) have carried out analysis of cinnamon samples collected from the fields with known growth factors of the plants and following the recommended procedures for preparation from harvesting through to drying and sampling for analysis. Quantitative determinations of bioactive compounds were carried out by the methodologies reported by Ejikeme et al. (2014) and Ezeonu and Ejikeme (2016) with slight modification in some cases. Variation of those phytochemicals on genotype, maturity, and agroecology is shown in Figs. 10.4, 10.5, and 10.6. Genotype dependence has been reported using two clonally propagated cinnamon varieties, 'Sri Gamunu' and 'Sri Vijaya,' maintaining the genetic factor for the bark and the leaf parts of plant materials. Alkaloids percentage in the bark of 'Sri Gamunu' is 7.9%, while that in the bark of 'Sri Vijaya' is 4.8%. In the case of leaf, generally alkaloids content is lower compared to the bark; alkaloids percentage in 'Sri Vijaya' (4.1%) is higher compared to 'Sri Gamunu' (2.1%). The bark part of 'Sri Gamunu' variety is the best between two varieties as a source of alkaloids.

Figure 10.5 shows the maturity dependence of phytochemicals for the bark and the leaf parts of Ceylon cinnamon using three maturity stages: 1.5–2 years, 2–2.5 years, and above 5 years, respectively. Alkaloids content in both bark and leaf is higher at the maturity of 2 years. In the bark, the percentage is about 8.3%, while in the leaf, it is about 4.1% at 2 years of maturity. It has been shown by Wijeweera et al. (2019, "unpublished data") that the best maturity stage for phytochemical abundance also depends slightly on the genotype. 'Sri Gamunu' gives the highest alkaloids percentage at the maturity level of 1.5–2 years, while that for 'Sri Vijaya'

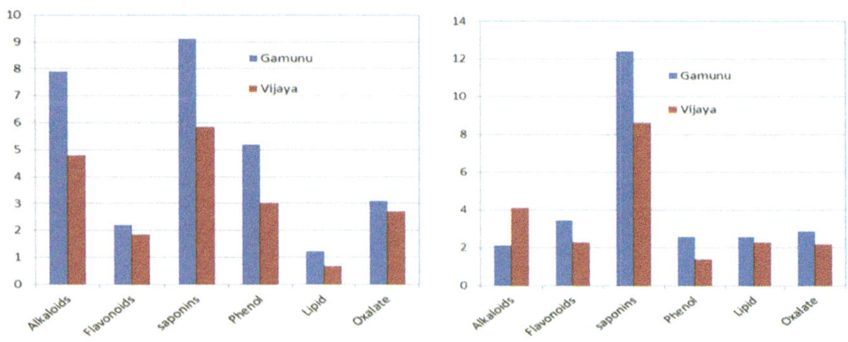

Fig. 10.4 Comparison in quantities of bioactive constituents in the bark (left) and the leaf (right) of cinnamon varieties 'Sri Gamunu' and 'Sri Vijaya'

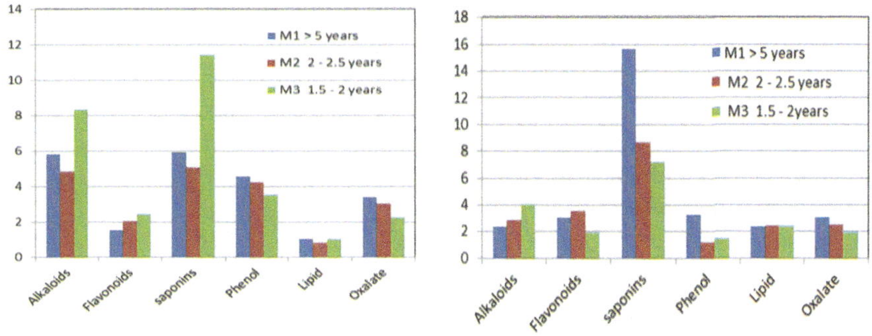

Fig. 10.5 Comparison in quantities of bioactive constituents in the bark (left) and the leaf (right) of Ceylon cinnamon harvested at three different maturity stages

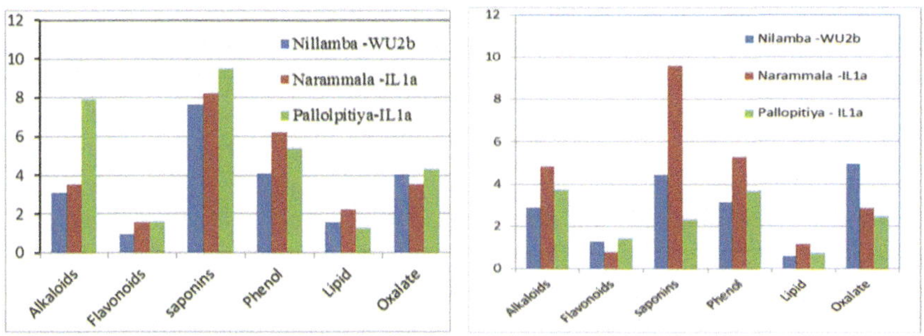

Fig. 10.6 Comparison in quantities of bioactive constituents in the bark (left) and the leaf (right) of *C. zeylanicum* 'Sri Gamunu' from three different locations in two agroecological zones

is at the maturity level of 2–2.5 years. Considering the genotype, maturity level of 2 years is the best harvesting period.

Wijeweera et al. (2019b) have also studied the agroecological dependence of quantity of bioactive compounds for the bark and the leaf samples of Ceylon cinnamon in two agroecological zones with two genotypes, 'Sri Gamunu' and 'Sri Vijaya.' Figure 10.6 shows the ecological dependence of bark constituents.

Alkaloids content in 'Sri Gamunu' variety is higher in Palolpitiya (7.9%) in IL1a agroecological zone, which is as twice asthe second highest location Narammala (3.6%) in the same agroecological zone. The lowest (3.1%) is reported from Nillamba. In the case of 'Sri Vijaya,' the highest and the second highest of alkaloids contents are from Narammala (4.8%) and Palolpitiya (3.7%), respectively, while the lowest is from Nillamba with 2.9%.

Four major parameters, genotype, maturity, part of the plant material, and agroecological factor, are shown to play a major role in alkaloids content of Ceylon cinnamon. The bark of 'Sri Gamunu' variety grown at Palolpitiya in the Intermediate Low Country Zone (IL1a) with about 2 years of maturity contains the highest alkaloids percentage.

The quantity of saponins in Ceylon cinnamon varies with the genotype, maturity of the plant, the part of the plant materials, and agroecological zone that the plant grows. Saponins are the most intense bioactive constituents in Ceylon cinnamon; their percentage in the bark of 'Sri Gamunu' is about 9%, while that in the bark of 'Sri Vijaya' is 4.9%. In the case of leaf, generally saponin content is higher compared to the bark of both genotypes; saponins percentage in 'Sri Gamunu' is about 12.4%, while that in 'Sri Vijaya' is 8.7%.

Saponin content in the bark is higher at the maturity of 2 years (11.4%) which is as twice as the other maturity stages. In leaf, the percentage is about 16% in plants maturing more than 5 years, while it is about 7% at the 2 years of maturity. It has been shown also that the most high yield maturity stage also depends heavily on the genotype. 'Sri Gamunu' gives the similar saponins percentage at all the three maturity levels, while for 'Sri Vijaya' the optimum is at the maturity level of 2 years. Considering the genotype, maturity level of 2 years is the best harvesting period.

Saponin content in 'Sri Gamunu' variety is higher in Palolpitiya (9.6%) in IL1a agroecological zone, and the second highest location is Narammala (8.2%) in the same agroecological zone. The lowest (7.7%) is reported from Nillamba. In the case of 'Sri Vijaya,' the highest saponin content is reported from Narammala (9.6%) which is as twice as the second highest reported from Nillamba (4.4%), and the lowest is from Palolpitiya (3.7%).

Similar to the alkaloids, for the quantity of saponins, four major parameters, genotype, maturity, part of the plant material, and ecological factor, are shown to play a major role. The bark of 'Sri Gamunu' variety grown at Palolpitiya in the Intermediate Low Country Zone (IL1a) with about 2 years of maturity contains the highest quantity of saponins.

Flavonoids are the least constituent present among four phytochemicals in the bark of Ceylon cinnamon. The flavonoid content in the bark part of 'Sri Gamunu' variety is about 2.2%, while that in the bark of 'Sri Vijaya' variety is about 1.8%. In leaf material, the flavonoid content is somewhat higher than the bark materials accounting for 3.4% in 'Sri Gamunu' and 2.3% in 'Sri Vijaya.' The highest flavonoid content in the bark is present at the 1.5–2 years of maturity (2.5%). In leaf, the percentage is about 3.6% at the 2–2.5 years of maturity. Similar to the other constituents, the maturity stage with the highest flavonoid content depends on the genotype. The bark of 'Sri Gamunu' gives the highest percentage at the maturity stage of 1.5–2 years, while that for 'Sri Vijaya' is at the maturity level of 2–2.5 years. Considering the genotype, maturity level of 2 years is the best harvesting period.

When considering the agroecological dependence of flavonoid content, the bark of 'Sri Gamunu' variety from both Palolpitiya and Narammala in IL1a agroecological zone has the highest content (1.6%), and for 'Sri Vijaya', while the highest yield is from Palolpitiya (1.4%), the bark from Nillamba (1.3%) also has a flavonoid content closer to Palolpitiya.

Four major parameters, genotype, maturity, part of the plant material, and ecological factor, are shown to play a major role in variation of flavonoid content. The bark part of 'Sri Gamunu' variety grown at Palolpitiya in the Intermediate Low

Country Zone (IL1a) with about 2 years of maturity contains the highest quantity of flavonoids.

Cinnamon has been shown to cure the metabolic syndrome associated with decreased antioxidant activity, insulin resistance, inflammation, increased glycation of proteins, and elevated glucose (Qin et al. 2010). Anderson et al. have reported that a class of water-soluble cinnamon polyphenol compounds displays insulin-potentiating, antioxidant, and related activities (Anderson and Roussel 2008). Polyphenols' presence in cinnamon inhibits the formation of glycation end products in the blood serum (Khan et al. 2003).

Understanding of variation in quantity of polyphenols in Ceylon cinnamon with the genotype, maturity of plants, part of plant materials, and the agroecology is important for an effective use of plant materials in drug industry. Both bark and leaf of 'Sri Gamunu' variety have higher content of polyphenols than those of 'Sri Vijaya'; the polyphenol content in bark of 'Sri Gamunu' and 'Sri Vijaya' is 5.2 and 3 mgTAE/gFW, respectively, while the polyphenol content in the leaf is 2.6 and 1.4 mgTAE/gFW, respectively.

The polyphenol content depends also on the maturity of the plant materials. The studies showed that the polyphenol content in the bark of plants with maturity of 2–2.5 years (4.6 mgTAE/gFW) and 5 years above (4.2 mgTAE/gFW), which have no significant difference, is higher than that in the bark of plants with maturity of 1.5–2 years (3.6 mgTAE/gFW). 'Sri Gamunu' variety has the higher polyphenol content than 'Sri Vijaya' variety.

The barks of 'Sri Gamunu' variety from both Narammala and Palolpitiya in IL1a agroecological zone have close polyphenol contents (6.2 and 5.4 mgTAE/gFW, respectively), and polyphenol content in 'Sri Gamunu' of Nillamba is about 4.1 mgTAE/gFW. In 'Sri Vijaya', while the highest yield is from Narammala (5.3 mgTAE/gFW), the bark from Palolpitiya (3.7 mgTAE/gFW) has the second highest.

Four major parameters, genotype, maturity, part of the plant material, and ecological factor, are shown to play a major role in variation of polyphenol content. The bark part of 'Sri Gamunu' variety grown at Palolpitiya in the Intermediate Low Country Zone (IL1a) is suitable for more effective use.

10.4.3 Coumarin Content in Ceylon Cinnamon

Among the compounds present in cinnamon, coumarin content in Ceylon cinnamon is significantly low when compared to that of cassia cinnamon (WHO Monograph 2004). However, it is difficult to distinguish Ceylon cinnamon from cassia when it is in powdered form, which is often traded using its common name, cinnamon/canella. Even though Ceylon cinnamon (*C. zeylanicum*) is considered as true cinnamon, cassia cinnamon (*C. cassia*) has a larger share in the world trade. Since coumarin is a carcinogenic and hepatotoxic compound, FAO/WHO has announced the threshold limit for coumarin content in foods as 2 ppm (GSFA 1995). In 1999, allowable human exposure to coumarin in a diet has been estimated and reported as approximately 0.02 mg/kg/day (Lake 1999). Further, the Theoretical Maximum Daily Intake (TMDI) of coumarin via food was estimated as 4.085 mg/day (0.07 mg/kg bw/day) (EFSA 2005; De Silva 1972).

In 2012, the Federal Institute of Risk Assessment (BfR) stated the bioavailability of coumarin in human body and reported that the total coumarin content in matrix of cinnamon bark is completely absorbed to the human body when cinnamon bark is consumed. The degree of absorption of coumarin is similar to that of the isolated coumarin. Further, consumers who frequently use large quantities of cinnamon as a condiment should therefore choose Ceylon cinnamon as it contains low coumarin content. Therefore, usage of cassia cinnamon as a regular supplement with meals was not encouraged, or the daily dosage of cinnamon was restricted in many countries due to the toxic effects of *C. cassia* on the liver and coagulation (BfR 036/2012).

Detection of coumarin can be carried out using GC or HPLC. A study carried out using reversed phase HPLC to quantify coumarin content in cinnamon bark from Sri Lanka and cassia indicated low coumarin content in Ceylon cinnamon (2–35 mg kg^{-1} or 0.0002–0.0035% w/w) and significantly high coumarin content (1000–10,000 mg kg^{-1} or 0.1–1.0% w/w) in cassia. In 2008, coumarin content had been detected in commercial Ceylon cinnamon bark and cinnamon bark oil using gas chromatography (Dayananda and Bandara 2008) and reported among the 38 cinnamon bark oil samples tested; maximum coumarin content detected was 100 ppm (mean 34.7 ± 0.16 ppm).

There are many reports (Table 10.8) on the coumarin content in cinnamon bark and cassia. The results revealed the presence of trace amount of coumarin in all the samples of Ceylon cinnamon bark, and they also indicated that the coumarin content in the cinnamon bark can increase with maturity. Nevertheless these values for Ceylon cinnamon did not exceed the toxic levels given in FAO/WHO. In the same studies, cassia bark oil and cassia bark samples tested had significantly high coumarin content, which was above the toxic levels reported. Further, the coumarin level in cassia can pose health risks if consumed in high quantities on a regular basis. Therefore, usage of cassia cinnamon as a regular supplement with meals was not encouraged due to the toxic effects on the liver and coagulation (Ransinghe et al. 2014; Dayananda and Bandara 2008; Weeratunga et al. 2013).

Table 10.8 Coumarin content in Ceylon cinnamon and cassia

Reference	Ceylon cinnamon bark oil	Ceylon cinnamon bark	Cassia Oil	Cassia bark
Vernin et al. (1994)	ND	ND	8.7	–
WHO monograph on selected medicinal plants (2004)	–	ND	–	0.45
Senanayake and Dill (1978)	–	–	15.3	–
Paranagama (1991)	0.66	–	–	–
Senanayake and Wijesekera (2005)	ND	–	8.7	–
Archer (1988)	–	ND	–	0.45

Wang et al. (2013) studied the coumarin content and other marker compounds in cinnamon-flavored food and food supplements in the United States using a validated UPLC-UV/MS method. The results indicated the presence of substantial amounts of coumarin. Yet, bark of the three species *C. aromaticum*, *C. loureiroi*, and *C. burmannii* could be used in the cinnamon-flavored food and food supplements in the United States.

Coumarin content detected in the powdered cinnamon samples collected from the Czech Republic was in the range 2650 to 7017 mg kg^{-1} indicating that cassia cinnamon is used as the source of cinnamon in the Czech retail market (Blahova and Syobodova 2012). In 2013, Ballin and Sorensen (2014) analyzed 74 food samples labeled with cinnamon from Denmark market with a validated UPLC-PDA method. It showed that fine bakery exceeded the EU limit for coumarin in almost 50% of the cases. One sample exceeded the EU limit for coumarin with more than a factor of 3.

Similarly coumarin content in cinnamon-flavored food and food supplements had been investigated from samples collected from Italy, Germany, Canada and India. In this study, 34 samples of cinnamon and 50 samples of cinnamon-containing foodstuffs had been collected from the Italian market. Quantitative determination of coumarin and cinnamaldehyde was performed by using HPLC with diode array detector (DAD). The results revealed that about 51% of cinnamon samples consisted of cassia, 10% were possibly a blend of cassia and Ceylon cinnamon, whereas only 39% were actually Ceylon cinnamon. The reported coumarin content in cinnamon-containing foods had exceeded the maximum level fixed in the European Flavorings Directive of 2 mg kg^{-1} (Silvia et al. 2008).

A study conducted in Germany on 47 powdered cinnamon samples collected from the retail market again confirmed high levels of coumarin in cassia cinnamon (Dayananda and Bandara 2008). In Canada a survey on coumarin content was carried out using 739 cinnamon samples obtained at the Canadian retail level. The results disclosed 63% of the samples with high content of coumarin. The results of this survey revealed coumarin concentrations in the samples ranged from 0.2 to 2170 ppm. Most of the spice mixes and cinnamon samples contained detected levels of coumarin (Blahova and Syobodova 2012).

In India, a study was carried out for the estimation of coumarin and other phenolic compounds using *C. zeylanicum* bark samples and samples collected from the retail market from south India. It was reported that coumarin contents in authentic bark samples were in the range of 12.3–143.0 mg/kg, whereas the coumarin contents in samples procured from the market were up to 3462.0 mg/kg. The high content of coumarin and cinnamaldehyde in cinnamon bark samples procured from the market suggested possible adulteration with *C. cassia* bark, which possesses substantial amounts of coumarin and cinnamaldehyde.

In another study conducted by Jayatilaka et al. (1995), it was reported that true cinnamon can be distinguished from cassia by the presence of eugenol and benzyl benzoate and the absence of coumarin and δ-cadinene, all minor components, referenced to the extracted amount of *trans*-cinnamaldehyde. Multivariate analysis of 16 semivolatile compounds commonly found in powdered cinnamon samples is used to classify samples of cinnamon purchased in 7 countries.

Acknowledgments The authors wish to thank the National Science Foundation (NSF) of Sri Lanka for the funding under research grant SP/CIN/2016/04 and the National Cinnamon Research and Training Center, Palolpitiya, Sri Lanka, for the support extended during research activities.

References

Abeywickrama KP (2009) Bioactivity and chemistry of cinnamon. In: Paranagama PA (ed) University of Kelaniya, Kelaniya, Sri Lanka, pp 9–23

Abraham K, Preiss-Weigert A (2010) Quantification of flavoring constituents in cinnamon: high variation of coumarin in cassia bark from the German retail market and in authentic samples from Indonesia. J Agric Food Chem 13:58

Anderson RA, Roussel AM (2008) Cinnamon, glucose and insulin sensitivity. In: Pasupuleti V, Anderson JW (eds) Nutraceuticals, glycemic health and type 2 diabetes. IFT Press Series, Wiley- Blackwell Publishing, pp 127–140

Archer AW (1988) Determination of cinnamaldehyde, coumarin and cinnamyl alcohol in cinnamon and cassia by HPLC. J Chromatogr 447:272–276

Ariyarathne HBMA, Weerasuriya SN, Senarath WTPSK (2018) Comparison of morphological and chemical characteristics of two selected accessions and six wild species of genus *Cinnamomum* Schaeff. Sri Lankan J Biol 3:11–23

Ballin NZ, Sorensen TA (2014) Coumarin content in cinnamon containing food products on the Danish market Published, www.elsevier.com/locate/foodcont

Blahova J, Syobodova Z (2012) Assessment of coumarin levels in ground cinnamon available in the Czech retail market. Sci World J. https://doi.org/10.1100/2012/263851

Dayananda KR, Bandara VS (2008) Coumarin content in Ceylon cinnamon (*Cinnamomum zeylanicum)* and its significance with Chinese cassia. Institute of Chemistry Ceylon, Proceedings of the international symposium on innovations in food science and technology and health-care for social well-being

De Silva S (1972) A report on the sample survey of the cinnamon industry of Sri Lanka. Ministry of Plantation Industries, Colombo

Dhanalaxmi RK, Vastrad J (2014) Phyto constituents: an analysis of cinnamon (*Cinnamomum verum)* leaf extracts. Asian J Home Sci 9:319–321

Ejikeme CM, Ezeonu CS, Eboatu AN (2014) Determination of physical and phytochemical constituents of some tropical timbers indigenous to Niger delta area of Nigeria. Eur Sci J 10:247–270

Ezeonu CS, Ejikeme CM (2016) Qualitative and quantitative determination of phytochemical contents of indigenous Nigerian softwoods. New J Sci 2016:5601327

Gellett E, Summons RE (1970) Alkaloids of the genus *Cinnamomum* II, the alkaloid of the bark of *Cinnamomum* sp. Aust J Chem 23:2095–2099

General standard for food additives; CODEX STAN 192 – 1995, Rev. 6-2005

Jayaprakasha GK, Rao LJ, Sakariah KK (1997) Chemical composition of the volatile oil from the fruits of *Cinnamomum zeylanicum* Blum. Flavor Frag J 12:331–333

Jayatilaka A, Poole SK, Poole CF, Chichila TPF (1995) Simultaneous micro steam distillation / solvent extraction for the isolation of semi-volatile flavor compounds from cinnamon and their separation by series coupled-column gas chromatography. Analytica Chimica Acta 302:147–162

Khan A, Safdar M, Alikhan M et al (2003) Cinnamon improves glucose and lipids of people with type 2 diabetes. Diabetes Care 26(1):3215–3218

Kumarathilaka DMHC, Senanayake SGJN, Wijesekara GAW, Wijesundara DSA, Ranawaka RAAK (2010) Extinction risk assessments at the species level: national red list status of endemic wild cinnamon species in Sri Lanka. Trop Agric Res 21:247–257

Lake BG (1999) Coumarin metabolism, toxicity and carcinogenicity: relevance for human risk assessment. BIBRA International, Surrey, UK, Food Chem. Toxicology

Laurentius SF, Wijesekera ROB, Devanathan MAV, Wijethunge MHC, Jayawardena AL, Ratnasingham K, Ponnachamy S, Rajendren S, Rajathurai P (1972) Sri Lanka Patent Application NG 6924

Liyanage T, Madhujith T, Wijesinghe KGG (2017) Comparative study on major chemical constituents in volatile oil of true cinnamon (*Cinnamomum zeylanicum* Blum.) and five wild cinnamon species grown in Sri Lanka. Trop Agric Res 28:270–280

Mallavarapu GR, Ramesh S (2000) Essential oil of the fruits of *Cinnamomum zeylanicum* Blum. J Essent Oil Res 12:628–630

New insights into coumarin contained in cinnamon (2012) Federal Institute of Risk Assessment; BfR 036/2012. https://www.bfr.bund.de/en/home.html

Paliwal R, Madungurum MA, Dahiru N (2018) Phytochemical analysis, physiochemical activity and antibacterial effects of *Cinnamomum zeylanicum* (Dalchini) extracts. Int J Eng Sci Res Technol. https://doi.org/10.5281/zenodo.1215434

Paranagama PA (1990) Analysis of Sri Lankan essential oils by gas chromatography and mass spectroscopy, MPhil thesis. University of Kelaniya, Kelaniya, Sri Lanka

Paranagama PA, Jayathilake GS, Jayawardene AL, Mubarak AM, Senanayeke UM, Wimalasena S (2001) A comparison of volatile constituents of cinnamon bark, leaf, root bark and fruit in Sri Lanka. J Natl Sci Found Sri Lanka 29:147–153

Qin B, Panickar K, Anderson R (2010) Cinnamon: potential role in the prevention of insulin resistance, metabolic syndrome, and type 2 diabetes. J Diabetes Sci Technol 4:685–693

Ransinghe P, Jayawardena R, Pigera S (2014) Evaluation of pharmacodynamics properties and safety of *Cinnamomum zeylanicum* (Ceylon cinnamon) in healthy adults a phase I clinical trial. BMC Complement Altern Med 14:309

Senanayake UM (1990) Essential oil product review. Ceylon Institute of Scientific and Industrial Research, Colombo

Senanayake UM, Wijesekera ROB (2005) Chemistry of cinnamon and cassia. In: Ravindran PN, Nirmal BK, Shylaja M (eds) Cinnamon and cassia, the genus *Cinnamomum*. CRC Press, London, pp 80–115

Senanayake UM, Lee TH, Wills RBH (1978) Volatile constituents of *Cinnamomum zeylanicum* oils. J Agric Food Chem 26:822–824

Silvia L, Fedrica A, Ettore C (2008) Coumarin and cinnamaldehyde in cinnamon marketed in Italy: a natural chemical hazard? Food Addit Contam 25:1297–1305

Sritharan R, Jacob VJ, Balasubramaniam S (1994) Thin layer chromatographic analysis of essential oils from *Cinnamomum* species. J Herbs Spices Med Plants 2:49–63

Taher M, Adibah F, Majid A, Sarmidi MR (2006) A proanthocyanidin from *Cinnamomum Zeylanicum* stimulates phosphorylation of insulin receptor in 3T3–L1 adipocytes. J Technol 44:385. https://doi.org/10.11113/jt.v44.385

The European Food Safety Authority (EFSA) (2005) Opinion of the scientific panel on food additives, flavorings, processing aids and material in contact with food (AFC) related to coumarin. EFSA J 245:1–14

Vernin G, Vernin C, Metzger J, Pujol L, Parkanyi C (1994) GC/MS analysis of cinnamon and cassia essential oils: a comparative study. In: Charalambous G (ed) Spices, herbs and edible fungi. Elsevier Science B V, Amsterdam

Wang Y, Avula B, Nanayakkara DNP, Zhao J, Khan IA (2013) Cassia cinnamon as a source of coumarin in cinnamon-flavored food and food supplements in the United States. J Agric Food Chem 61:4470–4476

Weeratunga HD, Premakumara GAS, Dayananda KR (2013) Investigation of coumarin levels in different grades of Ceylon cinnamon. Institute of Chemistry Ceylon Annual Sessions, Sri Lanka, 30(2):22

WHO monograph on selected Medicinal plants (2004)

Wijeweera A, Madhushika S, Hemalika S et al (2019a) Maturity dependence of bioactive constituents of Ceylon cinnamon: a qualitative and quantitative comparison. Braz J Pharm Sci (Submitted)

Wijeweera A, Madhushika S, Hemalika S et al (2019b) The ecological dependence bioactive constituents of Ceylon cinnamon: a qualitative and quantitative comparison. Braz J Pharm Sci (Submitted)

Chapter 11
Tapping into the Potential of Cinnamon as a Therapeutic Agent in Neurological Disorders and Metabolic Syndrome

Hanouf Ismail, Nalaka Wijekoon, Lakmal Gonawala, Stephanie Perera, Shamali Wasala, Yoonus Imran, Liraj Kodithuwakku, Nuwani Kodithuwakku, and Ranil De Silva

11.1 Cinnamon and Metabolic Syndrome

Cinnamon is one of the most important herbal drugs and has been widely used in Asia for more than 4000 years. As a folk medicine, cinnamon has been traditionally applied to the treatment of several disorders. Metabolic syndrome (MetS) is a multiplex risk factor that arises from abnormal adipose deposition and function with accompanying insulin resistance. The exact mechanism by which cinnamon elicits beneficial effects on MetS is yet to be identified, while several pathways are hypothesized based on research conducted on animal, human and animal cell models (Mollazadeh and Hosseinzadeh 2016). The targets suggested are insulin receptor, glucose transporter type 4 (GLUT-4), gluconeogenic pathway and proliferator activator receptor. The major active ingredients are hypothesized to be polyphenols present in cinnamon.

Several clinical trials and affirmation by recent systemic reviews and meta-analyses have suggested the beneficial effects of cinnamon consumption on diabetics and pre-diabetics on fasting blood sugar level, blood triglycerides and total cholesterol levels (Table 11.1). However, the data on the significance of its effects on haemoglobin A1c (HbA1c), low-density lipoprotein (LDL) and high-density lipoprotein (HDL), cholesterol levels, weight loss, blood pressure, waist circumference and insulin resistance are inconclusive (Blevins et al. 2007; Suppapitiporn et al. 2006; Vanschoonbeek et al. 2006).

The two major species of cinnamon used in research are *Cinnamomum zeylanicum* (Syn. *Cinnamomum verum*) and *Cinnamomum cassia.* However, given the high heterogeneity of data, the findings should be cautiously interpreted by clinicians despite having no serious side effects reported other than gastrointestinal disorders and allergic reactions of which majority were self-limiting (Tables 11.1 and 11.2).

H. Ismail · N. Wijekoon · L. Gonawala · S. Perera · S. Wasala · Y. Imran · L. Kodithuwakku
N. Kodithuwakku · R. De Silva (✉)
Interdisciplinary Centre for Innovation in Biotechnology and Neuroscience, University of Sri Jayewardenepura, Nugegoda, Sri Lanka
e-mail: ranil@sjp.ac.lk

R. Senaratne, R. Pathirana (eds.), *Cinnamon*,
https://doi.org/10.1007/978-3-030-54426-3_11

Table 11.1 Summary of studies of cinnamon effects on metabolic syndrome (MetS) conducted on subjects who meet one or more criteria of MetS or with risk factors for MetS. Abbreviations used are given in footnote

Reference	Use of anti-hypoglycaemic medication	Cinnamon				Outcome	Number of individuals
		Dosage	Duration	Type	Species		
Khan et al. (2003)	T2DM on sulfonylurea drugs, i.e. glibenclamide	Groups 1, 2 and 3 consumed 1, 3 or 6 g of cinnamon daily, respectively	40 days	Capsules	*Cinnamomum cassia*	Reduced the mean fasting serum glucose (18–29%), triglyceride (23–30%), LDL cholesterol (7–27%) and total cholesterol (12–26%) levels	60
Crawford (2009)	Type 2 diabetics on diabetic care and medication	1 g daily	90 days	Cinnamon capsules	*Cinnamomum cassia*	Lowered HbA1c	109
Mang et al. (2006)	Diabetes mellitus type 2 not on insulin therapy but treated with oral antidiabetics or diet	3 g daily	4 months	Cinnamon extract	N/A	Fasting plasma glucose level reduced	79
Stoecker et al. (2010)	With hyperglycaemia	250 mg twice per day	2 months	Dried water-extract cinnamon (CinSulin) capsule	N/A	Insulin concentrations and HOMA-IR (homeostasis model assessment-estimated insulin resistance) tended to be improved	137
Blevins et al. (2007)	77% of the subjects were taking diabetes medication	1 g daily	40 days	Cinnamon powder	*Cinnamomum cassia*	No significant change in fasting glucose, lipid, HbA1c or insulin levels	43

Reference	Use of anti-hypoglycaemic medication	Cinnamon				Outcome	Number of individuals
		Dosage	Duration	Type	Species		
Suppapitiporn et al. (2006)	Type 2 diabetic patients with their current treatment (metformin or sulfonylurea)	1.5 g daily	12 weeks	Cinnamon powder	*Cinnamomum cassia*	No significant difference in reducing fasting plasma glucose, HbA1c and serum lipid profile	60
Vanschoonbeek et al. (2006)	Postmenopausal patients with type 2 diabetes	1.5 g daily	6 weeks	Capsule	*Cinnamomum cassia*	No improvement in whole-body insulin sensitivity or oral glucose tolerance and did not modulate blood lipid profile	25
Gupta Jain et al. (2017)	Individuals with metabolic syndrome	3 g daily	16 weeks	Capsules	–	Significantly greater decrease in FBG, glycosylated haemoglobin, waist circumference and body mass index	116
Roussel et al. (2009)	Individuals with impaired fasting blood glucose with high BMI	250 mg two times per day	12 weeks	Aqueous extract of cinnamon (Cinnulin PF)	–	Reduced risk factors associated with diabetes and cardiovascular disease	22
Akilen et al. (2012)	T2D patients on hypoglycaemic agents and with an HbA1c more than 7%	2 g/day	12 weeks	Capsule	*Cinnamomum cassia*	Significantly reduces the HbA1c, SBP and DBP among poorly controlled type 2 diabetes patients	58

(continued)

Table 11.1 (continued)

Reference	Use of anti-hypoglycaemic medication	Cinnamon				Outcome	Number of individuals
		Dosage	Duration	Type	Species		
Allen et al. (2013)	T2D patients with their current treatment (not on insulin therapy)	1, 3, 6 g daily	40 days	Capsules	*Cinnamomum cassia*	Reduces serum glucose, triglyceride, LDL cholesterol and total cholesterol in people with T2D	60
Baker et al. (2008)	Patients with hyperglycaemia (fasting serum glucose >6.1 mmol/L or 2-h glucose >7.8 mmol/L)	250 mg two times per day	2 months	Dried water-extract of cinnamon (CinSulin) capsule	–	Reduced fasting insulin, glucose, total cholesterol and LDL cholesterol and enhanced insulin sensitivity of subjects with elevated blood glucose	137
Davis and Yokoyama (2011)	Subjects with T2D (fasting blood glucose ≥126 mg/dl)	3 glasses of black tea and either 3 g cardamom, cinnamon, ginger or 1 g saffron	8 weeks	Raw dried powder	*Cinnamomum verum*	Significantly beneficial effects on cholesterol, but not on measures of glycaemic control, oxidative stress and inflammation	204
Leach and Kumar (2012)	Non-insulin-dependent T2D, (HbA1c between 6% and 8% and FBG levels between 126 and 160 mg/dl)	3 g daily	8 weeks	Capsules	*Cinnamomum zeylanicum*	No significant differences were observed in glycaemic status indicators, lipid profile and anthropometric indicators between the groups at the end of intervention	37

Reference	Use of anti-hypoglycaemic medication	Cinnamon				Outcome	Number of individuals
		Dosage	Duration	Type	Species		
Akilen et al. (2010)	Subjects with T2D (with HbA1c more than 7% and on hypoglycaemic agents)	2 g daily	12 weeks	Capsules	*Cinnamomum zeylanicum*	No significant effect on glucose level, insulin response Ingestion of *C. zeylanicum* does not affect postprandial plasma glucose or insulin levels in human subjects	10
Khan et al. (2010)	Prediabetes and the metabolic syndrome	500 mg/day	12 weeks	Cinnulin PF®-specific water-soluble cinnamon extract	–	Decreases in FBG and SBP, increases in lean mass and decreases in body fat	22
Anderson et al. (2015)	Systematic literature search	1–6 g/day	40 days to 4 months	–	–	Decreases in FPG + decreases in HbA1c	6 clinical trials considered a total of 435 patients
Azimi et al. (2014)	Meta-analysis of 10 randomized controlled trials (RCT)	120 mg/day to 6 g/day	4–18 weeks	–	–	Decreases in FPG	543 patients
Vafa et al. (2012)	Systematic literature search	1–6 g/day		4 studies provided powder-filled capsules, while 1 study provided aqueous-filled capsules	*Cinnamomum cassia*	Did not change the results	5 RCTs 282 patients

(continued)

Table 11.1 (continued)

Reference	Use of anti-hypoglycaemic medication	Cinnamon				Outcome	Number of individuals
		Dosage	Duration	Type	Species		
Wickenberg et al. (2012)	Meta-analysis of clinical studies on people with type 2 diabetes and/or prediabetes	–	–	Cinnamon and cinnamon extract	–	Decreases in FPG	–
Ziegenfuss et al. (2006)	Literature search-active medication or no treatment in persons with either type 1 or type 2 diabetes mellitus	Mean dose of 2 g daily	4–16 weeks	Oral monopreparations of cinnamon	Predominantly *Cinnamomum cassia*	Effect of cinnamon on FBG level was inconclusive – -not statistically significant	10 RCTs involving a total of 577 participants
Mousavi et al. (2019)	Systematic review and dose-response meta-analysis of clinical trials that examined the effects of cinnamon supplements on obesity indices until September 2018	–	–	–	–	Cinnamon administration significantly decreased BW, BMI, WC and FM. Greater effects on BW were observed in subjects aged <50 years old and those with a baseline BMI of ≥30 kg/m^2	12 trials were included

Reference	Use of anti-hypoglycaemic medication	Cinnamon				Outcome	Number of individuals
		Dosage	Duration	Type	Species		
Maierean et al. (2017)	Meta-analysis of 13 randomized controlled trials	–	–	–	–	Significantly reduced blood triglycerides and total cholesterol concentrations without any significant effect on low-density lipoprotein cholesterol (LDL-C) and high-density lipoprotein cholesterol (HDL-C)	750 participants
Namazi et al. (2019)	Randomized clinical trials that examined the effects of cinnamon on at least fasting blood sugar until 31 February 2018	–	–	–	–	Significantly reduced fasting blood glucose. No significant change of the serum levels of insulin and insulin resistance, HbA1c, body weight, body mass index and waist circumference	18 studies

HbA1c haemoglobin A1c, *BMI* body mass index, *BW* body weight, *SBP* systolic blood pressure, *DBP* diastolic blood pressure, *FBG* fasting blood glucose, *FM* fat mass, *FPG* fasting plasma glucose, *IGT* impaired glucose tolerance, *N/A* not applicable, *WC* waist circumference, *T2D* type 2 diabetes

Table 11.2 Summary of studies of cinnamon effects on metabolic syndrome conducted on healthy subjects. Abbreviations used are given in footnote

Reference	Condition	Cinnamon				Outcome	Number of individuals
		Dosage	Duration	Type	Species		
Tang et al. (2008)	Healthy	3 g (6 capsules)/day	4 weeks	Capsule	–	No significant changes in fasting plasma glucose or lipids	11
Hlebowicz et al. (2009)	Healthy	3 g/week	2 months	Rice pudding with cinnamon	*Cinnamomum cassia*	Reduced postprandial serum insulin and increased GLP-1 concentrations without significantly affecting blood glucose, GIP, the ghrelin concentration, satiety or GER	15
Solomon and Blannin (2007)	Healthy	5 g of cinnamon	Once	–	–	Reduced total plasma glucose responses to oral glucose ingestion also improving insulin sensitivity as assessed by insulin sensitivity index	7
Beejmohun et al. (2014)	Healthy	1 g	Once	Capsule	*Cinnamomum verum*	Ceylon cinnamon hydro-alcoholic extract (CCE) may provide a natural and safe solution for the reduction of postprandial hyperglycaemia	18
Markey et al. (2011)	Healthy, young subjects	3 g	Once	Capsule	*Cinnamomum zeylanicum*	Did not alter the postprandial response to a high-fat test meal	9

GLP-1 glucagon-like peptide 1, *GIP* gastric inhibitory polypeptide, *GER* gastric emptying rate

11.1.1 The Interrelationship of Metabolic Syndrome and Neurodegenerative Diseases with Focus on Brain-Derived Neurotrophic Factor and the Role of Cinnamon

The brain-derived neurotrophic factor (BDNF) is found to be involved in the pathogenesis of MetS and neurodegenerative diseases (NDD). Cinnamon is a commonly used flavouring material, and its metabolite sodium benzoate (NaB) is a widely used food preservative and a FDA-approved drug for the treatment of urea cycle disorders in children (Modi et al. 2015; Brahmachari et al. 2007).

MetS causes decreased BDNF levels which in return decrease BDNF effective roles, which leads to serotonin signalling, altered dopaminergic system, increased inflammation, altered neurocognitive related roles and altered brain blood vessels status (Motamedi et al. 2017). NDD causes decreased BDNF levels which in return cause hyperphagia, glucose and lipid metabolic disorders, increased inflammation, hypertension and altered BDNF cardiovascular protective-related role which leads to diabetes, obesity and dyslipidaemia (Teillon et al. 2010).

It has been found that NaB dose-dependently induces the expression of BDNF and neurotrophin-3 (NT-3) in primary human astrocytes and neurons (Jana et al. 2013). It has also been found that after oral feeding, NaB enters into the brain and upregulate BDNF and NT-3 in vivo in the brain. Furthermore, oral administration of ground cinnamon was found to increase the level of NaB in blood and brain of mice and upregulate BDNF and NT-3 in vivo in the brain (Jana et al. 2013).

Activation of cAMP response element binding protein (CREB) seems essential for the transcription of most of the neurotrophic factors (Nair and Vaidya 2006). Therefore, for a drug to exhibit neurotrophic effect, it is almost mandatory to stimulate the activation of CREB. Evidence is found in research that NaB induces the activation of CREB via protein kinase A (PKA) (Jana et al. 2013). PKA has several key functions in the cell, including glycogen, sugar and lipid metabolism. Therefore, this suggests that cinnamon and NaB may also participate in glucose and lipid metabolism.

There are several advantages of NaB and cinnamon over other proposed anti-neurodegenerative therapies. Both NaB and cinnamon are fairly nontoxic. Cinnamon is metabolized to NaB. If in excess, NaB is excreted through the urine. Furthermore, cinnamon and NaB can be taken orally, the least painful route. Also, cinnamon and NaB are inexpensive compared to other existing anti-neurodegenerative therapies. After oral administration, NaB rapidly diffuses through the blood-brain barrier (BBB). Similarly, after oral administration of cinnamon, NaB is detected in the brain (Batshaw et al. 1988). Glycine toxicity is a problem in different neurological diseases because of movement disorders; glycine is one of the factors for inhibiting motor neurons. NaB is known to combine with glycine to produce hippurate, a compound that is readily excreted in the urine (Jana et al. 2013). These properties of cinnamon and NaB indicate that these compounds may be used for therapeutic intervention in NDDs and MetS as primary or adjunct therapy.

11.2 Neurological, Neuromuscular Diseases and Cinnamon

NDDs constitute a large group of pathological conditions, characterized by a progressive loss of neuronal cells, which compromise motor and/or cognitive functions. The most common NDDs are Alzheimer's disease (AD), amyotrophic lateral sclerosis (ALS), Parkinson's disease (PD) and Huntington's disease (HD). The causes of these pathologies are multifactorial and not fully understood, but it is well known that factors related to ageing and to the overproduction of free radicals and reactive oxygen species (ROS) lead to oxidative stress and cell death, which is extremely related (Sheikh et al. 2013). Whereas oxidative stress plays an unquestionable and central role in NDDs, the control of free radicals and ROS levels represents an interesting and promising strategy to delay neurodegeneration and attenuate the associated symptoms.

In vitro and in vivo studies, performed with extracts and fractions of plants and with isolated natural bioactive compounds, provide evidence of the role of these substances in the modulation of the cellular redox balance and in the reduction of the formation of ROS originating from oxidative stress, thereby demonstrating their great value as antioxidant agents and cellular protectors. NDDs, including AD, PD and HD, present a major health and financial burden to every health service organization in the world. The total number of AD patients globally is estimated to be well over 100 million by 2050 (Prince et al. 2013). However, AD is not the only disease for which the numbers will increase. A paper by Dorsey et al., "Projected number of people with Parkinson's disease in the most populous nations, 2005 through 2030", shows similar results. The number of PD patients will double or even increase above that for countries, such as China, India and Indonesia (Dorsey et al. 2007).

There have been 12 anti-Parkinsonian drugs, 6 drugs for multiple sclerosis and only 4 anti-Alzheimer drugs in the 30 years of research (Newman and Cragg 2012). It is important to note that none of the so-called anti-Alzheimer drugs slow down or prevent neuronal death and the malfunction of the human brain (Prince and Jackson 2009). Pharmacological treatments are only available for the partial treatment of the symptoms (Rafii and Aisen 2009). No new AD drugs have been approved by the Food and Drug Administration (FDA) since 2003 (Cummings et al. 2014). The numbers of drugs approved for other neurodegenerative diseases have not seen much of an increase over the last 8 years either. There was only a single new drug approved for ALS -Radicava in 2017 and HD- Austedo in 2017. A slightly higher number of drugs has been approved for PD -Duopa in 2015; Gocovri in 2017; Nuplazid in 2016; Rytary in 2015; Xadago in 2017 (Reviewed in Pohl and Kong Thoo Lin 2018). For other NDDs, such as prion diseases, e.g. Creutzfeldt-Jakob disease, or spinocerebellar ataxia (SCA), e.g. SCA 1, 3 or 6, no drug therapies to retard their progress are currently available (Aguzzi et al. 2018). Protein aggregations and their precursor stages have long been the target of several drug trials and significant research, for example, AD, PD and ALS, as well as HD and other polyglutamine diseases (e.g. SCA) (Ross and Poirier 2004), but despite this significant drug trials, there are no drugs to fight against the NDDs.

Neuromuscular diseases (NMD) refer to a heterogeneous group of diagnoses that share characteristics, such as loss of muscle strength, a neurodegenerative progression and chronicity of the pathology (Abresch et al. 2009; Pieterse et al. 2008). With a growing trend to seek out alternative therapies, it is not surprising that patients with NMDs are looking towards alternative therapies and nutraceuticals. If a nutraceutical could produce similar therapeutic benefits to available drugs for NMDs without adverse side effects, it could provide the majority of patients with an improved quality of life as well as reduce costs associated with recurrent hospital visits to monitor and treat drug-induced side effects (Woodman et al. 2016).

The development of a new drug is a complex, time-consuming and expensive process from the discovery of a new drug to its administration in a clinical setting takes approximately 12 years, with investments amounting to 1 billion US$ (Katiyar, Gupta, et al. 2012). Therapies that have been developed over the past 30 years or so reveal that ~40% of new chemical entities are natural products or are inspired by natural products. In Sri Lanka, the traditional medicine system, using local resources, has been practised for more than 5000 years in the treatment of various diseases, including those of complicated aetiology such as cancer (Kuruppu et al. 2019).

Ageing, one of the risk factors of many neurodegenerative diseases, was first associated with free radicals by Harman (1956). Since then, ROS have been shown to be the cause of oxidative stress and have been associated with a range of neurological disorders such as AD, PD, HD and Pick's disease. It is assumed that reducing oxidative stress within the human body could at least be part of future treatment options, targeting multiple drug targets, as most neurodegenerative diseases have been associated with several different pathways involved in the disease development (Patten et al. 2010; Knight 1998). It is well known that the production of ROS increases with age, while some of the endogenous defence mechanisms can decrease. If the balance between ROS and antioxidants is disturbed, the excessive amounts of ROS will damage cells by protein oxidation, DNA/RNA strand breakage, lipid peroxidation or the formation of advanced glycation end products (Moldogazieva et al. 2019). Redox dysregulation can be caused by the ineffectiveness of the endogenous antioxidant system to handle an increase in the production of free radicals (e.g. disease, mitochondrial dysfunction, exposure to environmental factors) or because of a decreased effectiveness of the endogenous antioxidant system itself. These additional stresses can cause damage to biological molecules that lead to rapid cell death, resulting in neurodegeneration, either by functional loss (ataxia) or sensory dysfunction (dementia) (Gandhi and Abramov 2012; Tang et al. 2005) In addition, inflammation, protein aggregations (e.g. amyloid in AD) and excessive presence of metal ions, such as iron (Fe^{2+}) and copper (Cu^{2+}), can cause oxidative stress (Sheikh et al. 2013).

Although the role of mitochondrial ROS is not completely understood, it is proposed that mitochondrial dysfunction causing excessive ROS production may be a prominent feature of several diseases (Görlach et al. 2015). Interactions among ROS and calcium signalling can be considered as bidirectional, wherein ROS can regulate cellular calcium (Ca^{++}), while calcium signalling is essential for ROS production. Thus, increased levels of Ca^{++} activate ROS-generating enzymes and

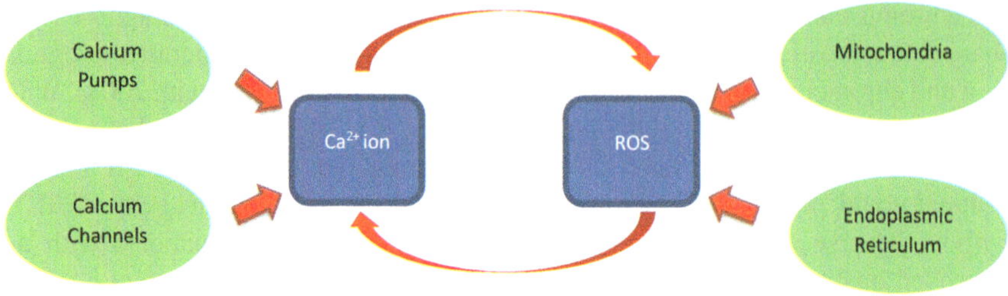

Fig. 11.1 The interplay between calcium and ROS. (Data for the image was collated from Görlach et al. 2015 and Gordeeva et al. 2003)

formation of free radicals (Gordeeva et al. 2003). Mitochondrial defects have been associated with a series of muscular dystrophies (Katsetos et al. 2013), and a substantial body of experimental work, derived from animal models, attests to a major role of mitochondria in the early process of muscle degeneration. Common mechanisms of mitochondria-related cell injury central to most MDs include reduction of mitochondrial mass and number, dysregulation of the mitochondrial permeability transition pore opening and defective respiratory chain activities (Lovering et al. 2005) (Fig. 11.1).

Natural product-based complementary and alternative medicine is becoming increasingly popular in Western countries and is a multibillion-dollar industry. As per the National Health Statistics Report 2016 by the National Institute of Health (NIH), $33.9 billion dollars were spent per year in the USA alone on complementary and alternative medicine (CAM), including nutraceutical products. More was spent on visits to complementary practitioners ($14.7 billion) followed by for purchases of natural product supplements ($12.8 billion) (NIH report 2016).

Research done over the past few decades has proven that cinnamon, a spice long used in traditional medicine, is a rich source of phytochemicals and recognized for decades as a therapeutic agent to improve memory and cognitive function. This part of the chapter aims to identify the role of cinnamon as a therapeutic approach against NDDs, according to current research. Developing an understanding of the pathology of these NDDs would aid in finding how cinnamon can be used as a treatment strategy. The ability of cinnamon to affect the nervous system will mostly depend on the metabolites ability to cross the BBB through the process of diffusion across the membrane and eventually enter the brain. It has been recommended that the capacity of cinnamon flavonoids and its metabolites to enter the brain upon crossing the BBB mostly depends on the extent of their lipophilicity (Lin et al. 2007). Despite this fact, there are certain animal studies which show the entry of glucuronides into the brain through BBB (Momtaz et al. 2018).

11.2.1 Role of Nrf2 and NF-κB in Neurodegenerative Disorders and Neuromuscular Diseases

Recent studies have highlighted that the finely regulated interplay between nuclear factor erythroid 2 p45-related factors 2 (Nrf2) and nuclear factor-kappa B (NF-κB) performs an important role in neuroprotection and the facilitation of a healthy neuronal environment. Dysregulation of this system has shown to result in chronic neurodegenerative disorders, such as AD, HD, ALS and PD (Esteras et al. 2016). The transcription factor Nrf2 is responsible for the regulation of genes that promote antioxidation, protection against electrophile damage, mitochondrial and protein homeostasis and prevention against inflammatory damage. In its resting state in the nucleus, it exists anchored to its repressor protein Keap1 (Kelch-like ECH-associated protein 1). During instances of cellular oxidative stress (OS), presence of polyphenols or other inducers, Keap1 is ubiquitinated and undergoes proteasomal degradation, which allows Nrf2 movement into the nucleus. It is then bound to the antioxidant response element (ARE) resulting in the downstream promotion of transcription of antioxidant, anti-inflammatory and autophagy functions within the cell.

NF-κB is a family of inducible transcription factors, which coordinates a wide range of genes relating to inflammatory and immune response. Apart from its role as an inflammatory mediator, it is also involved in the processes of cell death and cellular proliferation. Maggirwar et al. (1998) reported NF-κB facilitates the survival of sympathetic neurons, therefore exhibiting a role of NF-κB in neuronal death (Rahiman et al. 2015). NF-κB exists neutrally in the cytoplasm in conjunction with its repressor nuclear factor of kappa light polypeptide gene enhancer in B-cell inhibitor, alpha (IκBα). When phosphorylated and ubiquitinated, IκBα undergoes proteasomal degradation, releasing NF-κB and allowing its movement into the nucleus, similar to Nrf2. It is not just Nrf2 that takes the front seat in this relationship, as NF-κB too plays a pivotal role in mediating the interplay between NF-κB and Nrf2. This has been demonstrated by Yu et al. 2019 studies, where p65, a subunit of NF-κB, contributed to raising levels of Keap1 within the nucleus (Yu et al. 2011). This consequently causes a decline in Nrf2-ARE signalling by reducing the availability of free Nrf2 (Lattke et al. 2017). p65 can further inhibit Nrf2 by de-acetylation of histones and preventing it from forming a heterodimer with Nrf2, which is required for ARE binding – resulting in a drop in expression of ARE-related genes (Liu et al. 2019).

Neuroinflammation refers to the inflammatory processes that take place in the central nervous system (CNS) and involves the innate as well as the adaptive immune system. However, though the neuroinflammatory system has been evolved to protect the CNS from harmful pathogens, it could also cause severe debilitating damage due to intrinsic and extrinsic factors. Nrf2 is a well-established mediator of anti-inflammatory signalling, when activated, creating downstream effects on the transcriptional repression of pro-inflammatory cytokines, such as tumour necrosis factor-α (TN-α), interleukin-1 (IL-1), interleukin-6 (IL-6), microglia, macrophages, monocytes and astrocytes. However, this process is tightly regulated by Nrf2, by the inhibition of IκB kinase (IKK) and its subsequent inhibition of NF-κB, therefore attenuating the effects of neuroinflammation. NF-κB, which coordinates signal transduction and transcription factor regulation, is a key mediator of inflammation, cell survival and differentiation. Further, it is highlighted in the literature that the perturbation of the NF-κB signaling pathway is a common phenomenon in muscular dystrophies and that aberrant regulation of NF-κB could be a potential cause of the onset of muscular dystrophy. (Bhatnagar et al. 2009). On the other hand, inflammatory signaling, including NF-κB and cytokine (TNF-a, IL-1b) pathways, may be responsive to OS (Lingappan et al. 2018).

Oxidative stress created from the accumulation of ROS results in the production of protein aggregates, mitochondrial damage and neuronal loss. OS can affect some of the key diseases under the vast umbrella of neurodegenerative disorders, which include AD, PD, dementia, HD, ALS and SCA, through mechanisms including protein misfolding and mitochondrial dysfunction (Liu et al. 2017). By understanding the pathology of the diseases, an effective treatment strategy may be determined. OS is created when an imbalance occurs between the generation of ROS during respiration and the body's antioxidant combatting system. Therefore, finding a method to alleviate the effect of OS is crucial in improving the prognosis for neurodegenerative diseases. Evidence of OS such as lipid peroxidation nucleic acid oxidation as well as protein nitration has all been observed in the brains of AD and PD patients. Nrf2 is responsible for initiating an antioxidant response. However, studies have shown a difference in the pattern of Nrf2 availability within the different regions of the brain of patients suffering from AD and PD. There is a decreased nuclear Nrf2 abundance in the hippocampal neurons of AD patients, exhibiting more Nrf2 in the cytoplasm. In contrast, the substantia nigra of patients with PD show increased Nrf2 activity within the nucleus, demonstrating the expected defence response against OS. Due to the difference in cellular Nrf2 expression, it could be understood that there is a difference in how each disease affects the normal physiology of Nrf2 in the brain and therefore the pharmacological implications this might have in attempting to increase Nrf2 concentrations in the brain.

A study by Lastres-Becker et al. (2012) on mice where the microglial activation of Nrf2 was suppressed has shown to exhibit histological signs of mid to late stage of parkinsonism with neurodegeneration of nigral dopaminergic neuronal cells. Furthermore, the mice failed to express the genes responsible in the production of enzymes heme oxygenase-1 (HO-1) and NAD(P)H quinone dehydrogenase 1 (NQO1), key in defence against oxidative stress. This is not only just an example of the importance of Nrf2 activation on PD but also the interdependence of NF-κB and Nrf2 in facilitating inflammation (Sivandzade et al. 2019). Nrf2 has also been implicated in modulating neuroinflammation in ALS. Using murine models, a neuroprotective effect was noted when raising nuclear Nrf2 levels by the repression of activated microglial cells, recusing neurotoxicity (Dinkova-Kostova et al. 2018).

Aggregation of misfolded proteins is one of the major hallmarks of chronic NDDs. These misfoldings could originate from genetic aberrations, or from sporadic changes in proteostasis. Autophagy is the catabolic process of degradation of large protein aggregates as well as damaged cell organelles in order to maintain proteostasis and reduce organelle dysfunction. The pathogenesis of neurodegenerative disorders, such as AD, ALS, PD and HD, have all implicated a mechanism of faulty protein aggregate degradation. Since there is a key role played by autophagy dysregulation in cases of neurodegenerative disorders, it would prove important to understand the mechanism of autophagy, its regulators and the causes of its dysregulation in order to develop methods for its treatment. Nrf2 mediates autophagy via the induction of autophagosome cargo-protein p62/sequestosome-1 (p62/SQSTM1). In addition, p62, when phosphorylated, has a binding affinity to Keap1 protein, the negative regulator of Nrf2, and competes with Nrf2 for it, resulting in increased transcription of genes regulated by Nrf2. This simultaneously creates a positive feedback regulatory loop for p62/SQSTM1 itself (Jain et al. 2010). Nrf2 also regulates autophagy by mediating the expression of genes encoding proteins, such as serine/threonine-protein kinase (ULK1) (He et al. 2017) and autophagy-related 5 (ATG5), which play important roles in the assembly of the autophagosome (Grasso et al. 2018).

Autophagy is important in degrading amyloid precursor protein (APP), which when cleaved produces insoluble amyloid-β peptide found in senile plaques, a hallmark of AD. The downregulation of Nrf2 has been investigated in double transgenic mice expressing a chimeric mouse/hAPP (Mo/HuAPP695swe) and a mutant human γ-secretases presenilin 1 (PS1-dE9) 72. In comparison to mice with wild-type Nrf2, mice that are deficient for Nrf2 had increased cellular levels of APP and Aβ as well as increased neuroinflammation. In these Nrf2-deficient mice, histological stains revealed a build-up of multivesicular bodies, endosomes and lysosomes, all evidence of impaired autophagy (Tan et al. 2019). In cases of PD, Nrf2-mediated protection against neurodegeneration is activated, by the autophagic degradation of α-synuclein (Skibinski et al. 2016). Impaired autophagy also plays a role in neuromuscular disorders and myopathies such as congenital muscular dystrophy type 1(CMDT1) and Duchenne muscular dystrophy (DMD).

The abovementioned studies convey the importance of the Nrf2 and NF-κB interplay and demonstrate its potential as a possible therapeutic target. So far, for most NDDs, such as AD, there are no current disease-modifying therapies that act as a cure. However, the use of natural products in therapy could provide a solution to managing the symptoms of the disease with the least possible side effects, and cinnamon has proven to show some efficacy against the disease. Trans-cinnamaldehyde (TCA) exhibited neuroprotective effects in animal models, investigating ischaemic stroke. TCA was shown to possibly inhibit neuroinflammation by diminishing nitric oxide synthase (iNOS), COX-2 expression as well as the NF-κB signalling pathway. The NF-κB signalling pathway plays an important role in both acute neurodegenerative diseases like stroke and chronic neurodegenerative diseases like AD, where increased NF-κB activity results in the progression of the neurodegeneration process (Ho et al. 2013). Furthermore, studies have shown a strong correlation between amplified NF-κB activity and COX-2 transcription in superior temporal lobe gyrus of AD patients, adding to the pathology of the disease (Mattson and Camandola 2001).

Using a murine model, Jain et al. (2014) noted that when administered with *C. zeylanicum*, rats chemically induced to suffer amnesia with scopolamine showed a drop in OS markers in their brains and an improvement in cognitive function. Furthermore, a study by Modi et al. (2015) showed that oral feeding of *C. verum* powder and NaB suppressed the activation of p21rac (an NADPH oxidase complex) in transgenic mice, therefore decreasing OS in the brain by suppressing the production of ROS. Cinnamon has also shown to have a correlation with superoxide dismutase (SOD), an antioxidant enzyme which functions in the brain (Roussel et al. 2009). However, in a different study, Yu et al. (2000) showed that NF-κB serves a neuroprotective function in animal models of HD, where a lack of transcriptional factor p50 coding subunit for NF-κB results in cellular death. Therefore, cinnamon as therapy should be considered carefully before administration, and pharmacology of cinnamon should be identified so that it can be best-made use of as a therapy.

Nrf2/Keap1-ARE pathways were activated by *C. cassia* bark extract in experiments run by Wondrak et al. (2010), as a target for chemotherapy against colon cancers. Treatment with cinnamon extract increased cellular levels of Nrf2 and subsequently resulted in an antioxidant response including HO-1. NaB, a metabolite of cinnamon, has been shown to upregulate protein deglycase (DJ-1) (Wondrak et al. 2010), a neuroprotective protein responsible for ensuring cellular survival by managing oxidative stress, facilitating both of glutathione reductase and magnesium SOD through Nrf2-dependent pathways. Cinnamon and its metabolites, NaB, TCA and cinnamic acid, show potential in therapeutics against neurodegenerative disorders (Table 11.3, Ho et al. 2013; Jana et al. 2013; Yulug and Cankaya 2019) that would be valuable to tap into.

11.2.2 The Role of Nrf2 and NF-κB in Stroke

Stroke, a reduction in blood flow to the brain, is caused by a blockage in a cerebral artery by a clot or embolus (ischaemic stroke) or rupture of the blood vessel (haemorrhagic stroke). Both forms of stroke result in damage or death of neurons in the affected brain region, leading to the loss of brain function (Slemmer et al. 2008). In stroke, cerebral ischaemia triggers the pathological mechanisms, collectively known as the ischaemic cascade, causing rapid and irreversible neuronal injury within the ischaemic core. However, the surrounding hypoperfused brain tissue, known as the penumbra, can be salvaged if the flow is restored and/or efficacious therapies are applied. Notably, reperfusion from recanalized cerebral vessels can cause tissue injury due to cerebral oedema, brain haemorrhage and neuronal death (Jung et al. 2010).

The relative surplus of ROS, caused by excessive ROS generation and/or impaired ROS degradation, plays a key role in the pathological mechanisms of stroke followed by reperfusion. (Fraser 2011; Radermacher et al. 2013). The rapid increase of ROS production promotes ischaemia/reperfusion injury in many

Table 11.3 Cinnamon as a therapy against neurodegenerative disease

Model system	Condition	Outcome	Reference
(1-methyl-4-phenyl-1,2,3,6-tetrahydropyridine) MPTP mouse model	PD	Oral treatment of MPTP-intoxicated mice with cinnamon powder (*Cinnamomum verum*) and NaB reduced the expression of iNOS and protected Parkin/DJ-1 in the substantia nigra. These findings paralleled dopaminergic neuronal protection, normalized striatal neurotransmitters and improved motor functions by cinnamon in MPTP-intoxicated mice	Khasnavis and Pahan (2014)
(MPTP) mouse models	PD	Downregulated p62 in the substantia nigra of MPTP mice was increased by administration of CA. This study showed that blockage of autophagy using autophagy inhibitors protected the 1-methyl-4-phenylpyridinium (MPP+)-mediated death of BE(2)-M17 cells	Bea et al. (2018)
TCA efficacy on the 6-hydroxydopamine (6-OHDA)-induced dopaminergic degeneration in mice	PD	Trans-cinnamaldehyde (TCA) has a neuroprotective effect on dopaminergic neurons and that this effect may be associated with the inhibition of inflammatory responses	Pyo et al. (2013)
TH1 transfected SH-SY5Y	PD	Prevented H_2O_2-induced cell death in PD	Maiolo et al. (2018)
N/A	Stroke	Cinnamophilin is a novel thromboxane A2 receptor antagonist isolated from *Cinnamomum philippinensis*. Study reported that cinnamophilin confers protection against ischaemic damage in rat brains when administered at 80 mg/kg at different time intervals (2, 4 and 6 h) after insult. The effects were found to have a considerable effect (by 34–43%) on abridged brain infarction	Lee et al. (2009), Yu et al. (1994)
Fly model	AD	Cinnamon extract markedly inhibits the formation of toxic Ab oligomers and prevents the toxicity of Ab on neuronal PC12 cells	Frydman-Marom et al. (2011)
Primary hippocampal cells	AD	Extract from whole cinnamon effectively inhibited aggregation of human tau in vitro and that the inhibitory activity could be attributed to both a proanthocyanidin trimer and cinnamaldehyde	Peterson et al. (2009)

(continued)

Table 11.3 (continued)

Model system	Condition	Outcome	Reference
Primary mouse astrocytes and oligodendrocytes and primary human astrocytes	MS	Myelinogenic property of NaB and cinnamon (*Cinnamomum verum*)	Modi et al. (2015)
Human primary neurons		NaB increases the expression of neurotrophic factors (BDNF and NT-3) in brain cells and in vivo in the CNS of mice. Accordingly, oral administration of cinnamon increased the level of NaB in vivo in the CNS and upregulated CNS expression of neurotrophic factors. Furthermore, they have demonstrated that NaB utilizes the PKA-CREB pathway for the upregulation of neurotrophic factors	Jana et al. (2013)

different ways, including BBB disruption, inflammation, apoptosis and cellular necrosis (Kahles and Brandes 2012; Shi et al. 2015).

ROS regulates CNS inflammation via several mechanisms, of which interaction with NF-κB, a family of transcription factors that are master regulators of inflammation, seems to be of greatest importance (Rojo et al. 2014). Accumulating evidence suggests that neuronal cell death and expansion of damage to the penumbra after focal ischaemia are mediated by ROS-induced cell apoptosis. The two main pathways in cell-mediated apoptosis, the death receptor- and mitochondria-mediated pathways, are both affected by ROS through multiple mechanisms, such as apoptosis signal-regulating kinase 1 (ASK1)-Jun N-terminal protein kinase (JNK) signalling and permeability transition pore (PTP) proteins (Circu and Aw 2010). ROS causes cellular necrosis by lipid peroxidation, DNA damage, mitochondrial respiratory chain disruption and adenosine 5′-triphosphate (ATP) production (Olmez and Ozyurt 2012; Bolaños and Almeida 1999). Moreover, the aforementioned mechanisms of ischaemia/reperfusion injury induced by excessive ROS would interact with each other. For instance, NF-κB has been involved in the modulation of BBB integrity. The ROS-triggered NF-κB activation may enhance expression of intercellular cell adhesion molecule-1 (ICAM-1), resulting in cytoskeletal changes (Kim et al. 2008; EtienneManneville et al. 2000). NF-κB activation promotes tight junction disruption and MMP activation as well (Na et al. 2016; He et al. 2011).

The activation of Nrf2 appears to be influenced by several factors including cysteine residues, phosphorylation, autophagy and epigenetic factors. Under stressful conditions, the redox-stress sensitive cysteine residues in Keap1 are modified, allowing Nrf2 to be released from Keap1 (Sekhar et al. 2010). Phosphorylation of Nrf2 by various protein kinases has also been shown to affect Nrf2 activation. Phosphoinositide 3kinase (PI3K)/protein kinase B (Akt) signalling and extracellular regulated kinase (ERK) are involved in Nrf2 translocation. Inhibition of PI3K, Akt, or ERK attenuates Nrf2 activation (Nakaso et al. 2003; Li et al. 2006; Xu et al.

2006). Phosphorylation of Nrf2 by protein kinase C (PKC) promotes the release of Nrf2 from Keap1 (Bloom and Jaiswal 2013). Furthermore, p62, a selective autophagy substrate, interacts with the Nrf2-binding site on Keap1, resulting in stabilization and activation of Nrf2, which indicates that autophagy takes part in Nrf2 regulation (Komatsu et al. 2010; Lau et al. 2010). Additionally, epigenetic factors, such as microRNAs, histone acetylation and CpG methylation, have also been suggested as factors in the regulation of Nrf2 levels (Bryan et al. 2013; Narasimhan et al. 2012). Nrf2 seems to play an important role in the protection of brain cells against cerebral ischaemic injury. Loss of Nrf2 function increases the size of cerebral infarct and neurological deficits after an ischaemic event (Shah et al. 2007; Shi et al. 2015). This is further corroborated as recent studies highlight that the neuroprotection of many compounds after a stroke is mediated by Nrf2 signalling (Kunze et al. 2015; Shi et al. 2015; Ding et al. 2014; Zhang et al. 2014).

A number of studies have indicated that the expression of Nrf2 is largely increased in the acute phase of stroke (Yang et al. 2009; Thanaka et al. 2011; Dang et al. 2012; Srivastava et al. 2013). In a rat permanent middle cerebral artery occlusion (pMCAO) model, it has been demonstrated that Nrf2 mRNA and protein were upregulated beginning at 3 h after the ischaemic event with a peak of 24 h (Yang et al. 2009). Tanaka et al. (2011) found that in both peri-infarct and core infarct regions, Nrf2 expression began to increase at 2 h, peaked at 8 h and then decreased at 24 and 72 h of reperfusion in a mouse transient middle cerebral artery occlusion (tMCAO) model. Nrf2 protects the brain from ischaemia/reperfusion injury mainly by induction of its target antioxidant genes to defend against excessive ROS production and related disorders, such as BBB dysfunction and inflammation (Sandberg et al. 2014). Nrf2 activation initiates the transcription of a host of antioxidative genes. HO-1 and NQO1 expressions have been confirmed to increase after ischaemia, and genetic deletion of Nrf2 abolished the upregulation of HO-1 (Ding et al. 2015; Li et al. 2015). Activities of some Nrf2-target antioxidative genes were impaired after a stroke. Nrf2 activation restores their activities. It has been shown that the Nrf2 activator restored the stroke-induced decreased levels of SODs, glutathione peroxidase (GPx) and glutathione (GSH) (Ashabi et al. 2014; Ding et al. 2015). Furthermore, the upregulation of GSH induced by the Nrf2 activator was abolished in Nrf2-deficient mice (Shih et al. 2005). Loss of Nrf2 function increases the size of cerebral infarct and neurological deficits after an ischaemic event (Shah et al. 2007; Shih et al. 2005). This is further corroborated as recent studies highlight that the neuroprotection of many compounds after a stroke is mediated by Nrf2 signalling (Shah et al. 2014); Zhang et al. 2014). Thus, Nrf2 is a promising therapeutic target for the treatment of stroke.

It was reported recently that NF-κB could directly repress Nrf2 signalling at the transcription level. Pharmacological and genetic studies suggest that the absence of Nrf2 can exacerbate NF-κB activity leading to increased cytokine production, whereas NF-κB can modulate Nrf2 transcription and activity, having both positive and negative effects on the target gene expression (Wardyn et al. 2015). Nrf2 pathway inhibits the activation of NF-κB pathway by increasing antioxidant defences and HO-1 expression, which efficiently neutralizes ROS and detoxifies toxic chemicals and hence reduces ROS-mediated NF-κB activation. Nrf2 pathway also inhibits NF-κB-mediated transcription by preventing the degradation of IkB alpha. Similarly, NF-κB-mediated transcription reduces the Nrf2 activation, by reducing the ARE gene transcription, and decreases free CREB-binding protein (CBP), by competing with Nrf2 for Cys- and His-rich region 1 – KIX (CH1-KIX) domain of CBP, by competing with Nrf2 for CH1-KIX domain of CBP (Ganesh Yerra et al. 2013). Thus, the Keap1-Nrf2-ARE pathway and the crosstalk between Nrf2 and NF-κB are promising therapeutic targets for stroke and its complications.

Many electrophilic NRF2 activators are obtained from natural products, mainly non-nutrient phytochemicals, thus making them available to large populations at a low cost. Dietary supplementation with spices, such as turmeric, cinnamon and bixin, as well as the inclusion of whole foods, such as green tea, broccoli sprouts and other cruciferous vegetables, ensures constant exposure that cannot always be achieved with drugs due to low treatment adherence (Shu et al. 2010; Wondrak et al. 2010; Tao et al. 2016). Even though many studies have not been performed to investigate the relation of cinnamon and Nrf2 pathway in stroke, the present literature on risk factors of stroke and the effect of cinnamon proves that cinnamon has the potential as a promising therapeutic agent in stroke-associated brain cell death recovery by targeting the Nrf2 pathway (Zhou et al. 2014). Antioxidants reduce stroke risk. It has been well established in epidemiological trials that diets rich in fruits and vegetables help manage modifiable risk factors for stroke such as hypertension, thereby lowering stroke incidence (Kuklina et al. 2012; Aune et al. 2017; Tong et al. 2019; Wang et al. 2014)

11.3 The Association of Cinnamon and Cancer

Cancer is a global health burden, characterized by uncontrolled cell growth. In 2018, it has been estimated that the global cancer burden has risen to 18.1 million new cases and 9.6 million deaths. One in five men and one in six women worldwide develop cancer during their lifetime, and one in eight men and one in 11 women die from cancer (WHO 2018). With this issue burdening health authorities in almost every country, the prevention of cancer is one of the most significant public health challenges of the twenty-first century. Cinnamon has been widely studied for its potential use in both cancer prevention and treatment.

The process of angiogenesis, which is one of the hallmarks of cancer, is an important mechanism used by tumours to promote growth and metastasis. Angiogenesis is the formation of new blood vessels that develop fast from pre-existing vasculature to accommodate the increased nutrient and oxygen demand of a growing tumour (Nishida et al. 2006). The high oxygen demand results in a characteristic hypoxic state within the tumour, which is regulated primarily by the hypoxia-inducible factor-1 (HIF-1). In cancer cells, *HIF-1* induces the expression of vascular endothelial growth factor (VEGF), which fuels angiogenesis. Currently, there are several anti-VEGF agents which have been approved for cancer treatment; however, due to their severe side effects, such as bleeding and hypertension, their therapeutic application is limited. Lu et al. (2010) have shown that a cinnamon aqueous extract is a potent inhibitor of vascular endothelial growth factor receptor 2 (VEGFR2) kinase activity, directly inhibiting the kinase activity of purified VEGFR2 as well as mitogen-activated protein kinase- and Stat3-mediated signalling pathway in endothelial cells. Thus, the cinnamon extract was able to inhibit VEGF-induced endothelial cell proliferation, migration and tube formation in vitro and tumour-induced blood vessel formation in vivo (Lu et al. 2010). Another study has shown that essential oil of *C. zeylanicum* bark showed cytotoxicity with an IC50 value <20 μg/mL for Ras active (5RP7) and normal (F2408) fibroblasts cells. The study showed that 5RP7 cells were affected stronger than normal cells. Morphological observation of apoptotic cells indicated the induction of apoptosis at high concentrations of *C. zeylanicum* oil, especially in 5RP7 cells (Unlu et al. 2010). A study published in India has shown that treatment with aqueous extracts of cinnamon along with cardamom increased the activities of the detoxifying and antioxidant enzyme glutathione-s transferase (GST) with a concomitant reduction in lipid peroxidation levels in animals with colon cancer compared to controls. In living organisms, GSTs are a family of detoxifying enzymes, synthesized in the liver. Colon cancer is one of the leading causes of cancer deaths in many countries including in Sri Lanka. Thus, identifying chemopreventive substances, such as cinnamon, may help in arresting proliferation and enhancing apoptosis of colon cancer cells (Bhattacharjee 2007). In addition to extracts of cinnamon, a number of active compounds of cinnamon have also been reported to enhance the cytotoxicity of various cancer cell types. For instance, cinnamaldehyde has shown potent antiproliferative effects on liver cancer cells, such as PLC/PRF/5 cells, in a dose-dependent manner. The compound was also able to downregulate Bcl-2 and Mcl-1 expression and upregulate Bax protein in a time-response manner as a result of apoptosis (Wu et al. 2005). Cinnamon has shown to have inhibitory effects on tumour cell growth, by inhibiting angiogenesis and cell-cell adhesion. In conclusion, these data indicate the possibility that cinnamon and its constituents may find applications as potent anticancer agents.

Mitochondria are considered to be a target for preventing neurodegeneration, since they are highly reliant on respiratory task due to the bio-energetic requirements (Golpich et al. 2016). Mitochondrial 70 kDa heat-shock protein or mortalin, which plays a role in cell proliferation, stress response, tumorigenesis (Wadhwa et al. 2006), and neurodegeneration (Burbulla et al. 2014) and maintenance of the mitochondria (Golpich et al. 2016). The function of mortalin is altered in cellular stress conditions and its inducing the phosphorylation of p53 leads to apoptosis. The upregulation of mortalin can be induced by glucose deprivation, oxidative stress, Ca ionophore and hyperthyroidism (Wadhwa and Taira K Kaul 2002).

Several components of cinnamon can be used for the regulation of mortalin expression. Cinnamon reveals antioxidant, anti-inflammatory and anticancer properties and activities against neurological disorder (Rao and Gan 2014). Pleiotropic actions of the principal ingredients of cinnamon preserve neurons (Maiti et al.

2017). Procyanidins present in cinnamon have reduced the cellular oxidative damage in some studies on models (Vengoji et al. 2018).

Upregulation and/or maintenance of PD-related beneficial proteins, such as Parkin and DJ-1, in astrocytes during neurodegenerative insults may have therapeutic efficacy in PD. It is delineating that the levels of Parkin and DJ-1 decrease in activated glial cells and that treatment of activated glial cells with cinnamon metabolite NaB blocks such loss. Interestingly, NaB protects Parkin and DJ-1 in activated astrocytes via suppressing the production of NO and expression of iNOS. Similarly, in MPTP mouse model of PD, Khasnavis et al. (Khasnavis and Pahan 2014), found increased level of iNOS and decreased level of Parkin and DJ-1 in vivo in the nigra. Finally, it demonstrates that oral treatment of MPTP-intoxicated mice with cinnamon powder and NaB reduces the nigral expression of iNOS, blocks nigral loss of Parkin and DJ-1, protects the nigrostriatal axis and restores locomotor activities (Khasnavis and Pahan 2014). Furthermore, DJ-1 can bind to several chaperones, including HSP70, carboxy-terminus of HSP70-interacting protein (CHIP) and mitochondrial HSP70/mortalin/Grp75, and can help in the degradation of misfolded alpha-synuclein (SNCA) (Cardinale et al. 2014). Further, it has been shown that epigallocatechin gallate (EGCG) inhibits NF-κB activation and production of tumorigenic factors in cancer cells (Aggarwal and Gehlot 2009; Härdtner et al. 2012; Hoffmann et al. 2011; Hönicke et al. 2012), thereby blocking chronic inflammation, one of the hallmarks of cancer. The 10-year prospective cohort study by the group of Kazue Imai revealed that consumption of green tea significantly delayed cancer onset in humans, indicative of primary cancer prevention (Nakachi et al. 2000). In a randomized phase II clinical trial, the same amount of green tea prevented colorectal adenoma recurrence in polypectomy patients (Shimizu et al. 2008). Keeping in mind that targeting NF-κB should also reduce the expression of pro-tumorigenic HSPs and could thus be considered a form of therapeutic use of chaperones termed "chaperonotherapy", the use of suitable HSP70 (Mortalin) modulators in combination with radio-/chemotherapy is encouraged to improve the radiation sensitivity of tumours. This approach might be an attempt to supplement the hitherto existing conventional methods to disturb the concert between inflammation, malignant transformation and radio-chemoresistance of cancer cells. An intriguing novel approach aims to increase antitumour immune responses by targeting immune cells to tumours that specifically express Hsp70 on their cell surface or by blocking tumour-specific immune suppression mechanisms such as activation of myeloid suppressor cells (Goloudina et al. 2012).

11.4 Cinnamon and Gut Microbes

Gastrointestinal tract communicates with the central nervous system through the gut-brain axis to support neuronal development and maintenance, while gut dysbiosis manifests in neurological disease. There are three basic pathways through which the communication between the gut and the brain is mediated: direct neuronal communication, endocrine signalling mediators and the immune system (Westfall et al. 2017). Commensal microbiota produce a range of neuroactive molecules, such as serotonin, kynurenine, short-chain fatty acids (acetate, propionate and butyrate), melatonin, gamma-aminobutyric acid (GABA), catecholamines, histamine and acetylcholine (Lyte 2011). Dysregulation of the intestinal microflora leads to increased permeability of intestinal barriers, and the BBB resulting in increased penetration of products derived from microbial gut from the blood into the brain promotes neuroinflammation and, ultimately, neurodegeneration (Quigley 2017).

It is emerging that the regulation of microbiota composition can be realized using plant-derived bioactive molecules, such as polyphenols, that may help rescue neural signalling pathways impaired in neurodegenerative diseases. Several studies indicate that polyphenols are able to modulate cellular functions resulting in active neuroprotection. Examples are proanthocyanidins from cinnamon, resveratrol from grape and wine, curcumin from turmeric and epigallocatechin from green tea (Yoon and Baek 2005). Polyphenol absorption through the gut barrier can be increased after specific conjugation, such as methylation, sulfation and glucuronidation. The gut microbiota convert complex polyphenols into low-molecular-weight polyphenol metabolites found in blood and target organs that are readily absorbable (Lyte 2011).

11.5 Conclusion

There have been positive results when utilizing cinnamon as a therapeutic agent in slowing down the progression of neurological diseases, cancer and MetS, and this is an avenue that could be further explored. Understanding that natural product mixtures have a synergetic effect on the health of a person is also important to consider in order to effectively design clinical trials. Clinical trials could also tell us what factors would aid to amplify the neuroprotective and anticancer effects of cinnamon as it is important to consider that being able to maintain gut microbiota favourably would not only boost gastrointestinal health but also neuroprotection as well. Furthermore, a large number of existing research studies have demonstrated the anticancer effects of cinnamon, yet understanding the pharmacology and identifying the active component are crucial in its development as a drug, especially to test the pharmacodynamics, pharmacokinetics and toxicology. Though there have been many clinical trials exploring the efficacy of using cinnamon and its metabolites in the treatment of diseases of the MetS, the results show a conflicting outlook.

Acknowledgements The authors greatly appreciate the financial support received from the Ministry of Primary Industries, through National Science Foundation Grant Number SP/CIN/2016/02 and World Class University Grant Project, University of Sri Jayewardenepura (WCUP/PhD/19, WCUP/PhD/19B). We would like to acknowledge Anchala Kuruppu of Interdisciplinary Centre for Innovation in Biotechnology and Neuroscience, University of Sri Jayewardenepura, Sri Lanka, for her contribution to writing this chapter.

References

Abresch RT, Han JJ, Carter GT (2009) Rehabilitation management of neuromuscular disease: the role of exercise training. J Clin Neuromuscul Dis 11:7–21. https://doi.org/10.1097/CND.0b013e3181a8d36b

Aggarwal BB, Gehlot P (2009) Inflammation and cancer: how friendly is the relationship for cancer patients? Curr Opin Pharmacol 9:351–369. https://doi.org/10.1016/j.coph.2009.06.020

Aguzzi A, Lakkaraju AKK, Frontzek K (2018) Toward therapy of human prion diseases. Annu Rev Pharmacol Toxicol 58:331–351. https://doi.org/10.1146/annurev-pharmtox-010617-052745

Akilen R, Tsiami A, Devendra D, Robinson N (2010) Glycated haemoglobin and blood pressure-lowering effect of cinnamon in multi-ethnic Type 2 diabetic patients in the UK: a randomized, placebo-controlled, double-blind clinical trial. Diabet Med 27:1159–1167. https://doi.org/10.1111/j.1464-5491.2010.03079.x

Akilen R, Tsiami A, Devendra D, Robinson N (2012) Cinnamon in glycaemic control: systematic review and meta analysis. Clin Nutr 31:609–615. https://doi.org/10.1016/j.clnu.2012.04.003

Allen RW, Schwartzman E, Baker WL, Coleman CI, Phung OJ (2013) Cinnamon use in type 2 diabetes: an updated systematic review and meta-analysis. Ann Fam Med 11:452–459. https://doi.org/10.1370/afm.1517

Anderson RA, Zhan Z, Luo R, Guo X, Guo Q, Zhou J, Kong J, Davis PA, Stoecker BJ (2015) Cinnamon extract lowers glucose, insulin and cholesterol in people with elevated serum glucose. J Tradit Compl Med 18:332–336. https://doi.org/10.1016/j.jtcme.201503.005

Azimi P, Ghiasvand R, Feizi A, Hariri M, Abbasi B (2014) Effects of cinnamon, cardamom, saffron, and ginger consumption on markers of glycemic control, lipid profile, oxidative stress, and inflammation in type 2 diabetes patients. Rev Diabet Stud 11:258–266. https://doi.org/10.1900/RDS.2014.11.258

Ashabi G, Khalaj L, Khodagholi F et al. (2014) Pre-treatment with metformin activates Nrf2 antioxidant pathways and inhibits inflammatory responses through induction of AMPK after transient global cerebral ischemia. Metabolic Brain Disease 30:747-754. doi: 10.1007/s11011-014-9632-2

Aune D, Giovannucci E, Boffetta P et al. (2017) Fruit and vegetable intake and the risk of cardiovascular disease, total cancer and all-cause mortality—a systematic review and dose-response meta-analysis of prospective studies. International Journal of Epidemiology 46:1029-1056. doi: 10.1093/ije/dyw319

Baker WL, Gutierrez-Williams G, White CM, Kluger J, Coleman CI (2008) Effect of cinnamon on glucose control and lipid parameters. Diabetes Care 31:41–43. https://doi.org/10.2337/dc07-1711

Batshaw M, Hyman S, Coyle J, Robinson M, Qureshi I, David Mellits E, Quaskey S (1988) Effect of sodium benzoate and sodium phenylacetate on brain serotonin turnover in the ornithine transcarbamylase-deficient sparse-fur mouse. Pediatr Res 23:368–374. https://doi.org/10.1203/00006450-198804000-00006

Beejmohun V, Peytavy-Izard M, Mignon C, Muscente-Paque D, Deplanque X, Ripoll C, Chapal N (2014) Acute effect of Ceylon cinnamon extract on postprandial glycemia: alpha-amylase inhibition, starch tolerance test in rats, and randomized crossover clinical trial in healthy volunteers. BMC Complement Altern Med 14:351. https://doi.org/10.1186/1472-6882-14-351

Bhattacharjee TR (2007) Inhibition of lipid peroxidation and enhancement of GST activity by cardamom and cinnamon during chemically induced colon carcinogenesis in Swiss albino mice. Asian Pac J Cancer Prev 8:578–582

Blevins SM, Leyva MJ, Brown J, Wright J, Scofield RH, Aston CE (2007) Effect of cinnamon on glucose and lipid levels in non insulin-dependent type 2 diabetes. Diabetes Care 30:2236–2237. https://doi.org/10.2337/dc07-0098

Bolaños J, Almeida A (1999) Roles of nitric oxide in brain hypoxia-ischemia. Biochimica Et Biophysica Acta (BBA) – Bioenergetics 1411:415–436. https://doi.org/10.1016/s0005-2728(99)00030-4

Bryan H, Olayanju A, Goldring C, Park B (2013) The Nrf2 cell defence pathway: Keap1-dependent and -independent mechanisms of regulation. Biochem Pharmacol 85:705–717. https://doi.org/10.1016/j.bcp.2012.11.016

Burbulla LF, Fitzgerald JC, Stegen K, Westermeier J, Thost AK, Kato H, Rapaport D (2014) Mitochondrial proteolytic stress induced by loss of mortalin function is rescued by Parkin and PINK1. Cell Death Dis 5:e1180. https://doi.org/10.1038/cddis.2014.103

Brahmachari S, Pahan K (2007) Sodium benzoate, a food additive and a metabolite of cinnamon, modifies T Cells at multiple steps and inhibits adoptive transfer of experimental allergic encephalomyelitis. The Journal of Immunology 179:275–283. https://doi.org/10.4049/jimmunol.179.1.275

Bae W, Choi J, Jeong J (2018) The Neuroprotective Effects of Cinnamic Aldehyde in an MPTP Mouse Model of Parkinson's Disease. International Journal of Molecular Sciences 19:551. doi: 10.3390/ijms19020551

Bloom D, Jaiswal A (2003) Phosphorylation of Nrf2 at Ser40 by Protein Kinase C in Response to Antioxidants Leads to the Release of Nrf2 from INrf2, but Is Not Required for Nrf2 Stabilization/Accumulation in the Nucleus and Transcriptional Activation of Antioxidant Response Element-mediated NAD(P)H:Quinone Oxidoreductase-1 Gene Expression. Journal of Biological Chemistry 278:44675-44682. doi: 10.1074/jbc.m307633200

Bryan H, Olayanju A, Goldring C, Park B (2013) The Nrf2 cell defence pathway: Keap1-dependent and -independent mechanisms of regulation. Biochemical Pharmacology 85:705-717. doi: 10.1016/j.bcp.2012.11.0162

Bhatnagar S, Kumar A (2009) Therapeutic targeting of signaling pathways in muscular dystrophy. Journal of Molecular Medicine 88:155-166. doi: 10.1007/s00109-009-0550-4

Cardinale A, Chiesa R, Sierks M (2014) Protein misfolding and neurodegenerative diseases. Int J Cell Biol 2014:217371. https://doi.org/10.1155/2014/217371

Circu M, Aw T (2010) Reactive oxygen species, cellular redox systems, and apoptosis. Free Radic Biol Med 48:749–762. https://doi.org/10.1016/j.freeradbiomed.2009.12.022

Crawford P (2009) Effectiveness of cinnamon for lowering hemoglobin A1C in patients with type 2 diabetes: a randomized, controlled trial. J Am Board Fam Med 22:507–512. https://doi.org/10.3122/jabfm.2009.05.080093

Cummings JL, Morstorf T, Zhong K (2014) Alzheimer's disease drug-development pipeline: Few candidates, frequent failures. Alzheimers Res Ther 6:37. https://doi.org/10.1186/alzrt269

Davis PA, Yokoyama W (2011) Cinnamon intake lowers fasting blood glucose: meta-analysis. J Med Food 14:884–889. https://doi.org/10.1089/jmf.2010.0180

Ding S, Hou X, Yuan J, Tan X, Chen J, Yang N, Luo Y, Jiang Z, Jin P, Dong Z, Feng L, Jia X (2015) Wedelolactone protects human bronchial epithelial cell injury against cigarette smoke extract-induced oxidant stress and inflammation responses through Nrf2 pathway. Int Immunopharmacol 29:648–655. https://doi.org/10.1016/j.intimp.2015.09.015

Dinkova-Kostova A, Kostov R, Kazantsev A (2018) The role of Nrf2 signaling in counteracting neurodegenerative diseases. FEBS J 285:3576–3590. https://doi.org/10.1111/febs.14379

Dorsey ER, Constantinescu R, Thompson JP, Biglan KM, Holloway RG, Kieburtz K, Marshall FJ, Ravina BM, Schifitto G, Siderowf A, Tanner CM (2007) Projected number of people with Parkinson disease in the most populous nations, 2005 through 2030. Neurology 68:384–386. https://doi.org/10.1212/01.wnl.0000247740.47667.03

Ding Y, Chen M, Wang M et al. (2014) Posttreatment with 11-Keto-β-Boswellic Acid Ameliorates Cerebral Ischemia–Reperfusion Injury: Nrf2/HO-1 Pathway as a Potential Mechanism. Molecular Neurobiology 52:1430-1439. doi: 10.1007/s12035-014-8929-9

Dang J, Brandenburg L, Rosen C et al. (2012) Nrf2 Expression by Neurons, Astroglia, and Microglia in the Cerebral Cortical Penumbra of Ischemic Rats. Journal of Molecular Neuroscience 46:578-584. doi: 10.1007/s12031-011-9645-9

Esteras N, Dinkova-Kostova AT, Abramov AY (2016) Nrf2 activation in the treatment of neurodegenerative diseases: a focus on its role in mitochondrial bioenergetics and function. Biol Chem 397:383–400. https://doi.org/10.1515/hsz-2015-0295

EtienneManneville S, Manneville JB, Adamson P, Wilbourn B, Greenwood J, Couraud PO (2000) ICAM--coupled cytoskeletal rearrangements and transendothelial lymphocyte migration involve intracellular calcium signaling in brain endothelial cell lines. J Immunol 15:3375–3383. https://doi.org/10.4049/jimmunol.165.6.3375

Fraser PA (2011) The role of free radical generation in increasing cerebrovascular permeability. Free Radic Biol Med 51:967–977. https://doi.org/10.1016/j.freeradbiomed.2011.06.003

Gandhi S, Abramov AY (2012) Mechanism of oxidative stress in neurodegeneration. Oxidative Med Cell Longev 2012:428010. https://doi.org/10.1155/2012/428010

Ganesh Yerra V, Negi G, Sharma SS, Kumar A (2013) Potential therapeutic effects of the simultaneous targeting of the Nrf2 and NF-κB pathways in diabetic neuropathy. Redox Biol 1:394–397. https://doi.org/10.1016/j.redox.2013.07.005

Goloudina AR, Demidov ON, Garrido C (2012) Inhibition of HSP70: a challenging anti-cancer strategy. Cancer Lett 325:117–124. https://doi.org/10.1016/j.canlet.2012.06.003

Gordeeva AV, Zvyagilskaya RA, Labas YA (2003) Cross-talk between reactive oxygen species and calcium in living cells. Biochemistry (Mosc) 68:1077–1080. https://doi.org/10.1023/a:1026398310003

Görlach A, Bertram K, Hudecova S, Krizanova O (2015) Calcium and ROS: a mutual interplay. Redox Biol 6:260–271. https://doi.org/10.1016/j.redox.2015.08.010

Grasso D, Renna FJ, Vaccaro MI (2018) Initial steps in mammalian autophagosome biogenesis. Front Cell Dev Biol 23(146). https://doi.org/10.3389/fcell.2018.00146

Gupta Jain S, Puri S, Misra A, Gulati S, Mani K (2017) Effect of oral cinnamon intervention on metabolic profile and body composition of Asian Indians with metabolic syndrome: a randomized double -blind control trial. Lipids Health Dis 12:113. https://doi.org/10.1186/s12944-017-0504-8

Golpich M, Amini E, Mohamed Z et al. (2016) Mitochondrial Dysfunction and Biogenesis in Neurodegenerative diseases: Pathogenesis and Treatment. CNS Neuroscience & Therapeutics 23:5-22. doi: 10.1111/cns.12655

Härdtner C, Multhoff G, Falk W, Radons J (2012) (−)-Epigallocatechin- 3-gallate, a green tea-derived catechin, synergizes with celecoxib to inhibit IL-1-induced tumorigenic mediators by human pancreatic adenocarcinoma cells Colo357. Eur J Pharmacol 684:36–43. https://doi.org/10.1016/j.ejphar.2012.03.039

Harman D (1956) Aging: a theory based on free radical and radiation chemistry. J Gerontol 11:298–300. https://doi.org/10.1093/geronj/11.3.298

He Y, She H, Zhang T, Xu H, Cheng L, Yepes M, Zhao Y, Mao Z (2017) p38 MAPK inhibits autophagy and promotes microglial inflammatory responses by phosphorylating ULK1. J Cell Biol 217:315–328. https://doi.org/10.1083/jcb.201701049

Hlebowicz J, Hlebowicz A, Lindstedt S, Björgell O, Höglund P, Holst JJ, Darwiche G, Almér LO (2009) Effects of 1 and 3 g cinnamon on gastric emptying, satiety, and postprandial blood glucose, insulin, glucose-dependent insulinotropic polypeptide, glucagon-like peptide 1, and ghrelin concentrations in healthy subjects. Am J Clin Nutr 89:815–821. https://doi.org/10.3945/ajcn.2008.26807

Ho SC, Chang KS, Chang PW (2013) Inhibition of neuroinflammation by cinnamon and its main components. Food Chem 138:2275–2282. https://doi.org/10.1016/j.foodchem.2012.12.020

Hoffmann J, Junker H, Schmieder A, Venz S, Brandt R, Multhoff G, Falk W, Radons J (2011) EGCG downregulates IL-1RI expression and suppresses IL-1-induced tumorigenic factors in human pancreatic adenocarcinoma cells. Biochem Pharmacol 82:1153–1162. https://doi.org/10.1016/j.bcp.2011.07.063

Hönicke AS, Ender SA, Radons J (2012) Combined administration of EGCG and IL-1 receptor antagonist efficiently downregulates IL-1- induced tumorigenic factors in U-2 OS human osteosarcoma cells. Int J Oncol 41:753–758. https://doi.org/10.3892/ijo.2012.1498

He F, Peng J, Deng X et al. (2011) RhoA and NF-κB are involved in lipopolysaccharide-induced brain microvascular cell line hyperpermeability. Neuroscience 188:35-47. doi: 10.1016/j.neuroscience.2011.04.025

Jain A, Lamark T, Sjøttem E, Larsen KB, Awuh JA, Øvervatn A, McMahon M, Hayes JD, Johansen T (2010) p62/SQSTM1Is a target gene for transcription factor NRF2 and creates a positive feedback loop by inducing antioxidant response element-driven gene transcription. J Biol Chem 285:22576–22591. https://doi.org/10.1074/jbc.M110.118976

Jana A, Modi KK, Roy A, Anderson JA, van Breemen RB, Pahan K (2013) Up-regulation of neurotrophic factors by cinnamon and its metabolite sodium benzoate: therapeutic implications for neurodegenerative disorders. J Neuroimmune Pharmacol 8:739–755. https://doi.org/10.1007/s11481-013-9447-7

Jung JE, Kim GS, Chen H, Maier CM, Narasimhan P, Song YS, Niizuma K, Katsu M, Okami N, Yoshioka H, Sakata H, Goeders CE, Chan PH (2010) Reperfusion and neurovascular dysfunction in stroke: from basic mechanisms to potential strategies for neuroprotection. Mol Neurobiol 41:172–179. https://doi.org/10.1007/s12035-010-8102-z

Jain S, Sangma T, Shukla S, Mediratta P (2014) Effect ofCinnamomum zeylanicumextract on scopolamine-induced cognitive impairment and oxidative stress in rats. Nutritional Neuroscience 18:210-216. doi: 10.1179/1476830514y.0000000113

Kahles T, Brandes RP (2012) NADPH oxidases as therapeutic targets in ischemic stroke. Cell Mol Life Sci 69:2345–2363. https://doi.org/10.1007/s00018-012-1011-8

Khan A, Safdar M, Ali Khan MM, Khattak KN, Anderson RA (2003) Cinnamon improves glucose and lipids of people with type 2 diabetes. Diabetes Care 26:3215–3218. https://doi.org/10.2337/diacare.26.12.3215

Khan R, Khan Z, Shah S (2010) Cinnamon may reduce glucose, lipid and cholesterol level in type 2 diabetic individuals. Pakistan J Nutr 9:430–433. https://doi.org/10.3923/pjn.2010.430.433

Khasnavis S, Pahan K (2014) Cinnamon treatment upregulates neuroprotective proteins Parkin and DJ-1 and protects dopaminergic neurons in a mouse model of Parkinson's disease. J Neuroimmune Pharmacol 9:569–581. https://doi.org/10.1007/s11481-014-9552-2

Kim JM, Stewart R, Kim SW, Yang SJ, Shin IS, Kim YH, Yoon JS (2008) BDNF genotype potentially modifying the association between incident stroke and depression. Neurobiol Aging 29:789–792. https://doi.org/10.1016/j.neurobiolaging.2006.11.021

Knight JA (1998) Free radicals: their history and current status in aging and disease. Ann Clin Lab Sci 28:331–346

Komatsu M, Kurokawa H, Waguri S, Taguchi K, Kobayashi A, Ichimura Y, Sou YS, Ueno I, Sakamoto A, Tong KI, Kim M, Nishito Y, Iemura S, Natsume T, Ueno T, Kominami E, Motohashi H, Tanaka K, Yamamoto M (2010) The selective autophagy substrate p62 activates the stress responsive transcription factor Nrf2 through inactivation of Keap1. Nat Cell Biol 12:213–223. https://doi.org/10.1038/ncb2021

Kuklina EV, Tong X, George MG, Bansil P (2012) Epidemiology and prevention of stroke: a worldwide perspective. Expert Rev Neurother 12:99–208. https://doi.org/10.1586/ern.11.99

Kunze R, Urrutia A, Hoffmann A, Liu H, Helluy X, Pham M, Reischl S, Korff T, Marti HH (2015) Dimethyl fumarate attenuates cerebral edema formation by protecting the blood–brain barrier integrity. Exp Neurol 266:99–111. https://doi.org/10.1016/j.expneurol.2015.02.022

Kuruppu AI, Paranagama P, Goonasekara CL (2019) Medicinal plants commonly used against cancer in traditional medicine formulae in Sri Lanka. Saudi Pharm J 27:565–573. https://doi.org/10.1016/j.jsps.2019.02.004

Katsetos C, Koutzaki S, Melvin J (2013) Mitochondrial Dysfunction in Neuromuscular Disorders. Seminars in Pediatric Neurology 20:202-215. doi: 10.1016/j.spen.2013.10.010

Lastres-Becker I, Ulusoy A, Innamorato NG, Sahin G, Rábano A, Kirik D, Cuadrado A (2012) α-Synuclein expression and Nrf2 deficiency cooperate to aggravate protein aggregation, neuronal death and inflammation in early-stage Parkinson's disease. Hum Mol Genet 21:3173–3192. https://doi.org/10.1093/hmg/dds143

Lattke M, Reichel SN, Baumann B (2017) NF-κB-mediated astrocyte dysfunction initiates neurodegeneration. Oncotarget 8:50329–50330. https://doi.org/10.18632/oncotarget.18320

Lau A, Wang XJ, Zhao F, Villeneuve NF, Wu T, Jiang T, Sun Z, White E, Zhang DD (2010) A non-canonical mechanism of Nrf2 activation by autophagy deficiency: direct interaction between Keap1 and p62. Mol Cell Biol 30:3275–3285. https://doi.org/10.1128/MCB.00248-10

Leach MJ, Kumar S (2012) Cinnamon for diabetes mellitus. Cochrane Database Syst Rev 12:CD007170. https://doi.org/10.1002/14651858.CD007170.pub2

Li MH, Cha YN, Surh YJ (2006) Peroxynitrite induces HO-1 expression via PI3K/Akt-dependent activation of NF-E2-related factor 2 in PC12 cells. Free Radic Biol Med 41:1079–1091. https://doi.org/10.1016/j.freeradbiomed.2006.06.010

Lin LC, Wang MN, Tseng TY, Sung JS, Tsai TH (2007) Pharmacokinetics of (−)-epigallocatechin-3-gallate in conscious and freely moving rats and its brain regional distribution. J Agric Food Chem 55:1517–1524. https://doi.org/10.1021/jf062816a

Liu Z, Zhou T, Ziegler AC, Dimitrion P, Zuo L (2017) Oxidative stress in neurodegenerative diseases: from molecular mechanisms to clinical applications. Oxidative Med Cell Longev 2017(2525967). https://doi.org/10.1155/2017/2525967

Liu L, Locascio LM, Doré S (2019) Critical role of Nrf2 in experimental ischemic stroke. Front Pharmacol 10(153). https://doi.org/10.3389/fphar.2019.00153

Lovering RM, Porter NC, Bloch RJ (2005) The muscular dystrophies: from genes to therapies. Phys Ther 85:1372–1388. https://doi.org/10.1093/ptj/85.12.1372

Lu J, Zhang K, Nam S, Anderson RA, Jove R, Wen W (2010) Novel angiogenesis inhibitory activity in cinnamon extract blocks VEGFR2 kinase and downstream signalling. Carcinogenesis 31:481–488. https://doi.org/10.1093/carcin/bgp292

Lyte M (2011) Probiotics function mechanistically as delivery vehicles for neuroactive compounds: microbial endocrinology in the design and use of probiotics. BioEssays 33:574–581. https://doi.org/10.1002/bies.201100024

Lee E, Chen H, Hung Y et al. (2009) Therapeutic window for cinnamophilin following oxygen–glucose deprivation and transient focal cerebral ischemia. Experimental Neurology 217:74-83. doi: 10.1016/j.expneurol.2009.01.019

Li H, Yao W, Irwin M et al. (2015) Adiponectin ameliorates hyperglycemia-induced cardiac hypertrophy and dysfunction by concomitantly activating Nrf2 and Brg1. Free Radical Biology and Medicine 84:311-321. doi: 10.1016/j.freeradbiomed.2015.03.007

Lingappan K (2018) NF-κB in oxidative stress. Current Opinion in Toxicology 7:81-86. doi: 10.1016/j.cotox.2017.11.002

Maggirwar SB, Sarmiere PD, Dewhurst S, Freeman RS (1998) Nerve growth factor-dependent activation of NF-κB contributes to survival of sympathetic neurons. J Neurosci 18:10356–10365. https://doi.org/10.1523/jneurosci.18-24-10356.1998

Maierean SM, Serban MC, Sahebkar A, Ursoniu S, Serban A, Penson P, Banach M, Lipid and Blood Pressure meta-analysis Collaboration (LBPMC) Group (2017) The effects of cinnamon supplementation on blood lipid concentrations: a systematic review and meta-analysis. J Clin Lipidol 11:1393–1406. https://doi.org/10.1016/j.jacl.2017.08.004

Maiolo SA, Fan P, Bobrovskaya L (2018) Bioactive constituents from cinnamon, hemp seed and polygonum cuspidatum protect against H_2O_2 but not rotenone toxicity in a cellular model of Parkinson's disease. J Tradit Complement Med 8:420–427. https://doi.org/10.1016/j.jtcme.2017.11.001

Maiti P, Manna J, Dunbar G (2017) Current understanding of the molecular mechanisms in Parkinson's disease: targets for potential treatments. Transl Neurodegener 6:28. https://doi.org/10.1186/s40035-017-0099-z

Mang B, Wolters M, Schmitt B, Kelb K, Lichtinghagen R, Stichtenoth DO, Hahn A (2006) Effects of a cinnamon extract on plasma glucose, HbA, and serum lipids in diabetes mellitus type 2. Eur J Clin Investig 36:340–344. https://doi.org/10.1111/j.1365-2362.2006.01629.x

Markey O, McClean CM, Medlow P, Davison GW, Trinick TR, Duly E, Shafat A (2011) Effect of cinnamon on gastric emptying, arterial stiffness, postprandial lipemia, glycemia, and appetite responses to high-fat breakfast. Cardiovasc Diabetol 10:78. https://doi.org/10.1186/1475-2840-10-78

Mattson MP, Camandola S (2001) NF-κB in neuronal plasticity and neurodegenerative disorders. J Clin Invest 107:247–254. https://doi.org/10.1172/jci11916
Modi K, Jana M, Mondal S, Pahan K (2015) Sodium benzoate, a metabolite of cinnamon and a food additive, upregulates ciliary neurotrophic factor in astrocytes and oligodendrocytes. Neurochemical Research 40:2333–2347. https://doi.org/10.1007/s11064-015-1723-x
Mollazadeh H, Hosseinzadeh H (2016) Cinnamon effects on metabolic syndrome: a review based on its mechanisms. Iran J Basic Med Sci 19:1258–1270. https://doi.org/10.22038/ijbms.2016.7906
Momtaz S, Hassani S, Khan F, Ziaee M, Abdollahi M (2018) Cinnamon, a promising prospect towards Alzheimer's disease. Pharmacol Res 130:241–258. https://doi.org/10.1016/j.phrs.2017.12.011
Motamedi S, Karimi I, Jafari F (2017) The interrelationship of metabolic syndrome and neurodegenerative diseases with focus on brain-derived neurotrophic factor (BDNF): kill two birds with one stone. Metab Brain Dis 32:651–665. https://doi.org/10.1007/s11011-017-9997-0
Mousavi SM, Rahmani J, Kord-Varkaneh H, Sheikhi A, Larijani B, Esmaillzadeh A (2019) Cinnamon supplementation positively affects obesity: a systematic review and dose-response meta-analysis of randomized controlled trials. Clin Nutr 39:123–133. https://doi.org/10.1016/j.clnu.2019.02.017
Moldogazieva N, Mokhosoev I, Mel'nikova T et al. (2019) Oxidative Stress and Advanced Lipoxidation and Glycation End Products (ALEs and AGEs) in Aging and Age-Related Diseases. Oxidative Medicine and Cellular Longevity 2019:1-14. doi: 10.1155/2019/3085756
Na J, Lee K, Na W, Shin JY, Lee MJ, Yune TY, Lee HK, Jung HS, Kim WS, Ju BG (2016) Histone H3K27 demethylase JMJD3 in cooperation with NF-κB regulates keratinocyte wound healing. J Invest Dermatol 136:847–858. https://doi.org/10.1016/j.jid.2015.11.029
Nair A, Vaidya VA (2006) Cyclic AMP response element binding protein and brain-derived neurotrophic factor: molecules that modulate our mood? J Biosci 34:423–434. https://doi.org/10.1007/bf02704114
Nakachi K, Matsuyama S, Miyake S, Suganuma M, Imai K (2000) Preventive effects of drinking green tea on cancer and cardiovascular disease: epidemiological evidence for multiple targeting prevention. Biofactors 13:49–54. https://doi.org/10.1002/biof.5520130109
Nakaso K, Yano H, Fukuhara Y, Takeshima T, Wada-Isoe K, Nakashima K (2003) PI3K is a key molecule in the Nrf2-mediated regulation of antioxidative proteins by hemin in human neuroblastoma cells. FEBS Lett 546:181–184. https://doi.org/10.1016/s0014-5793(03)00517-9
Namazi N, Khodamoradi K, Khamechi SP, Heshmati J, Ayati MH, Larijani B (2019) The impact of cinnamon on anthropometric indices and glycemic status in patients with type 2 diabetes: a systematic review and meta-analysis of clinical trials. Complement Ther Med 43:92–101. https://doi.org/10.1016/j.ctim.2019.01.002
Narasimhan M, Patel D, Vedpathak D, Rathinam M, Henderson G, Mahimainathan L (2012) Identification of novel microRNAs in post-transcriptional control of Nrf2 expression and redox homeostasis in neuronal, SH-SY5Y cells. PLoS One 7:e51111. https://doi.org/10.1371/journal.pone.0051111
Newman DJ, Cragg GM (2012) Natural products as sources of new drugs over the 30 years from 1981 to 2010. J Nat Prod 75:311–335. https://doi.org/10.1021/np200906s
Nishida N, Yano H, Nishida T, Kamura T, Kojiro M (2006) Angiogenesis in cancer. Vasc Health Risk Manag 2:213–219. https://doi.org/10.2147/vhrm.2006.2.3.213
Olmez I, Ozyurt H (2012) Reactive oxygen species and ischemic cerebrovascular disease. Neurochem Int 60:208–212. https://doi.org/10.1016/j.neuint.2011.11.009
Patten DA, Germain M, Kelly MA, Slack RS (2010) Reactive oxygen species: stuck in the middle of neurodegeneration. J Alzheimers Dis 20:357–367. https://doi.org/10.3233/jad-2010-100498
Pieterse AJ, Cup EH, Knuijt S, Akkermans R, Hendricks HT, Van Engelen BGM, Van Der Wilt G, Oostendorp RAB (2008) Development of a tool to guide referral of patients with neuromuscular disorders to allied health services. Part two. Disabil Rehabil 30:863–870. https://doi.org/10.1080/09638280701403460

Pyo J, Jeong Y, Yeo S, Lee J, Jeong M, Kim S, Choi Y, Lim S (2013) Neuroprotective effect of trans-Cinnamaldehyde on the 6-Hydroxydopamine-induced dopaminergic injury. Biol Pharm Bull 36:1928–1935. https://doi.org/10.1248/bpb.b13-00537

Prince M, Guerchet M, Prina M (2013) Policy Brief: The Global Impact of Dementia 2013-2050 | Alzheimer's Disease International. In: Alz.co.uk. https://www.alz.co.uk/research/G8-policy-brief. Accessed 14 Oct 2020

Prince M, Jackson J (2009) World Alzheimer Report 2009: The Global Prevalence of Dementia | Alzheimer's Disease International. In: Alz.co.uk. https://www.alz.co.uk/research/world-report-2009. Accessed 14 Oct 2020

Pohl F, Kong Thoo Lin P (2018) The Potential Use of Plant Natural Products and Plant Extracts with Antioxidant Properties for the Prevention/Treatment of Neurodegenerative Diseases: In Vitro, In Vivo and Clinical Trials. Molecules 23:3283. doi: 10.3390/molecules23123283

Peterson D, George R, Scaramozzino F et al. (2009) Cinnamon Extract Inhibits Tau Aggregation Associated with Alzheimer's Disease In Vitro. Journal of Alzheimer's Disease 17:585-597. doi: 10.3233/jad-2009-1083

Quigley E (2017) Microbiota-brain-gut axis and neurodegenerative diseases. Curr Neurol Neurosci Rep 17:94. https://doi.org/10.1007/s11910-017-0802-6

Radermacher KA, Wingler K, Langhauser F, Altenhöfer S, Kleikers P, Hermans JR, Hrabě De Angelis M, Kleinschnitz C, Schmidt HH (2013) Neuroprotection after stroke by targeting NOX4 as a source of oxidative stress. Antioxid Redox Signal 18:1418–1427. https://doi.org/10.1089/ars.2012.4797

Rafii MS, Aisen PS (2009) Recent developments in Alzheimer's disease therapeutics. BMC Med 7:7. https://doi.org/10.1186/1741-7015-7-7

Rahiman A, Rajan N, Sreekumaran E (2015) Neuroprotective effect of vitex negundo against scopolamine induced cognitive impairment and oxidative stress in Wistar albino rats. Biosci Biotechnol Res Asia 12:301–307. https://doi.org/10.13005/bbra/2040

Rodrigo R, Libuy M, Feliú F, Hasson D (2013) Oxidative stress-related biomarkers in essential hypertension and ischemia-reperfusion myocardial damage. Dis Markers 35:773–790. https://doi.org/10.1155/2013/974358

Ross CA, Poirier MA (2004) Protein aggregation and neurodegenerative disease. Nat Med 10:10–17. https://doi.org/10.1038/nm1066

Roussel A, Hininger I, Benaraba R, Ziegenfuss TN, Anderson RA (2009) Antioxidant effects of a cinnamon extract in people with impaired fasting glucose that are overweight or obese. J Am Coll Nutr 28:16–21. https://doi.org/10.1080/07315724.2009.10719756

Rojo A, McBean G, Cindric M et al. (2014) Redox Control of Microglial Function: Molecular Mechanisms and Functional Significance. Antioxidants & Redox Signaling 21:1766-1801. doi: 10.1089/ars.2013.5745

Rao P, Gan S (2014) Cinnamon: A Multifaceted Medicinal Plant. Evidence-Based Complementary and Alternative Medicine 2014:1-12. doi: 10.1155/2014/642942

Sandberg M, Patil J, D'Angelo B, Weber S, Mallard C (2014) NRF2-regulation in brain health and disease: implication of cerebral inflammation. Neuropharmacology 79:298–306. https://doi.org/10.1016/j.neuropharm.2013.11.004

Sekhar K, Rachakonda G, Freeman M (2010) Cysteine-based regulation of the CUL3 adaptor protein Keap1. Toxicol Appl Pharmacol 244:21–26. https://doi.org/10.1016/j.taap.2009.06.016

Shah Z, Li R, Ahmad A, Kensler T, Yamamoto M, Biswal S, Doré S (2014) The Flavanol (−)-Epicatechin prevents stroke damage through the Nrf2/HO1 pathway. J Cereb Blood Flow Metab 34:735–735. https://doi.org/10.1038/jcbfm.2010.53

Sheikh S, Safia HE, Mir S (2013) Neurodegenerative diseases: multifactorial conformational diseases and their therapeutic interventions. J Neurodegener Dis:1–8. https://doi.org/10.1155/2013/563481

Shimizu M, Fukutomi Y, Ninomiya M, Nagura K, Kato T, Araki H, Suganuma M, Fujiki H, Moriwaki H (2008) Green tea extracts for the prevention of metachronous colorectal adenomas:

a pilot study. Cancer Epidemiol Biomark Prev 17:3020–3025. https://doi.org/10.1158/1055-9965.epi-08-0528

Sivandzade F, Prasad S, Bhalerao A, Cucullo L (2019) NRF2 and NF-κ interplay in cerebrovascular and neurodegenerative disorders: molecular mechanisms and possible therapeutic approaches. Redox Biol 21:101059. https://doi.org/10.1016/j.redox.2018.11.017

Skibinski G, Hwang V, Ando DM, Daub A, Lee AK, Ravisankar A, Modan S, Finucane MM, Shaby BA Finkbeiner S (2016) Nrf2 mitigates LRRK2- and α-synuclein–induced neurodegeneration by modulating proteostasis. Proc Natl Acad Sci U S A 114:1165–1170. https://doi.org/10.1073/pnas.1522872114

Slemmer JE, Zhu C, Landshamer S, Trabold R, Grohm J, Ardeshiri A, Wagner E, Sweeney MI, Blomgren K, Culmsee C, Weber JT, Plesnila N (2008) Causal role of apoptosis-inducing factor for neuronal cell death following traumatic brain injury. Am J Pathol 173:1795–1805. https://doi.org/10.2353/ajpath.2008.080168

Solomon TP, Blannin AK (2007) Effects of short-term cinnamon ingestion on in vivo glucose tolerance. Diabetes Obes Metab 9:895–901. https://doi.org/10.1111/j.1463-1326.2006.00694.x

Stoecker BJ, Zhan Z, Luo R, Mu X, Guo X, Liu Y, Guo Q, Zhou J, Kong J, Zhou Z, Cui B, Anderson R (2010) Cinnamon extract lowers blood glucose in hyperglycemic subjects. FASEB J 24:722

Suppapitiporn S, Kanpaksi N, Suppapitiporn S (2006) The effect of cinnamon cassia powder in type 2 diabetes mellitus. J Med Assoc Thail 89:200–205

Shi H, Jing X, Wei X et al. (2015) S-allyl cysteine activates the Nrf2-dependent antioxidant response and protects neurons against ischemic injuryin vitroandin vivo. Journal of Neurochemistry 133:298-308. doi: 10.1111/jnc.12986

Shah Z, Li R, Thimmulappa R et al. (2007) Role of reactive oxygen species in modulation of Nrf2 following ischemic reperfusion injury. Neuroscience 147:53-59. doi: 10.1016/j.neuroscience.2007.02.066

Srivastava S, Alfieri A, Siow R et al. (2013) Temporal and spatial distribution of Nrf2 in rat brain following stroke: quantification of nuclear to cytoplasmic Nrf2 content using a novel immunohistochemical technique. The Journal of Physiology 591:3525-3538. doi: 10.1113/jphysiol.2013.257964

Shih A (2005) A Small-Molecule-Inducible Nrf2-Mediated Antioxidant Response Provides Effective Prophylaxis against Cerebral Ischemia In Vivo. Journal of Neuroscience 25:10321-10335. doi: 10.1523/jneurosci.4014-05.2005

Shu L, Cheung K, Khor T et al. (2010) Phytochemicals: cancer chemoprevention and suppression of tumor onset and metastasis. Cancer and Metastasis Reviews 29:483-502. doi: 10.1007/s10555-010-9239-y

Tan Z, Hu B, Zang F, Wang J, Zhang X, Chen H (2019) Nrf2 drives oxidative stress-induced autophagy in nucleus pulposus cells via a Keap1/Nrf2/p62 feedback loop to protect intervertebral disc from degeneration. Cell Death Dis 10:510. https://doi.org/10.1038/s41419-019-1701-3

Tang LL, Wang R, Tang XC (2005) Huperzine a protects SHSY5Y neuroblastoma cells against oxidative stress damage via nerve growth factor production. Eur J Pharmacol 519:9–15. https://doi.org/10.1016/j.ejphar.2005.06.026

Tang M, Larson-Meyer DE, Liebman M (2008) Effect of cinnamon and turmeric on urinary oxalate excretion, plasma lipids, and plasma glucose in healthy subjects. Am J Clin Nutr 87:1262–1267. https://doi.org/10.1093/ajcn/87.5.1262

Teillon S, Calderon G, Rios M (2010) Diminished diet-induced hyperglycemia and dyslipidemia and enhanced expression of PPARα and FGF21 in mice with hepatic ablation of brain-derived neurotropic factor. J Endocrinol 205:37–47. https://doi.org/10.1677/joe-09-0405

Tong T, Appleby P, Bradbury K, Perez-Cornago A, Travis R, Clarke R, Key T (2019) Risks of ischaemic heart disease and stroke in meat eaters, fish eaters, and vegetarians over 18 years of follow-up: results from the prospective EPIC-Oxford study. BMJ 366:l4897. https://doi.org/10.1136/bmj.l4897

Tanaka N, Ikeda Y, Ohta Y et al. (2011) Expression of Keap1–Nrf2 system and antioxidative proteins in mouse brain after transient middle cerebral artery occlusion. Brain Research 1370:246-253. doi: 10.1016/j.brainres.2010.11.010

Tao S, Rojo de la Vega M, Quijada H et al. (2016) Bixin protects mice against ventilation-induced lung injury in an NRF2-dependent manner. Scientific Reports. doi: 10.1038/srep18760

Unlu M, Ergene E, Unlu GV, Zeytinoglu HS, Vural N (2010) Composition, antimicrobial activity and in vitro cytotoxicity of essential oil from *Cinnamomum zeylanicum* Blume (Lauraceae). Food Chem Toxicol 48:3274–3280. https://doi.org/10.1016/j.fct.2010.09.001

Vafa MR, Mohammadi F, Shidfar F, Sormaghi MS, Heidari I, Golestan B, Amiri F (2012) Effects of cinnamon consumption on glycemic status, lipid profile and body composition in type 2 diabetes patients. Int J Prev Med 3:8

Vanschoonbeek K, Thomassen BJ, Senden JM, Wodzig WK, van Loon LJ (2006) Cinnamon supplementation does not improve glycemic control in postmenopausal type 2 diabetes patients. J Nutr 136:977–980. https://doi.org/10.1093/jn/136.4.977

Vengoji R, Macha M, Batra S, Shonka N (2018) Natural products: a hope for glioblastoma patients. Oncotarget 9:22194–22219. https://doi.org/10.18632/oncotarget.25175

Wadhwa R, Taira K Kaul SC (2002) Mortalin: a potential candidate for biotechnology and biomedicine. Histol Histopathol 17:1173–1177. https://doi.org/10.14670/hh-17.1173

Wadhwa R, Taira K, Kaul SC (2006) Upregulation of mortalin/mthsp70/Grp75 contributes to human carcinogenesis. Int J Cancer 11:2973–2980. https://doi.org/10.1002/ijc.21773

Wang X, Ouyang Y, Liu J, Zhu M, Zhao G, Bao W, Hu FB (2014) Fruit and vegetable consumption and mortality from all causes, cardiovascular disease, and cancer: systematic review and dose-response meta-analysis of prospective cohort studies. BMJ 349:g4490. https://doi.org/10.1136/bmj.g4490

Wardyn J, Ponsford A, Sanderson C (2015) Dissecting molecular cross-talk between Nrf2 and NF-κB response pathways. Biochem Soc Trans (4):621–626. https://doi.org/10.1042/bst20150014

Westfall S, Lomis N, Kahouli I, Dia S, Singh S, Prakash S (2017) Microbiome, probiotics and neurodegenerative diseases: deciphering the gut brain axis. Cell Mol Life Sci 74:3769–3787. https://doi.org/10.1007/s00018-017-2550-9

WHO World Health Organisation (2018). Latest global cancer data. https://www.who.org. Accessed 27 July, 2018

Wickenberg J, Lindstedt S, Berntorp K, Nilsson J, Hlebowicz J (2012) Ceylon cinnamon does not affect postprandial plasma glucose or insulin in subjects with impaired glucose tolerance. Br J Nutr 107:1845–1849. https://doi.org/10.1017/S0007114511005113

Wondrak GT, Villeneuve NF, Lamore SD, Bause AS, Jiang T, Zhang DD (2010) The cinnamon-derived dietary factor cinnamic aldehyde activates the Nrf2-dependent antioxidant response in human epithelial colon cells. Molecules 15:3338–3355. https://doi.org/10.3390/molecules15053338

Wu S, Ng L, Lin C (2005) Cinnamaldehyde-induced apoptosis in human PLC/PRF/5 cells through activation of the proapoptotic Bcl-2 family proteins and MAPK pathway. Life Sci 77:938–951. https://doi.org/10.1016/j.lfs.2005.02.005

Woodman K, Coles C, Lamandé S, White J (2016) Nutraceuticals and Their Potential to Treat Duchenne Muscular Dystrophy: Separating the Credible from the Conjecture. Nutrients 8:713. doi: 10.3390/nu8110713

Xu S, Li N, Ning M et al. (2006) Bioactive Compounds fromPeperomiapellucida. Journal of Natural Products 69:247-250. doi: 10.1021/np050457s

Yang L, Calingasan NY, Thomas B, Chaturvedi RK, Kiaei M, Wille EJ, Liby KT, Williams C, Royce D, Risingsong R, Musiek ES, Morrow JD, Sporn M, Beal MF (2009) Neuroprotective effects of the triterpenoid, CDDO methyl amide, a potent inducer of Nrf2-mediated transcription. PLoS One 4:5757. https://doi.org/10.1371/journal.pone.0005757

Yoon J, Baek S (2005) Molecular targets of dietary polyphenols with anti-inflammatory properties. Yonsei Med J 46:585. https://doi.org/10.3349/ymj.2005.46.5.585

Yu Z, Zhou D, Cheng G, Mattson MP (2000) Neuroprotective role for the p50 subunit of NF-κB in an experimental model of Huntington's disease. J Mol Neurosci 15:31–44. https://doi.org/10.1385/jmn:15:1:31

Yu M, Li H, Liu Q, Liu F, Tang L, Li C, Yang X (2011) Nuclear factor p65 interacts with Keap1 to repress the Nrf2-ARE pathway. Cell Signal 23:883–892. https://doi.org/10.1016/j.cellsig.2011.01.014

Yulug B, Cankaya S (2019) Translational perspective: is cinnamon a suitable agent for cognitive impairment and Alzheimer's disease associated with brain trauma? Neural Regen Res 14:1372. https://doi.org/10.4103/1673-5374.253518

Yu S, Wu T, Teng C (1994) Pharmacological characterization of cinnamophilin, a novel dual inhibitor of thromboxane synthase and thromboxane A2 receptor. British Journal of Pharmacology 111:906-912. doi: 10.1111/j.1476-5381.1994.tb14824.x

Zhang L, Zhu Z, Liu J, Zhu Z, Hu Z (2014) Protective effect of N-acetylcysteine (NAC) on renal ischemia/reperfusion injury through Nrf2 signaling pathway. J Recept Signal Transduct Res 34:396–400. https://doi.org/10.3109/10799893.2014.908916

Ziegenfuss TN, Hofheins JE, Mendel RW, Landis J, Anderson RA (2006) Effects of a water-soluble cinnamon extract on body composition and features of the metabolic syndrome in pre-diabetic men and women. J Int Soc Sports Nutr 3:45–53. https://doi.org/10.1186/1550-2783-3-2-45

Chapter 12
Pharmacological Properties of Ceylon Cinnamon

G. A. S. Premakumara and W. P. K. M. Abeysekera

12.1 Traditional Health Claims and Medicinal Uses of Ceylon Cinnamon

Cinnamon as a medicinal ingredient is recorded in ancient medical texts, such as Susrutha Samhitha of Indian Ayurvedic system of medicine and ola leaf texts of Sri Lankan traditional medical systems, which date back to over 3000 years. It is recorded as an important ingredient in several multidrug formulations in the traditional medical texts in Sri Lanka, and those formulations have been used therapeutically to treat a number of aliments (Department of Ayurveda, Sri Lanka 2019). In the Unani system of medicine, cinnamon bark is used as a constituent of a formulation in the treatment of gastrointestinal ailments. Bark oil also forms a component of both Ayurvedic and Unani medicines of India (Ravindran et al. 2004). The bark of cinnamon is used for the treatment of dyspepsia, flatulence, diarrhoea, dysentery, vomiting, bronchitis, gangrene of the lungs, cramps of the stomach, toothache, paralysis of the tongue, phthisis and massive doses in the treatment of cancer and acute and chronic rheumatism (Jayaweera 1982).

G. A. S. Premakumara (✉)
Department of Basic Sciences and Social Sciences, Faculty of Nursing, University of Colombo, Colombo, Sri Lanka
e-mail: sirimal@dss.cmb.ac.lk

W. P. K. M. Abeysekera
Herbal Technology Section, Modern Research & Development Center, Industrial Technology Institute, Halbarawa, Malabe, Sri Lanka

R. Senaratne, R. Pathirana (eds.), *Cinnamon*,
https://doi.org/10.1007/978-3-030-54426-3_12

12.2 Pharmacological Properties of Ceylon Cinnamon

Ceylon cinnamon has been investigated and reported for a wide variety of biological activities worldwide. Those biological activities were reported under the scientific names of *Cinnamomum zeylanicum* and *Cinnamomum verum*. Some of the investigated biological activities included antioxidant (Rao and Gan 2014; Abeysekera et al. 2013a; Adisakwattana et al. 2011; Brewer 2011; Dudonne et al. 2009; Prasad et al. 2009; Mathew and Abraham 2006), anti-diabetic (Bandara et al. 2012; Ho and Chang 2012; Kumar et al. 2012; Ranasinghe et al. 2012; Adisakwattana et al. 2011; Sangal 2011; Ranilla et al. 2010), antilipidemic (Lopes et al. 2015; El-Desoky et al. 2012; Hassan et al. 2012; Javed et al. 2012; Ranasinghe et al. 2012; Gholamhoseinian et al. 2010), anti-inflammatory (Gunawardena et al. 2015; Vetal et al. 2013; Joshi et al. 2010; Mathew and Abraham 2006), antibacterial (Ranasinghe et al. 2013), antifungal (Ranasinghe et al. 2013), anti-nociceptive (Atta and Alkofahi 1998), anti-ageing (Bharti et al. 2013), immunomodulatory (Balekar et al. 2014; Niphade et al. 2009) and anticancer-related activity (Priyarani et al. 2010; Herdwiani et al. 2016). A variety of solvent and water extracts of bark, both bark and leaf essential oils and few extracts of leaf of Ceylon cinnamon were reported for above-mentioned biological activities. In most studies, bark extracts, bark and leaf essential oils and leaf extracts have been used. Reported biological activities were mainly based on a wide range of in vitro methods and in vivo animal models. Clinical studies on authenticated Ceylon cinnamon is limited to few trials, including the trial conducted by the author's team (Ranasinghe et al. 2017).

12.2.1 Antioxidant Properties of Ceylon Cinnamon

Ceylon cinnamon contains a very strong antioxidant activity and is a rich source of naturally occurring antioxidants. The reported antioxidant properties were from various aqueous and solvent extracts of cinnamon bark, bark and leaf essential oils and leaf. These antioxidant properties have been tested using standard bioassays, such as 2,2-diphenyl-1-picrylhydrazyl (DPPH), (2,2′-azino-bis(3-ethylbenzothiazoline-6-sulfonic acid)) (ABTS), oxygen radical absorbance capacity (ORAC), nitric oxide (NO) and superoxide; reducing power by ferric reducing antioxidant power assay (FRAP); metal iron chelation; lipid peroxidation and antioxidants with chain-breaking capacity and carbonyl compounds trapping (Abeysekera et al. 2017; Arachchige et al. 2017; Rao et al. 2014; Abeysekera et al. 2013a; Varalakshmi et al. 2012; Adisakwattana et al. 2011; Brewer 2011; Dudonne et al. 2009; Prasad et al. 2009; Mathew and Abraham 2006). Phenolics were the most abundant antioxidants in Ceylon cinnamon, particularly flavonoids and proanthocyanidins (Arachchige et al. 2017; Abeysekera et al. 2016a; Abeysekera et al. 2013a; Mateos-Martín et al. 2012; Prasad et al. 2009). Some nonphenolic antioxidants also have been reported from cinnamon extracts (Abeysekera et al. 2017; Lee et al. 2015; Brewer 2011; Rao et al. 2014). Individual antioxidant compounds identified in Ceylon cinnamon include vanillic, caffeic, gallic, protocatechuic, p-hydroxybenzoic, p-coumaric and ferulic acids, p-hydroxybenzaldehyde, trans-cinnamaldehyde, trans-cinnamic acid, eugenol, quercetin, quercetrin, kaempferol, catechin, epicatechin and type A and B procyanidin oligomers (Abeysekera et al. 2017; Lee et al. 2015; Brewer 2011; Rao and Gan 2014; Mateos-Martín et al. 2012). The presence of these antioxidants and antioxidant activities has been attributed to a variety of biological activities, such as anticancer, anti-inflammatory, anti-diabetic and lipid-lowering properties of Ceylon cinnamon (Fiedor and Burda 2014; Shebis et al. 2013; Brewer 2011; Pham-Huy et al. 2008; Blokhina et al. 2003; Paiva and Russell 1999).

12.2.2 Anti-diabetic Properties of Ceylon Cinnamon

Cinnamon including principle cinnamon species, *C. zeylanicum, C. aromaticum, C. burmannii* and *C. loureiroi*, have been reported to contain various anti-diabetic properties (Kim et al. 2016; Ho and Chang 2012; Kumar et al. 2012; Adisakwattana et al. 2011; Boğa et al. 2011; Peng et al. 2010; Ranilla et al. 2010; Dearlove et al. 2008; Peng et al. 2008). However, comprehensive and detailed studies related to anti-diabetic activity of cinnamon is mainly focused on *C. aromaticum* (cassia cinnamon or Chinese cinnamon) followed by *C. zeylanicum* (true cinnamon or Ceylon cinnamon) compared to other *Cinnamomum* species worldwide (Bandara et al. 2012). The anti-diabetic activity of Ceylon cinnamon has been established using various in vitro and in vivo models. Bark is the most investigated part for anti-diabetic activity. The in vivo anti-diabetic activity of bark has been shown to be mediated through various mechanisms (Bandara et al. 2012; Ho and Chang 2012; Kumar et al. 2012; Ranasinghe et al. 2012; Adisakwattana et al. 2011; Sangal 2011; Ranilla et al. 2010). These mechanisms included reduction of postprandial intestinal glucose absorption inhibiting the activity of pancreatic α-amylase and α-glucosidase enzymes, stimulation of glucose metabolism, glycogen synthesis, insulin release and cellular glucose uptake (membrane translocation of GLUT-4), inhibition of gluconeogenesis (affects key regulatory enzymes) and potentiating of insulin receptor activity (Bandara et al. 2012; Ranasinghe et al. 2012; Sangal 2011). Further, the in vivo studies confirmed that the anti-diabetic activity observed works through multiple mechanisms. Such mechanisms include reduction in fasting blood glucose and HbA1c, increase in circulating insulin levels, attenuation of weight loss associated with diabetes and reduction of LDL cholesterol and increase of HDL cholesterol (Bandara et al. 2012; Ranasinghe et al. 2012; Sangal 2011). In addition, bark of *C. zeylanicum* also demonstrated beneficial effects against diabetic neuropathy and nephropathy (Ranasinghe et al. 2012).

Only very few studies have been conducted to confirm the anti-diabetic activity of Ceylon cinnamon leaf. The leaf is reported to have anti-diabetic activity via the inhibition of α-amylase in vitro (Ponnusamy et al. 2011) and anti-hyperglycaemic activity in a rat model (Tailang et al. 2008). An author's work on Ceylon cinnamon leaf reported anti-amylase, anticholinesterases, anti-glycation and glycation reversing potential of leaf of Ceylon cinnamon in vitro (Arachchige et al. 2017). Clinical trials on anti-diabetic activity of cinnamon species are also limited. Bark of *C. aromaticum* has been reported clinically for its anti-diabetic activity (Bandara et al. 2012; Sangal 2011). However, all clinical studies on bark of *C. aromaticum* did not demonstrate beneficial effects on diabetes, and the findings of some studies are inconclusive particularly with regard to insulin sensitivity. Some studies on *C. aromaticum* did not demonstrate reduction of fasting blood glucose, HbA1c and lipid levels compared to control (Sangal 2011).

Studies so far conducted on anti-diabetic activity of *Cinnamomum* species have been focused mainly on the effect of cinnamon bark. Little attempt has been made to investigate the effect of leaf on the regulation of glycaemia. As in vitro anti-diabetic-related properties of leaf extracts, α-amylase inhibitory activity from *C. verum* (Ponnusamy et al. 2011), anti-glycation activity from *C. loureiroi* (Lee and Yoon 2010), anti-glycation and glycation reversing activity from *C. zeylanicum* (Arachchige et al. 2017) and acetylcholinesterase and butyrylcholinesterase inhibitory activities from *C. zeylanicum* (Arachchige et al. 2017; Dalai et al. 2014a) and *C. tamala* (Dalai et al. 2014b) have been reported. Further, ethanolic extract of leaf of *C. zeylanicum* (Tailang et al. 2008) and water extract of leaf of *C. tamala* (Chakraborty and Das 2010) were reported to have anti-hyperglycaemic activity in vivo. The authors have demonstrated anti-diabetic properties in both bark and leaf of Ceylon cinnamon in vitro in terms of anti-amylase and anticholinesterases inhibition and anti-glycation and glycation reversing activities. In general, bark showed higher anti-amylase and anti-butyrylcholinesterase activities compared to leaf, whereas the leaf extracts showed high anti-glycation and glycation reversing activities compared to bark (Arachchige et al. 2017).

12.2.3 Antilipidemic Properties of Ceylon Cinnamon

Ceylon cinnamon has been reported to possess remarkable lipid-lowering activity in both in vitro and in vivo models (Lopes et al. 2015; El-Desoky et al. 2012; Hassan et al. 2012; Javed et al. 2012; Ranasinghe et al. 2012; Gholamhoseinian et al. 2010). Reported lipid-lowering activity of Ceylon cinnamon was mainly from its bark (Lopes et al. 2015; El-Desoky et al. 2012; Hassan et al. 2012; Javed et al. 2012; Ranasinghe et al. 2012; Gholamhoseinian et al. 2010). Very few investigations report the effects of leaf (Abeysekera et al. 2014; Lin et al. 2011b) and essential oil (Ciftci et al. 2010; Dalkilic et al. 2009). The reported in vitro studies attributed its antilipidemic activity to anti-lipase, anti-HMG-CoA reductase, anti-cholesterol esterase, cholesterol micellization inhibition and bile acid-binding activities (Abeysekera et al. 2017; Lopes et al. 2015; Gholamhoseinian et al. 2010). In vivo investigations carried out on Ceylon cinnamon using rat models have shown reduction of total cholesterol and LDL cholesterol (El-Desoky et al. 2012; Hassan et al.

2012; Ranasinghe et al. 2012) and triglycerides (El-Desoky et al. 2012; Hassan et al. 2012). However, it is worthy to note that cinnamon simultaneously increased HDL cholesterol while reducing total cholesterol, triglycerides and LDL cholesterol (Hassan et al. 2012). Further, Javed et al. (2012) have shown that the lipid-lowering effect of methanolic extract of bark of *C. zeylanicum* at 0.75 g/kg of body weight was similar to the lipid-lowering effect of simvastatin at 0.6 mg/kg body weight in hyperlipidaemic albino rabbits.

12.2.4 Anticancer Properties of Ceylon Cinnamon

There are claims that cinnamon has been used for the treatment of cancers in traditional systems of medicine in India and Sri Lanka (Jayaweera 1982). This traditional claim has been proven by several researchers using a number of in vitro experiments worldwide (Herdwiani et al. 2016). The demonstrated in vitro anticancer-related activity of Ceylon cinnamon was mainly on cytotoxicity against different carcinoma cell lines (Herdwiani et al. 2016; Abeysekera et al. 2016b). From the different parts of this plant, particularly the bark, leaf and bark essential oils have been tested on breast carcinoma (MCF7, T47D), lung carcinoma (A-549), prostate carcinoma (PC-3), brain carcinoma (glioblastoma, T98G), hepatocarcinoma (HePG2), human cervix carcinoma (SiHa), human skin carcinoma (A431), human neuroblastoma (SK-N-MC), oral carcinoma (KB cells) and lymphoid leukaemia (L1210 cells) cell lines (Abeysekera et al. 2016b; Herdwiani et al. 2016; Varalakshmi et al. 2014; Sudan et al. 2013; Priyarani et al. 2010; Unlu et al. 2010; Zu et al. 2010).

The mechanisms of anticancer activity of Ceylon cinnamon included scavenging of free radicals (Sudan et al. 2013; Priyarani et al. 2010), inhibition of expression of pro-angiogenic factors (Vascular endothelial growth factor – VEGF) (Lu et al. 2010), cytotoxicity and growth inhibition (Sudan et al. 2013; Priyarani et al. 2010), modulation of enzyme activities in detoxification (Abeysekera et al. 2013b), inhibition of tumour angiogenesis via suppression of VEGFR2 signalling (Lu et al. 2010) and initiation of apoptosis mechanisms (Unlu et al. 2010). The reported *Cinnamomum* species having cytotoxicity against different carcinoma cell lines included *C. zeylanicum* (*C. verum*), *C. cassia*, *C. burmannii*, *C. tamala*, *C. osmophloeum*, *C. subavenium*, *C. kotoense*, *C. tenuifolium* and *C. esmophicum* (Herdwiani et al. 2016). The reported cytotoxicities were on various carcinoma cell lines, such as breast carcinoma (MCF7, MDA-MB-231, T47D), epidermoid carcinoma (A431), human cancer promyelocytic leukaemia (HL-60), human cervical carcinoma (SiHa), human colorectal carcinoma (HCT 116, HT 29 and SW 480), human epithelioid cervix carcinoma (HeLa), human glioblastoma multiform tumour (T98G), human leukaemia (K562), leukaemia rat embryonic fibroblast (5RP7), human liver cancer (Hep-1), human lymphoblast lung (U937), human melanoma (A375), human nasopharyngeal carcinoma (NPC) (NPC/HK1 and C666-1), human oral cancer (KB), lymphocytic leukaemia (L1210), human oral squamous cell carcinoma (SCC) (Ca9-22 and

SCC12), human prostate cancer (DU145 and PC-3) and hepatoma Hep G2 (Hep G2 and Hep3B) (Herdwiani et al. 2016; Rad et al. 2015; Priyarani et al. 2010).

Some of the active ingredients isolated/identified from *Cinnamomum* species, which have anticancer-related activity, included kotomolide A, isokotomolide A, kotomolide B, secokotomolide A, tenuifolide A, isotenuifolide A, tenuifolide B, secotenuifolide A, subamolide D, subamolide E, secusubamolide A, (7′S,8′R,8R)-lyoniresinol-9-O-(E)-feruloyl ester, 9,9′-di-O-feruloyl-(+)-5,5′-dimethoxy secoisolariciresinol ester, (7′S,8′R,8R)-lyoniresinol-9,9′-di-O-(E)-feruloyl ester, eugenol, trans-cinnamaldehyde, coumarin, cinnamic acid and cinnamyl alcohol (Herdwiani et al. 2016; Rad et al. 2015; Jaganathan and Supriyanto 2012; Ng and Wu 2011; Chen et al. 2010). Trans-cinnamaldehyde, cinnamic acid, cinnamyl alcohol, eugenol and procyanidin type A trimer and a tetramer are some of the identified individual antioxidants with anticancer-related properties of Ceylon cinnamon (Ng and Wu 2011; Lu et al. 2010; Jaganathan and Supriyanto 2012).

12.2.5 Anti-inflammatory Properties of Ceylon Cinnamon

Traditionally Ceylon cinnamon has been used for inflammatory diseases (Jayaweera 1982). This anti-inflammatory property claimed in traditional systems has been scientifically proven by several studies (Gunawardena et al. 2015; Vetal et al. 2013; Joshi et al. 2010; Mathew and Abraham 2006). The investigations were based on both in vitro and in vivo studies (Gunawardena et al. 2015; Vetal et al. 2013; Joshi et al. 2010; Mathew and Abraham 2006). In vitro studies on anti-inflammatory activity of Ceylon cinnamon were both on cell line- and non-cell line-based assays (Abeysekera et al. 2015, 2016c; Gunawardena et al. 2015; Varalakshmi et al. 2012; Joshi et al. 2010; Mathew and Abraham 2006). In vivo studies were based on rats as the animal model (Abeysekera et al. 2016c; Vetal et al. 2013; Joshi et al. 2010). Solvent and water extracts mainly from stem bark are the mostly investigated part for anti-inflammatory activity of Ceylon cinnamon (Abeysekera et al. 2016c; Gunawardena et al. 2015; Vetal et al. 2013; Joshi et al. 2010; Mathew and Abraham 2006). The authors have recently reported anti-inflammatory activity from bark and leaf extracts of Ceylon cinnamon using in vitro assays (Abeysekera et al. 2015).

The mechanisms of anti-inflammatory activity of Ceylon cinnamon included inhibition of production of nitric oxide (NO) and superoxide ($O_2^{\bullet-}$) biological radicals, inhibition of cyclooxygenases, inhibition of production of prostaglandins and other autacoids, free radical scavenging and antioxidant actions (Abeysekera et al. 2013a, 2016c; Gunawardena et al. 2015; Vetal et al. 2013; Varalakshmi et al. 2012; Joshi et al. 2010; Mathew and Abraham 2006). Ceylon cinnamon contains flavonoids, proanthocyanidins, saponins, tannins, catechins and hydroxycinnamic acids (Abeysekera et al. 2013a, 2017; Gunawardena et al. 2013). These phytochemical classes are reported to have anti-inflammatory activity and thus may be responsible for the reported anti-inflammatory activity of Ceylon cinnamon (Hussain et al. 2016; Gunawardena et al. 2013).

12.2.6 Antiarthritic Properties of Ceylon Cinnamon

Ceylon cinnamon has been tested for its antiarthritic properties by several workers. Ceylon cinnamon bark demonstrated to decrease the TNF-α levels in animal models of inflammation and rheumatoid arthritis. The administration of cinnamon extract has markedly improved serum TNF-α and C-reactive protein (CRP) levels in irradiated rats (Azab et al. 2011). Farideh Shishehbor et al. (2018) showed consumption of cinnamon improves the clinical symptoms and inflammatory markers in women with rheumatoid arthritis using *C. burmannii*. Further, the beneficial effects of cinnamon extract and cinnamon polyphenols on the reduction of serum levels of TNF-α, CRP and interleukin (IL)-6 and improved clinical signs have been reported in animal models (Liao et al. 2012; Azab et al. 2011). Administration of cinnamaldehyde (Liao et al. 2012) and a polyphenolic fraction of *C. zeylanicum* bark reduced paw oedema in animal models of inflammation and arthritis (Lee et al. 2011). Vetal et al. (2013) demonstrated antiarthritic activity from type A procyanidin from Ceylon cinnamon in a rat rheumatoid arthritis model.

12.2.7 Immunomodulatory Properties of Ceylon Cinnamon

Evidences suggest that cinnamon extract is useful in regulating the immune system (Cao et al. 2008). Several studies have demonstrated immunomodulatory properties from Ceylon cinnamon (Niphade et al. 2009; Balekar et al. 2014). Cell culture studies revealed the beneficial effect of cinnamon polyphenols. They are responsible in improving immune responses, and the effects are modulated by regulating the expression of proinflammatory and anti-inflammatory cytokine genes (Cao et al. 2008; Rathi et al. 2013; Bodhankar et al. 2013; Mohan and Thakurdesai 2013; Lee et al. 2011).

12.2.8 Antibiotic Properties of Ceylon Cinnamon

Many studies have conclusively demonstrated antibiotic properties of cinnamon bark, leaf and essential oils by testing against a range of pathogenic and food-borne bacteria and fungi (Shan et al. 2007; Trinh et al. 2015; Yap et al. 2015; Utchariyakiat et al. 2016). Wang et al. (2018) recently showed a strong anti-bacterial effect from *C. zeylanicum* bark essential oil against the oral pathogen *Porphyromonas gingivalis*, a Gram-negative anaerobic bacterium that causes chronic periodontitis. Firmino et al. (2018) showed 99.9% reduction in biofilm biomass from *C. zeylanicum* and cinnamaldehyde. The mechanism by which the cinnamon extracts, its essential oils and compounds inhibit bacteria were reported to be by insulting bacterial cell membrane integrity, changing lipid profile, inhibiting ATPases, cell division, membrane porins, motility, and biofilm formation and disturbing the quorum sensing in bacterium (Vasconcelos et al. 2018). Further, *C. zeylanicum* essential oil also demonstrated to possess synergistic activity with other antibiotics. This was demonstrated by Guerra et al. (2012) with antibiotics such as amikacin, gentamicin, imipenem and meropenem using *Acinetobacter baumannii*.

12.2.9 Anti-nociceptive and Analgesic Properties

Ceylon cinnamon bark oil has been used in the management of pain and aches from ancient times. In Sri Lanka, cinnamon bark oil is a common household remedy for the management of aches and pains, especially in severe toothaches and joint pains. The anti-nociceptive and analgesic effect of cinnamon extract has been demonstrated by several workers in acute and chronic pain in rat models (Arzi et al. 2011; Atta and Alkofahi 1998; Mohammad et al. 2016), and the effect is attributed to the major compound cinnamaldehyde present in cinnamon (Mohammad et al. 2016). The other major compound in cinnamon leaf oil, eugenol, is also a known painkiller and is a commonly used ingredient in dental procedures (Bó et al. 2013; Ohkubo and Shibata 1997; Park et al. 2011) and is reported to reduce neuropathic (Lionnet et al. 2010) and orofacial pain (Park et al. 2009). The nociception of cinnamon mediates via both central and peripheral actions (Jain and Gupta 2019). Bó et al. (2013) provided more information on the effect of eugenol on the acute pain, suggesting participation of glutamatergic and TNF-α pathways in its mechanism of action. Their studies provide further evidence for the involvement of opioid receptors in the anti-nociceptive action of eugenol (Bó et al. 2013). Other major compounds reported from cinnamon essential oils, caryophyllene (Katsuyama et al. 2013), linalool (Venâncio et al. 2011) and ρ-cymene (Quintans et al. 2013), too reported to contain anti-nociception through various central and peripheral mechanisms.

12.2.10 Cardioprotective Properties

A recent study reported the potential effects of two compounds, cinnamic aldehyde and cinnamic acid, isolated from *C. cassia* against myocardial ischaemia (Song et al. 2013), indicating that cinnamon also has the potential to be used to treat cardiovascular diseases. Presence of antihypertensive activity in cinnamon also can be attributed to its cardioprotective properties (Ranasinghe et al. 2013).

12.2.11 Antihypertensive Properties

Several studies have demonstrated antihypertensive properties of cinnamon (Ranasinghe et al. 2013). Wansi et al. (2007) showed antihypertensive effect in an ethanol extract of *C. zeylanicum*. In as early as 1975, Mastoshi and Shingo (1975) demonstrated cinnamaldehyde, the major constituent of cinnamon bark oil, causes hypotensive effects in guinea pigs and anesthetized dogs. They attributed this effect to possible vasodilatation activity. Examining the effect of Ceylon cinnamon on mean arterial blood pressure in normotensive rats, salt-loaded hypertensive rats, L-NAME hypertensive rats and spontaneously hypertensive rats, it has been demonstrated that immediately after IV administration, a significant blood pressure drop was evident in normotensive, salt-loaded hypertensive and L-NAME hypertensive rats, confirming the antihypertensive effect of *C. zeylanicum* (Nyadjeu et al. 2011).

12.2.12 Gastroprotective Properties

A number of studies have demonstrated gastroprotective effect from *C. zeylanicum* (Rafatullah et al. 2011; Alqasoumi 2012; Mohammed 2014; Ranasinghe et al. 2013). These studies have used aqueous suspensions of Ceylon cinnamon bark and also had looked at the possible mechanisms of action. Rafatullah et al. (2011) tested gastroprotection in Wistar rats and attributed the gastroprotection effect they observed to elevation of the non-protein sulfhydryls contents and the mucus coat in the stomach.

12.2.13 Skin Protection, Whitening and Anti-ageing Properties

Cinnamomum species were reported to have skin anti-ageing activity in several in vitro studies (Priani et al. 2014; Bharti et al. 2013; Chang et al. 2013; Takasao et al. 2012; Khan and Ahmad 2011; Lin et al. 2011a; Marongiu et al. 2007; Lee et al. 1999; Lee et al. 1997; Ramos et al. 1996). Species such as *C. zeylanicum* and *C. aromaticum* (*C. cassia*) have been studied in much detail for skin anti-ageing activity than other *Cinnamomum* species. *C. burmannii*, *C. tamala* and *C. osmophloeum* are the other *Cinnamomum* species reported to have skin anti-ageing activity via different in vitro studies. Reported skin anti-ageing activities of *Cinnamomum* species were on bark extracts and bark essential oils.

The mechanisms of skin anti-ageing activity of *Cinnamomum* species are diverse, and some of those included free radical scavenging and antioxidant action and attenuation of degradation of skin matrix through inhibition of matrix metalloproteinases, promoting collagen synthesis, photoprotection, anti-glycation activity, anti-inflammatory activity, improved skin elasticity and depigmentation (Priani et al. 2014; Abeysekera et al. 2013a; Bharti et al. 2013; Chang et al. 2013; Chou et al. 2013; Takasao et al. 2012; Khan and Ahmad 2011; Lin et al. 2011a; Marongiu et al. 2007; Mukherjee et al. 2011; Lee et al. 1997, 1999).

Phytochemical classes, such as catechins, flavonoids, hydroxybenzoic acids, hydroxycinnamic acids and proanthocyanidins, are responsible for the skin anti-ageing activity of *Cinnamomum* species, and to date very few individual compounds are known, and some of which included cinnamaldehyde, cis-2-methoxycinnamic acid, (E)-caryophyllene, eugenol and R-terpineol (Abeysekera et al. 2013a; Chang et al. 2013; Takasao et al. 2012; Khan and Ahmad 2011; Marongiu et al. 2007).

12.2.14 Other Pharmacological Properties of Ceylon Cinnamon

Ceylon cinnamon is reported for several biological activities other than the above-mentioned biological activities worldwide. Some of the investigated and reported biological activities of Ceylon cinnamon included anti-parasitic (Ranasinghe et al. 2013), anti-secretagogue and anti-gastric ulcer (Alqasoumi 2012), anti-nociceptive (Atta and Alkofahi 1998), hepatoprotective (Eidi et al. 2012) activity as well as positive effects on neurological disorders (Peterson et al. 2009) and memory performance (Mesripour et al. 2016), inhibitory effects on osteoclastogenesis (Tsuji-Naito 2008) and wound healing properties (Farahpour and Habibi 2012). Antiarthritic property has also been demonstrated from the ethanol and petroleum ether extracts of Ceylon cinnamon in a rat model (Mohamed et al. 2013).

12.3 Clinical Studies

There are about 12 randomized control trials on cinnamon reported so far, and all were mainly focused on evaluation of ability of cinnamon in the management of diabetes and dyslipidaemia. Allen et al. (2013) in their meta-analysis on ten randomized control trials conducted between 2003 and 2012 concluded that the consumption of cinnamon is associated with a statistically significant decrease in fasting plasma glucose, total cholesterol, LDL cholesterol and triglyceride levels and an increase in HDL cholesterol levels and no significant effect on HbA1c. The first clinical study on cinnamon is the trial conducted by Khan et al. (2003), and this study was on Chinese cinnamon, *C. cassia*. However, this study for the first time demonstrated clinically that intake of daily dose of cinnamon reduces serum glucose, triglyceride, LDL cholesterol and total cholesterol in people with type 2 diabetes (Khan et al. 2003). Though not comparable, the first US study conducted by Blevins et al. (2007) on *C. cassia* could not observe a significant reduction in fasting glucose or lipids in diabetic patients. Akilen et al. (2010) conducted a randomized, placebo-controlled, double-blind clinical trial on cinnamon in multi-ethnic type 2 diabetic patients in the UK and concluded that 2 g of cinnamon significantly reduces the HbA1c and blood pressure in diabetic patients. A randomized, placebo-controlled, double-blind study conducted in Germany using aqueous extract of *C. cassia* concluded the effect observed was moderate in poorly controlled diabetic patients (Mang et al. 2006). Lu et al. (2012) conducted a double-blind placebo-controlled study in a group of Chinese diabetic patients and concluded that *C. cassia* can help control blood sugar in diabetic patients. In another study conducted on postmenopausal diabetic patients in the Netherlands with *C. cassia,* supplementation failed to show beneficial effect on diabetic control (Vanschoonbeek et al. 2006). Most of the clinical studies so far reported either refer to *C. cassia* or do not clearly mention the cinnamon species used. The only clinical study so far that refers to authenticated Ceylon cinnamon *C. zeylanicum* obtained from the original plantations of true cinnamon is the trial conducted at the Medical Faculty of University of Colombo, Sri Lanka, on a cinnamon capsule developed from an aqueous fraction of Ceylon cinnamon stem bark. According to this study, Ceylon cinnamon reduces serum glucose, HbA1C, triglyceride, LDL cholesterol and total cholesterol in people with type 2 diabetes with no marked side effects (Ranasinghe 2019; Ranasinghe et al. 2017).

12.4 Toxicology of Ceylon Cinnamon

The US Food and Drug Administration has noted that the amount of cinnamon (only *C. zeylanicum* and *C. cassia*) bark and their essential oils in commonly found foods are non-toxic, well-tolerated and generally considered as safe (GRAS) substances. The recommended allowable daily intake (ADI) by the World Health Organization (WHO)/United Nations Food and Agriculture Organization (FAO) is 700 μg/kg of body weight when used as food additives (FAO/WHO Tech. Report 1984). Cinnamaldehyde is the main constituent in both bark and bark essential oil of *C. zeylanicum*. The LD_{50} value of cinnamaldehyde is reported as 132, 6110 and 2225 mg/kg body weight in mice by intravenous, intraperitoneal and oral administration, respectively. The LD_{50} value of bark essential oil is reported as 5.36 ml/kg body weight in mice by oral administration (Yadav et al. 1999). Eugenol is the main phytoconstituent in leaf and leaf essential oil of *C. zeylanicum*. It is reported to cause toxicity reactions, such as irritation, vomiting and ataxia, in dogs at dosages of nearly 0.5 g/kg body weight, and LD_{50} value was 1.8 ml (1.93 g/kg) in rats administered by stomach tube without subsequent aspiration (Sober et al. 1950). Cinnamon contains coumarins, which are naturally occurring plant compounds having carcinogenic and hepatotoxic properties (Medagama 2015). The maximum allowable limit of coumarin is 2 mg/kg for many foods and beverages by the Council of the European Communities (Medagama 2015). The coumarin content of Ceylon cinnamon was undetectable or minimum (0.017 g/kg) (Wang et al. 2013) compared to *C. cassia* (2880–4820 mg/kg by HPLC analysis) (Sproll et al. 2008) and other

Cinnamomum species [0.31 to 6.97 g/kg by ultra-performance liquid chromatography (UPLC) method] (Wang et al. 2013). Animal studies on Ceylon cinnamon too did not demonstrate any significant adverse effects or toxicity on the liver, the kidney and the pancreas in rats at the studied doses ranging from 100 to 2400 mg/Kg (Medagama 2015; Ranasinghe et al. 2012). However, few studies demonstrated elevated levels of alkaline phosphatase (ALP), serum glutamic oxaloacetic transaminase (SGOT) and serum glutamic-pyruvic transaminase (SGPT), effect on renal functions and significant rise in uric acid and blood urea upon administration of cinnamon (Medagama 2015).

12.5 Conclusion

Pharmacological effects of Ceylon cinnamon are diverse and carry a range of health benefits as a functional food, food additive, spice, beverage and health supplement. Its uses in traditional medicinal preparations have been justified through many scientific studies conducted on the claims or applications in those medical and health food systems. Clinical studies for the effects on diabetic and lipid control have proved its potential pharmaceutical and nutraceutical applications in the management of blood sugar and lipid profile in patients, without significant side effects. Many other beneficial effects of cinnamon are yet to be uncovered, and bioactive molecules are yet to be isolated and characterized.

References

Abeysekera WPKM, Premakumara GAS, Ratnasooriya WD (2013a) In vitro antioxidant properties of bark and leaf extracts of Ceylon cinnamon (*Cinnamomum zeylanicum* Blume). Trop Agric Res 24(2):128–138

Abeysekera WPKM, Premakumara GAS, Ratnasooriya WD (2013b). Glutathione S –transferase enzyme inhibitory activity of Ceylon cinnamon (*Cinnamomum zeylanicum* Blume). Proceedings of the annual research symposium of University of Colombo, Sri Lanka, p 185

Abeysekera WPKM, Premakumara GAS, Ratnasooriya WD (2014). *In vitro* lipase, cholesterol esterase and cholesterol micellization inhibitory activities of leaf and bark of *Cinnamomum zeylanicum* Blume (Ceylon cinnamon). Proceedings of the annual sessions of young scientists forum, Sri Lanka, p 1

Abeysekera WPKM, Premakumara GAS, Ratnasooriya WD (2015) Nitric oxide and superoxide radical scavenging activity of bark and leaf of Ceylon cinnamon (*Cinnamomum zeylanicum* Blume) in vitro. Abs. Proceedings of the 35th annual sessions of the Institute of Biology, Sri Lanka, p 50

Abeysekera WPKM, Premakumara GAS, Ratnasooriya WD (2016a) Proanthocyanidins and selected phenolic and non-phenolic compounds in bark and leaf of Ceylon cinnamon (*Cinnamomum zeylanicum* Blume). Proceedings of the 72nd annual sessions of Sri Lanka Association for the Advancement of Science, Sri Lanka, p 96

Abeysekera WPKM, Premakumara GAS, Ratnasooriya WD (2016b) Growth inhibition and cytotoxicity of leaf extracts of Ceylon cinnamon (*Cinnamomum zeylanicum*) on human breast carcinoma (MCF 7) and hepatocarcinoma ($HepG_2$) cell lines. Proceedings of the annual research symposium of University of Colombo, Sri Lanka, p 257

Abeysekera WPKM, Premakumara GAS, Ratnasooriya WD (2016c) Anti-inflammatory activity of bark of Ceylon cinnamon (*Cinnamomum zeylanicum* Blume). Abs. 9th international research conference of General Sir John Kotelawala Defense University, Sri Lanka, p 260

Abeysekera WPKM, Premakumara GAS, Ratnasooriya WD (2017) Bark extracts of Ceylon cinnamon possess antilipidemic activities and bind bile acids in vitro. Evid Based Complement Alternat Med 2017:7347219, 10 pages

Adisakwattana S, Lerdsuwankij O, Poputtachai U, Minipun A, Suparpprom C (2011) Inhibitory activity of cinnamon bark species and their combination effect with acarbose against intestinal α- glucosidase and pancreatic α- amylase. Plant Foods Hum Nutr 66:143–148

Akilen R, Tsiami A, Devendra D, Robinson N (2010) Glycated haemoglobin and blood pressure-lowering effect of cinnamon in multi-ethnic Type 2 diabetic patients in the UK: a randomized, placebo-controlled, double-blind clinical trial. Diabet Med 27(10):1159–1167

Alqasoumi S (2012) Anti-secretagogue and antiulcer effects of 'cinnamon' *Cinnamomum zeylanicum* in rats. J Pharmacogn Phytother 4:53–61

Arachchige SPG, Abeysekera WPKM, Ratnasooriya WD (2017) Anti-amylase, anti-cholinesterases, anti-glycation and glycation reversing potential of bark and leaf of Ceylon cinnamon (*Cinnamomum zeylanicum* Blume) in vitro. Evid Based Complement Alternat Med 2017:5076029, 13 pages

Arzi A, Sarkaki A, Aghel N, Nazari Z, Saeidnejad S (2011) Study of analgesic effect of hydroalcoholic extract of Cinnamon. Sci Med J 10(3):271–279

Atta AH, Alkofahi A (1998) Anti-nociceptive and anti-inflammatory effects of some Jordanian medicinal plant extracts. J Ethnopharmacol 60:117–124

Azab KS, Mostafa AHA, Ali EM, Abdel-Aziz MA (2011) Cinnamon extract ameliorates ionizing radiation-induced cellular injury in rats. Ecotoxicol Environ Saf 74(8):2324–2329

Allen RW, Schwartzman E, Baker WL, Coleman CI, Phung OJ (2013) Cinnamon Use in Type 2 Diabetes: An Updated Systematic Review and Meta-Analysis. Ann Fam Med 11(5):452–459 https://doi.org/10.1370/afm.1517

Balekar N, Bodhankar S, Mohan V, Thakurdesai PA (2014) Modulatory activity of a polyphenolic fraction of *Cinnamomum zeylanicum* L. bark on multiple arms of immunity in normal and immunocompromised mice. J Appl Pharm Sci 4(7):114–122

Bandara T, Uluwaduge I, Jansz ER (2012) Bioactivity of cinnamon with special emphasis on diabetes mellitus: a review. Int J Food Sci Nutr 63(3):380–386

Bharti P, Seasotiya L, Bai S, Malik A, Dalal S (2013) Anti-bacterial, anti-oxidant, collagenase inhibition, urease inhibition, activities and chemical composition of the essential oil and extracts of *Cinnamomum zeylanicum* bark. Int J Phytomedicine 5:36–45

Blevins SM, Leyva MJ, Brown J, Wright J, Scofield RH, Aston CE (2007) Effect of cinnamon on glucose and lipid levels in non–insulin-dependent type 2 diabetes. Diabetes Care 30(9):2236–2237

Blokhina O, Virolainen E, Fagerstedt KV (2003) Antioxidants, oxidative damage and oxygen deprivation stress. Ann Bot 91:179–194

Bó WD, Luiz AP, Martins DF, Mazzardo-Martins L, Santos ARS (2013) Eugenol reduces acute pain in mice by modulating the glutamatergic and tumor necrosis factor α (TNF-α) pathways. Fundam Clin Pharmacol 27:517–525

Bodhankar S, Mohan V, Thakurdesai P (2013) Ameliorative effects of a polyphenolic fraction of *Cinnamomum zeylanicum* L. bark in animal models of inflammation and arthritis. Sci Pharm 81(2):567–590

Boğa M, Hacıbekiroğlu I, Kolak U (2011) Antioxidant and anticholinesterase activities of eleven edible plants. Pharm Biol 49(3):290–295

Brewer MS (2011) Natural antioxidants: sources, compounds, mechanisms of action, and potential applications. Compr Rev Food Sci Food Saf 10:221–247

Cao H, Urban JF, Anderson RA (2008) Cinnamon polyphenol extract affects immune responses by regulating anti-and pro-inflammatory and glucose transporter gene expression in mouse macrophages. J Nutr 138(5):833–840. https://doi.org/10.1093/jn/138.5.833

Chakraborty U, Das H (2010) Antidiabetic and antioxidant activities of *Cinnamomum tamala* leaf extracts in Stz-treated diabetic rats. Global J Biotechnol Biochem 5(1):12–18

Chang CT, Chang WL, Hsu JC, Shih Y, Chou ST (2013) Chemical composition and tyrosinase inhibitory activity of *Cinnamomum cassia* essential oil. Bot Stud 54(10):1–7

Chen TH, Huang YH, Lin JJ, Liau BC, Wang SY, Wu YC, Jong TT (2010) Cytotoxic Lignan esters from *Cinnamomum osmophloeum*. Planta Med 76:613–619

Chou ST, Chang WL, Chang CT, Hsu SL, Lin YC, Shih Y (2013) *Cinnamomum cassia* essential oil inhibits α-MSH-induced melanin production and oxidative stress in murine B16 melanoma cells. Int J Mol Sci 14:19186–19201

Ciftci M, Simsek UG, Yuce A, Yilmaz O, Dalkilic B (2010) Effects of dietary antibiotic and cinnamon oil supplementation on antioxidant enzyme activities, cholesterol levels and fatty acid compositions of serum and meat in broiler chickens. Acta Vet Brno 79:33–40

Dalai MK, Bhadra S, Chaudhary SK, Chanda J, Bandyopadhyay A, Mukherjee PK (2014a) Anticholinesterase activity of *Cinnamomum zeylanicum* L. leaf extract. Tang [Humanitas Medicine] 4(2):11.1–11.6. 6 Pages

Dalai MK, Bhadra S, Chaudhary SK, Bandyopadhyay A, Mukherjee PK (2014b) Anticholinesterase potential of *Cinnamomum tamala* (Buch.-Ham.) T. Nees & Eberm. Leaves. Indian J Tradit Knowl 13(4):691–697

Dalkilic B, Ciftci M, Guler T, Cerci IH, Ertas ON, Guvenc M (2009) Influence of dietary cinnamon oil supplementation on fatty acid composition of liver and abdominal fat in broiler chicken. J Appl Anim Res 35(2):173–176

Dearlove RP, Greenspan P, Hartle DK, Swanson RB, Hargrove JL (2008) Inhibition of protein glycation by extracts of culinary herbs and spices. J Med Food 11(2):275–281

Department of Ayurveda, Sri Lanka 2019. http://www.ayurveda.gov.lk/Ayurveda%20in%20Sri%20Lankan.html. Accessed 21 Nov 2019.

Dudonne S, Vitrac X, Coutiere P, Woillez M, Merillon JM (2009) Comparative study of antioxidant properties and total phenolic content of 30 plant extracts of industrial interest using DPPH, ABTS, FRAP, SOD and ORAC assays. J Agric Food Chem 57:1768–1774

Eidi A, Mortazavi P, Bazargan M, Zaringhalam J (2012) Hepatoprotective activity of cinnamon ethanolic extract against CCL 4-induced liver injury in rats. Exp Clin Sci 11:495–507

El-Desoky GE, Aboul-Soud MAM, Al-Numair KS (2012) Antidiabetic and hypolipidemic effects of Ceylon cinnamon (*Cinnamomum verum*) in alloxan-diabetic rats. J Med Plant Res 6(9):1685–1691

FAO/WHO: Food Additive Series – Toxicological Evaluation of Certain Food Additives and Contaminants (1984), 28th Report, WHO Technical Report Series 19, 710, World Health Organization, 1986

Farahpour MR, Habibi M (2012) Evaluation of the wound healing activity of an ethanolic extract of Ceylon cinnamon in mice. Veterinarni Medicina 57:53–57

Fiedor J, Burda K (2014) Potential role of carotenoids as antioxidants in human health and disease. Nutrients 6:466–488

Firmino DF, Cavalcante TTA, Gomes GA, Firmino NCS, Rosa LD, De Carvalho MG, Catunda FEA Jr (2018) Antibacterial and antibiofilm activities of *Cinnamomum* Sp. essential oil and cinnamaldehyde: antimicrobial activities. Scientific World Journal:7405736, 9 pages. https://doi.org/10.1155/2018/7405736

Gholamhoseinian A, Shahouzehi B, Sharifi-far F (2010) Inhibitory effect of some plant extracts on pancreatic lipase. Int J Pharmacol 6(1):18–24

Guerra FQS, Mendes JM, De Sousa JP, Morais-Braga MF, Santos BH, Melo Coutinho HD, Lima E d O (2012) Increasing antibiotic activity against a multidrug-resistant *Acinetobacter* spp. by essential oils of *Citrus limon* and *Cinnamomum zeylanicum*. Nat Prod Res 26:2235–2238

Gunawardena D, Govindaraghavan S, Muench G (2013) Anti-inflammatory properties of cinnamon polyphenols and their monomeric precursors. In: Wastson RR, Preedy VR, Zibadi S (eds) Polyphenols in human health and disease, vol 1. Elsevier Science, Burlington, pp 409–425

Gunawardena D, Karunaweera N, Lee S, van Der Kooy F, Harman DG, Raju R, Bennett L, Gyengesi E, Sucher NJ, Munch G (2015) Anti-inflammatory activity of cinnamon (*C. zeylanicum* and *C. cassia*) extracts - identification of E-cinnamaldehyde and o-methoxy cinnamaldehyde as the most potent bioactive compounds. Food Funct 6(3):910–919

Hassan SA, Barthwal R, Nair MS, Haque SS (2012) Aqueous bark extract of *Cinnamomum zeylanicum*: a potential therapeutic agent for streptozotocin- induced type 1 diabetes mellitus (T1DM) rats. Trop J Pharm Res 11(3):429–435

Herdwiani W, Soemardji AA, Elfahmi TMI (2016) A review of cinnamon as a potent anticancer drug. Asian J Pharm Clin Res 9(3):8–13

Ho SC, Chang PW (2012) Inhibitory effects of several spices on inflammation caused by advanced glycation end products. Am J Plant Sci 3:995–1002

Hussain T, Tan B, Yin Y, Blachier F, Tossou MCB, Rahu N (2016) Oxidative stress and inflammation: what polyphenols can do for us?: a review. Oxid Med Cell Longev 2016:7432797, 9 pages

Jaganathan SK, Supriyanto E (2012) Antiproliferative and molecular mechanism of eugenol-induced apoptosis in cancer cells: a review. Molecules 17:6290–6304

Jain S, Gupta S (2019) Effects of *Cinnamomum zeylanicum* bark extract on nociception and anxiety like behavior in mice. Asian J Pharm Clin Res 12(9):236–241

Javed I, Faisal I, Rahman Z, Khan MZ, Muhammad F, Aslam B, Ahmed M, Shahzadi A (2012) Lipid lowering effect of *Cinnamomum zeylanicum* in hyperlipidaemic albino rabbits. Pak J Pharm Sci 25(1):141–147

Jayaweera DMA (1982) Medicinal plants (indigenous and exotic) used in Ceylon. Part III, National Science Council of Sri Lanka, p 117

Joshi K, Awte S, Bhatnagar P, Walunj S, Gupta R, Joshi S, Sabharwal S, Bani S, Padalkar AS (2010) *Cinnamomum zeylanicum* extract inhibits proinflammatory cytokine TNFμ: in vitro and in vivo studies. Res Pharm Biotechnol 2(2):14–21

Katsuyama S, Mizoguchi H, Kuwahata H, Komatsu T, Nagaoka K, Nakamura H, Bagetta G, Sakurada T, Sakurada S (2013) Involvement of peripheral cannabinoid and opioid receptors in β-caryophyllene-induced antinociception. Eur J Pain 17:664–675

Khan MSA, Ahmad I (2011) In vitro antifungal, anti-elastase and anti-keratinase activity of essential oils of *Cinnamomum*, *Syzygium* and *Cymbopogon* species against *Aspergillus fumigates* and *Trichophyton rubrum*. Phytomedicine 19:48–55

Khan A, Safdar M, Khan MMA, Khattak KN, Anderson RA (2003) Cinnamon improves glucose and lipids of people with type 2 diabetes. Diabetes Care 26(12):3215–3218

Kim CR, Choi SJ, Kwon YK, Kim JK, Kim YJ, Park GG, Shin DH (2016) *Cinnamomum loureirii* extract inhibits acetylcholinesterase activity and ameliorates trimethyltin-induced cognitive dysfunction in mice. Biol Pharm Bull 39:1130–1136

Kumar S, Brijeshlata, Dixit S (2012) Screening of traditional Indian spices for inhibitory activity of acetylcholinesterase and butyrylcholinesterase enzymes. Int J Pharm Bio Sci 3(1):59–65

Lee HS, Yoon JA (2010) Inhibitory activity of advanced glycation end products (AGE) formation of edible plants for development of anti-wrinkle ingredients. J Korean Soc Food Sci Nutr 39(2):186–192

Lee KT, Kim BJ, Kim JH (1997) Biological screening of 100 plant extracts for cosmetic use (I): inhibitory activities of tyrosinase and DOPA auto-oxidation. Int J Cosmet Sci 19:291–298

Lee KK, Kim JH, Cho JJ, Choi JD (1999) Inhibitory effects of 150 plant extracts on elastase activity, and their anti-inflammatory effects. Int J Cosmet Sci 21(2):71–82

Lee BJ, Kim YJ, Cho DH, Sohn NW, Kang H (2011) Immunomodulatory effect of water extract of cinnamon on anti-CD3-induced cytokine responses and p38, JNK, ERK1/2, and STAT4 activation. Immunopharmacol Immunotoxicol 33(4):714–722. https://doi.org/10.3109/08923973.2011.564185

Lee J, Lee DG, Park JY, Chae S, Lee S (2015) Analysis of the trans-cinnamic acid content in *Cinnamomum* spp. and commercial cinnamon powder using HPLC. J Agric Chem Environ 4:102–108

Liao JC, Deng JS, Chiu CS, Hou WC, Huang SS, Shie PH, Huang GJ (2012) Anti-inflammatory activities of *Cinnamomum cassia* constituents in vitro and in vivo. Evid Based Complement Alternat Med:429320, 12 pages

Lin CC, Yang CH, Wu PS, Kwan CC, Chen YS (2011a) Antimicrobial, anti-tyrosinase and antioxidant activities of aqueous aromatic extracts from forty-eight selected herbs. J Med Plant Res 5(26):6203–6209

Lin TY, Liao JW, Chang ST, Wang SY (2011b) Antidyslipidemic activity of hot-water extracts from leaves of *Cinnamomum osmophloeum* Kaneh. Phytother Res 25:1317–1322

Lionnet L, Beaudry F, Vachon P (2010) Intrathecal eugenol administration alleviates neuropathic pain in male Sprague-Dawley rats. Phytother Res 24:1645–1653

Lopes BP, Gaique TG, Souza LL, Paula GSM, Kluck GEG, Atella GC, Gomes ACC, Simas NK, Kuster RM, Ortiga-Carvalho TM, Pazos-Mouraa CC, Oliveira KJ (2015) Cinnamon extract improves the body composition and attenuates lipogenic processes in the liver and adipose tissue of rats. Food Funct 6:3257–3265

Lu J, Zhang K, Nam S, Anderson RA, Jove R, Wen W (2010) Novel angiogenesis inhibitory activity in cinnamon extract blocks VEGFR2 kinase and downstream signaling. Carcinogenesis 31(3):481–488

Lu T, Sheng H, Wu J, Cheng Y (2012) Cinnamon extract improves fasting blood glucose and glycosylated hemoglobin level in Chinese patients with type 2 diabetes. Nutr Res 32(6):408–412

Mang B, Wolters M, Schmitt B, Kelb K, Lichtinghagen R, Stichtenoth DO, Hahn A (2006) Effects of a cinnamon extract on plasma glucose, HbA1c, and serum lipids in diabetes mellitus type 2. Eur J Clin Investig 36(5):340–344

Marongiu B, Piras A, Porcedda S, Tuveri E, Sanjust E, Meli M, Sollai F, Succa P, Rescigno A (2007) Supercritical CO_2 extract of *Cinnamomum zeylanicum*: chemical characterization and antityrosinase activity. J Agric Food Chem 55:10022–10027

Mastoshi H, Shingo Y (1975) Pharmacological studies on Chinese cinnamon. II. Effects of cinnamaldehyde on the cardiovascular and digestive systems. Chem Pharm Bull (Tokyo) 23:941–947

Mateos-Martín ML, Fuguet E, Quero C, Pérez-Jiménez J, Torres JL (2012) New identification of proanthocyanidins in cinnamon (*Cinnamomum zeylanicum* L.) using MALDI-TOF/TOF mass spectrometry. Anal Bioanal Chem 402:1327–1336

Mathew S, Abraham TE (2006) Studies on the antioxidant activities of cinnamon (*Cinnamomum verum*) bark extracts, through various in vitro models. Food Chem 94(4):520–528

Medagama AB (2015) The glycaemic outcomes of cinnamon, a review of the experimental evidence and clinical trials. Nutr J 14:108

Mesripour A, Moghimi F, Rafieian-Kopaie M (2016) The effect of *Cinnamomum zeylanicum* bark water extract on memory performance in alloxan-induced diabetic mice. Res Pharm Sci 11(4):318–323

Mohammad HDR, Mohammad DQ, Farzane SMD, Mohammadreza NY, Seyyed MB (2016) Comparative effect of cinnamon essential oil, diclofenac and morphine on acute and chronic pain in mice. Int J Med Lab 3:92–103

Mohammed A (2014) Antiulcer effect of Cinnamomum zeylanicum bark in rats. Res J Biol Sci 9(5):182–187

Mohamed DA, Mahmoud EA, Abdel-Moniem S, Hassan M (2013) Anti-inflammatory and anti-arthritic activity of some spices extracts on adjuvant induced arthritis in rats. J Appl Sci Res 9(8):5303–5312

Mohan V, Thakurdesai P (2013) Ameliorative effects of a polyphenolic fraction of *Cinnamomum zeylanicum* L. bark in animal models of inflammation and arthritis. Sci Pharm 81(2):567–590

Mukherjee PK, Maity N, Nema NK, Sarkar BK (2011) Bioactive compounds from natural resources against skin aging. Phytomedicine 19(1):64–73

Ng LT, Wu SJ (2011) Antiproliferative activity of *Cinnamomum cassia* constituents and effects of pifithrin-alpha on their apoptotic signaling pathways in HepG2 cells. Evid Based Complement Alternat Med 2011:492148, 6 pages

Niphade SR, Asad M, Chandrakala GK, Toppo E, Deshmukh P (2009) Immunomodulatory activity of *Cinnamomum zeylanicum* bark. Pharm Biol 47(12):1168–1173

Nyadjeu P, Dongmo A, Nguelefack TB, Kamanyi A (2011) Antihypertensive and vasorelaxant effects of *Cinnamomum zeylanicum* stem bark aqueous extract in rats. J Complement Integr Med 2011:8

Ohkubo T, Shibata M (1997) The selective capsaicin antagonist Capsazepine abolishes the antinociceptive action of eugenol and guaiacol. J Dent Res 76:848–851

Paiva SAR, Russell RM (1999) Beta-carotene and other carotenoids as antioxidants. J Am Coll Nutr 18(5):426–433

Park CK, Kim K, Jung SJ, Kim MJ, Ahn DK, Hong SD, Kim JS, Oh SB (2009) Molecular mechanism for local anesthetic action of eugenol in the rat trigeminal system. Pain 144:84–94

Park SH, Sim YB, Lee JK, Kim SM, Kang YJ, Jung JS, Suh HW (2011) The analgesic effects and mechanisms of orally administered eugenol. Arch Pharm Res 34:501–507

Peng X, Ma J, Chao J, Sun Z, Chang RC, Tse I, Le ET, Chen F, Wang M (2010) Beneficial effects of cinnamon proanthocyanidins on the formation of specific advanced glycation end products and methylglyoxal-induced impairment on glucose consumption. J Agric Food Chem 58(11):6692–6696

Pham-Huy LA, He H, Pham-Huy C (2008) Free radicals, antioxidants in disease and health. Int J Biomed Sci 4(2):89–96

Ponnusamy S, Ravindran R, Zinjarde S, Bhargava S, Kumar AR (2011) Evaluation of traditional Indian antidiabetic medicinal plants for human pancreatic amylase inhibitory effect in vitro. Evid Based Complement Alternat Med 2011:515647, 10 pages

Prasad KN, Yang B, Dong X, Jiang G, Zhang H, Xie H, Jiang Y (2009) Flavonoid contents and antioxidant activities from *Cinnamomum* species. Innov Food Sci Emerg Technol 10:627–632

Priyarani M, Venkatesan J, Binilraj SS, Sasidharan I, Padmakumriamma KP (2010) Antioxidant and cytotoxic potential of acetone and methanolic extracts of fresh and dry barks of *Cinnamomum zeylanicum* : in vitro study. J Cell Tissue Res 10(1):2131–2138

Peng X, Cheng KW, Ma J, Chen B, Ho CT, Lo C, Chen F, Wang M (2008) Cinnamon bark proanthocyanidins as reactive carbonyl scavengers to prevent the formation of advanced glycation endproducts. J Agric Food Chem 56(6):1907–1911

Priani SE, Humanisya H, Darusman F (2014) Development of sunscreen emulgel containing *Cinnamomum burmannii* stem bark extract. Int J Sci Res 3(12):2338–2341

Peterson DW, George RC, Scaramozzino F, Lapointe NE, Anderson RA, Graves DJ, Lew J (2009) Cinnamon extract inhibits tau aggregation associated with alzheimer's disease in vitro. J Alzheimers Dis 17:585–597

Quintans JSS, Menezes PP, Santos MRV, Bonjardim LR, Almeida JRGS, Gelain DP, Araújo AAS, Quintans-Júnior LJ (2013) Improvement of p-cymene antinociceptive and anti-inflammatory effects by inclusion in β-cyclodextrin. Phytomedicine 20:436–440

Rad SK, Kanthimathi MS, Abd Malek SN, Lee GS, Looi CY, Wong WF (2015) *Cinnamomum cassia* suppresses caspase-9 through stimulation of AKT1 in MCF-7 cells but not in MDA-MB-231 cells. PLoS One 10(12):e0145216

Rafatullah S, Alqasoumi S, Al-Dosary M, Al-Yahya M, Al-Mofleh I (2011) Gastroprotective effect of a popular spice cinnamon "*Cinnamomum zeylanicum*" in rats. Eur J Pharmacol 668:e42. https://doi.org/10.1016/j.ejphar

Ramos MFS, Santos E, Bizarri CHB, Mattos HA (1996) Preliminary studies towards utilization of various plant extracts as anti-solar agents. Int J Cosmet Sci 18:87–101

Ranasinghe P, Galappaththy P, Constantine GR, Jayawardena R, Weeratunga HD, Premakumara S, Katulanda P (2017) *Cinnamomum zeylanicum* (Ceylon cinnamon) as a potential pharmaceutical agent for type-2 diabetes mellitus: study protocol for a randomized controlled trial. Trials 18:446, 8 pages

Ranasinghe P, Perera S, Gunatilake M, Abeywardena E, Gunapala N, Premakumara S, Perera K, Lokuhetty D, Katulanda P (2012) Effects of *Cinnamomum zeylanicum* (Ceylon cinnamon) on blood glucose and lipids in a diabetic and healthy rat model. Pharm Res 4(2):73–79

Ranasinghe P, Pigera S, Premakumara GAS, Galappaththy P, Constantine GR, Katulanda P (2013) Medicinal properties of 'true' cinnamon (*Cinnamomum zeylanicum*): a systematic review. BMC Complement Altern Med 13:275–285

Ranasinghe P (2019) Evaluation of *Cinnamomum zeylanicum* (Ceylon cinnamon) as a pharmaceutical agent for diabetes mellitus. PhD Thesis, Faculty of Medicine, University of Colombo, Sri Lanka

Ranilla LG, Kwon YI, Apostolidis E, Shetty K (2010) Phenolic compounds, antioxidant activity and in vitro inhibitory potential against key enzymes relevant for hyperglycemia and hypertension of commonly used medicinal plants, herbs and spices in Latin America. Bioresour Technol 101(12):4676–4689

Rao PV, Gan SH (2014) Cinnamon: a multifaceted medicinal plant: a review. Evid Based Complement Alternat Med 2014:642942, 12 pages

Rathi B, Bodhankar S, Mohan V, Thakurdesai P (2013) Ameliorative effects of a polyphenolic fraction of *Cinnamomum zeylanicum* L. bark in animal models of inflammation and arthritis. Sci Pharm 81(2):567–590. https://doi.org/10.3797/scipharm.1301-16

Ravindran PN, Babu KN, Shylaja M (eds) (2004) Cinnamon and Cassia: The genus *Cinnamomum*. CRC Press LLC, United States of America, pp 278

Shan B, Cai Y, Brooks JD, Corke H (2007) Antibacterial properties and major bioactive components of cinnamon stick (*Cinnamomum burmannii*): activity against foodborne pathogenic bacteria. J Agric Food Chem 55:5484–5490

Shishehbor F, Safar MR, Rajaei E, Haghighizadeh MH (2018) Cinnamon consumption improves clinical symptoms and inflammatory markers in women with rheumatoid arthritis. J Am Coll Nutr. https://doi.org/10.1080/07315724.2018.1460733

Sangal A (2011) Role of cinnamon as beneficial antidiabetic food adjunct: a review. Adv Appl Sci Res 2(4):440–450

Shebis Y, Iluz D, Kinel-Tahan Y, Dubinsky Z, Yehoshua Y (2013) Natural antioxidants: function and sources. Food Nutr Sci 4:643–649

Sober HA, Hollander F, Sober EK (1950) Toxicity of eugenol: determination of LD_{50} on rats. Proc Soc Exp Biol Med 73(1):148–151

Song F, Li H, Sun J, Wang S (2013) Protective effects of cinnamic acid and cinnamic aldehyde on isoproterenol-induced acute myocardial ischemia in rats. J Ethnopharmacol 150(1):125–130

Sproll C, Ruge W, Andlauer C, Godelmann R, Lahenmeier DW (2008) HPLC analysis and safety assessment of coumarin in foods. Food Chem 109(2):462–469

Sudan R, Bhagat M, Gupta S, Chitrarakha DT (2013) Comparative analysis of cytotoxic and antioxidant potential of edible *Cinnamomum verum* (bark) and *Cinnamomum tamala* (Indian bay leaf). Free Radicals Antioxid 3(2s):S70–S73

Tailang M, Gupta BK, Sharma A (2008) Antidiabetic activity of alcoholic extract of *Cinnamomum zeylanicum* leaves in alloxon induced diabetic rats. People's J Sci Res 1:9–11

Takasao N, Tsuji-Naito K, Ishikura S, Tamura A, Akagawa M (2012) Cinnamon extract promotes type I collagen biosynthesis via activation of IGF-I signaling in human dermal fibroblasts. J Agric Food Chem 60:1193–1200

Trinh NT, Dumas E, Thanh ML, Degraeve P, Amara BBC, Gharsallaoui A, Oulahal N (2015) Effect of a Vietnamese *Cinnamomum cassia* essential oil and its major component trans-cinnamaldehyde on the cell viability, membrane integrity, membrane fluidity, and proton motive force of *Listeria innocua*. Can J Microbiol 61(4):263–271

Tsuji-Naito K (2008) Aldehydic components of cinnamon bark extract suppresses RANKL-induced osteoclastogenesis through NFATc1 downregulation. Bioorg Med Chem 16:9176–9183

Unlu M, Ergene E, Unlu GV, Zeytinoglu HS, Vural N (2010) Composition, antimicrobial activity and in vitro cytotoxicity of essential oil from *Cinnamomum zeylanicum* Blume (Lauraceae). Food Chem Toxicol 48:3274–3280

Utchariyakiat I, Surassmo S, Jaturanpinyo M, Khuntayaporn P, Chomnawang MT (2016) Efficacy of cinnamon bark oil and cinnamaldehyde on anti-multidrug resistant *Pseudomonas aeruginosa* and the synergistic effects in combination with other antimicrobial agents. BMC Complement Altern Med 16:158

Vanschoonbeek K, Thomassen BJ, Senden JM, Wodzig WK, Van Loon LJ (2006) Cinnamon supplementation does not improve glycemic control in postmenopausal type 2 diabetes patients. J Nutr 136(4):977–980

Varalakshmi B, Vijaya anandh A, Vijayakumar K, Prasanna R (2012) In vitro antioxidant activity of *Cinnamomum zeylanicum* Linn bark. Int J Inst Pharm Life Sci 2(3):154–166

Vasconcelos NG, Croda J, Simionatto S (2018) Antibacterial mechanisms of cinnamon and its constituents: a review. Microb Pathog. https://doi.org/10.1016/j.micpath.2018.04.036

Venâncio AM, Marchioro M, Estavam CS, Melo MS, Santana MT, Onofre ASC, Guimarães AG, Oliveira MGB, Alves PB, Pimentel HC, Quintans-Jr LJ (2011) *Ocimum basilicum* leaf essential oil and (−)-linalool reduce orofacial nociception in rodents: a behavioral and electrophysiological approach. Braz J Pharmacog 21:1043–1051

Vetal S, Bodhankar SL, Mohan V, Thakurdesai PA (2013) Anti-inflammatory and anti-arthritic activity of type-A procyanidine polyphenols from bark of *Cinnamomum zeylanicum* in rats. Food Sci Human Wellness 2:59–67

Varalakshmi B, Anand AV, Karpagam T, Bai JS, Manikandan R (2014) In vitro antimicrobial and anticancer activity of *Cinnamomum zeylanicum* Linn bark extracts. Int J Pharm Pharm Sci 6(1):12–18

Wang Y, Zhang Y, Shi YQ, Pan XH, Lu YH, Cao P (2018) Antibacterial effects of cinnamon (*Cinnamomum zeylanicum*) bark essential oil on *Porphyromonas gingivalis*. Microb Pathog 116:26–32

Wang YH, Avula B, Nanayakkara N, Zhao J, Khan IA (2013) Cassia cinnamon as a source of coumarin in cinnamon-flavored food and food supplements in the United States. J Agric Food Chem 61(18):4470–4476

Wansi SL, Nyadjeu P, Ngamga D, Pami ENM, Nguelefack TB, Kamanyi A (2007) Blood pressure lowering effect of the ethanol extract from the stem bark of *Cinnamomum zeylanicum* (Lauraceae) in rats. Pharmacology online 3:166–176

Yadav P, Dubey NK, Joshi VK, Chansouria JPN, Yadav P (1999) Antidermatophytic activity of essential oil of *Cinnamomum* as herbal ointment for cure of dermatomycosis. J Med Aromat Plant Sci 21:347–351

Yap PS, Krishnan T, Chan KG, Lim SH (2015) Antibacterial mode of action of *Cinnamomum verum* bark essential oil, alone and in combination with piperacillin, against a multi-drug-resistant *Escherichia coli* strain. J Microbiol Biotechnol 25(8):1299–1306

Zu Y, Yu H, Liang L, Fu Y, Efferth T, Liu X, Wu N (2010) Activities of ten essential oils towards *Propionibacterium acnes* and PC-3, A-549 and MCF-7 cancer cells. Molecules 15(5):3200–3210

Chapter 13
Industrial Applications of Ceylon Cinnamon (*Cinnamomum zeylanicum* Blume) as Nutraceuticals and Cosmeceuticals

Tuley De Silva

13.1 Introduction

Cinnamomum verum Berchtold and Presl, a medicinal plant that belongs to the family Lauraceae, is a small tropical tree that was originally named as *Cinnamomum zeylanicum* Blume indicating its origin from Sri Lanka. It is also known as true cinnamon. The cinnamon of commerce is the dried inner bark of the tree. When it dries, it forms strips that curl into rolls, called cinnamon sticks or quills, that are pale brown in colour with a mild sweet hot taste and spicy-hot fragrance.

These sticks can be ground to form cinnamon powder. Bark oil and leaf oil are obtained by the steam distillation of the bark and the leaves of the tree, respectively. Cinnamon oleoresin is obtained by the solvent extraction of the bark.

Cinnamon has been used for centuries for culinary purposes as a very popular spice. It is also used as a flavour in a variety of cuisines, snacks, deserts, baked savoury foods and cereals and a flavour additive in wines and teas. The aroma and flavour of cinnamon are due to its essential oil and its principal component, cinnamaldehyde, as well as numerous other constituents, including cinnamyl acetate and eugenol.

13.1.1 Ayurveda

Ayurveda is one of the oldest healing systems that has a holistic approach for healing and promoting longevity. Holistic Ayurvedic remedies are directed at providing total health including physical, mental and spiritual health. From ancient times,

T. De Silva (✉)
R&D Center, Link Natural Products (Pvt) Ltd, Malinda, Kapugoda, Sri Lanka

R. Senaratne, R. Pathirana (eds.), *Cinnamon*,
https://doi.org/10.1007/978-3-030-54426-3_13

parts of the cinnamon tree and the oils obtained from the bark and leaves have been used in Ayurveda and indigenous systems of medicine in the treatment and prevention of ailments and diseases. The uses in Ayurveda and other traditional systems of medicine are recorded in ancient texts and present-day pharmacopoeia. Cinnamon has been used in Ayurveda to treat nausea, digestive disorders, painful menstrual problems, anaemia and respiratory conditions. It has also been used to improve blood circulation, control blood sugar levels and relieve headaches, flu and colds, particularly for people classed under the kapha Ayurvedic type. Its essential oil has been used in soothing nerves and muscle pains (Jakhetia et al. 2010). The bark powder has been used in tooth powders, and the oil has been used for medicinal purposes, such as toothaches and dental problems (Ayurveda Pharmacopoeia of Sri Lanka,1976.1979,1985).

The uses of cinnamon besides Ayurveda and other medical systems and culinary use are gaining popularity. These products that could be developed as industries are formulations for aromatherapy, nutraceuticals and cosmeceuticals. Development of these products can be a prospect for economic growth for developing countries like Sri Lanka endowed with a rich biodiversity supported by indigenous medical knowledge of the health effects of many indigenous plant species.

Some claimed health benefits of cinnamon indicated in indigenous medical systems are supported by scientific studies, thereby confirming that there have some medicinal benefits. These include the use of cinnamon to treat pain and dental problems, promote overall health and well-being and facilitate the healing process (Gruenwald et al. 2010; Rao and Gan 2014).

Recently published research and clinical studies have shown that cinnamon has other medicinal properties, such as antioxidant, anti-arthritis, antibacterial, antifungal and anti-inflammatory activities (Kumar et al. 2019). Its oil has been used in medicine as an astringent, a carminative and an antiseptic. Cinnamon oil reduces drowsiness, irritability and headaches (Singh et al. 2007). Cinnamon also gives relief from indigestion, nausea, vomiting and morning sickness, respiratory problems and menstrual discomfort.

There are reports of preliminary studies that cinnamon may prevent or delay diabetes and colon cancer. The US National Library of Medicine reports that cinnamon is effective in the treatment of diarrhoea, emesis, muscle cramps, infections and flu. Cinnamon has shown some effectiveness against Alzheimer disease (Peterson et al. 2009) and herpes infection (Benencia and Courreges 2000).

Cinnamon also is reported to lower the risk of developing cardiovascular disease, which is linked to metabolic syndrome, indicating the presence of usually three or more medical conditions, such as high blood pressure, obesity, high triglyceride levels, low HDL levels and high fasting levels of blood sugar (Mollazadeh and Hosseinzadeh 2016; Mottillo et al. 2010).

13.2 True Cinnamon

This chapter will be about products that can be obtained from true cinnamon (*C. zeylanicum*) and not from *Cinnamomum cassia*. There are chemical, physical and organoleptic differences between these two species, the major difference between true cinnamon and cassia being that cassia has high levels of a toxic compound called coumarin (1,2-benzopyrone). Coumarins have been reported to have anticoagulant, carcinogenic and hepatotoxic properties (Abraham et al. 2010). The European Food Safety Authority (EFSA) (EFSA 2008) recommends the tolerable daily intake (TDI) for coumarin to be 0.1 mg/kg body weight/day. The German Federal Institute for Risk Assessment (BfR) reported (2006) that 1 kg of cassia (CC) powder contained approximately 2.1–4.4 g of coumarin, which means 1 teaspoon of CC powder would have around 5.8–12.1 mg of coumarin. The BfR report specifically states that true cinnamon contains "hardly any" coumarin. The amount in cassia is above the tolerable daily intake (TDI) for coumarin recommended by the European Food Safety Authority (EFSA) (Abraham et al. 2010). Therefore, the EFSA discourages the regular, long-term use of *C. cassia* as a supplement due to its coumarin content (EFSA 2008).

13.3 Cinnamon Oil

The starting materials for use in industries would be cinnamon powder obtained from the inner bark and oils from the bark and the leaves obtained by steam distillation. The bark essential oil is brownish and the leaf oil is pale yellow in colour, while both are light and highly aromatic oils.

The chemical composition of the two oils is different. The bark oil contains a high content of cinnamaldehyde and a small percentage of eugenol, while the leaf oil has a small content of cinnamaldehyde and a high percentage of eugenol. Cinnamon oil is reported to possess many healing properties. The oils have demonstrated health benefits such as reducing blood sugar and blood cholesterol levels, fighting inflammation and infections and enhancing immunity. Cinnamon oil has been shown to reduce the symptoms of colds and flu, such as a runny nose, sneezing and sore throat, and provide relief from headaches and rheumatoid arthritis pains by its ability to enhance blood circulation (Stiles 2016). In addition, the oils have also demonstrated efficacy as antibacterial, astringent, antioxidant, anti-inflammatory, digestive and anthelminthic agents. These properties can be effectively developed into products as nutraceuticals and cosmeceuticals and aromatherapy formulations. They also can be used as blends with other essential oils, such as clove, orange, rosemary, vanilla and peppermint. They are often used as topical applications or diffused at home for aromatherapy. External applications of cinnamon oil include use as a shampoo for the hair, in chewing gums, as mouthwashes and as a toothpaste flavouring to freshen breath, as a cleansing agent to prevent infections and as a massage oil for relieving aches and pains. Cinnamon oil is also used in formulating perfumes.

13.4 Industrial Applications of Cinnamon

The uses of cinnamon besides Ayurveda and other medical systems and culinary use are gaining popularity. These products that could be developed as industries are formulations for aromatherapy, nutraceuticals and cosmeceuticals. Development of these products can be a prospect for economic growth for developing countries like Sri Lanka endowed with a rich biodiversity supported by indigenous medical knowledge of the health effects of many indigenous plant species. In order to benefit, the country has to invest on goal-oriented research and development activities to differentiate from competitors using the latest areas of research, such as gene expression, multiple receptor activities and antioxidant defences, that can be used to improve health by supporting functions of your body. Nutraceuticals have shown effects on immune status and susceptibility to cold and cold-related symptoms and other disease conditions. They also exhibit disease-modifying indications related to oxidative stress, including allergy, Alzheimer's disease, cardiovascular diseases, cell proliferation, obesity and hypoglycaemia (Vangalapati et al. 2012).

13.4.1 Aromatherapy

Aromatherapy is the medicinal use of aromatic substances to improve health and well-being. Currently, aromatherapy uses fragrant parts or oils of aromatic plants to get beneficial effects. This can be achieved even by taking a sniff of a spice like cinnamon or a fragrance of flowers. Other benefits achieved by the inhalation of specific fragrances include relief of depression, stress and anxiety, thereby helping you to relax and be calm. Some companies use selected fragrances to change the mood of workers by using ventilation systems to change the ambiance of the environment, and hospitals are testing the use of aromatherapy to help patients relax, so that other healing modalities can be carried out (Price and Price 2019; Ann 2016)

There is a large amount of published information on aromatherapy detailing the basic theory, use and different recipes of essential oils to be used in common conditions that can be treated ranging from mood changes, skin conditions to relief from certain ailments (Colletti and Cicero 2018). Aromatherapy is adaptable to be used in different ways to treat a wide range of physical and emotional difficulties (Keville and Green 2008; Purchon 2014).

Massaging aromatic oils into the skin is the most common and the safest way to use essential oils. Hence, aromatherapy remedies include massage oils, including cinnamon oil, as the aromatic compounds present in the oils have therapeutic benefits in addition to their fragrance properties. Aromatherapy properties of cinnamon are well known. Cinnamon has a sweet, spicy-hot fragrance and is used as a physical and emotional stimulant in aromatherapy. Research publications report that cinnamon lessens stress, drowsiness, irritability and headaches. In one study, the aroma of cinnamon in the room helped participants to concentrate and perform better. Cinnamon essential oil, as a component of recipes, is used in aromatherapy to diminish depression and fatigue and provide relaxation, to reduce joint and muscle pain and to strengthen immunity to relieve the symptoms of colds and the flu, lessen the pain associated with headaches and soothe nervous tension (Price and Price 2019).

When diffused throughout the home indoor environment, its aroma freshens and deodorizes while emitting its characteristic warm and relaxing fragrance that is known to have soothing, calming and tonic effects. External applications of cinnamon oil include use as a shampoo for the hair, in chewing gums, as mouthwashes and as a toothpaste flavouring to freshen breath, as a cleansing agent to prevent infections, and as a massage oil for relieving aches and pains.

Cinnamon oil helps to relax tightness of muscles, relieve painful joints and menstrual cramps. In addition, it increases blood circulation when used as a massage oil. The aroma of cinnamon increases the appetite and liven up the senses. Acne, burns, depression, dermatitis, fatigue, nausea, nerve and muscle pains and insect bites are some of the conditions best treated by aromatherapy using cinnamon. In addition, cinnamon possesses antibacterial, astringent, antioxidant, anthelminthic and antifungal properties, which can contribute to aromatherapy.

Cinnamon oil can be used as blends with other essential oils, such as clove, orange, rosemary, vanilla and peppermint. They are often used as topical applications or diffused at home for aromatherapy. Aromatherapy applications of cinnamon oil have shown to lessen depression, fatigue and relax the body. It also enhances circulation, which helps to reduce the pain associated with headaches. When diffused throughout the indoor environments, its aroma freshens and deodorizes, calms and gives a soothing effect. Its ability to reduce nervous tension helps to reduce the risk of memory loss.

Used cosmetically or topically in general, cinnamon essential oil is used to soothe dry skin and to relieve aches and pains in the muscles and joints. Its antibacterial properties help in alleviating acne, rashes and minor infections. Its anti-ageing properties are due the antioxidant compounds.

13.4.2 Nutraceuticals

Nutraceuticals are health or dietary supplements having a promoting or preventing effect on the health of human beings. They differ from drugs in that they cannot have claims on the treatment or prevention of diseases. Nutraceuticals are differentiated from foods in that they are defined as any food which provides health benefits, such as health promotion and prevention of symptoms of disease.

The term "nutraceutical" was first introduced in 1989, by Stephen L. DeFelice, Founder Chairman of the Foundation of Innovation Medicine. It indicates a combination of the words "nutrition" and "pharmaceutical" for products that range from vitamins, minerals, health and dietary supplements, herbal products to dietary enzymes. These, in addition to supplementing diets, help in preventing or relieving some health conditions, including healing of ailments.

Nutraceuticals are not rigorously tested or regulated as much as pharmaceutical drugs. The stringent requirement for a nutraceutical is safety, followed by labelling on claims and recommended use. In addition, good manufacturing practice (GMP) regulations have to be followed during production.

The results of a number of clinical studies done so far on some nutraceuticals support their effectiveness as well as their general safety. Nutraceuticals are permitted to make "structure/function claims" on how they affect the body. As an example, one can claim that, "calcium builds strong bones". But they cannot claim to treat or cure specific diseases such as osteoporosis.

Nutraceuticals are grouped in categories such as dietary supplements, functional foods and medicinal foods. A dietary supplement is a product that contains nutrients derived from food sources and are in dosage forms, such as capsule, powder or pill form. Functional foods are whole foods enriched or fortified, which provide more benefits than traditional nutrients and medicinal foods, which are taken under medical supervision intended for specific management of a disease.

Nutraceuticals have attracted a lot of interest due to their potential nutritional and restorative effects and safety. Furthermore, these could have some added effect in other biological actions, such as antioxidant defences, immune status and susceptibility to certain disease states, such as diabetes, obesity and gastrointestinal disorders (Gupta 2016; Cicero and Colletti 2018). The nutraceutical industry is a dynamic, innovative and an expanding business that opens novel opportunities to scientific research and discovery to develop products that meets the growing consumer interest in health-promotion. By tracing the consumer trends, industry could supply products to satisfy the consumer demands. The nutraceutical industry covers three main segments: functional foods, dietary supplements and herbal healthcare products.

The benefits of nutraceuticals may help us live longer, improve the healthiness from our diet, prevent particular medical conditions and may be perceived to be more "natural" than traditional medicine and less likely to produce unpleasant side effects. The growing dissatisfaction among the patients about the synthetic therapeutic agents as having toxic and side effects resulted in the demand for natural healthcare products.

Classification of dosage forms frequently encountered as nutraceuticals/health supplements can be solid, semisolid or liquid, either as unit or bulk formulations. Unit solid formulations include tablets, capsules, sachets, pills and lozenges and bulk dosage forms as powders. Semisolid will include creams, ointments, pastes, gels and balms. Liquid dosages can be solutions or multiphasic, such as emulsions and suspensions.

Among the solid dosage forms, the most convenient to use are the tablets, as they can be made to different shapes, thicknesses and sizes to be very reproducible and to be uniform. Tablets are also easy to protect from environmental conditions, as they can be coated, strip or blister packed and dispensed without direct handling.

13.4.3 Regulatory Requirements

Consumers largely determine the usefulness and benefits offered by nutraceuticals. Nutraceuticals are regulated as a category of food in many countries. In some countries, there is an explicit set of regulations for supplements (e.g. the USA, European Union). The Federal Food, Drug, and Cosmetic Act was amended by the Nutrition Labeling and Education Act of 1990 to include most foods, including dietary supplements to include nutrition labelling.

The Food and Drug Administration (FDA) of the USA regulates nutraceuticals under the Dietary Supplement Health and Education Act of 1994 (DSHEA), which has a different set of regulations than those covering the usual foods and drug products. This prohibits the marketing of products that are adulterated or misbranded without evaluating the safety and labelling of their products before marketing to ensure that they meet all the requirements of the DSHEA and FDA regulations. These products are only allowed to make what are called "structure/function claims" on how a nutraceutical affects the body such as "calcium builds strong bones" and not mention about treatment or cure specific diseases. Any health claims used on a label will have to conform to the 1990 Nutrition Labeling and Education Act (NLEA) in the USA. The DSHEA has granted the FDA authority to establish regulations regarding dietary supplement manufacturing, regulating health claims and labelling of dietary supplements.

In the European Union, food legislation is largely under the umbrella of the European Food and Safety Authority (EFSA). This legislation focuses on "food supplements", which are defined as concentrated sources of nutrients (e.g. proteins, minerals and vitamins) and other substances with a beneficial nutritional effect. The main EU legislation related to food supplements is Directive 2002/46/EC. In Canada and Australia, nutraceuticals are regulated more closely as a drug than food category.

In Latin America, the market entry requirements for nutraceuticals vary, with registration-based approaches employed in Colombia, Brazil, and Argentina and notification-based approaches in Mexico and Chile. In countries such as Brazil, China and Taiwan, the regulators require animal and/or human clinical studies as a requirement for the product registration. Whichever the market, the quality and safety of the product has to be assured and good manufacturing practices have to be followed.

13.5 Prevention of Diseases

Nutraceuticals have attracted immense interest due to their potential nutritional and therapeutic effects and their safety. They can be used to improve health, prevent chronic diseases and postpone the ageing process and for improving your overall health supporting functions. Nutraceutical supplements are used for the prevention of some diseases, such as diabetes, renal and gastrointestinal disorders as well as some infections. These also play useful roles in immune status and susceptibility to certain disease conditions. They also help disease-modifying indications related to oxidative stress including allergy, Alzheimer's disease, cardiovascular diseases, Parkinson's disease and obesity. Nutraceuticals also help in improving overall health supporting functions.

13.6 Cosmeceuticals

Cosmeceutical is a novel term, which describes cosmetic preparation that has pharmaceutical properties. Raymond Reed, Founder of US Society of Cosmetic Chemists, created the concept of "cosmeceutical", which was then popularized by an American dermatologist, Albert Kligman, in the late 1970s. Cosmeceuticals are external applications of combinations of products with both cosmetic effects and some health benefit mainly intended to enhance the beauty through ingredients that provide additional health-related function or benefit. Cosmeceuticals are the fastest-growing segment in the skin care market. It has no legal status as a separate category of cosmetics under the Food, Drugs and Cosmetics (FD&C) Act of the USA, which recognizes only three categories: drugs, cosmetics and soaps. These are not approved by the FDA for sale, nor are they intended to affect structure or function of the body.

Any cosmetic that contains active ingredients that are known to be beneficial to humans in some way can be called a cosmeceutical. An example is a cream that contains vitamin C, which is an antioxidant. Hence, careful labelling of cosmeceuticals is necessary to avoid claims about drug properties in order to bypass regulation. Cosmetics are intended to beautify, promote attractiveness, alter appearance or cleanse; they are not approved by the FDA for sale, nor are they intended to effect structure or function of the body.

Cosmeceuticals are products, usually creams and lotions, which are marketed as cosmetics but contain some kind of biologically active ingredient intended to benefit the user in some way. Classic examples of cosmeceuticals are anti-wrinkle creams and hair restorers, both of which purport to improve the appearance of the user while having an active effect on their body – a kind of marriage of cosmetic and medicinal treatment.

Cosmeceutical industry is expanding with advances in technology, specially to innovate products for the enhancement of beauty particularly aimed at keeping consumers looking young and healthy. This is also backed by the rising consumer demand toward herbal and organic products. Ayurveda has been using a large number of medicinal plants to enhance beauty and health. The development of new cosmeceuticals has accelerated due to the use of 3D bio printing technology, which can be tested on live human tissue at a much less cost than clinical trials.

Cosmeceutical industry is the quickest-growing segment of the natural personal care industry. The main interest has been for skin care, especially anti-ageing natural products. All cosmeceuticals claim to contain functional ingredients with either therapeutic, disease-fighting or healing properties. Cosmeceuticals are typically cosmetic-pharmaceutical mixes used to enhance the health and beauty of skin. Some cosmeceuticals are naturally derived, while others are synthetic, but all contain functional ingredients with either therapeutic, disease-fighting or healing properties. The growth of the cosmeceutical industry is very substantial due to the introduction of new categories that seek to meet the growing demands and needs of customers, such as products that delay visible signs of ageing, keeping consumers looking young and healthy. This trend is increasing customer preference toward green and organic cosmeceuticals.

Cosmeceuticals can be classified according to their intended use as facial, hair, spa treatments, sun care, body care, wellness, eye care and anti-ageing. Antibacterial, anti-inflammatory and antioxidant properties contribute toward cinnamon being effective against acne. Powdered cinnamon when rubbed on the skin exfoliates by removing the dead cells, dirt, germs and the sebum embedded in skin pores. Cinnamon promotes cell regeneration and blood circulation, thereby hastening the healing process of blemishes and scars. Similarly, cinnamon is effective in reducing cellulite particularly in women. Cinnamon has anti-ageing properties due to its richness in vitamins, minerals and antioxidants, which keep the skin cells well-nourished and healthy. Cracked heels can also be treated with cinnamon, which soften the cracked heels by sealing moisture and smoothening the rough areas. Lips can be plumped up and moisturized with cinnamon. Hair and scalp can be cleansed with cinnamon, which also helps in hair growth due to its nutritional value. Cinnamon also has an aroma that can refresh mouth and freshen breath by masking bad odour. Whenever the essential oil is used during topical applications, it needs to be mixed with a carrier vegetable oil or honey, as it alone can cause severe skin irritation (Sivamani et al. 2015).

13.7 Organic Cinnamon

Presently, a new trend is to label raw materials as organic to indicate that they have been grown without the use of harmful pesticides or prohibited synthetic fertilizers. Hence, there can be organic cinnamon collected from plants that have not been grown using seeds or cuttings genetically modified or subjected to ionizing radiation, using good cultivation practices that have the least impact on the environment and harvested using fair trade practices. These cultivations have to be protected from environmental pollution from other sources. Hence, a short procurement supply chain is preferable having direct contacts, so as to obtain the highest quality cinnamon. Nutrients in the soil of organic crops are maintained by farming practices, such as crop rotation and tilling. Composted manure is an approved fertilizer for organic crops. There are licensing agencies, which certify products as organic after following up cultivation practices from germination to harvesting. Besides the optimum growing conditions for cinnamon trees, the region is protected from pollution. Thus, it is necessary to control the whole cinnamon procurement supply chain from growing to the delivery of raw materials.

13.8 Manufacturing Process

If one uses cinnamon bark or powdered cinnamon in the development of nutraceuticals and cosmeceuticals, these have to be of good quality as the finished product always depend on the raw material. The quality process has to be maintained from the starting material to the finished product. The product has to be standardized to get uniformity of usage. Hence, specifications have to be drawn up for the raw material, intermediate products and the finished product. All these aspects need the usage of modern analytical instruments and up-to-date processing equipment. Being plant material, there is a greater tendency for them to undergo decomposition due to exposure to moisture, heat, light, air and handling. The choice of dosage form and packaging are important concerns that have to be considered. Thereafter, the product has to be subjected to stability testing as per international standards. Then, an expiration date has to be determined to be put on the label. The dosage has to be determined depending on the safety of the product and the period of usage. Interactions with food or other drugs have to be considered in order to establish safety of the product, although it has been used safely for centuries.

13.9 Research Areas

Some of the very promising activities for the development of drugs, nutraceuticals and cosmeceuticals are briefly described, so that scientific studies can be initiated to innovate products that could be of economic importance (Kawatra and Rajagopalan 2015).

13.9.1 Antioxidant Effects

Many of the health benefits of cinnamon are due to its strong antioxidant activity (Jayaprakasha et al. 2003). Antioxidants are capable of protecting living cells against the damage associated with free radicals, which are produced due to pollution, poor diet and stress. Antioxidants found in cinnamon can block the free radical damage and prevent diseases caused by them, such as ageing process, cardiovascular disease and certain forms of cancer (Lee et al. 2002; Dugoua et al. 2007). To date, 41 different protective compounds have been identified in cinnamon (Rao and Gan 2014). Cinnamon polyphenols possess free radicals scavenging activity, and detoxification processes and inhibition of certain enzymes. In a study of 26 spices that compared the antioxidant activity, cinnamon was proved to be higher in antioxidants than many other herbs and spices (Shan et al. 2005).

13.9.2 Anti-inflammatory Effects

Chronic inflammation can be responsible for the onset of numerous age-related medical conditions, such as arthritis and cardiovascular and neurodegenerative diseases (González-Gallego et al. 2007). Flavonoids have been shown to be effective in the treatment of inflammatory diseases (Garcia-Lafuente et al. 2009). Cinnamon contains flavonoids, which can be responsible for its anti-inflammatory activities (da Silveira et al. 2014). Another study on mice has shown that an ethanol extract of cinnamon was effective in the reduction of acute inflammation (Pan et al. 2010). A study on true cinnamon has shown that it has the most active anti-inflammatory effect, and the compounds responsible for this activity are E-cinnamaldehyde and o-methoxycinnamaldehyde (Gunawardena et al. 2015; Schink et al. 2018).

13.9.3 Antidiabetic Effects

Cinnamon is used in naturopathy to treat type 2 diabetes, but there is no scientific evidence using clinical trials. The American Diabetes Association published, in a 2003 edition of "Diabetes Care" publication, the findings of a study, which concluded that daily doses of 1–6 g of cinnamon can help to reduce blood glucose levels. Research using animals have shown that polyphenols present in cinnamon possess insulin-like activity (Anderson et al. 2004; Qin et al. 2003). Clinical studies have shown that cinnamon use leads to a reduction of sugar and lipid levels of people with type 2 diabetes (Khan et al., 2003). Studies have found a significant reduction in fasting glucose levels and improvements in haemoglobin A1c (HbA1c) in patients with type 2 diabetes after taking cinnamon products (Crawford et al. 2009). Published data on human clinical trials show some activities but need more trials to confirm to a statistically significant level (Allen et al. 2013; Davis et al. 2011; Blevins et al. 2007).

13.9.4 Antibacterial Activity

Cinnamon has shown to be effective in fighting pathogenic bacteria (Nabavi et al. 2015). A review of controlled clinical trials also refers to the antibacterial activity of cinnamon (Martin et al. 2003). In vitro experiments on *Helicobacter pylori*-infected gastric epithelial cells showed that cinnamon extract had high inhibitory activity (Zaidi et al. 2012). Antimicrobial activity has been displayed by cinnamaldehyde (He et al. 2019).

13.9.5 Hypolipidemic Effects

Hyperlipidaemia is a risk factor for atherosclerosis and cardiovascular diseases. Hypolipidemic effect of cinnamaldehyde in cinnamon is due to its ability to decrease fat absorption. Cinnamaldehyde administration to streptozotocin-induced diabetic male Wistar rats lowered the serum total cholesterol and triglyceride levels (Subash et al. 2007). Meta-analyses of controlled trials on the effect of cinnamon on the total cholesterol and triglycerides level conclude that there is a significant lowering effect (Maierean et al. 2017). Cholesterol- and lipid-lowering effects of cinnamon were shown in many studies. Khan et al. study on humans showed that cinnamon caused a reduction in triglyceride, total cholesterol and LDL-c cholesterol levels (Khan et al. 2003); hypolipidemic effects of cinnamon has been shown to be due to antioxidant enzyme activity in the liver (Javed et al. 2012; Lee et al. 2003).

13.9.6 Antifungal Effects

Candida infection is one of the causes of morbidity, as it has become resistant to antifungal drugs. There are promising signs of using cinnamon oil as an alternative. An in vitro study on candida isolated from the blood of patients was inactivated by a mix of cinnamon and olive oils (Goel et al. 2016). Cinnamon oil was found to be very effective in inhibiting biofilms of three strains of Candida (Farisa Banu et al. 2018).

Furthermore, cinnamaldehyde has demonstrated activity against *Aspergillus flavus* (Wang et al. 2019), and cinnamon extracts have halted mycelial growth and aflatoxin synthesis in *Aspergillus parasiticus* (Tantaoui-Elaraki et al. 1994).

13.9.7 Possible Cancer Protection Studies

A study that tested the antitumour activity of cinnamon extract on various tumour cells using in vitro methods and in living mice showed that the cinnamon extract suppressed cell tumour growth and caused tumour cell death (Kwon et al. 2019).

13.9.8 Hypotensive Effects

Three placebo-controlled, randomized clinical trials evaluated the use of cinnamon on blood pressure in prediabetic or type 2 diabetic patients and found a significant benefit towards the baseline pressure with cinnamon administration (Akilen et al. 2013; Wansi et al. 2007).

13.9.9 Analgesic Effects

Some of the antioxidants in cinnamon have anti-inflammatory effects, which can lower swelling and inflammation leading to reduction of pain. Studies have shown that cinnamon helps to relieve pain associated with muscle soreness, allergic reactions, menstruation and other age-related arthritis (Jaafarpour et al. 2015).

13.9.10 Cardiovascular Effects

Studies have shown cinnamon has the ability to improve heart health by lowering risk factors, such as high cholesterol and triglyceride levels and high blood pressure (Akilen et al. 2010; Rao and Gan 2014). Cinnamon also increases circulation and regenerates heart tissue to help fight heart attacks, heart disease and stroke (Rao and Gan 2014). Cinnamon showed cardiovascular protective activity due to its antioxidant, antidiabetic and anti-inflammatory activities and control of lipid levels being able to reduce metabolic syndrome problems (Ziegenfuss et al. 2006; Couturier et al. 2010; Shen et al. 2012). Another study on the cardiovascular activity of rats showed that the rats receiving cinnamon had better cardiac activity than the placebo group (Badalzadeh et al. 2014). The cardiovascular protective effects of cinnamon were shown in many studies (Ranasinghe et al. 2013).

13.9.11 Products for Neurodegenerative Diseases

Cinnamon has shown improvements for Alzheimer's and Parkinson's diseases in animal studies. Neurodegenerative diseases are characterized by progressive loss of the structure or function of brain cells. Two compounds isolated from cinnamon appear to inhibit the built-up of a protein called tau in the brain, which have been shown to cause Alzheimer's disease (Peterson et al. 2009; George et al. 2013; Khasnavis et al. 2014). Neuroprotective effects have been demonstrated in rats with glutamate-induced neuronal cell death (Shimada et al. 2000).

13.9.12 Anti-obesity Effects

One of the important causes of cardiovascular disease is obesity, which is due to an increase in oxidative stress (Wronkowitz et al. 2014) and insulin resistance. Cinnamaldehyde has been shown to increase insulin sensitivity, which results in reduced cumulative food intake and gastric emptying rates.

A study using the water extract of cinnamon showed reduction of insulin resistance, blood glucose and serum lipid levels in diabetes-induced mice (Sartorius et al. 2014), indicating that the anti-obesity effect of cinnamon depended on cardiovascular protection, insulin sensitivity and immune stimulating activity (Arslan et al. 2014).

13.9.13 Antipyretic and Analgesic Effects

Cinnamon had an antipyretic effect on mice models (Kurokawa et al. 1998). Analgesic effect has also been achieved on administration of cinnamaldehyde (Mohammadi et al. 2014; Ranasinghe et al. 2013; Ismael et al. 2017).

13.9.14 Oral Hygiene Uses

Cinnamon has beneficial effects on oral hygiene. It could protect tooth decay, caries, bad breath and mouth infections. The essential oils from cinnamon have been shown to be responsible to fight oral bacteria to bring about these beneficial effects (Gupta et al. 2011).

13.9.15 Wound Healing

Animal studies have demonstrated that an ethanol extract of cinnamon has wound healing effects attributable to its antioxidant activity (Kamath et al. 2003). Cinnamtannin B-1 is an A-type proanthocyanidin contained in *C. zeylanicum*, which is a potent antioxidant and protective agent against oxidative stress and apoptosis in human platelets. It has been shown to promote migration of mesenchymal stem cells and accelerates wound healing in mice (Fujita et al. 2015).

13.9.16 Food Preservation

Food spoilage occurs due to lipid oxidation changing the flavour and taste of food producing a bad odour. This is prevented by adding synthetic chemicals. Cinnamon extract has shown preservative action as a natural preservative (Tzortzakis 2009).

Because cinnamon oil has antibacterial and antioxidant activities, it can be used to prevent spoilage in many foods and vegetables without the need for chemicals (Shobana et al. 2000). A study undertaken to assess the efficacy of cinnamon oil against food spoilage bacteria and fungi has concluded that the oils have good protective action to preserve the food and vegetables (Bullerman et al. 1977). Cinnamon also possesses antityrosinase activities, which can be useful in stopping the discoloration of fruits and vegetables as they oxidize and begin to rot (Shan et al. 2005; Chang et al. 2013). Cinnamon oil has inhibited the growth of pathogenic microorganisms (Yossa et al. 2014; Zhang et al. 2014).

13.9.17 Skin

Cinnamon has played an important role in improving a healthy skin and even remedied some diseases. It smoothens the skin due to its ability to stimulate blood vessels. It acts as an exfoliating agent to stimulate the scalp and provide nourishment to hair follicles when massaged after dilution with a bland career oil. Cinnamon has also demonstrated its effect to soothe and soften the skin by removing the dead skin cells, bringing about shine of the skin. Other reported benefits of cinnamon include the maintenance and improvement of a healthy skin, including the texture, complexion, dryness and in dermatitis conditions.

Breakdown of collagen causes the skin to lose elasticity, thus increasing signs of ageing. Cinnamon increases collagen formation to prevent the skin from losing its elasticity and can be used to promote collagen biosynthesis; thus, creams with cinnamon products can act as anti-ageing creams (Takasao et al. 2012).

13.10 Global Market Trends

13.10.1 Cinnamon Market

The demand for cinnamon is rapidly increasing due to the wide coverage of benefits of cinnamon containing nutraceuticals and cosmeceuticals, which can provide relief for infections, allergies, skin ailments and other non-communicable diseases. A market research report on cinnamon, published by Grand View Research (GVR 2019), states that the global cinnamon market size in 2018 was USD 760.2 million and Ceylon cinnamon had a share of more than 35.0% in the global market in 2018 and will continue its lead through to 2025.

Growing consumer demand for natural healthy foods and ingredients is a key factor driving the market. In addition to the commonly known uses, the scope of cinnamon has expanded substantially due to extensive research in the field. The newly found beneficial effects against diseases, like diabetes, Alzheimer's and cardiovascular diseases, will further increase the demand for cinnamon products in the future. This segment is expected to exhibit the highest Compound Annual Growth Rate (CAGR) of 12% by 2023 (RMR 2018).

13.10.2 Global Nutraceuticals Market

Nutraceuticals market is an expanding business due to the advantages envisaged in using herbal products against synthetic chemicals with side effects to stay active and young-looking by taking anti-ageing nutraceuticals. Consumers are seeking safer less-toxic products. Consumers are now conscious about preventive aspects and look out for health-promoting products. More attention is now being made, which has caused the nutraceutical industries to undertake advanced research to develop new nutraceuticals as customer-made requirements. According to a market research engine report, the nutraceuticals market is expected to exceed US$ 359.9 Billion by 2024 at a CAGR of 7.8% annually between 2020 and 2025 (MRE Report 2019).

13.10.3 Global Market Analysis for Cosmeceuticals

In today's personal care industry, the market for cosmeceuticals has been reported as the fastest growing segment, especially in terms of innovative products, such as anti-ageing and skin care, including whitening and skin enrichment. The use of plant extracts has attributed to safer and low-cost products with much less reactions and side effects.

According to a GBI research's latest report on Cosmeceuticals (2018), cosmeceutical markets in the USA, the UK, France, Germany, Italy, Spain and Japan were estimated to reach $42.4 billion, following growth at a CAGR of 4.6%. Emerging markets, such as China, Latin America and India, are expected to add to these estimates in the future due to the higher demand for more efficient products and luxury brands. Cosmeceutical trade growth is further driven by digital marketing and the offer of personalized customer experiences and e-commerce. Specifically, antioxidants emerge as one of the most popular ingredients for skin care for their observed benefits to overall health. However, cosmeceuticals are not officially recognized by the FDA but should be cleared for their safety, as any product complaint about safety can end up in banning of the product. Although proof of the claimed therapeutic advantages is not demanded, the labelling of the products is well controlled so as not to make direct claims concerning results. Product efficacy is inspected by the US Federal Trade Commission (FTC), and cosmeceutical products that can be picked up with likely unrealistic expectations may be removed. Cosmeceutical companies must also follow the guidelines set by the FTC for product marketing, which control over stated or false product claims. Cosmeceuticals must also follow the specific regulations for GMP to ensure that products are of good quality.

According to Euromonitor International, the total global retail sales of natural products-based cosmeceuticals was US$2.98 billion in 2015 and has shown a CAGR of 4% during the past 5 years.

13.10.4 Global Aromatherapy Products Market

The demand for aromatherapy products is expanding as a result of the increase in medicinal applications of essential oils that are being proved to possess therapeutic effects for the treatment of numerous medical conditions and infections. The effects of aroma on relief of stress, insomnia and bringing about relaxation are factors contributing to this demand. People with acne problems often opt for therapeutic grade essential oils. Burn scars can also be cured with aromatherapy products. In addition, pain caused by burn may be treated with this therapy.

TMR (2020) predicts a marked increase in the demand for aromatherapy formulations. The global aromatherapy market size was USD 1.3 billion in 2018 and is projected to achieve a CAGR of 10.4% by 2026.

13.11 Conclusions

New products have to be launched into the market after considering pre-market resources and consumer affordability, making sure to guarantee product safety and quality in the marketplace. As described before, cinnamon is reported to possess many health-promoting and healing properties. Extensive research has to be conducted to develop nutraceutical and cosmeceutical products for these effects for the benefit of consumers. With so many benefits, the popularity of cinnamon herbal supplements and cosmetics will continue to flourish. Taking cinnamon as a daily supplement, in the right dose, will provide a significant impact on human health that will help a person look, feel better and live longer.

At present, consumers are mainly responsible for determining the usefulness and value offered by nutraceuticals. It is still necessary to bring in more regulation related to quality and safety of these products, so that the industry will benefit from risks of regulatory backlash. The growth of the industry will depend on consumer demand for health-promoting products from natural resources as synthetic products have created reasonable doubts about safety. Thus, nutraceutical and cosmeceutical industries represent vibrant and growing industries that offer novel opportunities to merge scientific discovery with growing consumer interest in health- augmenting products.

The major health benefits of cinnamon have been studied. Much more investigations are necessary to provide additional clinical evidence to develop protective products against cancer, cardiovascular and neurological disorders and ageing. Further research needs to be conducted to determine the effectiveness of the active compounds isolated from cinnamon and their therapeutic effects in the prevention and treatment of diseases.

References

Abraham K, Wöhrlin F, Lindtner O, Heinemeyer G, Lampen A (2010) Toxicology and risk assessment of coumarin: focus on human data. Mol Nutr Food Res 54:228–239

Akilen R, Tsiami A, Devendra D, Robinson N (2010) Glycated haemoglobin and blood pressure-lowering effect of cinnamon in multi-ethnic Type 2 diabetic patients in the UK: a randomized, placebo-controlled, double-blind clinical trial. Diabet Med 27(10):1159–1167

Akilen R, Pimlott Z, Tsiami A, Robinson N (2013) Effect of short-term administration of cinnamon on blood pressure in patients with prediabetes and type 2 diabetes. Nutrition 29(10):1192–1196. 23867208

Allen RW, Schwartzman E, Baker WL, Coleman CI, Phung OJ (2013) Cinnamon use in type 2 diabetes: an updated systematic review and meta-analysis. Ann Fam Med 11(5):452–459. doi:10.1370/afm.15173

Anderson RA, Broadhurst CL, Polansky MM, Schmidt WF, Khan A, Flanagan VP, Schoene NW, Graves DJ (2004) Isolation and characterization of polyphenol type-A polymers from cinnamon with insulin-like biological activity. J Agric Food Chem 52(1):65–70

Ann WV (2016) The complete book of essential oils and aromatherapy, revised and expanded: over 800 natural, nontoxic, and fragrant recipes to create health, beauty, and safe home and work environments. New World Library, Barnes & Noble Press. ISBN 178-1-57731-139-3

Arslan N, Sayin O, Tokgoz Y (2014) Evaluation of serum xenin and ghrelin levels and their relationship with nonalcoholic fatty liver disease and insulin resistance in obese adolescents. J Endocrinol Investig 37:1091–1097

Ayurvedic Pharmacopoeia (1976) Volume 1, Part1(1979) Vol 1 Part II (1985) volume 1 part III Department of Ayurveda, Sri Lanka

Badalzadeh R, Shaghaghi M, Mohammadi M, Dehghan G, Mohammadi Z (2014) The effect of cinnamon extract and long-term aerobic training on heart function, biochemical alterations and lipid profile following exhaustive exercise in male rats. Adv Pharm Bull 14(4):515–520

Benencia F, Courreges MC (2000) In vitro and in vivo activity of eugenol on human herpesvirus. Phytother Res 14(7):495–500

Blevins SM, Leyva MJ, Brown J, Wright J, Scofield RH, Aston CE (2007) Effect of cinnamon on glucose and lipid levels in non insulin-dependent type 2 diabetes. Diabetes Care 30(9):2236–2237.17563345

Bullerman LB, Lieu FU, Seier SA (1977) Inhibition of growth and aflatoxin production by cinnamon and clove oils, cinnamic aldehyde and eugenol. J Food Sci 42(4):1107–1109

Chang CT, Chang WL, Hsu JC, Shih Y, Chou ST (2013) Chemical composition and tyrosinase inhibitory activity of *Cinnamomum cassia* essential oil. Bot Stud 54(1):10. https://doi.org/10.1186/1999-3110-54-10

Cicero AFG, Colletti A (2018) Handbook of Nutraceuticles for Clinical Use. Springer. ISBN-10: 3319736418

Couturier K, Batandier C, Awada M, Hininger-Favier I, Canini F, Anderson RA, Leverve X, Roussel AM (2010) Cinnamon improves insulin sensitivity and alters the body composition in an animal model of the metabolic syndrome. Arch Biochem Biophys 501:158–161

Crawford P (2009) Effectiveness of cinnamon for lowering hemoglobin A_{1c} in patients with type 2 diabetes: a randomized, controlled trial. J Am Board Fam Med 22(5):507–512.19734396

da Silveira RDCS, Andrade LN, de Oliveira RDRB, De Sousa DP (2014) A review on anti-inflammatory activity of phenylpropanoids found in essential oils. Molecules 19(2):1459–1480. https://doi.org/10.3390/molecules19021459

Davis PA, Yokoyama W (2011) Cinnamon intake lowers fasting blood glucose: meta-analysis. J Med Food 14(9):884–889

Dugoua JJ, Seely D, Perri D, Cooley K, Forelli T, Mills E, Koren G (2007) From type 2 diabetes to antioxidant activity: a systematic review of the safety and efficacy of common and cassia cinnamon bark. Can J Physiol Pharmacol 85(9):837–847

European Food Safety Association (2008) Coumarin in flavourings and other food ingredients with flavouring properties. EFSA J 793:1–15

Farisa Banu S, Rubini D, Shanmugavelan P, Murugan R, Gowrishankar S, Karutha Pandian S, Nithyanand P (2018) Effects of patchouli and cinnamon essential oils on biofilm and hyphae formation by *Candida* species. J Mycol Med 20:S1156-5233(17)30378-5

Fujita K, Kuge K, Ozawa N, Sahara S, Zaiki K, Nakaoji K, Hamada K, Takenaka Y (2015) Cinnamtannin B-1 promotes migration of mesenchymal stem cells and accelerates wound healing in mice. PLoS One 10(12):e0144166. https://doi.org/10.1371/journal.pone.0144166

García-Lafuente A, Guillamón E, Villares A, Rostagno MA, Martínez JA (2009) Flavonoids as anti-inflammatory agents: implications in cancer and cardiovascular disease. Inflamm Res 58(9):537–552

George RC, Lew J, Graves DJ (2013) Interaction of cinnamaldehyde and epicatechin with tau: implications of beneficial effects in modulating Alzheimer's disease pathogenesis. J Alzheimers Dis 36(1):21–40. https://doi.org/10.3233/JAD-122113

Goel N, Rohilla H, Singh G, Punia P (2016) Antifungal activity of cinnamon oil and olive oil against Candida Spp. isolated from blood stream infections. J Clin Diagn Res 10(8):DC09–DC11. https://doi.org/10.7860/JCDR/2016/19958.8339

González-Gallego J, Sánchez-Campos S, Tuñón M (2007) Anti-inflammatory properties of dietary flavonoids. J Nutr Hosp 22(3):287–293

Gruenwald J, Freder J, Armbruester N (2010) Cinnamon and health. Crit Rev Food Sci Nutr 50:822–834

Gunawardena D, Karunaweera N, Lee S, van Der Kooy F, Harman DG, Raju R, Bennett L, Gyengesi E, Sucher NJ, Münch G (2015) Anti-inflammatory activity of cinnamon (*C. zeylanicum*) and *C. cassia* extracts - identification of E-cinnamaldehyde and o-methoxy cinnamaldehyde as the most potent bioactive compounds. Food Funct 6(3):910–919

Gupta RC (2016) Nutraceuticals efficacy, safety and toxicity. Academic Press. ISBN: 9780128021477

Gupta C, Kumari A, Pankaj Garg A, Catanzaro R, Marotta F (2011) Comparative study of cinnamon oil and clove oil on some oral microbiota. Acta Biomed 82(3):197–199

GVR (2019) Grand view research; cinnamon market size & share, global industry report, 2019-2025. ID GVR-3-68038-619-6

He Z, Huang Z, Jiang W, Zhou W (2019) Antimicrobial activity of cinnamaldehyde on *Streptococcus mutans* biofilms. Front Microbiol 10:2241. https://doi.org/10.3389/fmicb.2019.02241

Ismael AH, Muhsin HR, Sowdani KM (2017) The analgesic effect of Local Passiflora, Cinnamon and Chamomile plant ethanol extract in Swiss albino mice. Kerbala J Pharm Sci 12:109–119

Jaafarpour M, Hatefi M, Khani A, Khajavikhan J (2015) Comparative effect of cinnamon and Ibuprofen for treatment of primary dysmenorrhea: a randomized double-blind clinical trial. J Clin Diagn Res 9(4):QC04–QQC7. https://doi.org/10.7860/JCDR/2015/12084.5783

Jakhetia V, Patel R, Khatri P, Pahuja N, Garg S, Pandey A, Sharma S (2010) Cinnamon: a pharmacological review. J Adv Sci Res 1(2):19–23

Javed I, Imran F, Muhammad K, Faqir M, Bilal A, Mahmood A, Andleeb S (2012) Lipid lowering effect of *Cinnamomum zeylanicum* in hyperlipidaemic albino rabbits. Pak J Pharm Sci 25:141–147

Jayaprakasha GK, Rao LJM, Sakariah KK (2003) Volatile constituents from *Cinnamomum zeylanicum* fruit stalks and their antioxidant activities. J Agric Food Chem 51(15):4344–4348.12848508

Jayaratne K, Paranagama P, Abeywickrama K, Nugaliyadde L (2002) Inhibition of *Aspergillus flavus* Link and aflatoxin formation by essential oils of *Cinnamomum zeylanicum* (L.) and *Cymbopogon nardus* Rendle. Trop Agric Res 14:148–153

Kamath JV, Rana AC, Chowdhury AR (2003) Pro-healing effect of *Cinnamomum zeylanicum* bark. Phytother Res 17:970–972

Kawatra P, Rajagopalan R (2015) Cinnamon: Mystic powers of a minute ingredient. Pharm Res 7(Suppl 1):S1–S6. https://doi.org/10.4103/0974-8490.157990

Keville K, Green M (2008) Aromatherapy: a complete guide to the healing art paperback, 2nd edn. Crossing Press. ISBN-10: 1580911897

Khan A, Safdar M, Khan MMA, Khattak KN, Anderson RA (2003) Cinnamon improves glucose and lipids of people with type 2 diabetes. Diabetes Care 26(12):3215–3218

Khasnavis S, Kalipada P (2014) Cinnamon treatment upregulates neuroprotective proteins Parkin and DJ-1 and protects dopaminergic neurons in a mouse model of Parkinson's disease. J Neuroimmune Pharmacol 9:569–581

Kumar S, Kumari R, Mishra S (2019) Pharmacological properties and their medicinal uses of *Cinnamomum*: a review. J Pharm Pharmacol 71(12):1735–1761

Kurokawa M, Kumeda CA, Yamamura J, Kamiyama T, Shiraki K (1998) Antipyretic activity of cinnamyl derivatives and related compounds in influenza virus-infected mice. Eur J Pharmacol 348(1):45–51

Kwon H, Hwang J, So J, Lee C, Sahoo A, Ryu J, Jeon WK, Ko BS, Lee SH, Park ZY, Im S (2019) Correction to: cinnamon extract induces tumor cell death through inhibition of NFκB and AP1. BMC Cancer 19:1113. https://doi.org/10.1186/s12885-019-6342-5

Lee KG, Shibamoto T (2002) Determination of antioxidant potential of volatile extracts isolated from various herbs and spices. J Agric Food Chem 50(17):4947–4952

Lee JS, Jeon SM, Park EM, Huh TL, Kwon OH, Lee MK, Choi MS (2003) Cinnamate supplementation enhances hepatic lipid metabolism and antioxidant defense systems in high cholesterol-fed rats. J Med Food 6(3):183–191, PubMed PMID: 14585184.

Maierean SM, Serban MC, Sahebkar A, Ursoniu S, Serban A, Penson P, Banach M, Lipid and Blood Pressure Meta-analysis Collaboration (LBPMC) Group (2017) The effects of cinnamon supplementation on blood lipid concentrations: a systematic review and meta-analysis. J Clin Lipidol 11(6):1393–1406.28887086

Martin KW, Ernst E (2003) Herbal medicines for treatment of bacterial infections: a review of controlled clinical trials. J Antimicrob Chemother 51(2):241–246

Mohammadi A, Mohammad-Alizadeh-Charandabi S, Mirghafourvand M, Javadzadeh Y, Fardiazar Z, Effati-Daryani F (2014) Effects of cinnamon on perineal pain and healing of episiotomy: a randomized placebo-controlled trial. J Integr Med 12(4):359–366

Mollazadeh H, Hosseinzadeh H (2016) Cinnamon effects on metabolic syndrome: a review based on its mechanisms. Iran J Basic Med Sci 19(12):1258–1270

Mottillo S, Filion KB, Genest J, Joseph L, Pilote L, Poirier P, Rinfret S, Schiffrin EL, Eisenberg MJ (2010) The metabolic syndrome and cardiovascular risk, a systemic review and meta-analysis. J Am Coll Cardiol 56:1113–1132

MRE (2019) Market research engine report on Nutraceuticals. Id: FBNM819. https://www.marketresearchengine.com/nutraceuticals-market

Nabavi SF, Lorenzo AD, Izadi M, Sobarzo-Sánchez E, Daglia M, Nabavi SM (2015) Antibacterial effects of cinnamon: from farm to food, cosmetic and pharmaceutical industries. Nutrients 7(9):7729–7748. https://doi.org/10.3390/nu7095359

Pan MH, Lai CS, Ho CT (2010) Anti-inflammatory activity of natural dietary flavonoids. Food Funct 1(1):15–31. https://doi.org/10.1039/c0fo00103a

Peterson DW, George RC, Scaramozzino F, LaPointe NE, Anderson RA, Graves DJ, Lew J (2009) Cinnamon extract inhibits tau aggregation associated with Alzheimer's disease in vitro. J Alzheimers Dis 17(3):585–597

Price S, Price L (2019) Aromatherapy for health professionals, 5th edn. Elsevier Health Sciences. ISBN-10: 070207473X

Purchon N, Cantele L (2014) The complete aromatherapy & essential oils handbook for everyday wellness. Robert Rose Publishers. ISBN: 0778804860/978-0778804864

Qin B, Nagasaki M, Ren M, Bajotto G, Oshida Y, Sato Y (2003) Cinnamon extract (traditional herb) potentiates in vivo insulin-regulated glucose utilization via enhancing insulin signaling in rats. Diabetes Res Clin Pract 62(3):139–148.14625128

Ranasinghe P, Pigera S, Premakumara GS, Galappaththy P, Constantine GR, Katulanda P (2013) Medicinal properties of 'true' cinnamon (*Cinnamomum zeylanicum*): a systematic review. BMC Complement Altern Med 13:275

Rao PV, Gan SH (2014) Cinnamon: a multifaceted medicinal plant. Evid Based Complement Alternat Med 2014:642942. https://doi.org/10.1155/2014/642942

RMR (2018) Research and markets report ID: 4744639 global cinnamon market 2019–2023

Sartorius T, Peter A, Schulz N, Drescher A, Bergheim I, Machann J, Schick F, Siegel-Axell D, Schurmann A, Weigert C, Haring HU, Hennige AM (2014) Cinnamon extract improves insulin sensitivity in the brain and lowers liver fat in mouse models of obesity. PLoS One 9(3):e92358. https://doi.org/10.1371/journal.pone.0092358

Schink A, Naumoska K, Kitanovski Z, Kampf CJ, Nowoisky JF, Thines E, Pöschl U, Schuppan D, Lucas K (2018) Anti-inflammatory effects of cinnamon extract and identification of active compounds influencing the TLR2 and TLR4 signaling pathways. Food Funct 9(11):5950–5964

Shan B, Cai YZ, Sun M, Corke H (2005) Antioxidant capacity of 26 spice extracts and characterization of their phenolic constituents. J Agric Food Chem 53(20):7749–7759

Shen Y, Jia LN, Honma N, Hosono T, Ariga T, Seki T (2012) Beneficial effects of cinnamon on the metabolic syndrome, inflammation, and pain, and mechanisms underlying these effects - a review. J Tradit Complement Med 2(1):27–32. https://doi.org/10.1016/s2225-4110(16)30067-0

Shimada Y, Goto H, Kogure T, Kohta K, Shintani T, Itoh T, Terasawa K (2000) Extract prepared from the bark of *Cinnamomum cassia* Blume prevents glutamate-induced neuronal death in cultured cerebellar granule cells. Phytother Res 4(6):466–468.10960905

Shobana S, Naidu KA (2000) Antioxidant activity of selected Indian spices. Prostaglandins Leukot Essent Fatty Acids 62(2):107–110

Singh G, Maurya S, DeLampasona MP, Catalan CA (2007) A comparison of chemical, antioxidant and antimicrobial studies of cinnamon leaf and bark volatile oils, oleoresins and their constituents. Food Chem Toxicol 45:1650–1661

Sivamani RK (2015) In: Jagdeo JR, Elsner P, Maibach HI (eds) Cosmeceuticals and active cosmetics, 3rd edn. CRC Press, ISBN 10 1482214164

Stiles KG (2016) Cinnamon leaf essential oil research studies prove effectiveness plus+ how to use guide & recipes. New world library nook book. Barnes & Noble Press. ISBN-10: 1577311396

Subash Babu P, Prabuseenivasan S, Ignacimuthu S (2007) Cinnamaldehyde – a potential antidiabetic agent. Phytomedicine 14(1):15–22

Takasao N, Tsuji-Naito K, Ishikura S, Tamura A, Akagawa M (2012) Cinnamon extract promotes type I collagen biosynthesis via activation of IGF-I signaling in human dermal fibroblasts. J Agric Food Chem 60(5):1193–1200

TMR (2020) Transparency market report; aromatherapy market – global industry analysis, size, trends, growth, trends, and forecast 2019–2027. Rep Id: TMRGL74346

Tzortzakis NG (2009) Impact of cinnamon oil-enrichment on microbial spoilage of fresh produce. J Innov Food Sci Emerg Technol 10:97–102. https://doi.org/10.1016/j.ifset.2008.09.002

Vangalapati M, Satya N, Prakash DV, Avanigadda S (2012) A review on pharmacological activities and clinical effects of cinnamon species. Res J Pharm, Biol Chem Sci 3(1):653–663

Wang P, Ma L, Jin J, Zheng M, Pan L, Zhao Y, Sun X, Liu Y, Xing F (2019) The anti-aflatoxigenic mechanism of cinnamaldehyde in *Aspergillus flavus*. Sci Rep 9:10499. https://doi.org/10.1038/s41598-019-47003-z

Wansi SL, Nyadjeu P, Ngamga D, Mbuyo EPN, Nguelefack TB, Kamanyi A (2007) Blood pressure lowering effect of the ethanol extract from the stembark of *Cinnamomum zeylanicum* (Lauraceae) in rats. Pharmacol Online 3:166–176

Wronkowitz N, Romacho T, Sell H, Eckel J (2014) Adipose tissue dysfunction and inflammation in cardiovascular disease. Front Horm Res 43:79–92. https://doi.org/10.1159/000360560

Yossa N, Patel J, Macarisin D, Millner P, Murphy C, Bauchan G, Lo YM (2014) Antibacterial activity of Cinnamaldehyde and Sporan against *Escherichia coli*O157:H7 and *Salmonella*. J Food Process Preserv 38(3):749–757

Zaidi SF, Muhammad JS, Shahryar S, Usmanghani K, Gilani AH, Jafri W, Sugiyama T (2012) Anti-inflammatory and cytoprotective effects of selected Pakistani medicinal plants in *Helicobacter pylori*-infected gastric epithelial cells. J Ethnopharmacol 141(1):403–410

Zhang H, Zhou W, Zhang W, Yang A, Liu Y, Jiang Y, Huang S, Su AJ (2014) Inhibitory effects of citral, cinnamaldehyde, and tea polyphenols on mixed biofilm formation by foodborne *Staphylococcus aureus* and *Salmonella enteritidis*. J Food Prot 77(6):927–933. https://doi.org/10.4315/0362-028X.JFP-13-497

Ziegenfuss TN, Hofheins JE, Mendel RW, Landis J, Anderson RA (2006) Effects of a water-soluble cinnamon extract on body composition and features of the metabolic syndrome in pre-diabetic men and women. J Int Soc Sports Nutr 3(2):45–53. https://doi.org/10.1186/1550-2783-3-2-45

Chapter 14
Public-Private Partnership in Growth and Development of Cinnamon Industry in Sri Lanka

Kumudini Gunasekare, Sarada de Silva, and Ravindu Runage

14.1 Introduction

Cinnamon is the prime spice of Sri Lanka. The first records of cinnamon from Sri Lanka extend back into 1500 B.C. Sri Lanka holds a monopoly of Ceylon cinnamon (*Cinnamomum zeylanicum* Blume) in the export market, and this is the only monopoly the country holds. Statistically it is over 90% (Anon 2010). Ceylon cinnamon is exported to over 70 countries, bringing US$ 259 million turnover to the country in 2018 from all cinnamon products (Anon 2018). During the last decade, Sri Lanka has dominated the world market in cinnamon exports, followed by Indonesia, China, and Vietnam. However, these countries produce cassia in huge quantities, whereas Sri Lanka produces pure cinnamon from *C. zeylanicum* (Thibbotuwawa et al. 2017).

Even though a national branding attempt was made in 2011 with the launch of "Pure Ceylon Cinnamon" along with its very own Lion logo, the cinnamon industry still has a long way to go. With the government setting a target of achieving US$ 1 billion foreign exchange from cinnamon along with other spices and allied products by 2020, the industry is now looking at aggressive growth with the support of the state organizations and private sector collaborations. Aiming to empower industry stakeholders and achieve recognition as an international brand synonymous with Sri Lanka's unique identity, a long-term strategy is envisaged to take cinnamon to the world, by inspiring stakeholders of the industry – cinnamon growers, processors,

K. Gunasekare (✉)
Agriculture Sector Modernization Project, Ministry of Plantation Industries & Export Agriculture, Level 16, "Suhurupaya", Battaramulla, Sri Lanka

S. de Silva
B. Darsin De Silva & Sons Private Limited, Colombo, Sri Lanka

R. Runage
Rathna Producers, Kamburupitiya, Sri Lanka

R. Senaratne, R. Pathirana (eds.), *Cinnamon*,
https://doi.org/10.1007/978-3-030-54426-3_14

exporters, and cinnamon-related product manufacturers in the country – with a vision to stimulate the growth of cinnamon exports in Sri Lanka.

Along with the relevant state partners, the private sector leads the way forward in the cinnamon industry. It is driven by many players such as cultivators (small farmers, proprietors, and large companies), processors, collectors, retailers, dealers, and exporters. This section of the community represents the private sector.

14.2 Private Sector

Private sector means basically the business sector in its diverse modalities. This sector usually covers all entities involved in private profit-seeking operations in fields, such as cultivation/production, processing, distribution, and provision of services, trade, and exports. These entities come in diverse scales and behaviors, from micro and small to medium and large enterprises, from small farms and shops owned by individuals, families, or cooperatives, all the way to giant transnational corporations. There are also gray areas and hybrids between private and public sectors, as in the case of state-controlled corporations with significant private equity and of jointly owned (public-private) firms as well as in the case of cooperatives.

14.3 Private Sector Entities in Cinnamon Industry

The private partners can be local or international and may include businesses or investors with technical or financial expertise. The stakeholders of the private sector involved in cinnamon at the local level in Sri Lanka are as follows: The Spice Council Sri Lanka (TSC), the Spices and Allied Products Producers and Traders Association (SAPPTA), the Cinnamon Training Academy (CTA), the Ceylon Cinnamon Association (CCA), the Cinnamon Cultivators Association (CinCA), and the Ceylon Cinnamon GI Association.

14.3.1 Involvement of Private Sector in the Cinnamon Industry

The cultivation of cinnamon is carried out mainly by smallholders and few large plantations. The high cost of production, low volumes, and poor quality and food safety standards have led the industry to lose competitiveness in major export markets. Cinnamon can be harvested twice a year, but approximately only 25% is harvested twice a year, while 65% is harvested once a year, and the balance is neglected or not harvested in time due to the acute shortages of cinnamon harvesters and processor, leading to yield loss (Macan-Marker 2017).

A lack of knowledge and skills in processing cinnamon safely and hygienically has restricted the ability of the industry to supply cinnamon according to the international food safety standards. The long-standing social stigma attached to cinnamon peelers has resulted in workers leaving to other occupations and discouraged the young generation from taking up this work. Currently, the industry is lacking over 35,000 skilled workers to carry out the harvesting and producing of cinnamon according to the export market demand.

14.4 Public Sector

The relevant state partners include relevant ministries and public sector institutions, such as the National Cinnamon Research and Training Center of the Department of Export Agriculture (under the Ministry of Plantation Indutries & Export Agriculture), Export Development Board (under the Ministry of Development Strategies and International Trade), Sri Lanka Standards Institute (SLIS – under the Ministry of Science and Technology), Department of Commerce (DOC – under the Ministry of Industry and Commerce), Sri Lanka Accreditation Body (SLAB – under the Ministry of Science and Technology), and Industrial Technology Institute of Sri Lanka (under the Ministry of Science and Technology). These entities play different roles and in certain instances their mandates are overlapping.

14.5 National Drive Toward PPP

Following a cabinet decision, the national private-public partnership (PPP) unit was set up under the Ministry of Finance. The prime objective of this unit is to fulfill national economic objectives of securing investments with private sector engagement for planning, implementation, financing, and maintenance of certain infrastructure assets owned by the state agencies. Given the importance of investment inflow for a sustainable economy, the initiative to set up the PPP unit is supported by the World Bank. It is entrusted with the task of developing new infrastructure services despite short-term fiscal constraints and providing value for money through efficiencies in procurement, construction, and operation. The aim is to improve service quality and innovation through the use of private sector expertise and performance incentives. This was part of an initiative to promote PPP for infrastructure development projects in Sri Lanka and restructuring of some of the state-owned enterprises.

14.6 Definitions for PPP

PPPs are defined by various agencies in different manner, and there is no universally accepted definition for it. Some definitions of PPP are given below:

According to the Department of Economic Affairs of the Ministry of Finance, India, PPP means an arrangement between a government or statutory entity or government-owned entity on one side and a private sector entity on the other, for the provision of public assets and/or related services for public benefit, through investments being made by and/or management undertaken by the private sector entity for a specified period of time, where there is substantial risk sharing with the private sector and the private sector receives performance-linked payments that conform to specified predetermined and measurable performance standards (Anon 2016). Local Government Procurement Agency of the UK describes PPP as a generic term for the relationships formed between the private sector and public bodies often with the aim of introducing private sector resources and/or expertise in order to help provide and deliver public sector assets and services. The term PPP is used to design, build, finance, and operate (DBFO) service contracts and formal joint venture companies (Anon 2015).

As described by the Department of Environment and Local Government of Ireland, PPP is a partnership between the public sector and private sector for the purpose of delivering a project or a service traditionally provided by the public sector. PPPs come in a variety of different forms, but at the heart of every successful project is the concept that better value for money may be achieved through the exploitation of private sector competencies and the allocation of risk to the party best able to manage it (Anon 2000).

FAO's definition of an agri-PPP or a PPP for agribusiness development is a formalized partnership between public institutions and private partners designed to address sustainable agricultural development objectives, where the public benefits anticipated from the partnership are clearly defined, investment contributions and risk are shared, and active roles exist for all partners at various stages throughout the PPP project life cycle (FAO 2016).

One commonality that emerges from the above definitions is that PPPs are long-term contractual arrangements between the public and private sectors for the delivery of public services in an efficient manner. However, there is no clear legal definition so far for the concept of PPP (Anon 2016).

In practice, agri-PPPs may involve either formal (contractual) or informal (collaborative) arrangements and tend to favor simpler, less complete contract modalities, such as memoranda of understanding, when compared to traditional PPPs for infrastructure (FAO 2016).

14.7 Importance of PPP Initiatives

PPPs involve three main features, namely, risk transfer, long-term contracts, and partnership agreement. The governments around the world tend to adopt this approach, owing to three main benefits:

- Ability of developing new infrastructure services despite short-term fiscal constraints
- Value for money through efficiencies in procurement, construction, and operation

- Improved service quality and innovation through the use of private sector expertise and performance incentives

Due to inadequacy of government and foreign funds, particularly for providing economic infrastructure, it is compelled to look for private sector participation for financing.

PPP is an approach that is used for attracting private capital strategically for identified sectors/subsectors, where the "risk factor" is crucial with regard to the return on investment. The efficient use of scarce public resources is a critical challenge for governments. Governments face an ever-increasing need to find sufficient financing to develop and maintain the infrastructure required to support agriculture-based industries. Combined with most governments' limited financial capacity, it has become imperative to mobilize private sector capital for infrastructure investment. Structured correctly, a PPP may be able to mobilize previously untapped resources from the local, regional, or international private sector, which is seeking investment opportunities. In this context, PPP can be considered as an efficient tool for greater efficiency.

PPPs in agriculture are also an important mechanism for harnessing technology, resources, skills, expertise, and market access to improve the livelihoods of, mainly, the resource-poor smallholders. For many years, PPPs focus has mainly been on large infrastructure projects. However, PPP interventions in agriculture, and particularly to that of small farmers, are relatively recent. There is now growing interest in PPPs' ability to create transformational change in the agriculture sector, including the cinnamon industry.

14.8 Motivation for Engaging in PPPs

The three main aspects that motivate governments to enter into PPPs are:

1. To attract private capital investment (often to either supplement public resources or release them for other public needs)
2. To increase the efficiency and use of available resources
3. To reform sectors through reallocation of roles, incentives, and accountability

14.9 Prerequisites for Successful Public-Private Partnership

Effective PPPs recognize that both the public and private sectors have certain advantages, relative to the other, in performing specific tasks. A strong PPP allocates the tasks, obligations, and risks among the public and private partners in an optimal way. The private partners can be local or international and may include businesses or investors with technical or financial expertise. It enables the government to fulfill its responsibilities in the efficient delivery of socio-economic goods and services by ensuring efficiency, effectiveness, accountability, quality, and outreach services.

Increasingly, PPPs may also include non-government organizations (NGOs) and/or community-based organizations (CBOs).

PPPs allow the government to pass operational roles to efficient private sector operators while retaining and improving focus on core public sector responsibilities, such as regulation and supervision. Properly implemented, this approach should result in a lower aggregate cash outlay for the government and better and cheaper service to the consumer. This should hold true even if the government continues to bear part of the investment or operational cost since government's cost obligation is likely to be targeted, limited, and structured within a rational overall financing strategy.

Another main requirement for a successful PPP is the proper guidance in the design phase of PPP projects. It is important to ensure transparency in the selection of private partners, risk sharing, and mitigation mechanisms to protect small farmers, as well as conflict resolution strategies that have often been overlooked (OECD 2014).

The success or failure of agri-PPPs is highly dependent on the enabling environment and the governance strategy designed to support the implementation of these partnerships. The legislation and regulation concerned with land access, enforceability of contract farming agreements and protection of intellectual property, and agricultural insurance to support SMEs are critical for the successful implementation of agribusiness PPPs (FAO 2016).

14.10 Unique PPP Endeavors to Elevate the Ceylon Cinnamon Subsector

Over the last two decades, the quality of Ceylon cinnamon exported from Sri Lanka has notably deteriorated coupled with the industry facing emerging competition, lack of a workforce, quality issues, and a threat of losing a global market share. High-level support from the government, in collaboration with strong championship by the private sector, is therefore seen as real drivers to transform the industry in order to reach global spice destinations.

The following engagements highlight the PPP interventions in developing and promoting the cinnamon industry both locally and internationally:

14.10.1 Training and Capacity Building of Cinnamon Value Chain Actors

The Cinnamon Training Academy (CTA) Limited is a PPP initiative with limited liability incorporated in Sri Lanka in 2006 and registered under the Companies Act. The shareholders of the company are members of the CCA and TSC, who are dedicated to the development of the cinnamon industry. CTA was financially supported by the Government of Sri Lanka and UNIDO/WTO-STDF, IFEAT, and 15 leading cinnamon producers and exporters. The primary objective of CTA is to "provide service to train value chain actors in the cinnamon industry for increasing the production capacity and enhance quality and standards compliance capacity to elevate the industry to a national symbol of quality and excellence." As a result, the Cinnamon Training Academy now provides National Vocational Qualification (NVQ) Level 3 and Level 4 for cinnamon factory and field officers. Awareness raising and National Vocational Qualification training on food safety and hygiene practices have reached over 1000 people. It will also serve to retain and attract more workers by promoting new job opportunities and enhancing working conditions through delivering systematic training with a nationwide accepted certification. The CTA is now equipped to implement the National Vocational Qualification (NVQ) for various occupations, namely, those of harvesters, processors, factory officers, and supervisors under cinnamon factory and field operations.

14.10.2 Ceylon Cinnamon: A Roadmap Toward Its Protection as Geographical Indication

The Sri Lanka Exports Development Board (EDB) linked up with the cinnamon industry stakeholders to obtain a geographical indication (GI) for "Ceylon cinnamon" in the European Union (EU). This covers five districts, namely, Galle, Matara, Kalutara, Ratnapura, and Hambantota. TSC and UNIDO have elaborated a project proposal in the field of food safety training and certification that was granted support by the Standards and Trade Development Facility (STDF). The implementation of the STDF-UNIDO-TSC project on cinnamon foresees an activity devoted to the registration of Ceylon cinnamon as geographical indication. The Ministry of Development Strategies and International Trade is the implementing authority of the project.

14.10.3 Enhancing the Compliance and Productive Capacities and Competitiveness of the Cinnamon Value Chains in Sri Lanka

TSC works with the other public sector institutions, viz., the Department of Export Agriculture (DEA), under the Ministry of Primary Industries, and the EDB, under the Ministry of Development Strategies and International Trade, to make interventions to address the core issues of the industry. TSC has been working along with the DEA to develop production and research focused on production-related aspects. TSC and the Department of Commerce jointly lobbied for promoting true cinnamon against cassia. This was made possible through enhancing the compliance, productive capacities, and competitiveness of the cinnamon value chain in Sri Lanka, which was designed with the financial support of STDF and the technical support of UNIDO in 2011. It provided support to the cinnamon industry stakeholders to enhance the competitiveness of their product and equip them to face tight competition at the global trade level, thereby increasing the share of the value-added cinnamon from Sri Lanka to the world market.

14.10.4 Branding of Pure Ceylon Cinnamon

At the insistence of CCA and TSC, the bifurcation of cinnamon and cassia in World Customs Organization Harmonized Code (H.S. Code) in 2007 has helped the cinnamon industry. The role played by the Department of Customs and the Department of Commerce is much appreciated by the cinnamon industry.

The branding effort of Pure Ceylon Cinnamon and launch of the Lion logo was a result of an initiative taken by individuals and organizations in the industry, from both the public and private sector. The Government of Sri Lanka also helped protect the botanical name of cinnamon and the launch of the Ceylon Cinnamon Lion logo. As a result, the industry is now able to enhance global market position of Ceylon cinnamon.

14.10.5 The Spices and Concentrates Strategy

The Spices and Concentrates Strategy was developed as part of the National Export Strategy (NES) of Sri Lanka, under the aegis of the Ministry of Development Strategies and International Trade (MoDSIT) and the EDB, with the financial assistance of the EU, as part of the "EU-Sri Lanka Trade-Related Assistance." The tangible potential of the spice sector in Sri Lanka has been recognized by making it a priority sector in the NES. The NES for the spice sector was formulated through consultations between private and public sector stakeholders in the industry. The overall aspiration of the NES is to position Sri Lanka as a trade hub in the region driven by investment and innovation. Thus, the sector strategy for spices seeks to propel the industry to greater heights by establishing Sri Lanka as the key exporter of spices in the region.

14.10.6 Adoption of Good Agricultural Practices (GAP)

One of the main barriers to exporting cinnamon to EU countries is the presence of pesticide residues and other substances/adulterates, with ensuing food safety concerns. Certification for Good Agricultural Practices (GAP) is one of the effective strategies to address this issue and also to overcome many other related issues that arise from cultivation to processing. State institutions such as Sri Lanka Standards Institute in collaboration with Department of Export Agriculture and with the private sector stakeholders have prepared the draft national standards for GAP for cinnamon, pepper, and coffee relating to production and processing so as to ensure a legally compliant, environmentally sound, socially acceptable, and economically viable quality product. This is yet to be implemented.

14.10.7 Increasing Trade Competitiveness of Small- and Medium-Sized Enterprises in Regional and European Union Markets

The Government of Sri Lanka has placed the export growth high on its development agenda. To increase the competitiveness of Sri Lankan small- and medium-sized enterprises (SME) in regional and EU markets, the Ministry of Industry and Commerce of the Government of Sri Lanka, through its Department of Commerce, approached EU Delegation with a request for trade-related technical assistance. It aims to increase the competitiveness of Sri Lankan small- and medium-sized enterprises (SME) in regional and EU markets and to support inclusive, trade-led growth by focusing on export strategy, trade policy and facilitation, national quality infrastructure (NQI), and value chains in the spices, food, and information technology (IT). Within the framework of the project, UNIDO facilitates the strengthening of NQI services to meet the quality, sanitary, and phytosanitary standards at home and abroad and enhances the value chain performance of SMEs that depend on these services. Specific attention is given to strengthening the core NQI pillars, which constitute standardization, metrology, accreditation, and conformity assessment, and to enhancing export compliance along the spices and processed food value chains (https://www.unido.org accessed on 30.10.2019).

Since 2012, a strong public-private partnership has brought together the Sri Lankan government, UNIDO, and the Spice Council – the apex body representing the cinnamon industry – which has helped to mobilize additional assistance. The partnership focused on boosting the productive capacities and competitiveness of the cinnamon value chain and on increasing exports to high-end markets.

14.11 International Experiences in Agribusiness PPPs

To improve understanding of both the potential benefits and the challenges of agri-PPPs, the FAO has gathered 70 case studies from 15 developing countries along with evidence from field-based support to PPP initiatives in Central America and Southeast Asia (FAO 2016). This publication provides a wealth of practical information. Its primary objective is to draw lessons that can be used to provide guidance on how to establish effective partnerships with the private sector to mobilize support for agribusiness development. From the 70 cases investigated, a typology of four common project types was identified: (i) partnerships that aim to develop agricultural value chains; (ii) partnerships for joint agricultural research, innovation, and technology transfer; (iii) partnerships for building and upgrading market infrastructure; and (iv) partnerships for the delivery of business development services to farmers and small enterprises.

14.12 Challenges and Opportunities

To continue to add value and intensify market development for the cinnamon and the spice sector in general, public and private industry representatives identified the following as the most pressing issues that should be addressed rapidly through PPP engagements:

1. Limited availability of high-yielding planting stock and low use of associated techniques cause slow productivity growth and low adoption of improved varieties.
2. Insufficient availability of skilled and unskilled labor, which is exacerbated by limited training opportunities.
3. Implementation of standards and codes of conduct at the production and processing levels.
4. Limited links between the public and private sectors.
5. Poor dissemination of trade information and limited promotion and inadequate brand promotion of cinnamon in destination markets.
6. Limited use of contract farming methods between producers, processors, and exporters.
7. Limited postharvest storage infrastructure, leading to high postharvest losses.
8. Low technology and mechanization adoption.
9. Traceability for organics and fair-trade segments.
10. Lack of guidance in the design phase of PPP projects.

Learning from the issues and challenges, there are many opportunities that the cinnamon industry can foster PPP as outlined below:

1. Data gathering channels and farmer registration efforts are ineffective due to limited outreach by the DEA to spice producers, especially those who operate a home garden. These small growers do not have the opportunities to take advantage of DEA offered benefits. It is essential to identify all spice growers and farmers to provide them with the support they need. This would also link farmers, growers, and processors for channeled information dissemination and so develop a relational value chain to empower small farmers. The need for DEA to register all producers, collectors, and dealers is a must in order to develop this industry by proper networking production and trade.

2. The Spices and Concentrates Strategy developed as part of the NES of Sri Lanka needs to be implemented with proper monitoring mechanisms backed with the consistent financial support essential to drive the industry with the strong footing in R&D.
3. To develop and deploy a mobile information service that provides regular real-time information led by the private sector. A mobile application could be introduced that would be easily accessible to provide daily information to all stakeholders, especially producers. This application could provide essential information in real time and gather and disseminate regular information, such as weather forecasts and patterns, market prices, demand and supply data, and harvest and postharvest information. This must be an entirely private sector initiative, supported by telecom service providers.
4. Globally, buyers are requesting that agriculture suppliers provide more tangible, transparent, and effective food processing and production safeguards and protocols to ensure product quality and safety. In North America and Europe, buyers are increasingly reluctant to accept products from suppliers without a food safety program and proof of certification from either an independent third-party verifier or certification body. These certifications or audits and market requirements cover GAP, GMP, phytosanitory, hygiene, food security, pest management, and environmental topics. As such, it is important to foster a nationwide pool of certified or accredited GAP and GMP consultants or service providers. This pool of certified and accredited consultants can be trained through funding from the private sector as a profitable business venture.

14.13 Way Forward

Recommendations to enhance the future of the cinnamon industry through strong PPPs rely on many factors. Primarily it is imperative to create an enabling environment to promote PPPs in research and development, extension, and advisory as well as fostering a strong business climate to promote the Ceylon cinnamon industry and elevate its global status. In achieving this, there are many prerequisites to meet such as identifying a common interest and create a win-win situation, creating robust institutional and management arrangements, transparency, effective communications, investing on value chain developments and addressing gaps in value chains, promoting R&D for evidence-based decision in the promotion of Ceylon cinnamon, and protecting cinnamon using geographical indication and adding value.

Another important aspect that both state and private sector entities should intervene in is to ensure the quality and safety of cinnamon products to meet buyer requirements in the EU and in other high-end markets. In this connection, it would be ideal if the state sector can take initiatives to establish one-stop shop where business community can access facilities at one place for laboratory analysis with state-of-the-art facilities, guiding the private sector in achieving these standards and certifications.

The European market currently provides excellent opportunities for trading in cinnamon, due to continuing rises in import levels and favorable prices. Europe is mainly a cassia market, but opportunities for Ceylon cinnamon are prominently apparent in specific countries and segments. The United Kingdom, Italy, and Belgium import a much larger share of Ceylon cinnamon than the European average (CBI – Ministry of Foreign Affairs – https://www.cbi.eu/market-information/spices-herbs/cinnamon/). There are also good prospects for sustainable suppliers and those supplying processed cinnamon with organic and other sustainable category certification systems, which Sri Lanka needs to explore in order to derive higher margins. Although sustainable cinnamon is still a niche market, the demand for products certified for compliance with sustainability standards is increasing. Organic and fair-trade cinnamon have been on the market for some time; however, the introduction of Rainforest Alliance-certified cinnamon still needs to be explored. A major challenge for the market for certified sustainable cinnamon is that it is sold at a higher price to cover some or all of the costs, and this is something that needs more attention. Nonetheless, there is an ongoing debate in the sector concerning the best way forward in implementing sustainability in the mainstream market. The option of third-party certification is still under debate. As mentioned above, self-verification is expected to become more common in the future in the mainstream market, and the Sri Lankan cinnamon industry needs to play a more proactive role in this so as to facilitate the stakeholders.

Governmental and nongovernmental organizations in developed countries often have programs and subsidies available for investments in sustainability. Therefore, private sector entities should look for partners in the promotion of sustainability with the aid of foreign funds.

Smallholder farmers will find it difficult to operate independently in this field. A certain scale of production is often required to make certification economically feasible. There may be foreign companies willing to invest in training farmers in the country of origin and helping them to obtain certification. Large processors and exporters can also play a leading role in this process and attract capital.

Limited innovations in processing also hamper the growth of the industry. Cinnamon harvesting is labor intensive and can account for up to 60% of the total cost of production. Machines are being developed both by the state and private sector, and proper collaboration between the two parties will increase the application of these machines that can reduce processing costs without compromising the quality of cinnamon product. As the quality of cinnamon is also judged by its appearance (broken or entire quills), hand-peeled cinnamon could be aimed at the high-end markets.

To be able to bear the initial cost of investment in machinery and training, PPPs and private investments, along with some public funding, can be a means of support. A cost sharing or matching grant approach with a financial mix consisting of bank loans, grants, and the industry's own resources needs to be facilitated and promoted.

Insurance services are also needed to safeguard the farmers, and intervention of private financial institutions to provide insurance services would be a way forward. Private financial institutions can play heterogeneous roles but may often provide credit and insurance services to farmers. In several cases, nonfinancial institutions, such as public- or donor-supported programs, private foundations, and farmer cooperatives, could also offer financial services. Similarly, nonprofit organizations could help improve the delivery and postharvest quality of products by providing growers with production loans to purchase agricultural inputs.

It is also necessary to implement and monitor activities in accordance with the NES in the area of cinnamon and concentrates and to align the monitoring of progress and collaborative efforts between the private sector and state institutions to achieve stated targets with a vision to promote and increase export-led, value-added products using cinnamon. All in all, these strategies and activities need to be supported by a long-term vision and by leadership of inclusive and sustainable industrial development built on a triple bottom line of social, economic, and environmental prosperity.

References

Anon (2000) Public-private partnership assessment guidance note four. Department of the Environment and Local Government, Dublin, p 2000

Anon (2010) Annual highlights of the spice council. Spice Council of Sri Lanka, Colombo

Anon (2015) OECD Review of public governance of PPP in the United Kingdom. Organization for Economic Co-operation and Development (OECD), GOV/PGC/SBO(2015)8, 02-Dec-2015 https://www.oecd.org/governance/procurement/toolbox/search/oecd-review-public-governance-ppp-uk.pdf

Anon (2016) PPP guide for practitioners. Department of Economic Affairs Ministry of Finance Government of India. https://www.pppinindia.gov.in/documents/20181/33749/PPP+Guide+for+Practitioners/

Anon (2018) Export Performance Indicators, Export Development Board – https://www.srilanka-business.com/pdf/Export_Performance_Indicators_of_Sri_Lanka_2009_2018.pdf accessed on July 2019

FAO (2016) In: Rankin M, Gálvez Nogales E, Santacoloma P, Mhlanga N, Rizzo C (eds) Public–private partnerships for agribusiness development – a review of international experiences, Rome

Macan-Marker M (2017) Labour shortage hits Sri Lanka's cinnamon industry. Asian Review, June 2017

OECD (2014) Public–private partnerships for agricultural innovation: 6th meeting of the Food Chain Analysis Network, 13–14 October 2014. Organisation for Economic Co-operation and Development (OECD), Paris

Thibbotuwawa M, Wijayasiri J, Hirimuthugodage D (2017) Analysis of cinnamon, pepper and cardamom value chains in Sri Lanka. Institute of Policy Studies of Sri Lanka, p 16

Chapter 15
Value Creation and Food Products of Cinnamon

Renda K. C. Jeewanthi, Achini M. De Silva, and Tharaka Weddagala

15.1 Value Creation

Competitive advantage facilitates to position the products and services in markets while securing healthy profits. In order to construct competitive advantage, the firm or value chain should name the capabilities and resources that are superior to competitors. Figure 15.1 shows the conceptual model on value creation and profit generation. Value must be defined by the downstream of the value chain, markets, and customers, and defined value should be operationalized through the upstream of the value chain. Therefore, production and value addition nodes of the value chain must understand the market/customer wants and market trends. Resources, processors, and capabilities are the key building blocks of the value creation. In general, resources and processes can be the same, but outcomes are different in different firms due to their own capabilities. The present status of the Ceylon cinnamon value-added exporters shows similar pattern where capable firms are in better positions.

Value creation in the cinnamon industry is vital for sustainable value chain upgrading. Creating value in the cinnamon value chain occurs through the addition of actual or perceived value to a target customer for a superior product. New concepts, new products, enhanced product characteristics, brand names, labels, packaging, unique customer experiences, etc. may all create additional value for cinnamon-based products. The two forms of the value creation are the addition of actual value and the addition of perceived value.

R. K. C. Jeewanthi (✉) · A. M. De Silva · T. Weddagala
Faculty of Agricultural Sciences, Sabaragamuwa University, Belihuloya, Sri Lanka

R. Senaratne, R. Pathirana (eds.), *Cinnamon*,
https://doi.org/10.1007/978-3-030-54426-3_15

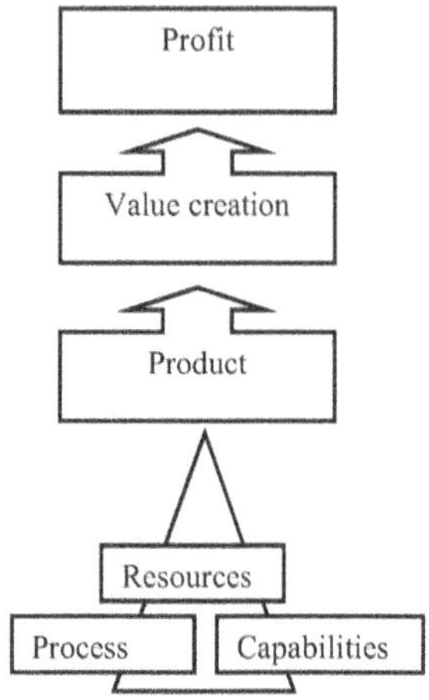

Fig. 15.1 Value creation model

15.1.1 Addition of Actual Value

Actual value adding options available range from cut, ground cinnamon to nutraceuticals, health care, personal care, and home care products. Figure 15.1 shows the product flowchart in the market that is already or soon to be available. Actual value addition in cinnamon industry is currently at very low levels. Figure 15.2 explains the present status of the export product range. The centuries-old Ceylon cinnamon export business is yet to change its strategies to cater for the end consumer. Still the industry heavily exports raw materials to the global market place. Value chain sustainability lies within the hands of actual value addition to the queen of spice to earn better returns. Value chain actors are involved in limited value adding to the buyers.

15.1.2 Addition of Perceived Value

Perceived value addition is based on market or consumer requirements. Products or services are developed to cater the end user requirements. Further, perceived value addition aims to fulfill the desires of consumers, and the expectations vary from country to country and market to market. Market and customer segmentation are essential to design the perceived value packs. Existing examples from Ceylon cinnamon industry range from safety and quality certified products (GMP/HACCP/Organic/BRC/FSSC/Fair Trade, etc.), labeling (antioxidants, coumarin content, free from irradiation, ethical labor, environmental friendly (Rainforest Alliance certified), etc.), packaging (protect taste, color, aroma, etc.), chemical fingerprint (separate Ceylon cinnamon from cassia), health benefits (antidiabetic, anti-inflammatory, antifungal, and antibacterial, richness of antioxidant components, etc.) (Fig. 15.3).

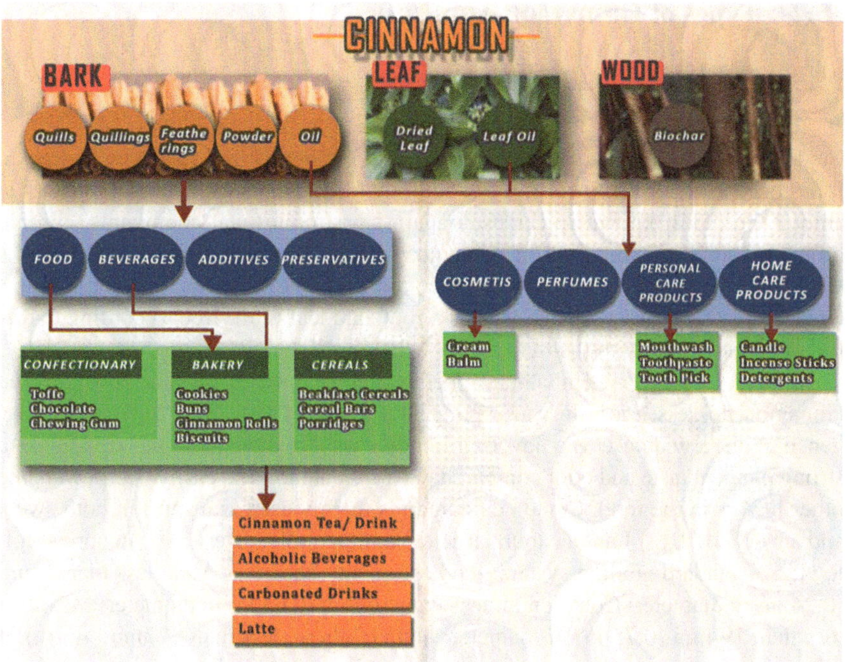

Fig. 15.2 Product flowchart of Ceylon cinnamon

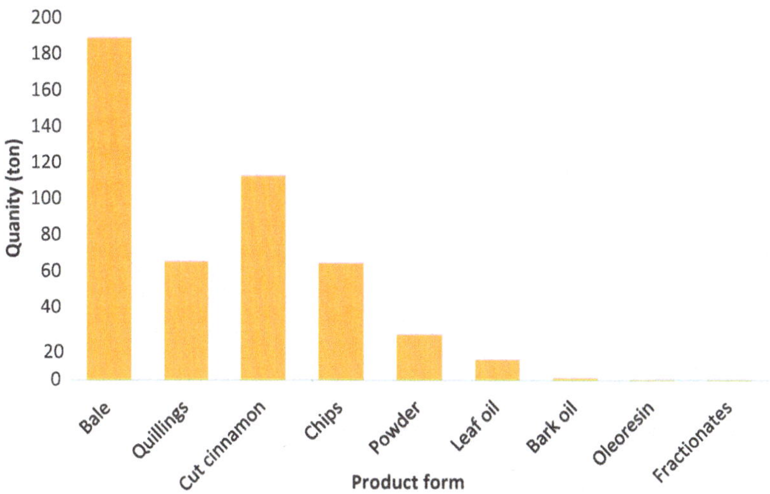

Fig. 15.3 Composition of Ceylon cinnamon export basket: volume of export products. (Source: Harindra 2017)

15.2 Types of Cinnamon Application

15.2.1 Products of Food Industry

Ceylon cinnamon is one of the earliest known commercialized spices in Sri Lanka. It has been used for many purposes in addition to medicinal uses. Cinnamon bark is among the oldest known spices used against gastrointestinal complaints, chronic bronchitis, and inflammation of eyes in Ayurvedic medicine for over 6000 years (Qais et al. 2019). Products of this spice are available as quills, powder, and oil. Cinnamon oil is extracted from the bark, root, and leaf of the tree, according to different applications, with their chemical variations of the oil.

In the food industry, cinnamon is being used as quills, sticks, or powder. True cinnamon possesses a mild warm pungent and sweet taste with a pleasant odor; hence, it has a wide use as a flavoring in numerous food preparations. Cassia has a strong pungent taste and odor compared to true cinnamon and easily overpowers the other flavors in the food. Ceylon cinnamon is mainly used as a spice in both sweet and savory food. Sri Lanka exports true cinnamon quills under four categories, following a standard grading system: the best quality named as Alba, less than 6 mm (0.24 in) in diameter; Continental, less than 16 mm (0.63 in) in diameter; Mexican, less than 19 mm (0.75 in) in diameter; and Hamburg, less than 32 mm (1.3 in) in diameter. These groups have subgroups depending on quill diameter and the number of quills per kilogram for specific grades. As an example, Mexican is divided into M00 000 special, M000000, and M0000.

15.2.1.1 Cinnamon as a Spice

The sweet taste with an aromatic, spicy fragrance of cinnamon bark leads to its use as a spice. A wide range of curry powders are used in Sri Lankan, Indian, and other traditional cuisines. Cinnamon is the second most important spice in the United States and Europe, second only to black pepper (Jayaprakasha et al. 2007). In the food industry, ground spice is essential for flavoring baked products, steamed puddings, candies, and desserts. Cinnamon is also added to pickles, sauces, soups, confectionaries, and canned fruits, such as Dutch pears or stewed rhubarb.

15.2.1.2 Bakery Products with Cinnamon

In the food industry, ground spice is needed for flavoring baked products. It is a famous application in many bakery products including cookies, breads, cinnamon sugar twists, cakes, pies, quick breads, and especially cinnamon rolls. The cinnamon roll has especially attracted the attention of bakers, whose methods, varying from area to area, for manufacturing it in different ways have been patented through decades (Table 15.1). The cinnamon roll is also named a cinnamon bun, cinnamon

Table 15.1 Patents of developing cinnamon roll as a commercial food

Year	Product	Patent no.	Description
1940	Apparatus for making cinnamon rolls	US2217896A	Treating the dough and rolling the treated sheet into a spiral form for 5 cinnamon rolls
1942	Method for making pastry	US2305712A	Making pastry using a chilled or frozen cylindrical mass of dough in an airtight container
1964	Guillotine-type cinnamon roll cutter	US3158057A	Epicycle gearing, connected to the cutter blade A cutter for dividing a ribbon dough into a plurality of segments
1976	Machine for making cinnamon rolls	US3930733	A machine containing a form into which dough is forced into a spiral shape
2002	Frozen uncooked cinnamon roll	US6468569	Attain the qualities of freshly prepared cinnamon roll when thawed and baked
2006	Method for reduced fat cinnamon roll	US2006/0188631A1	Pregelatinized chemically modified resistant starch cinnamon roll
2007	Assembly line technique for slicing cinnamon roll	US2007/0178205A1	Assembly line technique for pull-apart food products such as cinnamon rolls having heat-sensitive thick pastes
2007	Formula and process for producing frozen sheeted dough	US2007/0218167A1	High diameter, better flavor and taste
2011	A portable cinnamon roll	US8057834	Forming raw dough in the shape of a sphere having a central pocket; placing cinnamon, frosting, and sugar within the central pocket; and cooking the raw dough with cinnamon, frosting
2011	A developed dough composition	CA2615571A1	Extending the shelf life of bakery products, a developed dough composition
2012	Cinnamon roll kit	US20120231115A1	Packaged kit featuring premeasured ingredients to easily and expediently create a scrumptious selection of pecan caramel cinnamon rolls

swirl, and cinnamon snail in different countries. It is a sweet roll served commonly in Northern Europe (mainly in Scandinavia) and North America. In Denmark, it is known as *kanelsnegl* (cinnamon snail), and in Finland, it is known as *korvapuusti*. A *Pågens classic* is the name for cinnamon roll in Scandinavia. Cinnamon rolls having a sugar icing have been a staple article of home preparation and commerce for many years. The rolls have been made in a variety of sizes and recipes in two basic ways. One is as frozen cinnamon rolls for sale by retailers, while the other involves the preparation of a fresh cinnamon roll from basic ingredients. These rolls are made commercially or at home, and their flavor and aroma can be of a very high quality. A cinnamon roll is desirably cooked in a convection oven, at a temperature of 310 °F for about 17–19 min (Hansen et al. 2003).

Multiple cinnamon rolls are formed and combined in a well-known pop-open fiberboard tube container. Before the consumption, the cinnamon rolls are baked, iced, and served after taken out of the container. These rolls are classically made from hard dough using an extrusion process, which is then combined with cinnamon, rolled, and frozen into the finished product. The cinnamon roll after preparation is small, dense, and thinly iced using the icing supplied with the tube. Waldemar (1942), US Patent No. 2,305,712, revealed a method of making pastry using a chilled or frozen cylindrical mass of dough in an airtight container. The package can be stored until it is needed for the preparation of the roll. A portable cinnamon roll was introduced in 2011, by US patent 8,057,834, forming raw dough in the shape of a sphere having a central pocket; placing cinnamon, frosting, and sugar within the central pocket; and cooking the raw dough with cinnamon, frosting, and sugar until it is fully prepared for consumption (Kwitek 2011).

Cinnamon is commonly used in tea as powder, sticks, or extract. Cinnamon tea is naturally low in calories and regulates blood sugar levels, which could lead to consuming fewer calories and helping with weight control. Cinnamon has been used as an ingredient of traditional tea all over the world for centuries and, with its unique health benefits, has been handed down from generation to generation in traditional cultures. Mexicans, for example, use cinnamon as a tea with Blessed thistle, flower of borage, eucalyptus leaves, or everlasting flowers with honey. Also they use cinnamon with cocoa as a hot chocolate or latte type beverage. Normally they use cinnamon sticks dipped in a hot water jar. In Turkey people use cinnamon with the beverage called "Salep." Salep is a well-known indigenous Turkish drink made by boiling salep powder and milk with sugar and served with cinnamon sprinkled on top (Dogan and Kayacier 2004). The Korean version of cinnamon tea is known as "gyepi cha" and is often mixed with ginger tea. The Chilean version is "te con canela" and is often in the form of cinnamon sticks and is mixed with regular tea leaves. Cinnamon-flavored tea, as well as teas flavored with cardamom, is consumed as a hot tea in Bangladesh, India, and Pakistan.

15.2.1.3 Cinnamon Beverages

Tea with Cinnamon

The current beverage industry focuses on new, strong, or traditional flavors after being exposed to and responding to many food cultures. Cinnamon beverage types appear in the arena because of its ability to satisfy current health benefit requirements as well as the appeal of its flavor. There are a number of health targeting beverages being introduced including teas (Korean Patent 1,020,160,013,228) and other refreshing beverages. Table 15.2 shows recently patented cinnamon beverage types (with other herbal types) that claim a variety of health benefits. With the nanotech applications in the food industry, cinnamon seems to be recognized as a health booster. As an example, in the ready-to-drink industry, a novel chocolate beverage is being developed with colloidal cinnamon nanoparticles with the aim of

Table 15.2 Recently patented health benefits targeted cinnamon beverage types

Name	Cinnamon incorporation	Other main ingredients	Patent number (year)	Health claim
Cinnamon oolong tea beverage	0.007–0.008% of a cinnamon extract	Tea powder, saccharic acid, honey, sodium bicarbonate, essence, and water	CN109043009 (2018)	Weight loss
Weight loss beverage	0.395 g of cinnamon bark per serving	Oolong tea (Taiwanese tea, Min Lan tea), inulin, cacao powder, vanilla essential oil frankincense powder, ground nutmeg, Stevia (Rebaudioside A)	US20130344215A1 (2012)	Weight loss
Cinnamon oolong tea	Cinnamon extract	Black tea extract, white tea extract, guarana extract, oolong tea extract, green maté extract, choline; rooibos tea extract, yerba maté extract, grape pomace extract	US 7989009 B2 (2011)	Weight loss
Cinnamon tea	Boiled cinnamon	*Longan Arillus*, jujubes, honey	1020160013228 (2016)	Not specified
Pure natural beverage	Dried branches, leaves	Water	CN103445255 (2013)	Easy menstruation
Cinnamon oolong tea	Dried cinnamon bark 23%	Oolong tea	CN103504084B (2014)	Reduce blood sugar
Herbal tea	Not specified	Gymnema, bilberry (*Vaccinium myrtillus*), ginseng (*Panax ginseng*), fenugreek, marshmallow, bitter gourd, autumn crocus, bay laurel, colocynth, prickly pear	US08993008 (2015)	Control of diabetes

improving antioxidant activity and the physical stability of the drink (Muhammad et al. 2019). This cinnamon nanoparticle addition has improved the total phenolic content up to 40% and antioxidant activity up to 60%, evidencing the health boosting activity of cinnamon.

Oolong teas are considered as healthier tea types than ordinary black and green teas. These are semi-oxidized teas or semi-fermented teas. After oxidation and initial partial shaping like a ball, oolongs are heated or roasted to halt oxidation and shaped again. This process delivers darker aromas and strong flavors. Cinnamon has a constructive role in modern oolong tea industry and able to find novel applications (Li 2013, China Patent 103,445,255).

15.2.1.4 Cinnamon Alcoholic Beverages

There is historical evidence of cinnamon being used in brewing 5000 years ago, by Sumerians in Mesopotamia (Keersmaecker 1996). Cinnamon is a popular flavoring in numerous alcoholic beverages, such as "cinnamon liqueur," which is popular in Europe (Willard 2013). Fireball Cinnamon Whisky is a good example, which is among the top selling whisky brands in the United States in 2018. Cinnamon schnapps is another famous liqueur flavored with cinnamon. It is a distilled spirit, clear, with added red-colored liqueur. This spirit is also high proof, ranging from 40% to 50% alcohol by volume (80–100 proof). The actual flavor of cinnamon schnapps varies according to the brand. Commonly schnapps use the cinnamon accompanied with red hot candies. A few brands, e.g., Hiram Walker, use a more traditional cinnamon spice flavor with less heat (Graham 2019). Within the last decade, the United States and Europe have introduced many beer types that use cinnamon in the brewing process (Table 15.3). Some of them have already failed to perform in the market, while some products have been popular among the consumers.

Different types of rum are available with cinnamon in the market. In 2015, Bacardi launched a rum named spirit drink with exotic cinnamon flavors, which is made in torched oak barrels aged for a minimum of 1 year. *Orchata Cinnamon Cream rum was introduced by* Chila in the Virgin Islands that added dairy cream with cinnamon. There are many similar products throughout the world. Vodka types also are being introduced with cinnamon flavors added to the traditional flavors. Cinna-Sugar Twist, Cinnabon vodka, and Zinamom vodka are the few examples.

Peligroso's tequila is the first tequila released in the market with natural cinnamon extracts; it was launched in March 2013. This flavor is slightly spicy, has a nice balance of agave and warming cinnamon, and is a delight to sip straight and mix into cocktails. Mexico, the major importer of true cinnamon, announced Soltado, a new *tequila* infused with spicy serrano peppers and *flavored* with a *cinnamon.* Jose Cuervo Cinge cinnamon-flavored tequila is another famous cinnamon *tequila type*.

Cinnamon is also a famous ingredient in cocktails throughout the world. Bourbon cider cocktail with cinnamon and ginger, cinnamon toast crunch cocktail, apple cinnamon old-fashioned, caramel cinnamon martini, apple pie moonshine cocktail, homemade apple pie moonshine, pumpkin beer cocktail, spiced Manhattan cocktail, slow cooker cider wassail, and hot buttered hazelnut whisky are some of the examples of cinnamon-based cocktails.

15.2.1.5 Cinnamon Applications in Chewing Gum

Cinnamon flavors are very sensitive to certain ingredients in chewing gum compositions. This was the reason for the complications with the final products of cinnamon chewing gums, causing changes in texture, color, aroma, and other sensory quality characteristics. In sugarless cinnamon-flavored gums, certain sweeteners such as aspartame are unstable with the aldehyde-containing oils present in cinnamon. Researchers have conducted several trials to protect sweeteners such as aspartame

Table 15.3 Cinnamon-based beer types currently in the market

Name	Origin	Alcohol by volume (ABV)	Ingredients	Other remarks
Cigar City Hunahpu's Imperial Stout	Tampa, Florida, United States	11.0%	Cinnamon, cacao nibs, vanilla, and ancho and pasilla chilies	Ale
Perennial Abraxas – Barrel-Aged	St. Louis, Missouri, United States	11.0%	Rye, cacao nibs, ancho chilies, vanilla beans, and cinnamon	Imperial stout aged 11 months in Rittenhouse
Epic Big Bad Baptista	Salt Lake City, Utah, United States	11.7%	Cinnamon, vanilla, Mexican coffee roasted by blue copper, and solstice chocolate cacao nibs	Inspired by a traditional Mexican coffee
Cycle Trademark Dispute (Yellow)	Saint Petersburg, Florida, United States	No	Coffee, cinnamon	
Westbrook Mexican Cake Imperial Stout – Reserva	Mount Pleasant, South Carolina, United States	10.5%	Van winkle barrels, bourbon barrel-aged maple syrup, Tahitian vanilla, Valrhona cocoa nibs, cinnamon, and fresh habaneros	
Fremont Rusty Nail	Seattle, Washington, United States	13.2%	An imperial oatmeal stout with pale barley, smoked barley, brewer's licorice, cinnamon bark	Aged in Heaven Hill barrels for close to a year
Cycle Wednesday – 2016	Saint Petersburg, Florida, United States		Cinnamon hazelnut	
Bottle Logic Reaction State	Anaheim, California, United State	11.8%	Maple syrup barrel-aged, Tahitian-vanilla-bean-boosted, cinnamon-finished	The first level II release from Stasis Project 2017
Three Floyds Dark Lord – ChemTrailMix	Munster, Indiana, United States	15.0%	Cinnamon and pink peppercorns	
Cycle Greatest Hits &+	Saint Petersburg, Florida, United States	12.0%	Dried ancho, guajillo and pasilla peppers, as well as cinnamon and vanilla	
Bottle Logic Reaction State	Anaheim, California, United States	11.8%	Maple syrup barrel-aged, Tahitian-vanilla-bean-boosted, cinnamon-finished	

(continued)

Table 15.3 (continued)

Name	Origin	Alcohol by volume (ABV)	Ingredients	Other remarks
Perennial Abraxas	St. Louis, Missouri, United States	10.0%	Ancho chili peppers, cacao nibs, and cinnamon sticks	Ale
Magic Rock Bourbon Barrel Bearded Lady Dessert Edition	Huddersfield, West Yorkshire, England	10.5%	Chocolate, vanilla, cinnamon	
Nøgne Ø At Gale Force	Grimstad, Norway	15.0%	Vanilla and cinnamon in whisky	Use cognac barrels
Westbrook Mexican Cake Imperial Stout	Mount Pleasant, South Carolina, United States	10.5%	Cocoa nibs, vanilla beans, cinnamon sticks, and fresh habanero pepper	
Perennial Vanilla Abraxas	St. Louis, Missouri, United States	10.0%	Ancho chili peppers, cacao nibs, and cinnamon sticks with vanilla beans	
Wrecking Bar Mexican Siberius Maximus Russian Imperial Stout	Atlanta, Georgia, United States	12.5%	Smoked serranos, chipotle Morita peppers, cinnamon bark, toasted cocoa nibs, and vanilla beans	
Prairie Christmas Bomb! – Barrel-Aged	Tulsa, Oklahoma, United States	13.0%	Cinnamon, vanilla, chocolate, spaceship earth coffee, ancho chilies, and whiskey	
Cycle Rare DOS Double Barrel	Saint Petersburg, Florida, United States	11.0%	Cinnamon, hazelnut	

through coating formulations. In 1984, Warner-Lambert Company (US Patent 4,597,970) introduced cinnamon as a sweetener to chewing gum; however, discoloration occurs during storage of the product. The same company was granted a patent for a stable cinnamon-flavored chewing gum in 1988, comprising a gum base, cinnamon flavor, and a sweetener delivery system (Cherukuri et al. 1988/ US Patent 4,722,845). Recent studies disclosed more health benefits of cinnamon-based chewing gums. Commercial sugar-sweetened cinnamon chewing gum has claimed to have the benefit of preventing halitosis by reducing volatile sulfur compounds producing anaerobes in the oral cavity (Zhu et al. 2011). Current chewing gum products are introduced with novel techniques such as multiple layers filled in cinnamon and its flavors (Kabse et al. (2015) US patent application 0264958A1).

15.2.1.6 Cinnamon as a Natural Food Preservative

Food deterioration is caused by lipid rancidity. The antioxidant properties of polyphenols of cinnamon inhibit lipid rancidity and have been used as a natural food preservative for many centuries. Novel experiments still focus on cinnamon as a preservative in many food applications.

Almond paste consists of ground almonds and glucose syrup in the proportions 90/10 (wt/wt). This solid paste is available in two kinds of almond varieties in Morocco as cooked and non-cooked pastes. Faid et al. in 1995 found that adding 0.25% of cinnamon in almond paste has the ability to reduce yeast levels and extends its shelf life for 3 months without deterioration. The cinnamon extract incorporation into butter showed low levels of peroxide value, free fatty acid value, and low microbial count compared to butter without a preservative and potassium sorbate added butter. Antioxidant activity of cinnamon extends the shelf life of butter. Vidanagamage et al. (2016) showed that incorporation of 3% cinnamon extract can be used to formulate an antioxidant-rich butter, and it can be used as a natural preservative for preparation of butter.

Jo et al. (2015) developed anti-insect food packaging film containing cinnamon oil encapsulates. This is recommended to the food industry since this film does not affect the sensory qualities of the food products. For food preservation, plastic packaging material is innovated as a coated inner wall with cinnamon and garlic paste. It has been experimented considering the higher antioxidant and antimicrobial activity of cinnamon (El-Baroty et al. 2010). Garlic and cinnamon have proved useful as a natural food preservative for fish at room temperature, fried fish, and deep fried fish, since the activity index of the garlic and cinnamon is more than 0.5 (Ranjan et al. 2012). The antioxidant mechanism of chitosan-based coating with cinnamon is explained by Ojagh et al. (2010), whose experiments confirmed that this coating on fish enables its high quality to be retained longer, hence extending the shelf life of fish. Hu et al. (2015) report how cinnamon essential oil can be encapsulated into chitosan nanoparticles of three sizes. These nanoparticles improved the antioxidant and antimicrobial properties of pork and acted as a natural preservative for meat. The nanoparticles with larger size prolonged the shelf life of the chilled pork by maintaining the quality of the meat and associated meat products. In addition, this chitosan-based cinnamon coating activates defense-related enzymes, improves free radical scavenging activity, and the permeability and integrity of cell membrane in fruits. Xing (2016) disclosed the suitability of cinnamon as a natural preservative for fruits as well. The combination of nisin and cinnamon accelerates death of *Salmonella typhimurium* and *E. coli* O157:H7 in apple juice and enhances the safety of the product (Yuste and Fung 2002).

Vazirian et al. (2015) examined the preservative effects of cinnamon oil in cream-filled cakes. By analyzing the sensory quality of food preserved with cinnamon oil, they concluded that cinnamon is a natural preservative of food, especially of cream-filled cakes and pastries. Kordsardouei (2013) observed the preservative effects of cinnamon for cakes and suggested cinnamon as a natural food preservative sufficiently powerful to replace synthetic preservatives in foods. Researchers continuously have focused on medicinal plants for extracting natural antioxidants that can replace synthetic kinds that might be carcinogenic or toxic (Whysner et al. 1994).

15.2.1.7 Cinnamon Chocolate

It is believed that the chocolate originates from Mexico where the Mayas, Incas, and Aztecs cultivated the cacao tree. Since the sixteenth century, a cinnamon application with chocolate has been referred to (Lanza et al. 2011). Current studies incorporating cinnamon essential oil to chocolate mostly aim to improve the sensory qualities of chocolate, especially the aroma. Recent research showed that bicycle (3.1.1. heptane, 6-methyl), 8-methyl-6,8-Nonadien-2-one, bicycle (3.1.hexan-2-ol, 2-methyl), copaene, and 1-butanol, 3-methyl-benzoate occurred as a result of interaction between cinnamon and cocoa mass during conching (Albak and Tekin 2015). Further, this study evidenced that combining dark chocolate with cinnamon improved the fruity and flowery aroma of chocolate. Much of the research has been done on the use of cassia, not true cinnamon, in chocolate. Future studies and applications are needed in this area.

15.2.2 Personal and Home Care Products of Cinnamon

Personal products include mouthwash, toothpaste, toothpicks, dental floss, and deodorants. Dental caries and periodontal diseases are caused by microorganisms, and there is a need to apply an antibacterial agent to protect against them. Cinnamon has a strong history of being a natural component of dental care products due to antibacterial, anti-inflammatory, and antifungal activities in its essential oils. Recently many researchers caused alarm about the cinnamon application for mouthwash and toothpastes but without considering the type of cinnamon (Tremblay and Avon 2008; De Groot 2017). There is no proven evidence with true cinnamon of any harm, while cassia has been shown to have several disadvantages. Recent research with mouthwash containing true cinnamon extract (50%) has not been reported as having any harmful effects (Gupta et al. 2015). Oral hygiene targeted cinnamon toothpastes containing bark or bark oil are continually being introduced into the market by many product developers (Telrandhe et al. 2017; Akotakar et al. 2018).

Consumers are looking for more natural, aluminum-free, and silicone-free deodorants. Cinnamon is building good reputation in future deodorant products in the industry; this is proved by the number of current patent applications in the United States (Sturgis and Britt US patents 2019/0000747A1, Sturgis and Jones 2019/0000736A1). Detergents, candles, and incense sticks, among many other products, are available in the current market owing to cinnamon's natural and unique aroma. There is, therefore, considerable evidence for the traditional belief that cinnamon incense gives protection to people.

Acknowledgments We are grateful to Mr. Sarada De Silva, Mr. Vijith Jayathilake, and Mr. D.A. Perera for their dedicated and expeditious help in arranging field work as well as linking up the research team with overseas markets. We like to thank the participants in our focus groups for their time and willingness to share their experiences with us. We thank the project coordinators of the special project on cinnamon, Prof. Ranjith Senaratne and Prof. K.D.N. Weerasinghe, and our contact persons at the National Science Foundation for their enormous efforts in the development of cinnamon research. Finally, but not the least, we wish to convey our sincere thanks to the Ministry of Primary Industries, for opening up the research grant scheme on Ceylon cinnamon and the grant for value chain development (SP/CIN/2016/05) under which this research work was accomplished.

References

Akotakar AM, Thenge RR, Patil AV, Ghonge AB, Bhaltadak MB (2018) Formulation and comparative standardization of toothpaste. Int J Pharm Sci Res 3:12–15

Albak F, Tekin AR (2015) Effect of cinnamon powder addition during conching on the flavor of dark chocolate mass. J Food Sci Tech 52:1960–1970

Bhatia S, Morad MM (2007) US Patent 2007/0,218,167A1

Cherukuri SR, Mansukhani G, Jacob KC (1988) US Patent 4,722,845

De Groot A (2017) Contact allergy to (ingredients of) toothpastes. Dermatitis 28:95–114

De Keersmaecker J (1996) The mystery of lambic beer. Sci Am 275:74–80

Dogan M, Kayacier A (2004) Rheological properties of reconstituted hot salep beverage. Int J Food Prop 7:683–691

Dunker MJ, Wendland MJ (2002) US Patent 6,468,569

El-Baroty GS, El-Baky HA, Farag RS, Saleh MA (2010) Characterization of antioxidant and antimicrobial compounds of cinnamon and ginger essential oils. Afr J Biochem Res 4:167–174

Faid M, Bakhy K, Anchad M, Tantaoui-elaraki ABDE (1995) Almond paste: physicochemical and microbiological characterization and preservation with sorbic acid and cinnamon. J Food Prot 58:547–550

Fonken ME (1940) US Patent 2,217,896

Graham C (2019) Cinnamon Schnapps profile and drink recipes. https://www.thespruceeats.com/cinnamon-schnapps-profile-drink-recipes-760258. Accessed 20 Sept 2019

Gupta D, Nayan S, Tippanawar HK, Patil GI, Jain A, Momin RK, Gupta RK (2015) Are herbal mouthwash efficacious over chlorhexidine on the dental plaque? Pharm Res 7:277

Hansen LM, Anderson BR, Lorence MW, Reinke JD (2003) US Patent 6,589,583

Harindra M (2017) Branding as a secure vehicle for Ceylon cinnamon to win the international market. An unpublished dissertation, Sabaragamuwa University of Sri Lanka

Hartman JB (1964) US Patent 3,158,057

Holly S (1976) US Patent 3,930,733

Hu J, Wang X, Xiao Z, Bi W (2015) Effect of chitosan nanoparticles loaded with cinnamon essential oil on the quality of chilled pork. LWT- Food Sci Technol 63:519–526

Jayaprakasha GK, Negi PS, Jena BS, Rao LJM (2007) Antioxidant and antimutagenic activities of *Cinnamomum zeylanicum* fruit extracts. J Food Compos Anal 20:330–336

Jo HJ, Park KM, Na JH, Min SC, Park KH, Chang PS, Han J (2015) Development of anti-insect food packaging film containing a polyvinyl alcohol and cinnamon oil emulsion at a pilot plant scale. J Stored Prod Res 61:114–118

Kabse K, Norton CR, Zeller BL, Elejalde CC, Kamakura Y, Ueminami A (2015) US Patent application 2015/0,264,958A1

Kandaswami C, Stohs S, Coyle C (2011) US Patent 7,989,009

Kim DS (2016) South Korean Patent 1,020,160,013,228

Kirksey N, Fitzgerald S (2007) US Patent 2007/0,178,205A1
Kordsardouei H, Barzegar M, Sahari MA (2013) Application of *Zataria multiflora* Boiss. and *Cinnamomum zeylanicum* essential oils as two natural preservatives in cake. Avicenna J Phytomed 3:238
Kwitek BJ (2011) US Patent 8,057,834
Lanza CM, Mazzaglia A, Pagliarini E (2011) Sensory profile of a specialty Sicilian chocolate. Ital J Food Sci 23:36–44
Li H (2013) China patent 103,445,255A
Liotta DC, Petros JA, Wey SJ, Karr JF, Pohl J (2000) US Patent 6,153,596
Muhammad DRA, Gonzalez CG, Doost AS, Van de Walle D, Van der Meeren P, Dewettinck K (2019) Improvement of antioxidant activity and physical stability of chocolate beverage using colloidal cinnamon nanoparticles. Food Bioproc Tech 12:976–989
Ojagh SM, Rezaei M, Razavi SH, Hosseini SMH (2010) Effect of chitosan coatings enriched with cinnamon oil on the quality of refrigerated rainbow trout. Food Chem 120:193–198
Qais FA, Khan MS, Althubiani AS et al (2019) Understanding biochemical and molecular mechanism of complications of glycation and its management by herbal medicine. In: Mohd SAK, Iqbal A, Debprasad C (eds) New look to phytomedicine. Academic Press, Massachusetts, pp 331–366
Ranjan S, Dasgupta N, Saha P, Rakshit M, Ramalingam C (2012) Comparative study of antibacterial activity of garlic and cinnamon at different temperature and its application on preservation of fish. Adv Appl Sci Res 3:495–501
Roy S, Ray MM, Weber JL (2011) Canada Patent 2,615,571A1
Sturgis DA, Britt LM (2019) US Patent 2019/0,000,747A1
Sturgis DA, Jones SD (2019) US Patent 2019/0,000,736A1
Telrandhe R, Deshmukh P, Gunde M (2017) Formulation and evaluation of herbal toothpaste: compared with marketed preparation. Int J Pharm Drug Anal 5:406–410
Tremblay S, Avon SL (2008) Contact allergy to cinnamon: case report. J Can Dent Assoc 74:445–461
Vazirian M, Alehabib S, Jamalifar H, Fazeli MR, Najarian TA, Khanavi M (2015) Antimicrobial effect of cinnamon (*Cinnamomum verum* J. Presl) bark essential oil in cream-filled cakes and pastries. Res J Pharmacogn 2:11–16
Vidanagamage SA, Pathiraje PMHD, Perera ODAN (2016) Effects of cinnamon (*Cinnamomum verum*) extract on functional properties of butter. Procedia Food Sci 6:136–142
Waldemar K (1942) US Patent 2,305,712
Warner-Lambert Company (1984) US Patent US 4,597,970
Whysner J, Wang CX, Zang E, Iatropoulos MJ, Williams GM (1994) Dose response of promotion by butylated hydroxyanisole in chemically initiated tumours of the rat forestomach. Food Chem Toxicol 32:215–222
Willard H (2013) 11 cinnamon-flavored liquors for the holidays. http://www.thedaily meal.com/11-cinnamon-flavored-liquors-holidays/121613. Accessed 10 Sept 2019
Woo K, Sukh B, Clodualdo Mt, Lianfu Z, Edward T, George K, Shishir R et al (2006) US Patent 2006/0,188,631A1
Xing Y, Xu Q, Yang S, Chen C, Tang Y, Sun S, Zhang L, Li X (2016) Preservation mechanism of chitosan-based coating with cinnamon oil for fruits storage based on sensor data. Sensors 16:1111
Young DG, Schreuder M (2012) US Patent 20,130,344,215A1
Yuste J, Fung DYC (2002) Inactivation of *Listeria monocytogenes Scott* A 49594 in apple juice supplemented with cinnamon. J Food Prot 65:1663–1666
Zhang P, Zhang S, Wang D, Qiu Z, Wang Y, Yi G, Zhang X (2014) China patent CN103,504,084B
Zhu M (2018) China patent 109,043,009A
Zhu M, Carvalho R, Scher A, Wu CD (2011) Short-term germ-killing effect of sugar-sweetened cinnamon chewing gum on salivary anaerobes associated with halitosis. Int J Clin Dent 22:23–26

Chapter 16
Cinnamon Value Chain Dynamics and Enhancement

Achini M. De Silva and Mohamed Esham

16.1 Introduction

16.1.1 Value Chain

The chain of interconnected actors and their functions together delivering a bundle of attributes to the customer is the wholesome meaning of the concept (Kaplinsky and Morris 2001). A value chain comprises a series of activities that create and build value at every step. Porter (1985) introduced the concept of value chain comprising a set of activities performed by the firms of an industry that deliver a valuable product to the market. The value chain ideology is based on process view of the organization where a manufacturing organization operates as system made up of sub-systems (Cambridge University 2013). The total value delivered by the company is the sum of the value built up through the company (Porter 1985).

Agriculture value chains enwrap the flow of products, knowledge and information among key actors including input suppliers, farmers (or growers), transporters, sellers and consumers. The cinnamon value chain is created by stakeholders involved in input supplying, growing, processing, value addition and marketing of cinnamon-based products. Each actor in the value chain is an important player, and each needs to understand others' needs for better product performance and coordination of the chain for improved efficiency, customer satisfaction and profitability (Gwabu 2015). Furthermore, the value chain perspective helps to understand business relationships that connect the chain, increase efficiency and improve productivity (Webber and Labaste 2010). Value chains aim to produce value-added products or services for a target market, by transforming resources and utilizing the infrastructure within the business ecosystem (Trienekens 2011). Opportunities and

A. M. De Silva (✉) · M. Esham
Department of Agribusiness Management, Faculty of Agricultural Sciences, Sabaragamuwa University, Belihuloya, Sri Lanka

R. Senaratne, R. Pathirana (eds.), *Cinnamon*,
https://doi.org/10.1007/978-3-030-54426-3_16

constraints for value chain development are related to local, regional and international market access and market orientation (Grunert et al. 2005); available human, natural, social and financial resources together with physical infrastructures (Porter 1990); and regulative, cognitive and normative institutions (Scott 1995a, b). In general, value chain actors connect through vertical and horizontal linkages forming a network structure (Trienekens 2011).

Value addition and creation in each node is a key function of the value chain actors. Governance of value chain is determined by the authority and power relationships among the actors that determine resource allocation within a chain (Williamson 1999; Rindfleisch and Heide 1997; Gibbon et al. 2008; Gereffi 1994). Value chain improvement requires players/participants to do something different by changing their behaviour or market conditions or creating new market opportunities. Changing market conditions is supported by access to lump sums of money for investment; adoption of new technologies, new products or production processes; adding new functions that add value; or selling through new market channels and the development of new business relationships. The value chain is now increasingly used as a methodology for identifying appropriate points to provide opportunities for the stakeholders to obtain productive work in conditions of equity, freedom, security and dignity.

16.2 Value Chain Analysis: A Case of Ceylon Cinnamon

Ceylon cinnamon (*Cinnamomum zeylanicum* Blume) is among the first traded spices of the world, led the global market for many centuries aided by monopoly in the industry and was known for its medicinal, chemical and aromatic properties. Pharmaceuticals, baked food and confectionary, cosmetics and perfumes were the main global market segments for cinnamon. Historically, Sri Lanka has retained an unparalleled position among countries supplying cinnamon to the world market (Rupasena et al. 2007). However, the country's cinnamon products are exported with minimum value addition causing huge losses in relation to potential export earnings. So far, Sri Lankan producers have not really tapped into the global market, although the country is catering to 40 international market destinations. Sri Lanka contributes over 80% of world's production of true cinnamon (Fig. 16.1) with an annual production of approximately 16,000 MT (EDB 2018).

Cinnamon grown and produced in Sri Lanka has acquired an established reputation and demand in the international market owing to its unique fragrance, quality, colour, flavour and aroma as well as its remarkably low amounts of coumarin. Cinnamon ranks first as key export earner for Sri Lanka among the spice basket (54% of export earnings from total spice exports and 8% of total agricultural exports) while securing employment opportunities for about 400,000 people (EDB 2018).

Figure 16.2 shows the conceptual framework for the cinnamon value chain analysis. The general view of the conceptual framework is divided into three pillars;

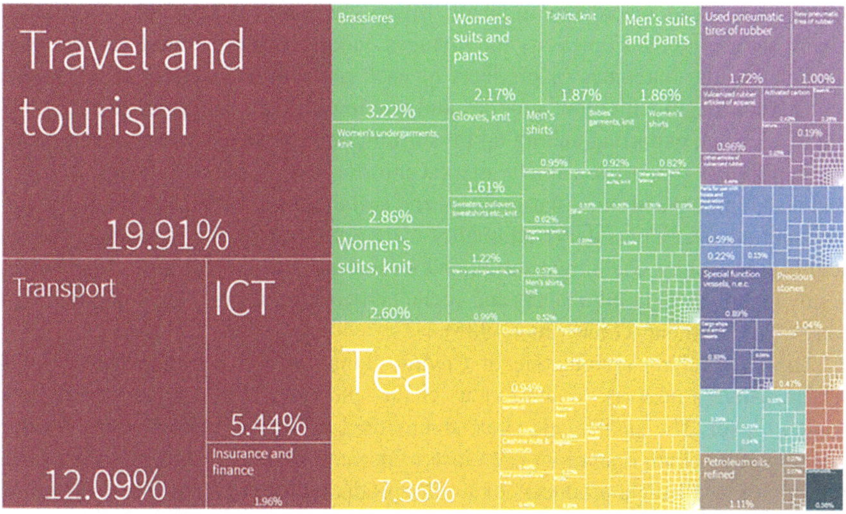

Fig. 16.1 Sri Lanka's exports in 2017 by products, Harvard atlas of complexity. (Source: Centre for International Development, Harvard University)

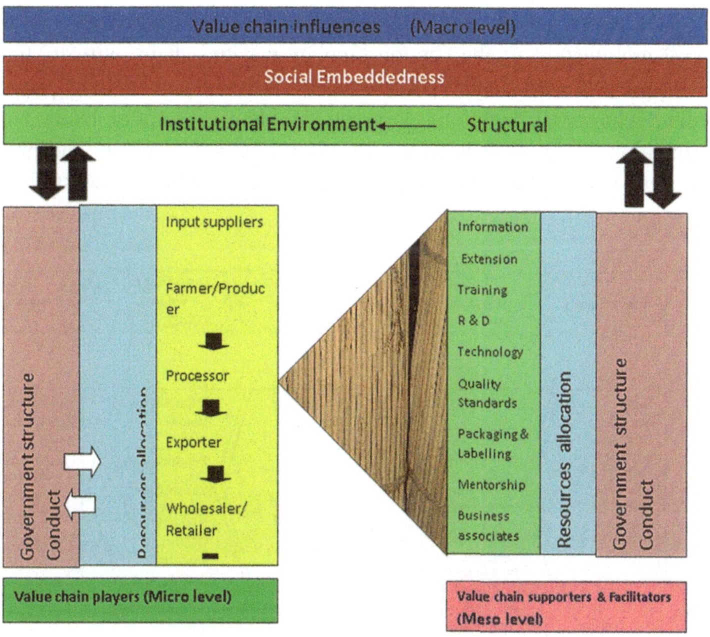

Fig. 16.2 Value chain analysis framework

micro, meso and macro environment. Micro environment consists of the value chain, actors and their functions and linkages. Value chain actors perform functions to generate the outcomes or products of each node. Key functions of cinnamon value chain begin with input supplies and moving through the production, trading, value addition, exporting ending up in consumption. Meso environment represents the supportive functions, namely, market information, financial services, research and development activities, partnerships and collaboration with institutions, promotion and advocacy on safety and quality management. Macro environment factors identify the policy, enabling a sustainable business environment. Institutional environment, value chain influencers and social embeddedness are main components of macro environment (Granovetter 1985).

The value chain architecture of Ceylon cinnamon has a fragile but complex nature where a network of short value chains end at local middlemen (collectors/traders) level. The small-scale short chains feed the longer export-oriented value chains. Value chain operations are close to perfect market conditions with a large number of small-scale producers of similar products that are price takers. A limited number of locally based collectors govern the chain through price decisions and volume demands. In contrast, export-oriented cinnamon value chains are longer with several nodes and are governed by exporters. Value chain membership varies with type of business operation or product flow through the chain. Limited value addition is a common feature of the Ceylon cinnamon industry. Figure 16.3 describes the general architecture of the Ceylon cinnamon value chain, with its functions, actors and products. Value chain comprises two hypothetical halves, upstream and downstream, where each function is governed by identical actor or actors. Upstream of the cinnamon value chain begins with input suppliers, growers and processors those who are responsible for primary production activities. Downstream of the

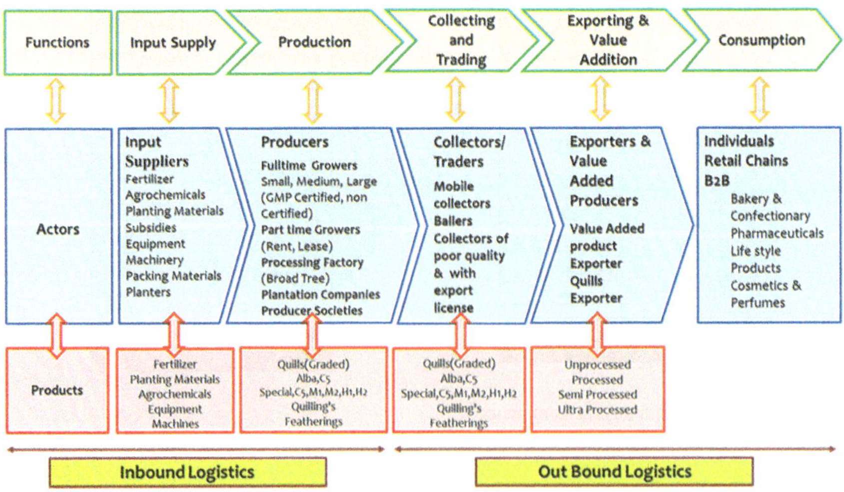

Fig. 16.3 Ceylon cinnamon value chain physical flow

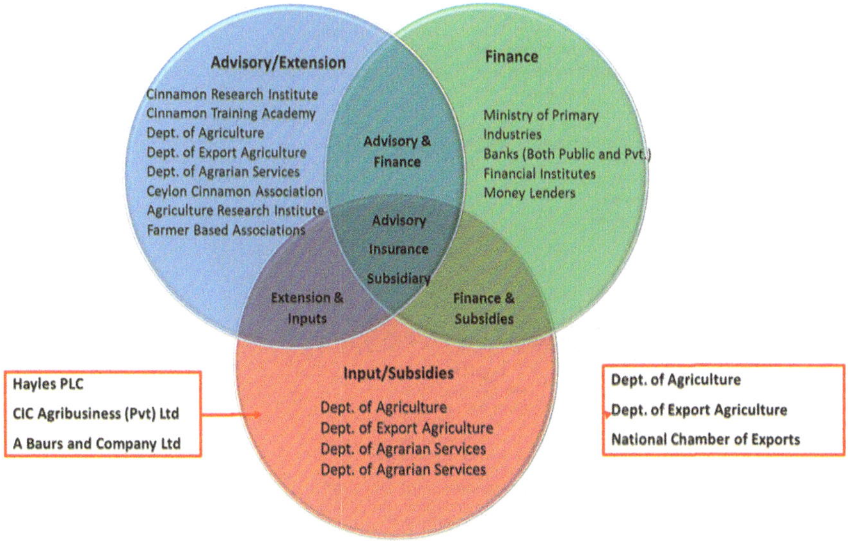

Fig. 16.4 Stakeholder map of the upstream of cinnamon value chain in Sri Lanka

cinnamon value chain represent the traders and collectors, exporters, secondary processors, retailers and end users.

The meso environment of the cinnamon value chain (Figs. 16.4 and 16.5) concerns its stakeholder composition. Institutional stakeholders facilitate value chain functions through financing, advisory and extension services, input supply subsidies for upstream and logistics, export operation, insurance, inventory management, legal and regulatory provisions and providing the manufacturing infrastructure for the downstream. In general, both government and private sector institutions were leading the supportive services arm. Common features were overlapping mandates, fragile organizations, poor institutional networks as well as weak collaborations with small-scale producers.

The cinnamon value chain begins at the input supplier node, and the suppliers are responsible for providing inputs such as planting material, fertilizer, agro-chemicals, tools and equipment, credit and finance for the shorter value chains and logistics, packaging, financing, testing, etc., for the longer export-oriented chains.

Cinnamon growers or planters represent an important upstream node, dominated by male land owners (76%) followed by female growers (24%). In general, land owners belong to the middle-aged to aged categories (age 51–60 years, 29%; age 41–50 years, 24%; age > 60 years, 17%). Experience filtered through generation after generation has helped them to develop a unique plantation management system where owners share the harvest with cinnamon processors (peelers). Aging cinnamon planters, successors living abroad or far from the cinnamon business, absence of successors and land fragmentation are critical issues that hinder future expansion. In general, cinnamon plantations can be categorized into three

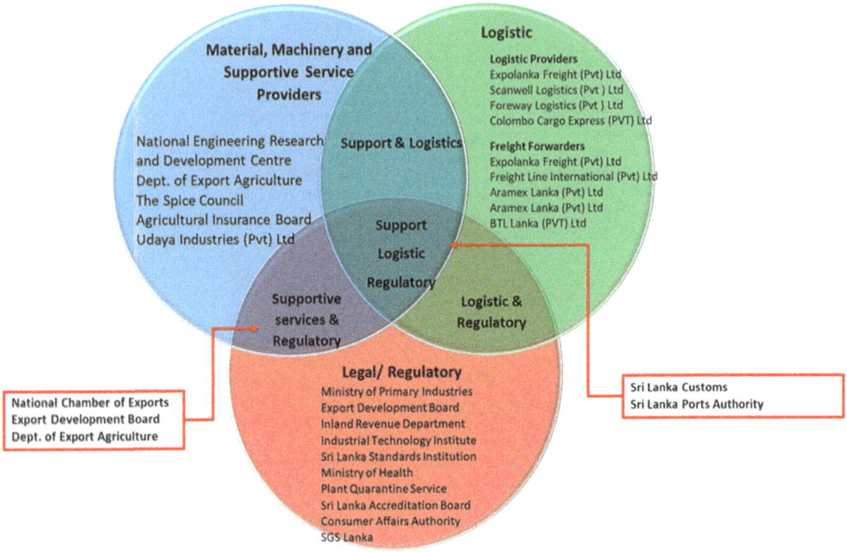

Fig. 16.5 Stakeholder map of the downstream of cinnamon value chain

sub-categories: small scale (<0.8 ha or 2 acres), medium scale (0.8–2 ha or 2–5 acres) and large scale (>2.4 ha or >6 acres). Composition of the cinnamon plantations is skewed with about 71% small-scale operations and 9% large-scale operations (Figs. 16.6 and 16.7).

Functions, responsibilities, attitudes, value and behaviour of growers are decisive in producing quality outcome. Figure 16.8 identifies the essential requirements (percentage of cinnamon growers) of the cinnamon growers based in key cinnamon-growing areas of the country: Galle, Matara, Rathnapura and Kalutara districts. Quality and availability of services varied across the cinnamon-growing areas. Extension services, training, financial support, knowledge on value addition and creation, market intelligence, subsidies and governance issues were considered as important aspects. Value chain services free from intermediaries were a prime concern of the Kalutara district, and Rathnapura demanded extension services, while Matara demanded information access.

16.2.1 Cinnamon Processors

The art of cinnamon processing (peeling) is carried out by artisans who are groomed and trained for many years. From generation to generation, the skills of peeling cinnamon bark have been passed down from father/mother to son/daughter. Cinnamon quill is unique to Sri Lanka and bears the identity of the country of origin. Quill making; only a selected artisans have perfected the traditional methods of peeling

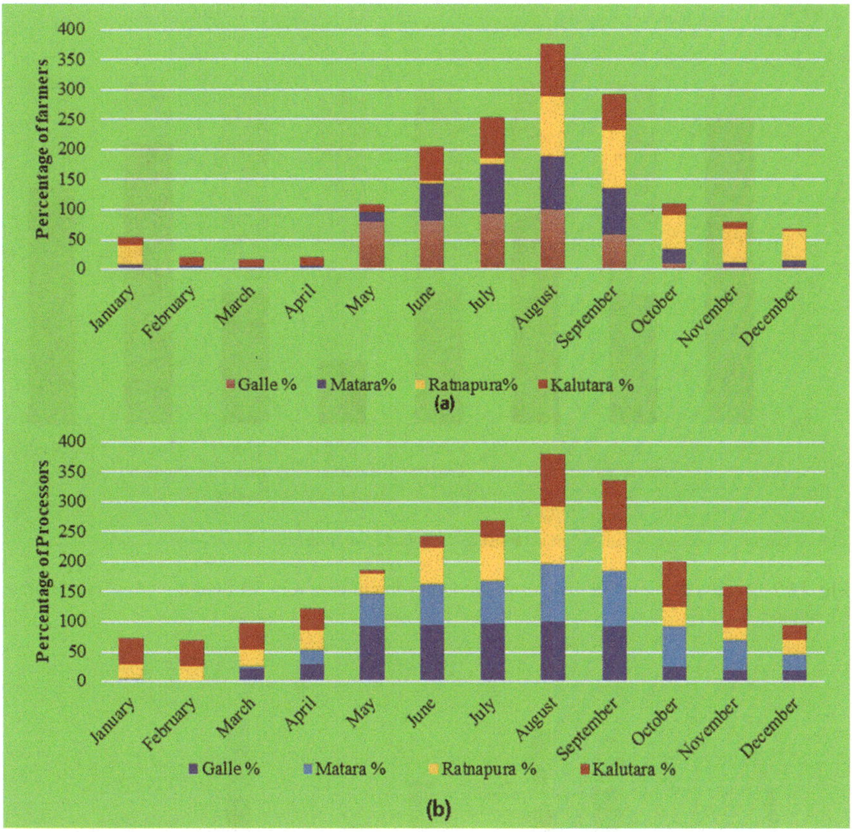

Fig. 16.6 Monthly income variability among growers (**a**) and processors (**b**) in main cinnamon-growing districts of Sri Lanka

through and deftly rolling the cinnamon into quills which are then dried. The processing of cinnamon complies with the standards (GMP/HACCP/ISO 9000 series, BRC, FSSC and organic) for the food and beverage and pharmaceuticals.

The art of cinnamon production is the result of skill and technique unique to the cinnamon processors (peelers) of Sri Lanka. The cinnamon processor's role at the production node is a decisive element in the entire value chain. Non-availability of essential numbers of peelers (about 15,000 cinnamon processors are employed in the sector currently, but the industry requires 350,00 processors for the optimum level of cinnamon processing) hinders the expansion of cinnamon industry as well as diminishes the earning power of the value chain. Cinnamon processors are organized into different systems to perform their duties: In the Kalli system, two to five members, particularly from the same family circle, are organized as a group to perform cinnamon processing with equal opportunity available to each member and income sharing equally among the members (Samarawickrama 2017; Amadoru 2017). In contrast is the line system, where cinnamon processors perform duties on

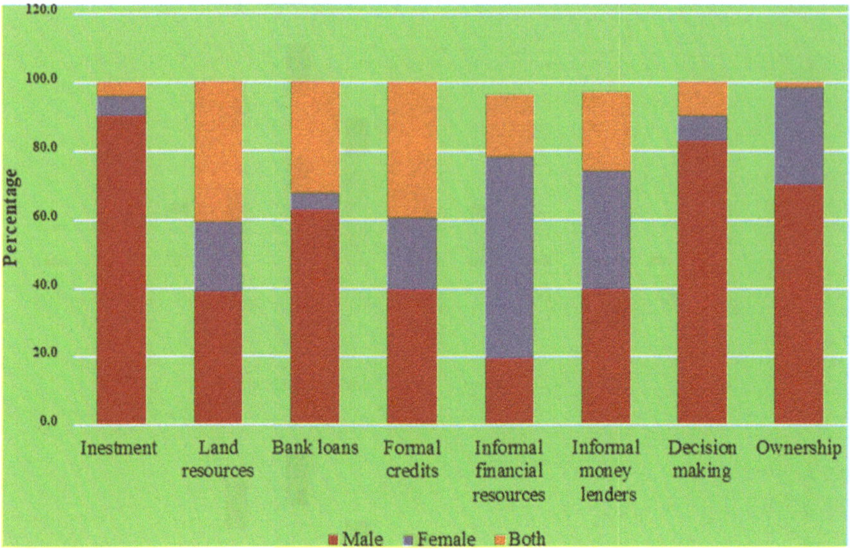

Fig. 16.7 Access and control profiles of resources in cinnamon in Sri Lanka – growers. (Source: Chandimal 2017)

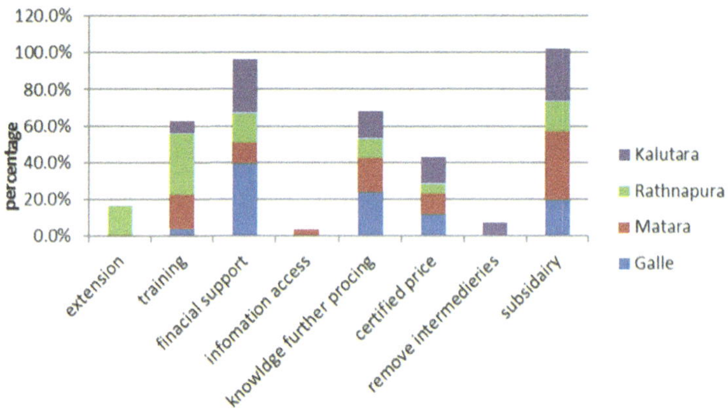

Fig. 16.8 Essentials of grower node in key cinnamon-growing areas

a daily paid basis and processors are trained for specialized tasks of quill making. Figure 16.9 identifies the important human resources management issues of the cinnamon value chain and responses (%) of cinnamon processors of four main districts of cinnamon cultivation in Sri Lanka. Lack of job security ranked first, followed by poor working conditions and lack of freedom. Value chain upgrading and promotion essentially requires strategic interventions to mitigate the human resources issues of the value chain. Figure 16.10 identifies the main concerns of processors (% responses) where financial security, pension scheme and subsidies ranked as the top most priorities.

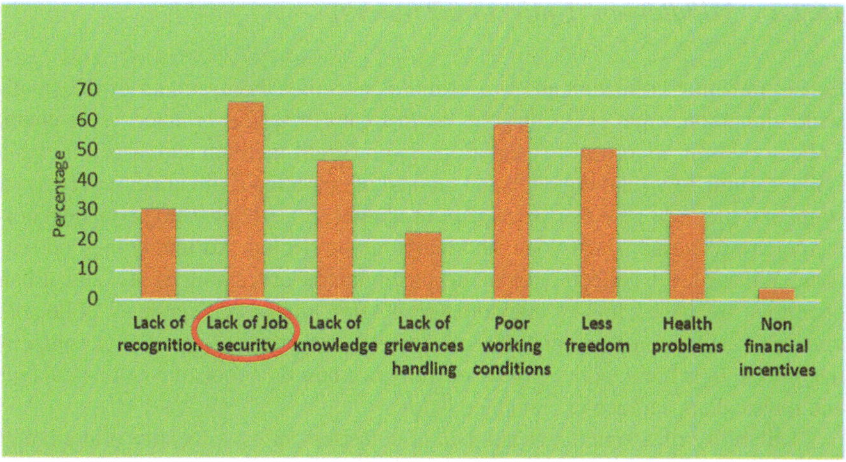

Fig. 16.9 Critical issues of human resources management in cinnamon value chain. (Source: Amadoru 2017)

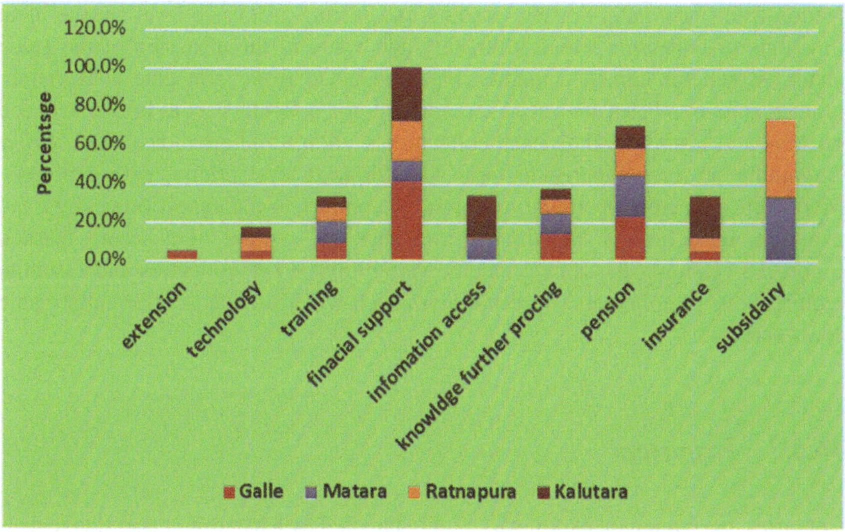

Fig. 16.10 Essentials of cinnamon processors. (Source: Amadoru 2017)

16.2.2 Middlemen (Collectors/Traders)

Local cinnamon collectors act as a bridge connecting the upstream of the small-scale cinnamon value chain into the downstream of export-oriented value chain. Collectors differ in terms of scale of operation (small-medium and large), location (village, rural, town and regional) and method of operation (individuals and contract suppliers; they can be either permanent or temporary). About 4000 permanent collectors are scattered island-wide, and temporary collectors work only in good seasons. Village-level collectors exist in all cinnamon-processing areas, and urban operators are based in key cinnamon-processing towns. Village-level collectors maintain regular contact with cinnamon growers and processors and offer competitive prices. In general, collectors decide the price based on moisture content, colour and physical appearance of cinnamon bales.

Availability of a significant number of collectors in each locality offers better bargaining power for both growers and processors. Personal contact with collectors, quality and regular supplies usually ensure better prices for growers. Mixed bales (different cinnamon grades bundled together to make a bale) are the common product form at farm gate level, and collectors use their own transport facilities for collection from farm gate. Sorting, grading and storing at a basic level are the main functions of collectors which ultimately add value to the next step of the value chain. In general, mixed cinnamon bales collected from farm gate convert into graded cinnamon bales are priced accordingly, and collectors act as bridge between growers and exporters. Individual collectors at village and urban level have better bargaining power compared to farm gate for next-level pricing with exporters. Moreover, collectors operating as contract suppliers to the export processors and exporters have limited bargaining power but are assured of a secure market. Figure 16.11 shows the annual income variability (%) of collectors and essential requirements of collectors (% responses) which will facilitate to offer better value chain services.

16.2.3 Exporters

Centuries ago cinnamon export was mainly handled by a handful of traditional business families and the monarchy. Export segment of the present cinnamon market landscape of is governed by a limited number of traditional players with few innovative players from non-traditional cinnamon families. Land fragmentation, migration of traditional large-scale cinnamon growers and absence of successors are the main issues.

Present status of cinnamon exporters composed of 266 registered exporters where 130 are occasional exporters, 48 regular exporters and 88 inactive members (EDB 2018). Family-owned cinnamon export business is managed by members of the family and in generally headed by the male members. Collaborative

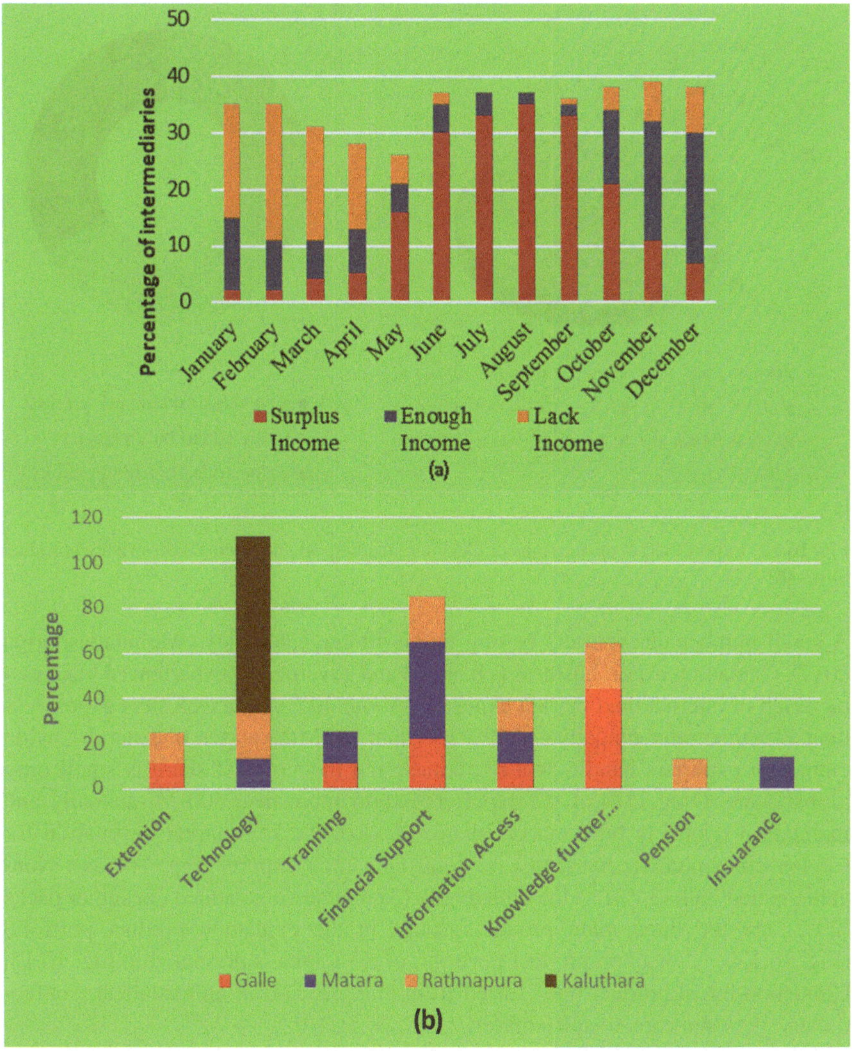

Fig. 16.11 (**a**) Annul income variability of collectors. (**b**) Essentials of cinnamon collectors

interventions with export processing and/or value addition with local or foreign expertise are rare in Ceylon cinnamon export. Family-owned and archaic management strategies, each with its own international market efforts and linkages, are common in the business landscape. Joint efforts for upgrading and promotion in the cinnamon value chain are ignored on most occasions. Individual efforts and strategic interventions have opened up several opportunities, and industry-wide upgrading is rare (Thanthirige 2011).

A majority of the exporters (42%) belong to the 41- to 50-year age category; middle (41–50 years) to old (>60 years) age groups represent 91% of the exporters.

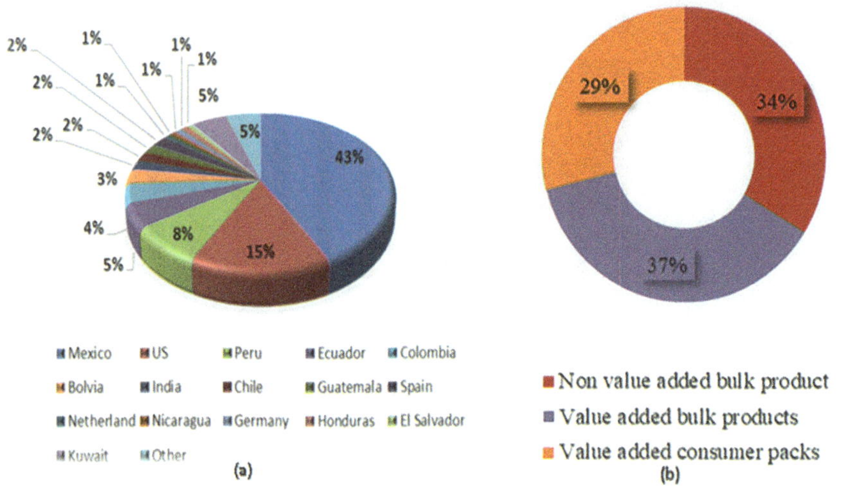

Fig. 16.12 Export market destinations of Ceylon cinnamon (**a**) and main export-oriented product forms (**b**)

A small number of exporters belong to 31–40 age categories (Warnakulasooriya 2017). Cinnamon export business is handled and governed by experienced businessmen rich in international market exposure. About 40% of the exporters export less than 50 Mt/exporter annually and 49% less than 100 Mt/exporter. In general, Ceylon cinnamon exporters handle small volumes, and their market share is small on a global scale. Only 11% of the exporters export more than 500 Mt annually and maintain a relatively high market share. Mexico, the key market destination for Ceylon cinnamon, receives 48% of total cinnamon exports (Fig. 16.12a). Non-value-added bulk (34%), value-added bulk (37%) and value-added consumer packs (29%) are the three main product groups in the cinnamon product portfolio (Fig. 16.12b). Cinnamon product portfolio of Sri Lanka is presented in Fig. 16.13. Cinnamon-based product lines varied from pharmaceutical, nutraceutical, confectionary, personal care, to spice mixes.

Ethno domination is observed in the Ceylon cinnamon value chain where over 90% all value chain actors are Sinhalese Buddhists. Portuguese rulers handed over the cinnamon business to low country people and British declared a separate cast, Salagama, for the cinnamon growers. For centuries, the cinnamon business was managed by the traditional Sinhalese families. Moreover, Cinnamon plantations, processing and value addition ventures are based in the southern part of the country where the plantation owners are Sinhala people.

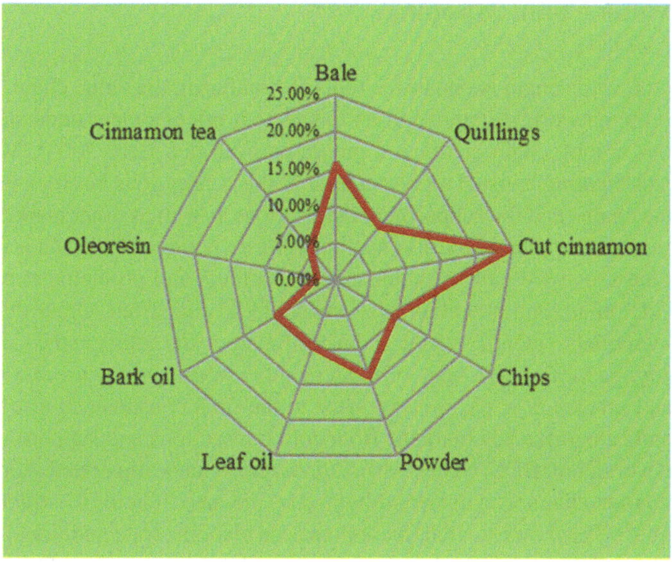

Fig. 16.13 Composition of Ceylon cinnamon export basket

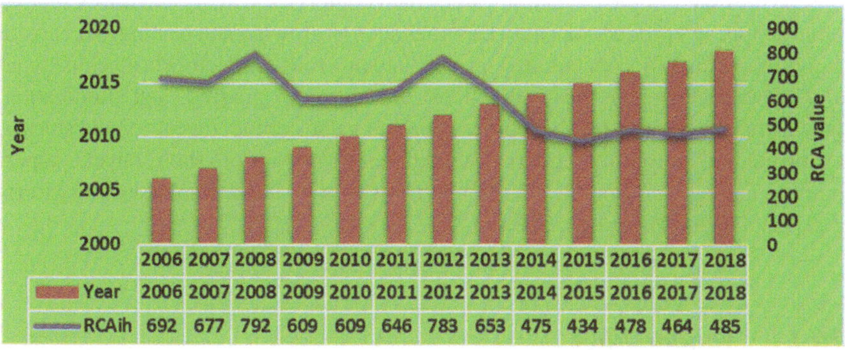

Year	2006	2007	2008	2009	2010	2011	2012	2013	2014	2015	2016	2017	2018
RCAih	692	677	792	609	609	646	783	653	475	434	478	464	485

Fig. 16.14 Revealed comparative advantage index of Ceylon cinnamon exports. (Source: Rajapaksha 2019)

16.3 Export Competitiveness

Export competitiveness was measured by using revealed comparative advantage (RCA), and the RCA index (RCAI) for cinnamon from 2006 to 2018 in Sri Lanka is shown in Fig. 16.14. Competitiveness of Ceylon cinnamon varied with the time. The peaks in 2008 and 2012 are due to favourable international market conditions and foreign exchange rates and drawbacks faced by its competitor, cassia cinnamon (Rajapaksha 2019). RCAI was lowest in recent years, 2014–2018, and the market downturn is due to low level of supplies, unfavourable foreign exchange rates, safety and quality issues and the price of cinnamon being higher than that of cassia.

16.4 Value Chain Mapping

Mapping the value chain is one of the tools for analysing the value chain. The map visualizes the network structure and its linkages. It offers a clear understanding of connections among chain actors and processes (Van den Berg 2004). Mapping a chain means creating a visual representation of the connections between businesses in the value chains and the supporting organizations as well as other market players. Comprehensive mapping of the all-encompassing activities in the chain is important to gain a basic overview of the value chain. Conflicts and problems at each node could be identified, and potential solutions could be suggested. Other than that, it was very essential to identify strong and weak linkages and actors in the value chain.

By using the value chain map, one can readily understand the process by which product and service flow past several steps to the final consumer. It also serves to identify and categorize key markets. Both core transactions and supporting organizations (government, BDS, NGOs, associations, etc.) can be illustrated. Additionally different market channels will be highlighted by the map. It helps to explore market opportunities. Based on research, these maps can also illustrate additional information on the relevance of individual market channels and the nature of relationships (e.g. number of competitors, size of the market, number of workers, working conditions, value chain governance, etc.).

The upstream of the cinnamon value chain comprises primary processes such as input supply and production based activities such as growing and peeling. Midstream activities are mainly concentrated on middlemen's role. Collectors bridge upstream and downstream and main functions are sorting, grading and fumigation. Downstream activities involved processing and value adding, inventory and storing, exporting and freight forwarders. International actors are off-shore buyers, supermarkets and retailers and the consumers (Neilson et al. 2014; Kaplinsky and Morris 2001; Dolan and Humphrey 2000).

16.4.1 Volume Flow

Volume flow illustrates volumes of value chain such as number of members, number of jobs, volume of production of each member and how to flow out the volume across the value chain. Figure 16.15 shows the volume of processed cinnamon handled by each actor of the value chain. Majority, small- and medium-scale producers are dealing with collectors, and about 70% of produce passes on to collector node, while large-scale producers are bypassing the collector node and supply their produce directly to exporters. Negligible portion of small- and medium-scale producers (8%) and majority of large-scale producers (20%) are able to eliminate the middlemen node which secures better returns to them. Collectors facilitate the smooth function of the value chain where 80% of produce handled by them transfer to the exporters. About 14% of produce ended up in local market for insider value addition

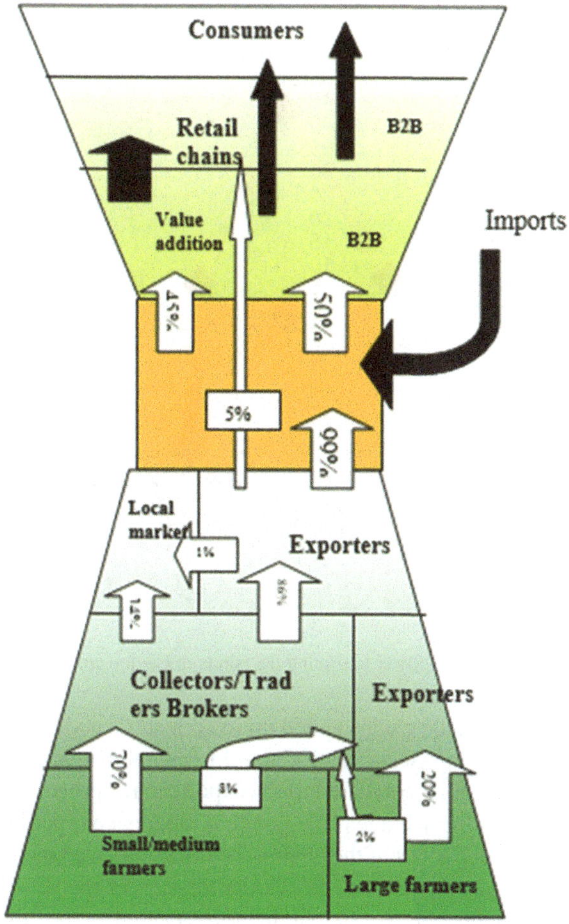

Fig. 16.15 Cinnamon value chain map, volume flow

and product development for local market, i.e. spice mixes, cinnamon tea, cosmetics, candles, etc. Fifty percent of export volume ended up in business-to-business operations overseas, and 45% goes for further value addition beyond the national boundaries.

16.4.2 Value Flow: Price

Value flow of the cinnamon value chain describes the prices at different nodes of the value chain. Farm gate prices vary season to season as well as region to region based on physical quality of the produce and demand conditions. Price bargaining ability of the value chain actors varies according to the value chain and their position.

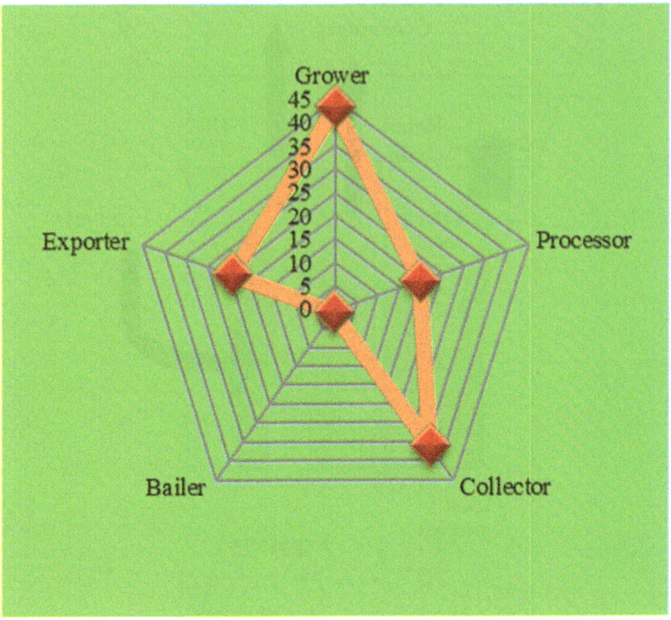

Fig. 16.16 Level of bargaining power in pricing decisions in Ceylon cinnamon

Small-scale cinnamon value chains ended up in collector's node and behave differently compared to the large-scale value chain. In general, bargaining power of collectors and growers are high compared to the exporters and processors (Fig. 16.16).

Figure 16.17 shows the average price (grade disaggregated) per kilogram of cinnamon quills at different nodes of the chain. The prices received by the grower were calculated as the average of the farm gate prices throughout the year. Fixed or certified price is absent in this industry so that the prices earned by the other actors in the chain were evaluated by a survey of the data.

16.4.3 Flow of Relationships and Linkages

Figure 16.18 shows strong relationships with unbroken lines and weak relationships with dotted lines. Cinnamon quills and leaves value chain relationships are included in map. The quills chain consists of a number of nodes and more complex than the leaf value chain. Growers always were related with the mobile or permanent collectors and not the bailers placed at higher levels of value chain and exporters (Weddagala 2018). A majority of the growers establish spot mark relationships with the collectors.

Business and social relationships are common among cinnamon value chain actors, and connection is strong among ancillary members. Social and business

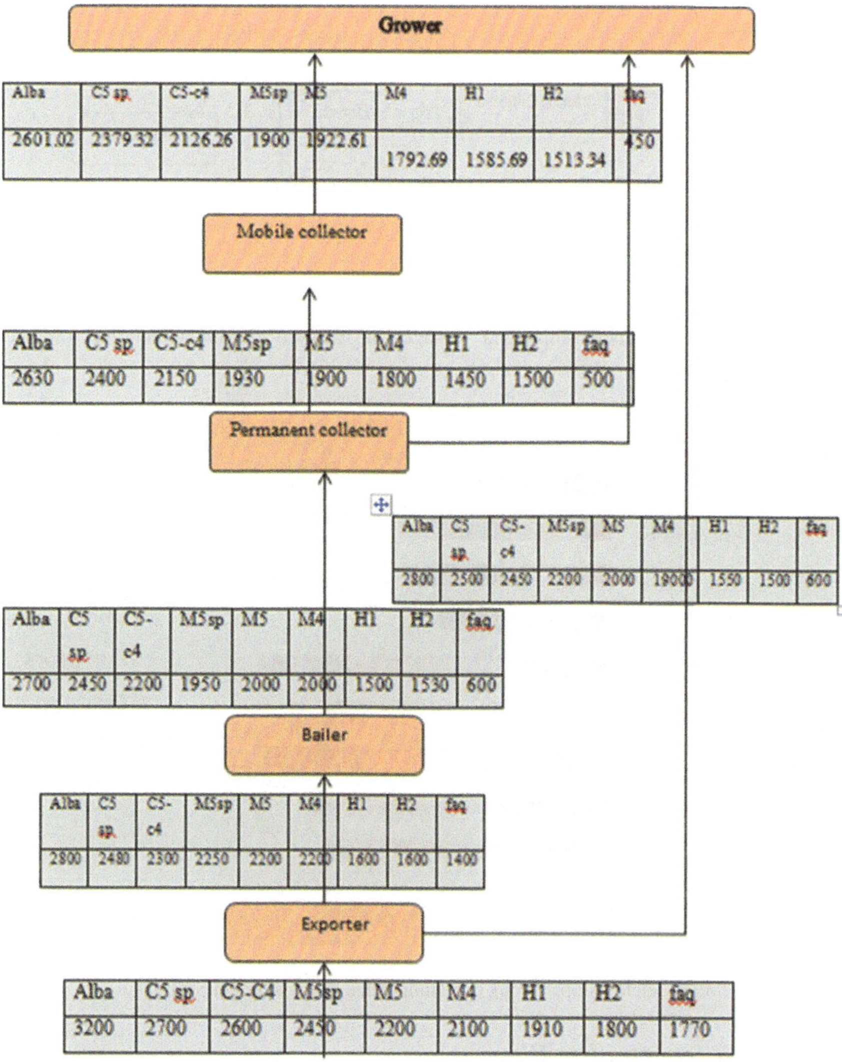

Fig. 16.17 Price flow of Ceylon cinnamon. (Source: Jayasinghe 2017; all values in Sri Lankan Rupees, 1US$ = 180 LKR)

relationships between growers and mobile collectors are stronger compared to growers and exporters. Only a handful of large-scale growers are maintaining relationships with exporters. Growers and peelers are weakly linked with the downstream members such as balers and exporters, and this acts as a barrier to fair and better prices for upstream members (Fig. 16.19).

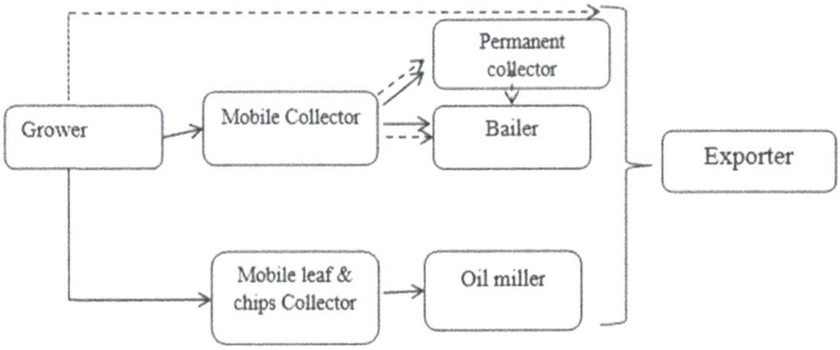

Fig. 16.18 Map of relationships and linkages in Ceylon cinnamon industry. (Source: Jayasinghe 2017)

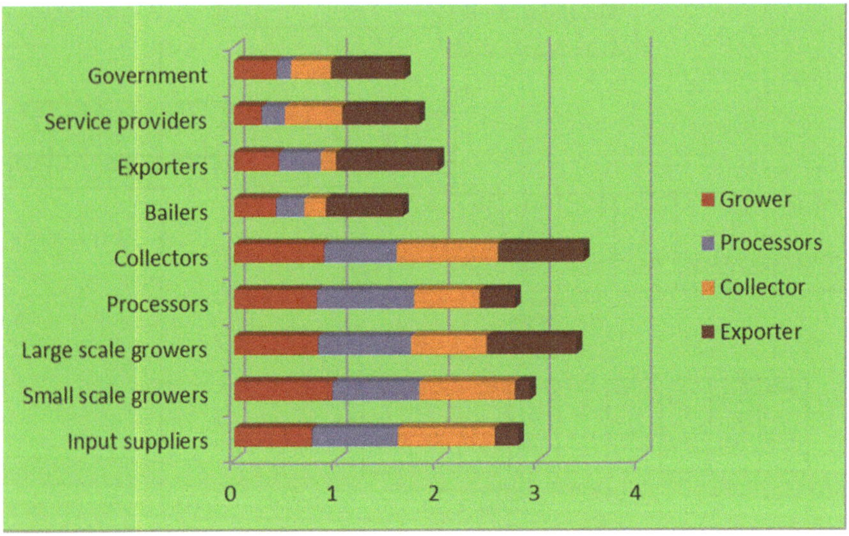

Fig. 16.19 Level of relationships among cinnamon value chain actors in Sri Lanka

16.4.4 Safety Quality Flow

Compliance on safety and quality standards is mandatory to enter high-value foreign markets, such as the USA, the EU, Canada, Japan, the UK, etc. Figure 16.20 shows the present status (number of value chain members' compliance with safety and quality standards) of safety and quality standards and certifications practised in Ceylon cinnamon value chain. GMP, ISO 9000 series and 22,000 series and HACCP are the most common standards maintained in cinnamon value chain (Fig. 16.20). USDA and EU organic certified cinnamon exporters available in considerable amount and limited amount of exporters maintain some volunteer standards based on buyers' requirements.

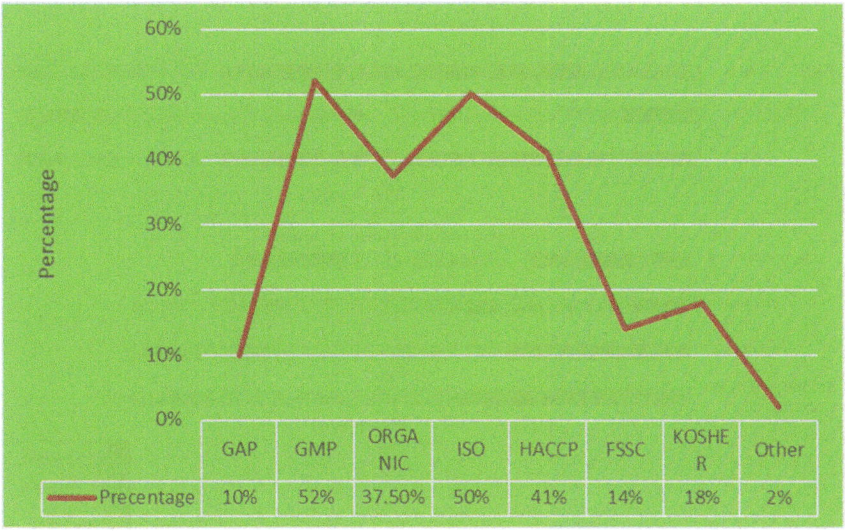

Fig. 16.20 Present status of the safety and quality compliance of Ceylon cinnamon industry

Level of cinnamon grower's awareness and compliance on safety and quality standards and certificates are shown in Fig. 16.21. In general, majority of the cinnamon growers were aware on Good Manufacturing Practices (GMP) and Hazard Analysis Critical Control Point (HACCP) but unaware on fair trade and GAP (Good Agricultural Practices) certificates compared to rest. Limited knowledge and training hinder their compliance on safety and quality standards. Ground-level mechanism for awareness and knowledge sharing was absent, and poor compliance of upstream value chain on safety and quality standards was the ultimate outcome. Cinnamon processors are playing an important role in safety and quality maintenance of processed cinnamon. But their level of awareness as well as the compliance on safety and quality standards shows similar to cinnamon growers (Fig. 16.22). Therefore, safety and quality maintenance of cinnamon upstream was weak, and value chain upgrading is vital for sustainability of the industry. Application of knowledge and training on safety and quality management in cinnamon processing is essential for cinnamon processors to produce safety and quality certified products.

Figure 16.23 shows the level of awareness and compliance on safety and quality standards of cinnamon collectors. They were unaware of current international market requirements; knowledge on BRC (British Retail Consortium), HACCP and ISO22000 were poor. Lack of traceability system negatively affected export market success, while cinnamon collectors demanded training, application knowledge and technology for safety and quality maintenance.

The upstream of the cinnamon value chain is unaware of most safety and quality standards and certifications. Therefore, compliance level is lower than that of other actors in the value chain. In contrast, exporters' awareness of safety and quality

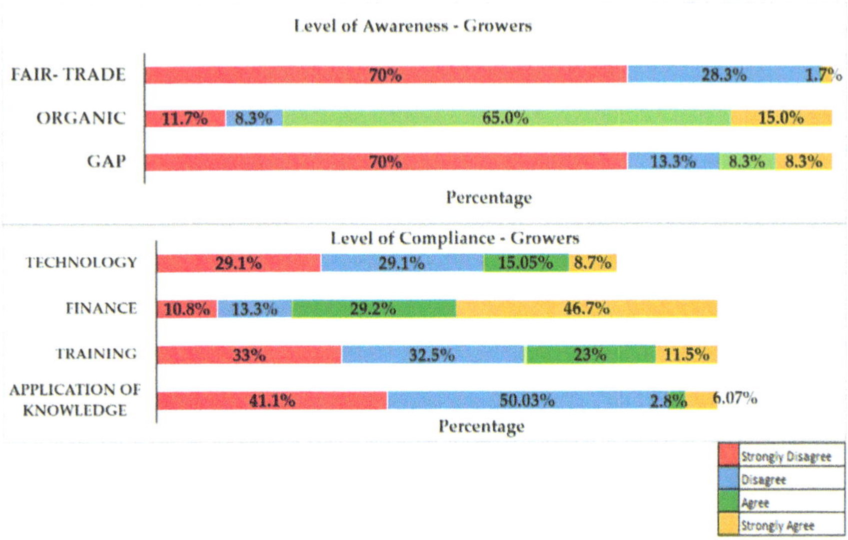

Fig. 16.21 Level of awareness and compliance on safety and quality standards: Growers. (Source: Madushani 2018)

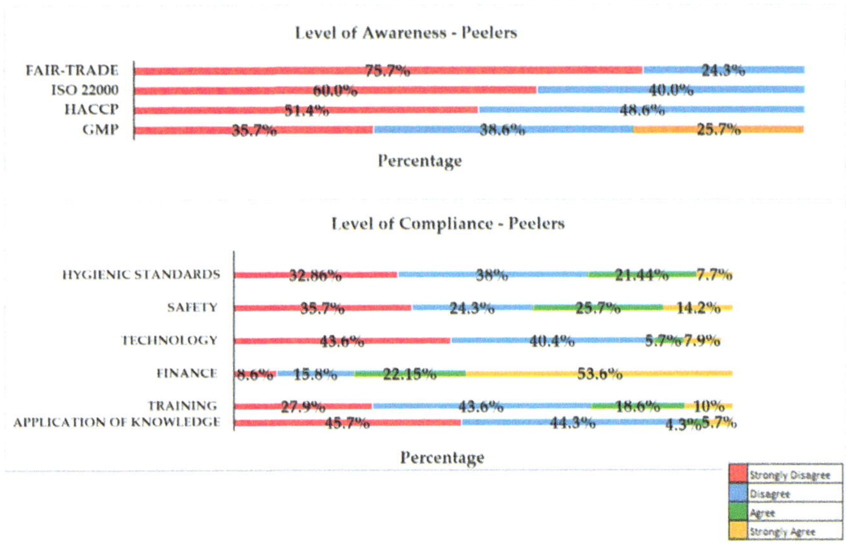

Fig. 16.22 Level of awareness and compliance on safety and quality standards: Processors. (Source: Madushani 2018)

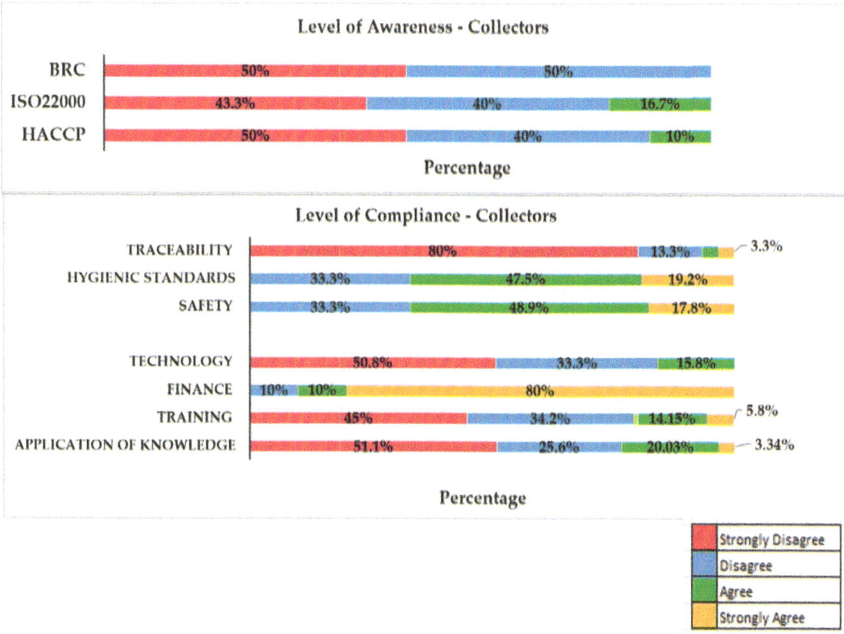

Fig. 16.23 Level of awareness and compliance on safety and quality standards: Collectors. (Source: Madushani 2018)

standards and their compliance levels are high. Figure 16.24 shows the present status of the cinnamon value chain on safety and quality issues.

Value chain innovation is vital for export market success. Innovative practices of both upstream and downstream of cinnamon value chain help to meet the market requirements as well as attract buyers. Downstream, mainly the cinnamon exporters based on long years of experience on cinnamon exports were able to develop market-oriented innovative practices. Value addition and special promotional methods were the most common strategies followed (Figs. 16.25 and 16.26).

16.5 Influence of Internal and External Factors on the Value Addition and Innovation Perspective of Ceylon Cinnamon Exporters and Growers

Political neglect and lack of priority make the cinnamon industry less attractive to the new generation; and poor research intervention and lack of R&D and innovation are among the factors that further weaken it. Also there is no proper programme for value addition of the cinnamon industry. The Ministry for Primary Industries and the Department of Export Agriculture as governmental arms responsible for

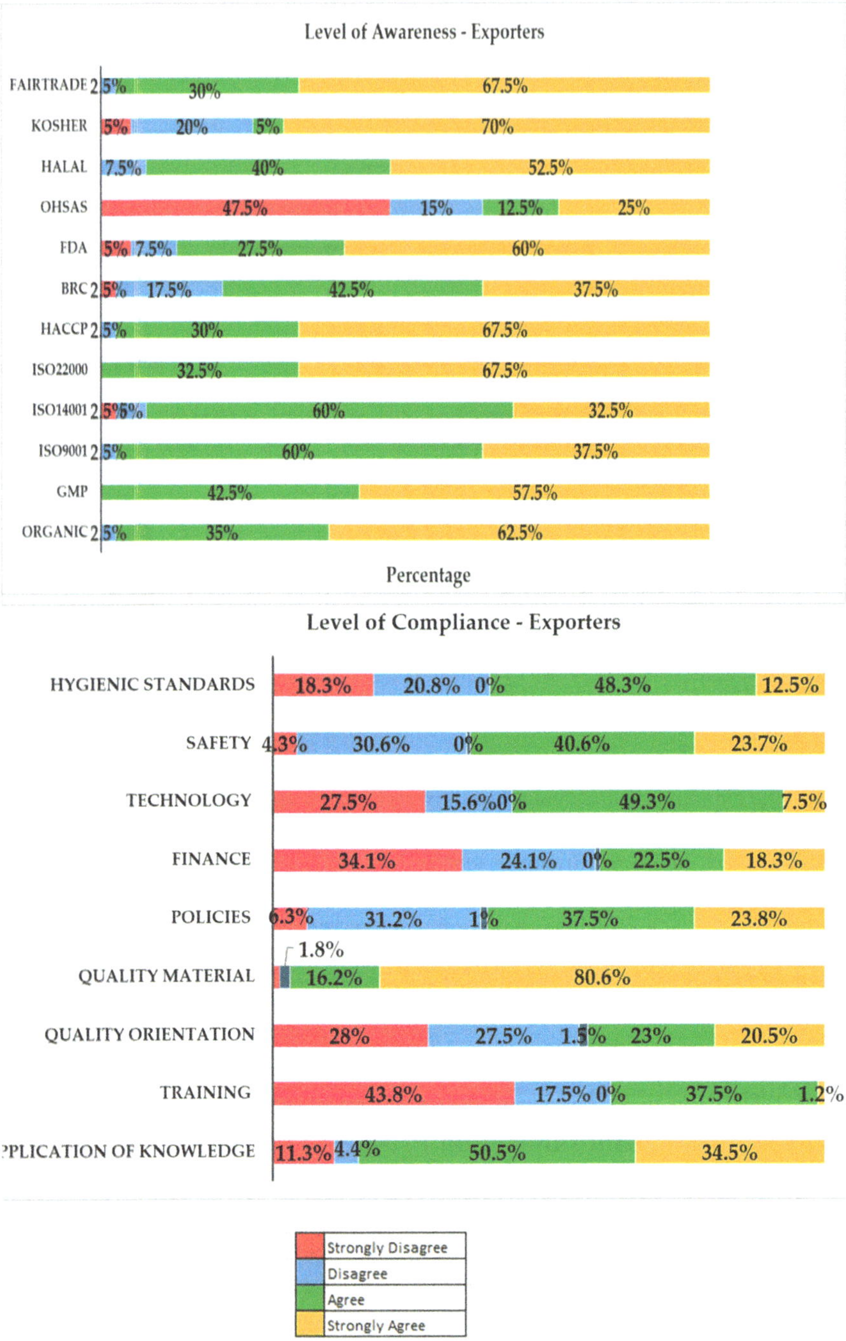

Fig. 16.24 Level of awareness and compliance on safety and quality standards: Exporters. (Source: Madushani 2018)

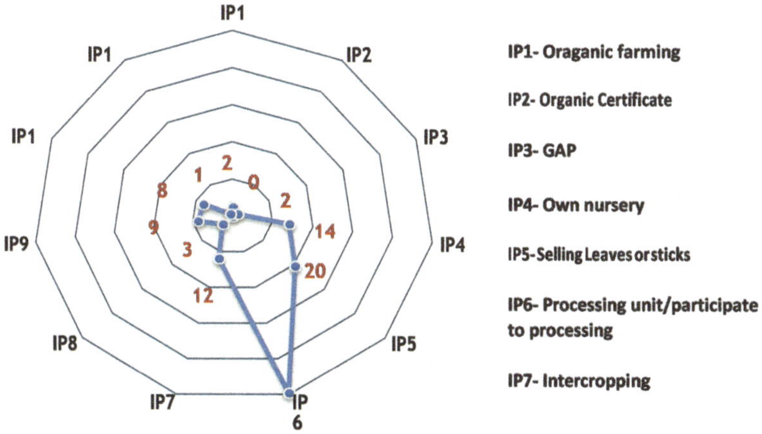

Fig. 16.25 Innovative practices aiming cinnamon value chain actors: Growers. (Source: Silva 2018)

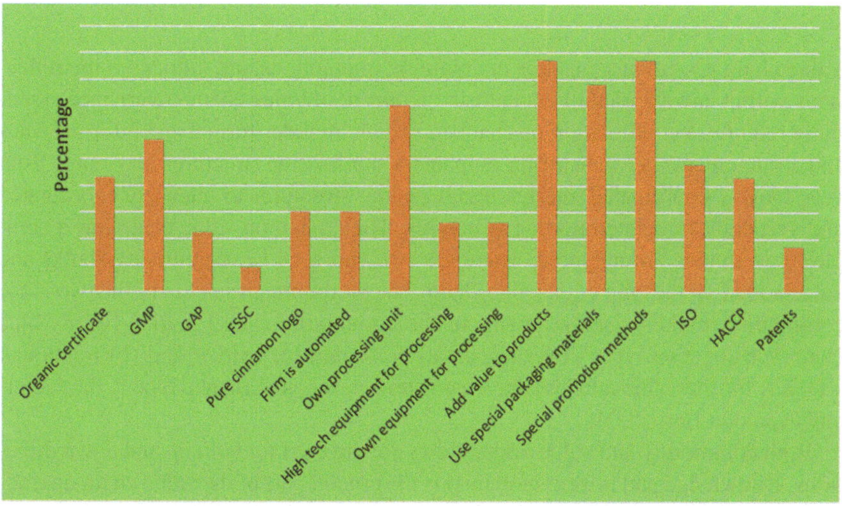

Fig. 16.26 Innovative practices of downstream of the cinnamon value chain. (Source: Silva 2018)

cinnamon industry have plans for certification of cinnamon cultivation and processing for GMP and GAP. Availability of the technology has significant influence on value addition. Also the study of Nguyen (2009) highlighted the technology transfer factor; there are no technology transfer mechanisms such as importing essential machinery from overseas, particularly Taiwan and China.

Cultural factors have significant influence on value addition and innovation. Export-oriented value-added products are specifically targeted in certain countries and regions such as South America. Technological factors in value addition and innovation include access to and availability of machinery, equipment, tools and systems. Access to modern production and processing technology, knowledge and knowhow has been limited and less exposed to cinnamon growers than to exporters. Value-added exporters and large-scale companies benefit from better technological facilities and access pathways (Figs. 16.23 and 16.24). Determinants of value-added production in the cinnamon industry (Jayathilaka 2015) also explain the plight of small- and medium-scale cinnamon companies.

Lack of research and limited intervention of both public and private sector companies in machinery production hinder the mechanization of cinnamon industry and retard its efficiency. Low level of value addition and innovation of small- and medium-scale operators were prevalent, and the issues need to be addressed immediately. Government policy, institutional facility and R&D are significant to value addition for exporters. Seven percent of the exporters obtained patents for their products, while 30% obtained the pure cinnamon logo. Most of the firms are automated and use high-tech equipment in manufacturing.

16.6 Governance Mechanism of the Cinnamon Value Chain

Value chain governance ensures maintaining interactions and influences the reflections within and external to the industry. Concept of value chain governance essentially requires a coordination mechanism, which includes formal and informal rules and arrangements between the participants. Coordination structures may range from very loosely coordinated, market-based trading structures to intensely coordinated and vertically integrated trading structures. The cinnamon value chain has a complex governance arrangement. There are two sub-sets, i.e. commodity (bulk) and value added, in the cinnamon value chain. The predominant sub-set of the two is the commodity market-based segment. In the commodity (bulk) market-based value chain sub-set, two types governance can be found, (a) market and (b) relational, whereas for the value-added value chain sub-set, the governance type is the hierarchy (Van den Berg 2004).

Both commodity and value-added flows are directed for both up and down directions. Knowledge and information map is the crucial part of the value chain upgrading. Then the responsible people can fulfil the deficiencies of the speed, access, direction and amount of the knowledge and information flow of the value chain (Van den Berg 2004). Two categories of knowledge sources could be identified, as the traditional and modern. Both peelers and growers have plenty of traditional knowledge. The traditional knowledge about this industry operations are gained through the family, own observation or other personal expertise in the sector. Modern knowledge passes to the value chain members through training programmes, workshops, exhibitions, etc. organized by the Department of Export Agriculture, Export Development Board, Cinnamon Training Academy, etc.

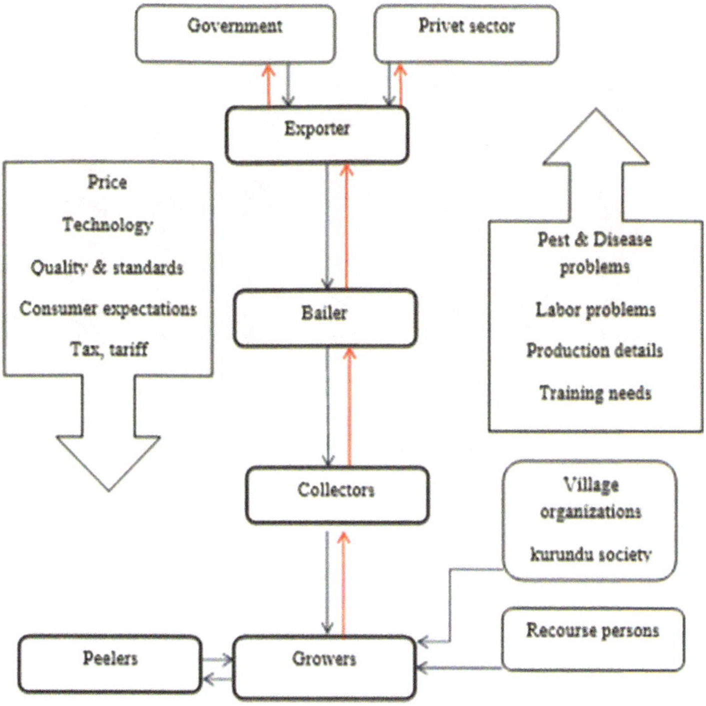

Fig. 16.27 Information flow of cinnamon value chain

16.7 Information Flow

Figure 16.27 shows that the information flow in the cinnamon value chain is a two-way communication process. At the top of the chain, the government and the private sector should be involved to provide information, such as safety and quality standards, tax, tariffs, technology, etc. From bottom to top training needs, problems in the chain members, production details, etc., flow upwards.

16.8 Gender in Cinnamon Value Chain

Gender issues fundamentally shape the totality of production, distribution and consumption within an economy but have often been overlooked in value chain development. From production to processing to disposal, gendered patterns of behaviour condition men's and women's jobs and tasks, the distribution of resources and benefits derived from income-generating activities in the chain and the efficiency and competitiveness of value chains in the global market (Rubin and Manfre 2014). Men and women occupy distinct roles all along the cinnamon value chain, and

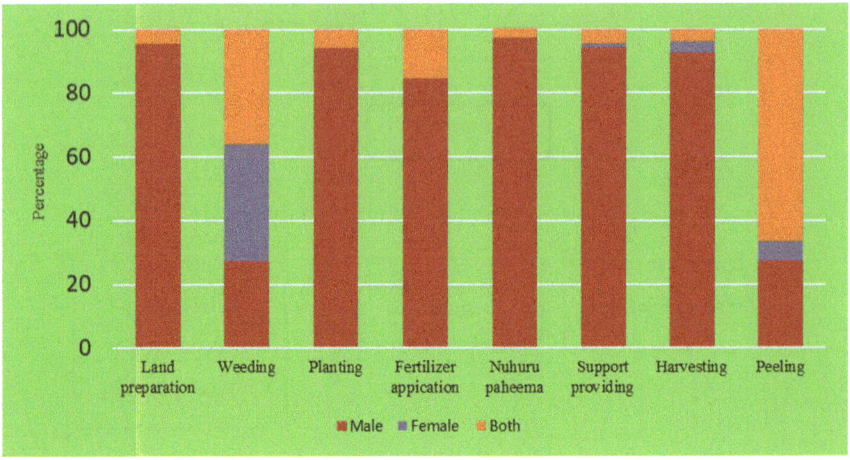

Fig. 16.28 Gender participation in cinnamon value chain activities

regardless of the location or level of industry development, female workers are consistently over-represented in low-skill, low-paid, low-valued positions and remain mostly absent at the other end of the value chain. Figure 16.28 shows the gender participation in cinnamon value chain in Sri Lanka, and male dominancy is common across the value chain activities.

16.9 Micro Perspective for the Cinnamon Industry

The cinnamon industry has the potential not only to lead in existing international markets and domestic markets but also to enter into new markets around the world. The SWOT analysis was used for the entire cinnamon industry to identify the weaknesses and threats and explore strengths and opportunities. This micro perspective was done by using the information gathered through two important nodes (exporters, consumers) of value chain actors of the cinnamon industry through focus groups and the interviewer administrative questionnaires.

16.10 Conclusions

Ceylon cinnamon or true cinnamon (*C. zeylanicum*) is native to Sri Lanka, and its characteristic chemical fingerprint, especially rich in antioxidants, ultra-low levels of coumarin and anti-microbial properties rank it as king of spices. Cassia cinnamon, a direct substitute, invaded the global cinnamon market through its economic advantage and has a 92% market share (EDB 2017). Centuries-old Ceylon cinnamon export business has been moving at a snail's pace, burdened by its inherited problems and issues. Cinnamon ranks first in export earnings of the spice sector of Sri Lanka and lies in the fourth place among plantation crops while contributing very much for the country's economy.

Fragmented cinnamon value chain comprised of scattered upstream with large number of participants link to lengthier, compressed downstream of limited numbers bridge through independent intermediaries. Ceylon cinnamon quill, a unique quill making process limited to Sri Lanka requires experienced and skilful processors' touch which currently act as main barriers to expand the industry. Seasonal income, poor recognition, lack of healthy and safe working environment and mechanization and innovative practices move youth away from the cinnamon-processing industry which is currently the highest-earning agricultural sector. Moreover, cinnamon is the only plantation industry that shares its profits with labourers. Plantation management and processing knowledge passes through generation to generation, while limited attention has been paid to technological, economical and agribusiness concepts. On the other hand, traditional upstream ignore the safety and quality standards and certifications demanded by the high-end international marketplace. Technology, training and application of knowledge on safety and quality management were the prime requirements of upstream of the cinnamon value chain to shift to quality culture. In contrast, compliance of downstream on safety and quality matches with global marketplace securing better returns to the downstream actors. Market opportunities are available in high-end value additions, niche markets, new marketplaces, etc. Mechanization is one of the prime concerns in efficiency and productivity enhancement of traditional cinnamon processing. Research and development focus along the cinnamon value chain is in a weak position where private sector is leading the area. Resistance to change together with attitudes, values and behaviours hinders the industry growth, low level of technology adaptation and less attractive to youth.

Fragmented and scattered cinnamon value chain with inherited poor coordination mechanism separated the upstream from downstream. Upstream has no clear idea on size, shape and worth of end users and their requirements. Value chain-wide intervention on R&D, market research, mechanization and adaptation to appropriate technology, introduction of modern human resources management into cinnamon value chain, crop improvement and high-end value addition are prime needs of Ceylon cinnamon industry. Moreover, value chain integration and establish string links with institutional partners are essential for value chain upgrading.

Acknowledgements We are grateful to Mr. Sarada De Silva, Mr. Vijith Jayathilake and Mr. D.A. Perera for their dedicated and expeditious help in arranging filed work as well as linking up the research team with overseas markets. We like to thank the participants in our focus groups for their time and willingness to share their experiences with us. We thank the project coordinators of special project on cinnamon, Prof. Ranjith Senaratne, Prof. K.D.N. Weerasinghe and our contact persons at the National Science Foundation for their enormous efforts in the development of cinnamon research. Finally, but not the least, we wish to convey our sincere thanks to the Ministry for Primary Industries, for opening up the research grant scheme on Ceylon cinnamon and the grant for value chain development (SP/CIN/2016/05) under which this research work was accomplished.

References

Amadoru CK (2017) Capturing wealth from human resources: determinants of the upstream value chain actors of the Ceylon Cinnamon Industry, an unpublished dissertation, Dept. of Agribusiness Management, Faculty of Agricultural Sciences, Sabaragamuwa University of Sri Lanka

Cambridge University (2013) Value chain definitions and characteristics, Institute for Sustainable Leadership, Cambridge University. Accessed 15 Mar 2017. https://www.cisl.cam.ac.uk/education/graduate-study/pgcerts/pdfs/Value_Chain_Definitions.pdf

Chandimal MWE (2017) Scanning the environment and measure its influence: a case of value addition and innovation in Ceylon cinnamon value chain, an unpublished dissertation, Dept. of Agribusiness Management, Faculty of Agricultural Sciences, Sabaragamuwa University of Sri Lanka

Dolan C, Humphrey J (2000) Governance and trade in fresh vegetables: the impact of UK supermarkets on the African horticulture industry. J Dev Stud 37(2):147–176

EDB (2017) Annual report, export development board, Nawam Mawatha, Colombo, Sri Lanka. Accessed 5 May 2018. https://www.srilankabusiness.com/buyers/export-performance-report.html

EDB (2018) Annual report, export development board, Nawam Mawatha, Colombo, Sri Lanka, Accessed 10 June 2019. https://www.srilankabusiness.com/buyers/export-performance-report.html

Gereffi G (1994) The organisation of buyer-driven global commodity chains: how U.S. retailers shape overseas production networks. In: Gereffi G, Korzeniewicz M (eds) Commodity chains and global capitalism. Praeger, Westport, pp 95–122

Gibbon P, Bair J, Ponte S (2008) Governing global value chains. Econ Soc 37(3):315–338. https://doi.org/10.1080/03085140802172656

Granovetter M (1985) Economic action and social structure: the problem of embeddedness. Am J Sociol 91(3):481–510

Grunert KG, Jeppesen LF, Jepersen KR, Sonne A-M, Trondsen T, Hansen K, Young JA (2005) Market orientation of value chains: a conceptual framework based on four case studies from the food industry. Eur J Mark 39(5/6):428–455. https://doi.org/10.1108/03090560510590656

Gwabu C (2015) Importance of agriculture value chain, the national. Accessed 15 Jan 2017. https://www.thenational.com.pg/importance-of-agriculture-value-chain/

Jayasinghe HDS (2017) Ceylon cinnamon value chain mapping and governance, an unpublished dissertation, Dept. of Agribusiness Management, Faculty of Agricultural Sciences, Sabaragamuwa University of Sri Lanka

Jayathilaka W (2015) Determinants of value added production in cinnamon industry, Uva Wellassa University of Sri Lanka

Kaplinsky and Morris (2001) A handbook for value chain research, vol 113. IDRC, Ottawa

Madushani KP (2018) Standards and certifications as a means for competitive advantage in marketing, a case of Ceylon Cinnamon Industry, an unpublished dissertation, Dept. of Agribusiness Management, Faculty of Agricultural Sciences, Sabaragamuwa University of Sri Lanka

Neilson J, Pritchard B, Yeung HW (2014) Global value chains and global production networks in the changing international political economy: an introduction. Rev Int Polit Econ 21(1):1–8. https://doi.org/10.1080/09692290.2013.873369

Nguyen TUH (2009) Information technology adoption in SMEs: an integrated framework, Int J Entrep Behav Res. https://doi.org/10.1108/13552550910944566

Porter ME (1985) Competitive advantage: creating and sustaining superior performance. Free Press, New York

Porter ME (1990) The competitive advantage of nations. Simon & Schuster, USA

Rajapaksha RHN (2019) Export competitiveness, value chain characteristics, firm behavior, & policy

Rindfleisch A, Heide JB (1997) Transaction cost analysis: past, present, and future applications. J Mark 61(4):30–54. https://doi.org/10.2307/1252085

Rubin D, Manfre C (2014) Promoting gender-equitable agricultural value chains: issues, opportunities, and next steps. Gender Agric. https://doi.org/10.1007/978-94-017-8616-4_12

Rupasena L, Rathnayake B, Ravichandran T (2007) Structure, conduct and performance analysis of cinnamon, pepper, coconut and tea market. Hector Kobbekaduwa Agrarian Research and Training Institute, Colombo

Samarawickrema S (2017) Assessment of the Kalli- and line system in the Sri Lankan cinnamon sector

Scott WR (1995a) Institutions and organizations. Sage, London

Scott WR (1995b) Institutions and organizations. Ideas, interests and identities, management 17(2):136. https://doi.org/10.3917/mana.172.0136

Silva MPMM (2018) Entrepreneurial orientation and innovation: a study of Ceylon cinnamon value chain, an unpublished dissertation, Dept. of Agribusiness Management, Faculty of Agricultural Sciences, Sabaragamuwa University of Sri Lanka

Thanthirige J (2011) Technology and marketing issues in cinnamon industry in Sri Lanka. University of Moratuwa, Moratuwa

Trienekens JH (2011) Agricultural value chains in developing countries: a framework for analysis. Int Food Agribus Manag Rev 14(2):51–82

Van den Berg M (2004) Making value chains work better for the poor: a toolbook for practitioners of value chain analysis: Asian Development Bank

Warnakulasooriya WMS (2017) Marketing of Ceylon cinnamon: managing tangible and intangible assets, an unpublished dissertation, Dept. of Agribusiness Management, Faculty of Agricultural Sciences, Sabaragamuwa University of Sri Lanka

Webber CM, Labaste P (2010) Building competitiveness in Africa's agriculture: a guide to value chain concepts and applications, World Bank. Accessed 5 Mar 2017. http://siteresources.worldbank.org/INTARD/Resources/Building_Competitiveness_in_Africa_Ag.pdf

Weddagala WMTB (2018) Development of a traceability system for an agriculture produce: a case of Ceylon cinnamon (*Cinnamomum zeylanicum*), an unpublished dissertation, Dept. of Agribusiness Management, Faculty of Agricultural Sciences, Sabaragamuwa University of Sri Lanka

Williamson OE (1999) Strategy research: governance and competence perspectives, Strategic Strateg Manag J 20(12):1087–1108

Chapter 17
Planning and Strategic Policy Interventions for Building a Globally Competitive Cinnamon Industry in Sri Lanka

Buddhi Marambe, Jeevika Weerahewa, and A. P. P. Disna

17.1 Introduction

The history of cinnamon (*Cinnamomum zeylanicum* Blume) dates back to about 2800 BC, where it can be found referenced as *kwai* in Chinese writings. Cinnamon is even mentioned in the Bible when Moses used it as an ingredient for anointing oil in ancient Rome. It was burned in Roman funerals perhaps partly as a way to ward off the odour of dead bodies.

Cinnamon was a precious spice in the West during the fourteenth to fifteenth centuries, and its primary use was to preserve meat, retarding the growth of bacteria. The quest for cinnamon was a major factor, which led to the exploration of the world in the fifteenth century. By that time, the real cinnamon was produced in only one place, namely, in Ceylon or Sri Lanka. Anyone who had the control of the supply flow would have made profits immensely. Portuguese traders made their way to Ceylon in the fifteenth century, enslaved the natives and took the control of the trade from Arabs. Soon the Dutch displaced the Portuguese and gained the control of the cinnamon monopoly. It was the Dutch who made a massive effort to boost the production by domestication of the crop and expanding the extent in the areas they had the control. Because of that effort, cinnamon cultivation was moved to western and southern coastal belts of the island. Since 1815, the British took the control of the island, and cinnamon trade was moved to their hands. By this time the relative

B. Marambe (✉)
Department of Crop Science, Faculty of Agriculture, University of Peradeniya, Peradeniya, Sri Lanka
e-mail: bmarambe@pdn.ac.lk

J. Weerahewa
Department of Agricultural Economics and Business Management, Faculty of Agriculture, University of Peradeniya, Peradeniya, Sri Lanka

A. P. P. Disna
Department of Export Agriculture, Getambe, Peradeniya, Sri Lanka

R. Senaratne, R. Pathirana (eds.), *Cinnamon*,
https://doi.org/10.1007/978-3-030-54426-3_17

importance of spices in the world market was declining due to the emerging plantation crop sectors of tea and rubber, which restricted the further expansion of cinnamon. The best historical evidence about the cinnamon trade in Sri Lanka is found in 'Up country-Dutch agreement (Hanguranketha agreement)' signed on 14th February 1766 between the Sri Lankan king Sri Keerthi Sri Rajasinghe and the Dutch government. By this agreement the King had permitted the Dutch to cut and peel cinnamon in certain forest areas of Sri Lanka, and the Dutch agreed to protect the Kingdom from foreign invasion (DEA 2019).

Cinnamon is a versatile spice, which can be added to any food item such as salads, confectionaries, beverages, soups, stews and sauces. Cinnamon drink made by immersing pieces of bark in hot water is popular among Latin American countries. Cinnamon-flavoured tea is becoming popular. It is also used as a common ingredient in Chinese and Ayurvedic medicine. Cinnamon leaf and bark oils are used to flavour food products, in perfume industry and in pharmaceutical industry. People in ancient Sri Lanka used cinnamon for many purposes, such as medicine, spice, perfume material and soft drink. It is considered as a 'miracle cure' for a raft of health issues ranging from obesity and diabetes to heart disease and arthritis (http://www.dailymirror.lk/business/features/30034-us-research-study-spells-boom-for-lankas-cinnamonindustry.html). At present cinnamon is widely used in bakery products, pharmaceutical preparations and cosmetics worldwide. It is also used in the manufacture of biscuits, breads, confectioneries, chocolate, flavoured coffee and tea, fruit preservatives, sauces, gherkin, pickles, sherbet, punch and mulled wine. Mexico is the largest importer of Sri Lankan cinnamon. Mexicans boil the bark in water and use the extract as a beverage similar to tea.

Cinnamon has unique medicinal properties. Throughout history, cinnamon has been linked to anti-clotting actions, antimicrobial activity, boosting brain function and aiding digestion. Perhaps cinnamon is most notable for its ability to reduce blood sugar, cholesterol and triglyceride levels in people with type 2 diabetes. Cinnamon extract has shown a moderate effect on reducing fasting plasma glucose concentration. An intake of 1, 3 or 6 grams of cinnamon per day reduced serum glucose, triglyceride, LDL cholesterol and total cholesterol in people with type 2 diabetes (Khan et al. 2003).

17.2 Cinnamon Situation Analysis: Production and Uses

Cinnamon is an evergreen tree, maintained as a low bush, and thrives well in a warm and humid climate, with a rainfall of over 1500 mm. In Sri Lanka, cinnamon seems to have originated in the central hills where seven wild species of cinnamon occur in Kandy, Matale, Belihuloya, Haputale, Horton plains and the Sinharaja forest range. Later, the crop has gradually shifted to the north and areas around Colombo and then moved to the Southern Province.

Presently, cinnamon cultivation is concentrated along the coastal belt from Negombo (Western province) to Matara (Southern province) and has also made inroads into Kalutara (Western province) and Ratnapura (Sabaragamuwa proince). In terms of districts, cinnamon is grown in the low-country wet zone in Galle, Matara, Ratnapura, Kalutara and Gampaha districts, parts of Matale and the Hambantota districts, where the required climatic conditions are prevalent. The cultivated extent of cinnamon in Sri Lanka has increased by 24% from 27,034 ha in 2003 to 35,589 ha in 2018. Cinnamon production increased by 21.9% from 18,945 mt in 2016 to 23,109 mt in 2018. The export earnings from cinnamon increased by 69% from US dollars 136 million in 2012 to US dollars 230 million in 2018. This progress was driven by high volumes and average export prices (DOA 2018; EAS 2013, 2018).

As the land area for new planting is limited, productivity of cinnamon has to be stepped up to meet the ever-increasing international export demand. One of the critical components in augmenting crop productivity is soil fertility management with balanced nutrition. The yields reported in the majority of cinnamon cultivations in Sri Lanka are relatively low. Many have reported a decline in soil fertility coupled with reduced crop productivity, resulting from poor soil and crop management practices. The trend in production of cinnamon in Sri Lanka over a period of 50 years is illustrated in Fig. 17.1. The annual cinnamon production in Sri Lanka has increased by more than threefold over a period of 56 years from 5,200 mt in 1961 to 17,255 mt in 2017. There are eight cinnamon species in Sri Lanka, and only *C. zeylanicum* is grown commercially. Traditionally, there were several types of cinnamon, categorized based on the taste of the bark, i.e. *Pani-Miris Kurundu* was the best with sweet-pungent taste, and *Pani Kurundu, Naga Kurundu, Weli Kurundu, Peiris Kurundu, Sewel Kurundu* and *Thiththa Kurundu* are the rest. Currently ten cinnamon accessions have been identified based on yield and quality performances, and the best two lines among them, named as *Sri Vijaya* and *Sri Gamunu*, have

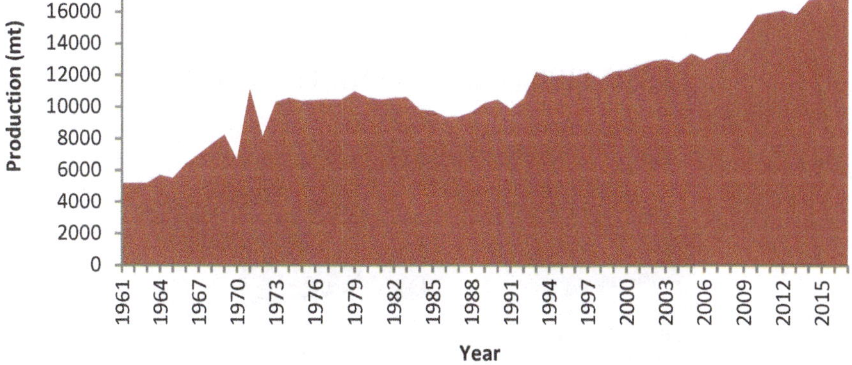

Fig. 17.1 Trends in production of cinnamon in Sri Lanka from 1961 to 2017. (Source: FAOSTAT - various issues)

been released for cultivation. Other selections are under evaluation in different agro-climatic zones (DEA – Unpublished report).

17.3 Cinnamon Situation Analysis: Trade

World cinnamon market comprises two commercial types as true cinnamon (*C. zeylanicum*) and cassia cinnamon (*Cinnamomum cassia*). More than 90% of the true cinnamon is supplied from Sri Lanka, while the rest is mainly coming from Madagascar and Seychelles. However, a large bulk of world supplies of cinnamon comprised cassia cinnamon, which is mainly exported by South East Asian countries. The Sri Lankan cinnamon industry has suffered heavily in the international markets mainly due to the stiff competition from *C. cassia* (Cassia; Fig. 17.2), which is less than 25% of the price of true cinnamon in the global trade.

In the global context, Sri Lanka is ranked fourth in production with a world share of 8% in 2017 (Fig. 17.3). Sri Lanka is also ranked fourth in the total cinnamon area harvested with a world share of 11.12% (31,598 ha) and ranked fifth in the average yield (5,461 hectogram per ha⁻; http://www.factfish.com/statistic/cinnamon%2C%20yield).

C. cassia, called Chinese cassia or Chinese cinnamon, is an evergreen tree originating in southern China and widely cultivated in China and in southern and eastern Asia (India, Indonesia, Laos, Malaysia, Taiwan, Thailand and Vietnam). It is one of the several species of *Cinnamomum* that are used primarily for their aromatic bark, which is used as a spice. In the USA, Chinese cassia is often sold under the culinary

Fig. 17.2 The Ceylon cinnamon vs cassia cinnamon. (Source: https://cinnamonbenefits.weebly.com/ceylon-vs-cassia.html)

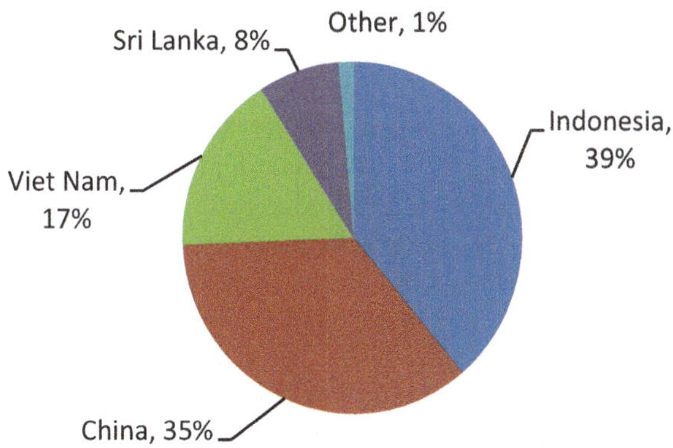

Fig. 17.3 World share of production of cinnamon in 2017. (Source: FAOSTAT - various issues)

name of 'cinnamon'. Chinese cassia is a close relative to Ceylon cinnamon (*C. zeylanicum*), Saigon cinnamon (*C. loureiroi*, also known as 'Vietnamese cinnamon') and Indonesian cinnamon (*C. burmannii*). In all four species, the dried bark is used as a spice. Chinese cassia's flavour is less delicate than that of Ceylon cinnamon. For this reason, it is less expensive and is sometimes called *bastard cinnamon* (http://en.wikipedia.org/wiki/Cinnamomum_cassia), in which the bark is thicker and more difficult to crush and has a rougher texture than that of Ceylon cinnamon. Owing to those differences in products, Sri Lankan cinnamon (Ceylon cinnamon) has been classified under a separate Harmonized System (HS) code in the international trade.

Sri Lankan spice volumes (including cinnamon) for export, despite their high quality, are facing the risk of rejection in tough international market segments, such as North America, due to the lack of various essential cleaning and sterilization process facilities at pre-export stages. Spices are exported to India duty-free, under the Indo-Sri Lanka Free Trade Agreement (ISFTA). However, export of cinnamon to India is limited as only retail packs are permitted under the ISFTA. Sri Lanka exports cinnamon bark products in various forms, such as quills, cut quills and pieces of bark in different sizes, as well as in crushed and ground forms. A small quantity of high-tech value-added products, such as ingredients, are exported. The preparation of a cinnamon quill is a unique processing method for Sri Lankan cinnamon and dates back to centuries. Usually quills are made at a diameter of 42 inches (106.7 cm), but current international demand is for quills that are in different sizes.

Sri Lanka also exports sizable quantities of cinnamon bark oil and leaf oil. The world demand for cinnamon leaf oil has been around 120–150 mt per annum in recent years, which is met almost entirely by Sri Lanka. The USA and Western Europe are the largest markets for cinnamon leaf oil. Imports into France and the UK have fallen in the last few years. Hong Kong is a significant importer, although most of the oil is re-exported. In 2017, earnings from exports of cinnamon, which accounted for 50.62% of earnings from the spice sector, were US dollars 202.2 million, which is an increase by 27.4% compared to 2016. Earnings from export of cinnamon to Mexico (30.81%), Peru (30.18%), Colombia (46.8%) and the USA (5.4%) have increased, while earnings from exports to India (−2.30%) have dropped (EDB 2018; EAS 2017–2018).

The exports of cinnamon quills, leaf oil and bark oil over the past 5 years are illustrated in Fig. 17.4. Cinnamon quills are the major cinnamon export product. However, the value-added product extracts, namely, cinnamon leaf and bark oils, are exported only in relatively low quantities but clearly have secured higher amount of foreign exchange due to the high value of the products. The statistics clearly indicate that export of high quality value-added products is a must if Sri Lanka is to achieve its cherished goals in the spice crop sector by 2025.

Sri Lanka exports approximately 17,084 t of cinnamon products to countries in North and South America and several countries in Europe. Sri Lanka is ranked No 1 on cinnamon exports to Mexico, the USA, Colombia, Peru and Germany with a market share of 28.2%. The export growth (quantity) of Ceylon cinnamon to these countries has been 7% over the period 2014–2018. The top ten destinations of Ceylon cinnamon directly exported from Sri Lanka, in terms of value of foreign exchange earned, are illustrated in Fig. 17.5, clearly indicating that Mexico and the USA are the major cinnamon importers at global scale. Sri Lanka has strongly established itself as a major contributor of providing high-quality true cinnamon products.

The statistics clearly indicate that Sri Lanka's share of 28.2% of the global cinnamon market in 2018 in monetary value has the potential to expand into many unexplored regions such as the middle-eastern countries and other expanding areas in terms of global imports including the European Union (Fig. 17.6). The major exporters of cinnamon in the global scale are depicted in Fig. 17.7, indicating that Sri Lanka occupies the top position. India is ranked 11th in the list of major exporters of cinnamon in US $ terms probably due to the export of low valued and high-coumarin-containing *C. cassia*.

The foreign exchange generated from the quantities of Ceylon cinnamon exported to selected destination countries is depicted in Figs. 17.8, 17.9, and 17.10. The export quantity of Ceylon cinnamon to the EU has fluctuated over the past 5 years but has shown a marginal declining trend (Fig. 17.8). Correspondingly the foreign exchange earnings have varied (Fig. 17.8). However, the overall cinnamon imports to the EU have increased over the past 5 years with the EU paying significantly higher amounts for the cinnamon imports from countries such as Indonesia. The statistics showed the enormous potential that exists in the EU for export of cinnamon, provided that the required quantity being produced by meeting the required quality standards and certification.

As in the case of cinnamon exports quantity to the EU, the quantity to the USA also has decreased from 2013 to 2015 but has increased in 2016 with a marginal decrease in 2017. However, the foreign exchange generated through exports of Ceylon cinnamon to the USA has been significantly higher, but the earnings have shown a corresponding decline in 2015, owing to the lower quantity exported. The overall cinnamon imports to the USA have increased over the past 5 years (Fig. 17.9), with a marginal reduction in 2015. The statistics show that currently the USA

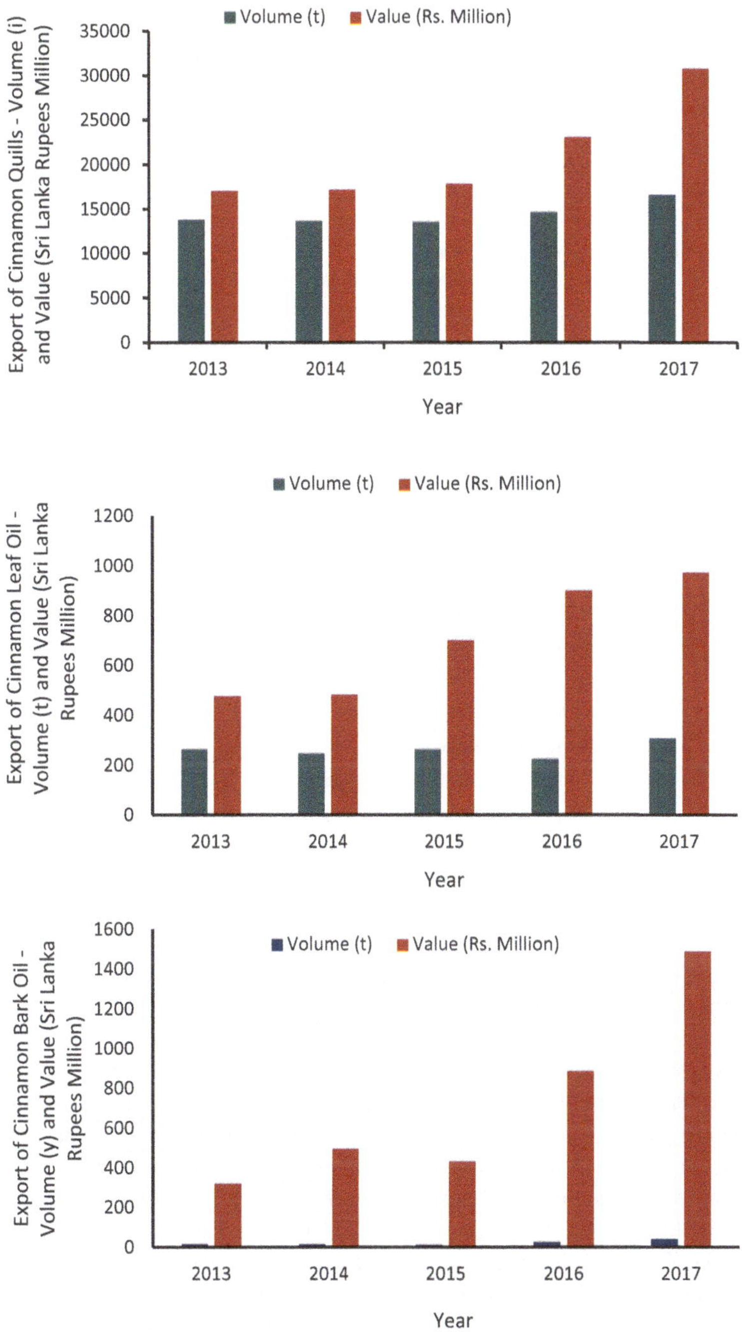

Fig. 17.4 Export volumes and value of different cinnamon products from Sri Lanka over a five-year period (US$ 1 = Rs 181.5). (Source: DEA 2018)

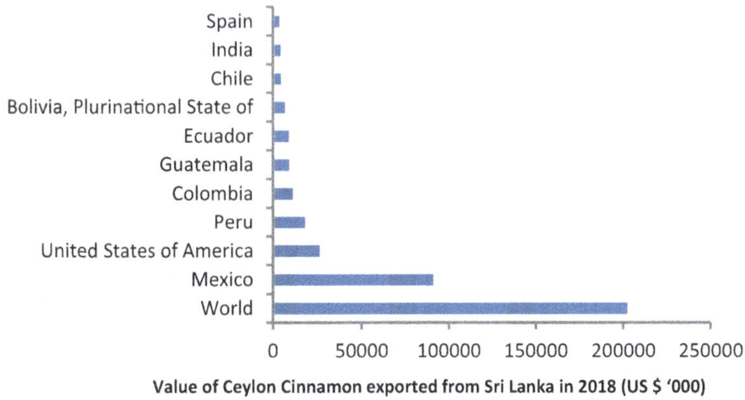

Fig. 17.5 The total exports and top ten destination countries of Ceylon cinnamon exported from Sri Lanka in terms of value in 2018. (Source: ITC 2018)

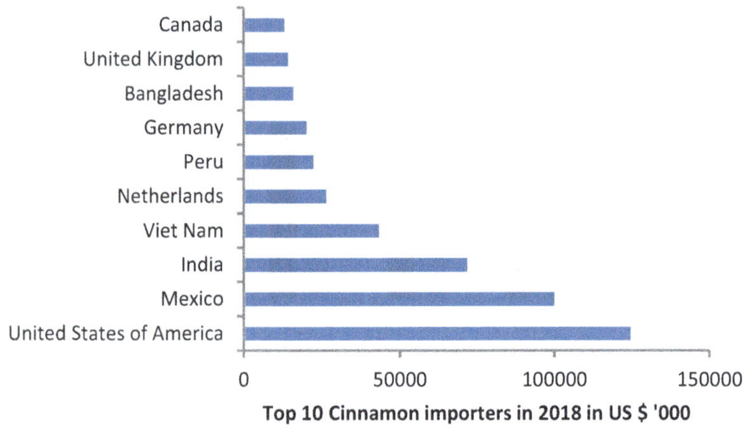

Fig. 17.6 Top ten global importers of cinnamon in terms of import value in 2018. (Source: ITC 2018)

imports vast quantities of *C. cassia*, and thus, there is a need to expand and explore potential for export of Ceylon cinnamon to the USA, with the required quantity and quality standards. Sri Lanka has clearly dominated the cinnamon export trade to Mexico (Fig. 17.10), proving approximately 86% of the requirements of Mexico.

The total Ceylon cinnamon exports from Sri Lanka to the global market are presented in Fig. 17.11, together with that of the market competitors India and Indonesia. These two countries are mainly marketing the low-quality and high-coumarin-containing *C. cassia* instead of the true cinnamon *C. zeylanicum*.

Despite the lower prices fetched by *C. cassia* compared to *C. zeylanicum*, the cinnamon exports especially from Indonesia dominates the global market in terms of the quantity exported. This also opens up an enormous opportunity for Sri Lanka to further penetrate to the cinnamon export market with a powerful and effective

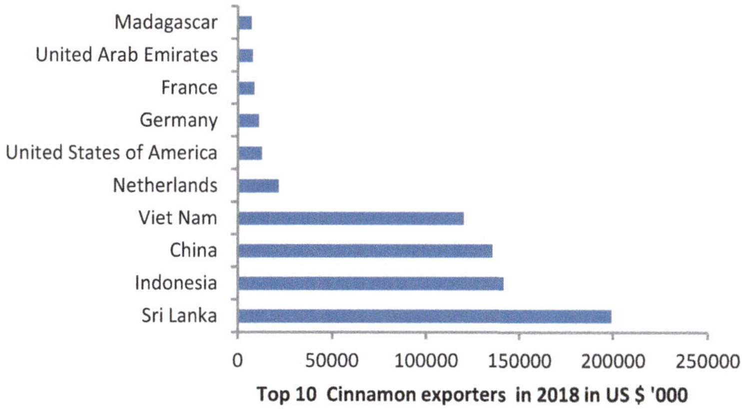

Fig. 17.7 Top ten exporters of cinnamon in terms of export value in 2018. (Source: ITC 2018)

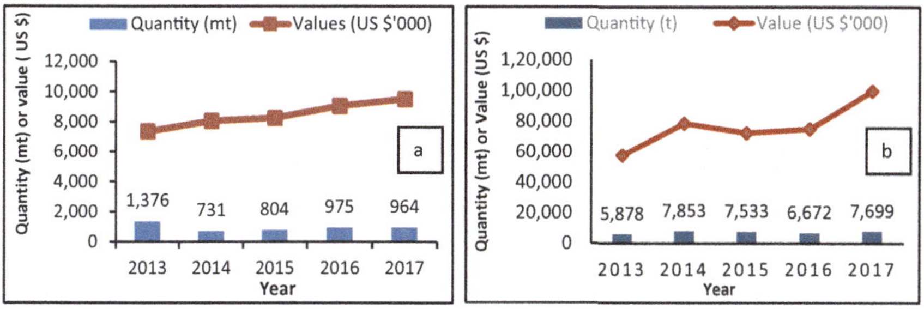

Fig. 17.8 Export quantity and earnings from cinnamon to the EU (**a**) and total cinnamon imports to the EU (**b**). (Source: ITC 2018)

propaganda and awareness campaign of the quality of true cinnamon produced and exported from the country.

In general, the quantities of Ceylon cinnamon exported has increased marginally over the past 5 years, while the foreign exchange generated has increased exponentially due to price escalations observed for cinnamon at global scale. The results are encouraging and signify the need to carry out promotional campaigns and establish and/or further strengthen the position of Ceylon cinnamon at different market places in the international trade.

The average international market prices for Ceylon cinnamon in the past 6 years have increased from US $ 4.95 in 2013/2014 and has stabilized at US $ 7.75 per pound (0.453 kg) of cinnamon by the year 2018/2019 (Fig. 17.12). The international market prices for cinnamon reached the highest level during 2018–2019 compared to the price fetched during the past 6 years, showing better signs for Sri Lanka's cinnamon market. Indian and Indonesian cinnamon, mainly comprising *C. cassia*, are marketed at extremely lower prices (Fig. 17.13), indicating severity of competition on the cinnamon exported from Sri Lanka.

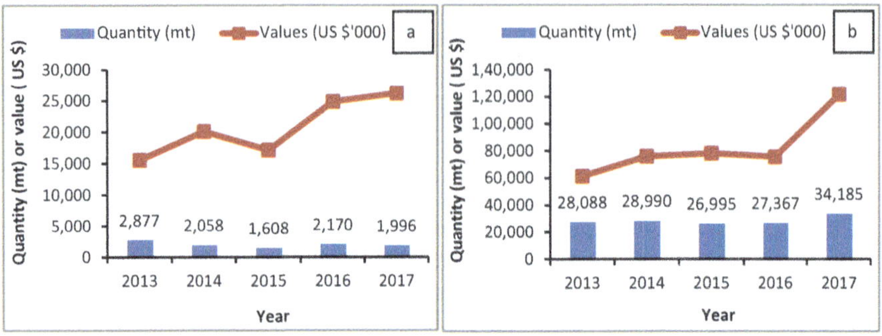

Fig. 17.9 Export quantity and earnings from Ceylon cinnamon to the USA (**a**) and total cinnamon imports to the USA (**b**). (Source: ITC 2018)

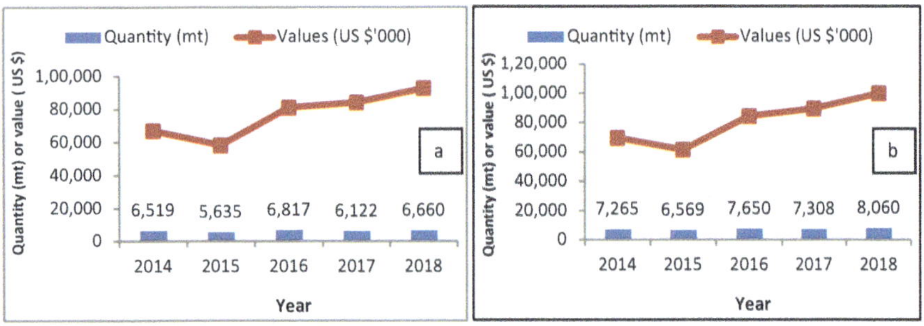

Fig. 17.10 Export quantity and earnings of Ceylon cinnamon to Mexico (**a**) and total cinnamon imports to Mexico (**b**). (Source: ITC 2018)

The price variations of cinnamon observed in Sri Lanka are illustrated in Fig. 17.14. A decline in prices in the local markets has been observed during the period 2013–2014 followed by a sharp increase since 2015.

The domestic farm-gate prices of cinnamon in Sri Lanka showed a steady increase during the period 2009–2015 (IPS 2017). Further, high prices were received by producers at the farm gate (1246.1 LKR/kg) in 2015, and it illustrates that the international prices determine the farm-gate prices to a great extent. Cinnamon prices did not exhibit clear seasonality, but somewhat low prices were seen in July–August period, with a peak only in July. Though cinnamon peeling is done throughout the year, a larger part of harvest comes into the market in July–September and could be the reason for the lower prices observed in the said period.

The bark of Ceylon cinnamon has been reported to have only traces of coumarin, whereas barks from other three cassia species, especially Vietnam cinnamon and Indonesian cinnamon, have been found recently to contain substantial amounts of coumarin (http://www.ft.lk/2013/05/23/us-research-shows-massive-potential-for-sri-lankan-cinnamon/). Coumarin is a known toxin to the liver, acts as an anticoagulant and is known to cause cancer in rodents. Coumarin has been banned in the USA as a food additive since 1954. The European health agencies already recognize the adverse side effects of coumarin, and the EU regulations specify a Tolerable Daily

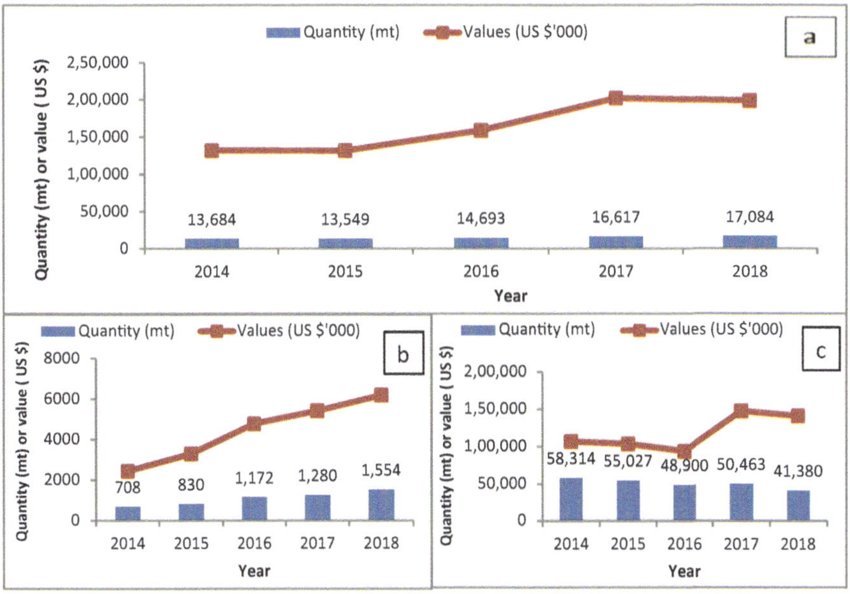

Fig. 17.11 Comparison of quantity and value of Ceylon cinnamon (*C. zeylanicum*) exports from (**a**) Sri Lanka to the global market and cinnamon exports from India (**b**) and Indonesia (**c**). (Source: ITC 2018)

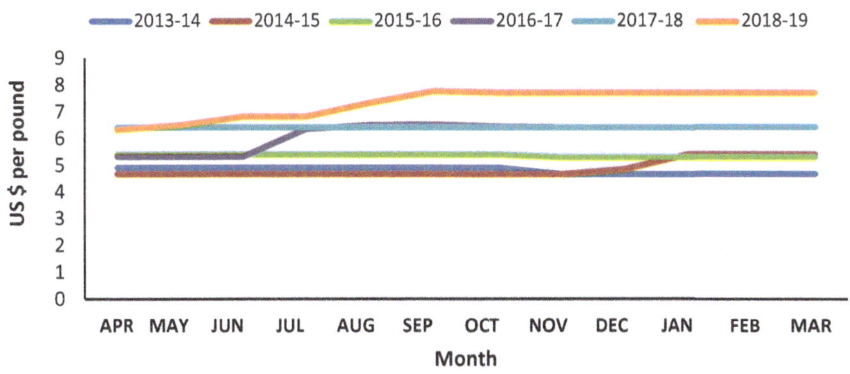

Fig. 17.12 Price variations observed in the world market for *C. zeylanicum* (at New York Prices). (Source: ISB 2019)

Intake (TDI) for coumarin of 0.1 mg per kg of body weight per day. However, setting such limits does not ensure compliance. Recent tests by a leading independent consumer protection group warned that coumarin levels in a variety of cookies, cereals and rice puddings sold in Germany were up to 20 times the European legal limit (Leelarathne 2013).

Cinnamon is exported in raw form as well as in value-added forms, such as cut pieces, powder and crushed form. Cinnamon is considered a unique product

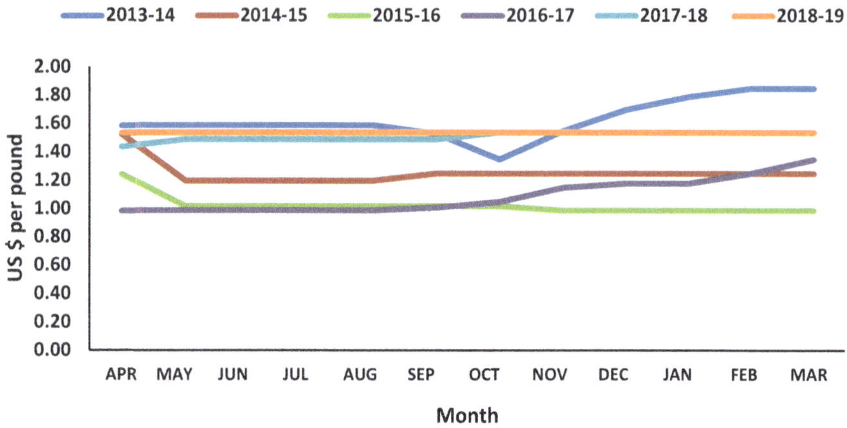

Fig. 17.13 Price variations observed in the world market for *C. cassia* (at New York Prices). (Source: ISB 2019)

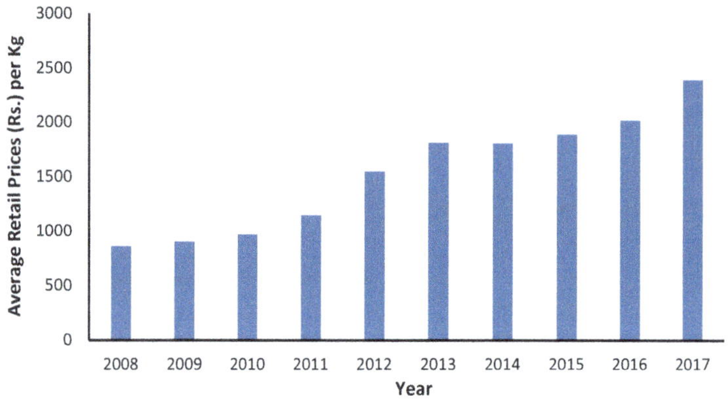

Fig. 17.14 Variations observed in prices of cinnamon at the Sri Lankan markets. (Source: CBSL 2018)

exported from Sri Lanka, and it has unique characteristics, which can be branded under the protection of the Geographical Indications (GI). Accordingly Ceylon cinnamon was introduced to the international market as a branded product, namely, *Pure Ceylon Cinnamon*, which reflects a combination of several intrinsic characteristics of *C. zeylanicum*.

Sri Lanka is far behind other countries in the export of value-added spices, including organic produce. Apart from improvements in distillation practice, the greatest advances in productivity and quality will come from breeding programmes aimed at producing superior germ plasm for planting. Some progress has already been made in identifying mother plants, which give high yields of oil and high contents of cinnamaldehyde and eugenol in the bark and leaves (Dr. A.P. Heenkende – Personal communication).

17.4 Standards for Cinnamon Exports

In Sri Lanka, the latest regulations were tabled in the Parliament in early July 2013 to standardize and improve the quality of the centuries-old cinnamon industry. This was by shifting the focus from popular export crops, namely, tea, coconut and rubber, to the value-added cinnamon in the international market. The regulations published in the Gazette Extraordinary 1813/15 focused on quality standards of cinnamon products, i.e. prohibiting the export of crushed or ground cinnamon, organic cinnamon, cut cinnamon quills, cinnamon in retail packs of 1 kg or less, cinnamon featherings and cinnamon chips or cinnamon in any other form without the SLS 81:2010 certification specified for Ceylon cinnamon. According to the new regulation, the Government of Sri Lanka expects all cinnamon exporters to furnish a certificate of conformity to the Sri Lanka Standards (SLS) stipulated for cinnamon issued by either a laboratory accredited for testing of the particular article or a laboratory recognized by the Sri Lanka Standards Institution (SLSI) in addition to the certificate of clearance issued by the Director General of SLSI. However, the expected procedure was not materialized due to some issues. Whiie recognizing the difficulty of meeting the quality standards as a major problem faced by Sri Lanka's cinnamon exporters, the 2019 budget of the Sri Lankan government proposed that the cinnamon exporters should obtain quanlity control certification for their products, at the point of export. Those exporters who fail such certification tests will be supported to improve their operations within 12 months, as part of the Export Market Access programme of the Export Development Board (GOSL 2018).

At the international level, the ISO 6539:2014 specifies requirements for whole or ground (powdered) cinnamon including recommendations for storage and transport of the Sri Lankan, Madagascan and Seychelles types of *C. zeylanicum*. It is important to note that the requirements for cassia (Chinese type, Indonesian type and Vietnamese type) are given separately in ISO 6538:1997.

The cinnamon industry has done extremely well during the recent past, with world market prices reaching the highest levels. However, with the prices increasing, the quality has suffered drastically, where the producers and dealers have lowered their quality to profit from the price boom. Exporters have been unable to meet their commitments as the superior grades were not produced. Problems have continued to exist with the difficulty in obtaining peelers for the processing of cinnamon quills and weak linkages that they have with producers.

17.5 Quality Requirements of Cinnamon

17.5.1 Product and Process Standards Relevant to Cinnamon

Cinnamon is sold generally as quills. Further, cinnamon is also exported as quillings, featherings and chips. The quills shall be of light brown colour, well formed and well-dried. The occurrence of reddish-brown patches on the quills, which may become dark brown with time, is known as foxing. These are defects, and the value of quills gets depreciated depending on the amount of foxing. The commercial specifications of cinnamon are shown in Table 17.1 and the chemical requirements in Table 17.2.

The SLSI has specified the required standards for Ceylon cinnamon (SLS 81: 2010); the focus is mainly on the physical appearance of the products. According to the specifications, there are two main grades: 'coarse' and 'fine'. The fine grade is named as 'Alba', of which the market price is significantly higher compared to other grades. The intermediate grade is 'continental', and the coarse grades are 'Mexican' and 'Hamburg'. These are categorized according to the diameter of quills, number of whole quills per kilogram and the occurrence of reddish-brown patches on the surface of the quills (the extent of foxing). The existing international standards applicable for cinnamon are the European Spice Association (ESA) specifications of quality for herbs and spices, the American Spice Trade Association (ASTA) cleanliness

Table 17.1 Commercial specifications of cinnamon according to Sri Lanka Standards

Parameter	Description
Colour	Pale brown to slightly reddish colourGround cinnamon – yellowish to reddish brown in colour
Odour	Characteristic fresh aroma
Flavour	Delicate and sweet flavour characteristic to Ceylon cinnamon. It shall be free from foreign flavour; including mustiness
Moisture	Not more than 15% for quills and 12% for other grades
Volatile oil	Minimum 1% for quills and 0.7% for other grades on dry basis
Shelf life	Minimum of 1 year
Packaging	Packaged in clean, sound, dry packages; made of jute, cloth, paper or polyethylene bags

Source: SLSI 81:2010

Table 17.2 Chemical requirements of cinnamon (Source: SPC 2019)

Characteristics	SLSI[a] 81:2010	ESA[b] requirements	ISO-6539:1997 requirements	SLSI/ISO test methods
Ash	5.0% w/w (max)	7% w/w (max)	5% on dry basis max	SLS 186: Part 5 ISO 928
Acid insoluble ash	1.0% w/w (max)	2% w/w (max)	1% (m/m) dry basis max	SLS 186: Part 3 ISO 930
H_2O %	15.0 w/w (max)	15% w/w (max)	14% (m/m) max	SLS 186: Part 4 ISO 939
Volatile oil	1.0 ml/ 100 g (min)	0.45 w/w (min)	ml/100 g on dry basis min.Whole cinnamon – 1% Ground cinnamon – 0.7%	SLS 186: Part 11 ISO 6571

[a]*SLSI* Sri Lanka Standards Institution
[b]*ESA* European Spice Association

Table 17.3 Cleanliness specifications and permissible pesticide residues (the ASTA[a] cleanliness specifications, effective from 28th August 1990)

Whole insect dead by count (No./lb)	Mammalian excreta by mg/lb	Other excreta by mg/lb	Mould % by weight	Insect infested % by weight	Extraneous matter % by weight
2	1	2.0	1.00	1.00	0.50

Source: SPC 2019

[a]*ASTA* American Spice Trade Association

Table 17.4 Defect action level for cinnamon by the US Food and Drug Administration (Source: SPC 2019)

Cinnamon whole	Mould (MPM[a] – V32)	Average of 5% or more pieces by weight is mouldy
	Insect filth (MPM – V32)	Average of 5% or more pieces by weight is insect infested
	Mammalian excreta (MPM – V32)	Average of 1 mg or more mammalian excreta per pond
Cinnamon ground	Insect filth (AOAC 968'38B)	Average of 400 or more insect fragments per 50 grams
	Rodent filth (AOAC[b] 968.38)	Average of 11 or more rodent hairs per 50 grams

[a]*MPM* Macroanalytical Procedures Manual

[b]*AOAC* Official Method of Analysis of the Association of Official Analytical Chemists

Table 17.5 Maximum levels for heavy metals tolerated in spices and herbs under the European Spice Association requirements. (Source: SPC 2019)

Heavy metal	Max. level (mg/kg)
Arsenic	5.00
Copper	20.0
Lead	10.0
Zinc	50.0

specifications, good agricultural practices (GAPs), the Hazard Analysis and Critical Control Points (HACCP), ISO 22000:2005 and fair trade. Table 17.3 highlights the cleanliness specifications, and the permitted defect level and maximum tolerated levels of heavy metals are presented in Tables 17.4 and 17.5, respectively.

17.5.1.1 Importance of Complying with Quality Standards

The quality determines the value of cinnamon exports, which helps in reaching the target export quantities and export earnings. Problems are encountered due to quality failures of exported cinnamon in the international market, and there is a risk of losing the international market for Sri Lankan cinnamon. The quality also requires high-level protection of consumer/human health.

17.5.2 Branding of Ceylon Cinnamon

Repositioning *Pure Ceylon Cinnamon* in the global market and brand promotion to make it the global brand is an important step while highlighting the main characteristics of Ceylon cinnamon and its comparative advantage over cassia. The EDB of Sri Lanka has completed the registration of *Pure Ceylon Cinnamon* trademark (Fig. 17.15) in Europe and the USA and obtained the certificates of registration from both destinations.

The EDB currently works with the stakeholders in the cinnamon industry to obtain GI certification for *Pure Ceylon Cinnamon* in the EU covering five districts, namely, Galle, Matara, Kalutara, Ratnapura and Hambantota. The United Nations Industrial Development Organization (UNIDO) and the EDB are also assisting the cinnamon industry in registering GI for Ceylon cinnamon with the European Commission. With these efforts, Ceylon cinnamon would be Sri Lanka's first with GI. The UNIDO also plays a major role in supporting the EDB and the cinnamon industry towards the legal framework for GI registration, preparation of the supporting documents for the application to the EC and promotion of Ceylon cinnamon GI. To supplement this initiative, the Department of Export Agriculture is promoting implementation of good agricultural practices (GAP) for cinnamon growers and good manufacturing practices (GMP) for cinnamon processors.

The UNIDO, along with the EDB, has led the industry in laying the groundwork for the formation of the Ceylon Cinnamon Geographical Indication Association (CCGIA) to organize all value chain actors in order to promote CCGIA and to harmonize the practices of value chain actors under a common framework. Those involved in the cinnamon supply chain need to be registered and obtain membership of the CCGIA to export Ceylon cinnamon to the EU under the '*Pure Ceylon Cinnamon*' name. The responsibility of the CCGIA is to implement an internal control mechanism to monitor its members meeting the specifications stated in the GI specification document. In this regard, the cooperation and active participation of the cinnamon industry stakeholders, including growers, manufacturers, traders and exporters, will be obtained by the CCGIA. To initiate the groundwork for the

Fig. 17.15 *Pure Ceylon Cinnamon* trademark

registration of GI, UNIDO has supported the Intellectual Property (IP) Office in Sri Lanka to amend the Intellectual Property Act to register GI in Sri Lanka. The policy support given in this regard is as follows:

- The development of traditional GI products should be empowered through capacity building and the involvement of local people in all steps of decision-making from production to marketing.
- The development of traditional products should be a multi-stakeholder effort, but the government needs to facilitate and support existing initiatives through the allocation of budgets.
- Exploration of potential markets and promotion of traditional products in the European Market region, which is not fully tapped. Currently, the most important market links for traditional products are with the tourism sector.
- Further strengthening of the capacity building to enable local products (traditional products) to comply with the GI procedures to facilitate their reach to the international markets.
- The private sector should also play an important role in promoting traditional knowledge, but the interests of the local people must be assured throughout the value chain and allow all eligible stakeholders to benefit from the GIs.

A nationally accredited vocational training programme has been developed in Sri Lanka to promote internationally recognized food safety certification. As a result, the Cinnamon Training Academy now provides National Vocational Qualifications (NVQ) for the cinnamon factory and field officers. Awareness raising and NVQ level training on food safety and hygiene practices have reached over 1000 people. Six cinnamon processing centres have been upgraded to date, allowing them to obtain GMP certification. More cinnamon peelers and processors (including women) have joined the sector due to the certified vocational training and decent working conditions. Social marketing has helped to mitigate the social stigma associated with cinnamon peeling and promote career opportunities in the sector. (http://www.standardsfacility.org/sites/default/files/STDF_PG_343.pdf). The 2019 budget proposals of the Government of Sri Lanka have recognized the importance of strengthening the Cinnamon Training Academy established in Kosgoda to train peelers and have allocated the necessary funds. The Department of Export Agriculture has also established a cinnamon training centre at Palolpitiya, Thihagoda, to train stakeholders in the cinnamon industry. Since 2017, it has conducted training programmes on hygienic cinnamon peeling, GAP, GMP and cinnamon value-added products. Over 2500 people have been trained since 2017 under these programmes.

17.6 Way Forward

17.6.1 Strategies to Increase Cinnamon Production

The current production of cinnamon in Sri Lanka is not adequate to cater to the increasing global demand. The world demand for both Ceylon cinnamon and cassia cinnamon shows an increasing trend. However, in comparison to the supply of cassia cinnamon, the supply of Ceylon cinnamon to the world market is around 10% of the total demand. As the cinnamon export from Sri Lanka is around 17,000 mt per annum, there is a need to increase the volume by several folds within a short period if the country wants to become a major player in the cinnamon world trade. Currently, cinnamon is mainly cultivated in a few districts, but there is a potential to expand cinnamon cultivation into other areas in the wet and intermediate agroclimatic zones. Before 2000, cinnamon was largely concentrated in the Southern province of Sri Lanka, and later it was expanded to Western and Sabaragamuwa provinces due to the new cultivation programmes introduced by the Department of Export Agriculture.

As the land area for new planting is limited, the productivity of the crop has to be increased to meet the ever-growing international demand for cinnamon. The national productivity of cinnamon is around 500 kg/ha, while there is a potential to increase to 750–1000 kg/ha with proper crop management practices and gap filling in cultivations. Another critical component in augmenting crop productivity is the soil fertility management with balanced supply of plant nutrients. Majority of low-yielding cinnamon cultivations in Sri Lanka have reported to have declined soil fertility due to poor soil and crop management practices. Investments in this area are highly necessary coupled with effective promotional and awareness programmes.

17.6.2 *Need for Quality Assurance*

Conducting awareness programmes on the importance of meeting quality standards is crucial from the producer level to the exporter level across the value chain of cinnamon production. At the producer level, it is vital to introduce techniques and methods to make peeling, rolling and quilling processes easier and provide proper training for peelers to produce quality products. The pressing need for establishing 'central processing units' in the main growing areas has been emphasized by stakeholders in the cinnamon industry. Moreover, it is useful to modernize 'in-house processing sheds' to maintain the quality of the products. It is also necessary to promote producers to adhere with GAPs.

At the collector level, a priority would be to promote sound storage and transport systems. Towards this end, awareness should be increased on conditions of warehouses, which should be covered and well protected from the rain, sun and excessive heat. Moreover, Sri Lanka has to strengthen existing mechanisms and introduce new technologies for quality checking.

When considering the exporter level, increasing their linkages with the producers and developing appropriate methods and programmes for processing are required. It is essential to introduce concessionary terms or incentive schemes for exporters who wish to obtain and maintain standard certificates, to safeguard Sri Lanka's position as the supplier of true cinnamon to the world. As highlighted earlier, the 2019 budget proposals of the Government of Sri Lanka have taken steps towards ensuring quality of cinnamon exports. However, more needs to be done along the cinnamon value chain to ensure the quality of exported cinnamon.

17.6.3 Government Support to Obtain GI

The Government of Sri Lanka should consider implementing several actions in order to develop the enabling environment that will support sustainable GIs. Identification of potential products and stakeholders or inter-professional bodies ready to cooperate is a current need to develop the cinnamon industry in Sri Lanka. Further, activities leading to raise consumer awareness through the media, launching pilot projects to develop newly registered GIs, particularly those products meant for export and provide temporary support such as investment, would benefit the industry. Assessment of the economic and rural development potential for the registration of new GIs and including GIs in a national strategy for tourism and export promotion of the country's products should be done by the Government of Sri Lanka to support the cinnamon industry (Hirimuthugoda 2019).

17.6.4 Diversification of Processing Methods

The Sri Lankan way of processing cinnamon as quills is a centuries-old unique method. There is a price advantage attached to quills in premium markets. However, due to the higher price, cinnamon is not competitive in markets with average consumers. Avoiding the high labour-intensive traditional peeling process and harvesting cinnamon from more matured trees will help Sri Lanka to supply higher quantities of cinnamon at a competitive price to the world market. In order to revive the national production, the country needs to realize the issues related to adhering to a peeling technology that has impeded the national cinnamon production. Cassia producers have managed to capture more than 85–90% of the world market by supplying cinnamon as powder or as pieces of bark for a cheaper price. In order to recuperate the national cinnamon industry, Sri Lanka should move away from the high-cost traditional peeling process and develop a simple, low-cost peeling method.

Introduction of a more cost-effective, simple method of peeling will help new growers to emerge from other parts of the wet and intermediate zones of the country. Traditional peelers will not be required for these new areas as they can adopt a simple low-cost peeling method. However, sustainability of such approaches should be examined by industry experts, scientists and academia.

17.7 Strategic Policy Planning for Cinnamon

The SWOT analysis carried out for cinnamon (MMECP 2014), which was updated by the authors for this publication, is presented in Annex 1. Based on the analysis conducted by the authors in this chapter and the SWOT analysis, the proposed strategic plan for improving the cinnamon industry under eight different thrust areas is presented in Table 17.6.

Annex 1

Cinnamon: SWOT Analysis for the Industry

Strengths

- An overarching policy aiming at improvement of spice crop production exports
- A line department devoted for generating technical know-how and information dissemination on the export agriculture sector in including cinnamon
- A statutory board to look into the interests of the export commodities of Sri Lanka
- Presence of a cinnamon research centre under the purview of the Department of Export Agriculture
- Availability of research facilities and services at the Department of Export Agriculture, Industrial Technology Institute (ITI) and universities
- Incentive schemes for new planting, productivity improvement and post-harvest assistance for cinnamon implemented by the Department of Export Agriculture
- Global market leader in Ceylon cinnamon production and exports
- Presence of a grower and cinnamon peeler training carried out by the Department of Export Agriculture
- Presence of an organized private sector to support the cinnamon industry (i.e. the Spice Council, SAPPTA, Cinnamon Cultivators' Association – CinCA)
- Involvement of the banking sector to promote Ceylon cinnamon (e.g. NDB)
- Interest shown by the private sector to prepare sectoral development plans for cinnamon
- Private sector taking initiatives for capacity building in terms of cinnamon peeler training programmes with financial assistance from the state and a UN organization
- Private sector showing interest on capacity building of cinnamon growers and processors of GMP and quality standards
- Presence of a competent research and extension staff at the Department of Export Agriculture
- Presence of SLS standards SLS81:2010 for cinnamon
- Presence of ISO standards for *Cinnamomum zeylanicum* (ISO 6539:2014) separately to that of *Cinnamomum cassia* (ISO 6538:1997)
- Specific microclimatic zones and soil conditions conducive for cinnamon cultivation
- Presence of a branding system with *Ceylon cinnamon*
- Ceylon cinnamon being popular as *Cinnamomum zeylanicum*
- Research published at international level on health benefits and quality of Ceylon cinnamon
- The bark of Sri Lanka cinnamon having only traces of coumarin
- Accepted traditional knowledge of over 4000 years in the Ayurvedic system
- Freight advantage due to close proximity to international trade routes

Weaknesses

- Absence of a proactive, well-directed and coordinated lobbying and promotional mechanism at global scale
- Smallholder cultivation dominating the sector with low commercial orientation
- Limited efforts made to establish the scientific name *Cinnamomum zeylanicum* for Ceylon cinnamon
- Major markets limited to Mexico, the USA, Columbia, Peru and Germany
- Lack of proactive promotional activities in the international arena
- Weak research prioritization mechanisms aiming at market needs
- Limited efforts to identify research priorities in consultation with the stakeholders
- Limited attention paid on updating online databases maintained by the Department of Export Agriculture on cultivation, production and export of cinnamon
- Imperfect markets and use of oligopoly power in bargaining
- Limited cultivation of the cinnamon crop to some districts
- Inadequate skilled labour especially for peeling
- Limited attention paid for quality standards and sanitation during cinnamon processing
- Social cast enigma in the job of cinnamon peeling, leading to lower attraction of the youth
- Inadequate availability of trainers to train personnel at all levels of cinnamon production and processing
- Traditional labour-intensive methods used for processing
- Seasonal variations in the crop production
- Poor maintenance of smallholder farms and poor post-harvest processing, leading to low-quality products
- Inadequate supply of quality products
- Weak linkage between the industry and research institutions, leading to retardation of innovation
- Weak linkage among producers, processors and exporters
- Inadequate capacity in the extension services
- Ineffective implementation of government incentive schemes
- Poor response from the plantation sector for cinnamon cultivation
- Most of the processors/exporters having no access to land or no direct involvement in cultivation of cinnamon

Opportunities

- Growing demand for Ceylon cinnamon at the global scale
- Separate ISO standards established for *Cinnamomum zeylanicum* to that of *C. cassia*
- Establishment and acceptance of Ceylon cinnamon brand logo at the EU and the USA

- Health consciousness of people worldwide
- High potential for the presence of coumarin-free germ plasm in Sri Lanka
- Considered as a 'miracle cure' for a raft of health issues ranging from obesity to heart disease and arthritis
- Research published at international level, confirming health benefits and quality of Ceylon cinnamon
- International Botanical Congress as the formal accepting authority of the scientific name of Ceylon cinnamon as *C. zeylanicum*
- Barks from the substituted cinnamon product cassia, having substantial amounts of coumarin
- 'Spice Gardens Promotion' and 'Spice Park' projects implemented by the Government of Sri Lanka
- Potential for expansion of cultivation in the wet and intermediate zones of Sri Lanka
- Potential for intercropping with other established perennial crop production systems in different agroecosystems
- Land opening up for cultivation in northern and eastern provinces
- Changing dietary pattern for more spicy food at global scale
- Trend towards natural flavours and fragrances and natural healthcare in the developed markets
- Move towards a holistic marketing approach from a trading mentality
- Vast unskilled labour force, making the smallholder outgrower system viable
- Relaxation of regulations to attract foreign investment in areas of value-added industries of cinnamon

Threats

- Low-cost cassia from India, Vietnam and Indonesia, competing with high-valued true Ceylon cinnamon
- India developing strains of cinnamon, of which Sri Lanka has a major share in the world market
- Ceylon cinnamon being traded by other countries such as India by playing the role of the middleman
- Low cost of production of Indian essential oils, resulting in higher Indian exports undercutting Sri Lankan cinnamon oil
- Ever-increasing standards by the developed countries, which act as a trade barrier for Sri Lankan cinnamon products to enter those markets
- Low standards of cinnamon supplies due to weaknesses in the coordination, awareness and infrastructure

Table 17.6 Strategic plan for development of cinnamon sub-sector in Sri Lanka

1. Thrust area: expansion of cultivation and introduction of new varieties		
1.1 Strategy	Facilitate expansion of cultivated extent of cinnamon	
	1.1.1. Programmes	Produce good-quality planting materials of cinnamon
		Promote new cultivation systems and practices for cinnamon
		Encourage private and public land for cinnamon cultivation
		Develop programmes for intercropping cinnamon with other major crops
		Develop programmes to introduce productive traditional organic practices and biodynamic applications for cinnamon
		Promote home garden cultivations of cinnamon at village level
		Develop cluster village system for cinnamon cultivations
		Expand cinnamon cultivation into non-traditional areas
		Introduce attractive incentive schemes for medium- and large-scale plantation entities to move into cinnamon cultivation
		Expand the existing incentive schemes to support cinnamon cultivation by small-scale producers
1.2 Strategy	Promote introduction of new cinnamon varieties	
	1.2.1 Programmes	Develop new high-yielding, high-quality, pest-and-disease-resistant and drought-tolerant varieties of cinnamon suitable to different agroecological regions
		Promote surveying, identification, multiplication and cultivation of no-coumarin cinnamon accessions
2. *Thrust area: improvement of productivity*		
2.1 Strategy	Introduction of productivity improvement techniques, methods and practices	
	2.1.1 Programmes	Perform surveys to identify productivity level of the cinnamon sector
		Develop programmes to increase the production of cinnamon, considering the climatic changes
		Ensure adoption of good agronomic practices and methods for cinnamon
		Ensure adoption of good pest and disease control practices for cinnamon
		Promote indigenous/traditional cultivation practices
		Develop an appropriate mechanization programme for cinnamon cultivation
		Develop soil conservation and land management practices
		Develop programmes to ensure soil fertility and conservation in cinnamon plantations
		Develop programmes for optimum utilization of fertilizers
		Develop programmes to introduce productive harvesting systems
		Develop a motivation scheme for cinnamon stakeholders

(continued)

Table 17.6 (continued)

2.2 Strategy	Promote adoption of good agricultural practices (GAP) for cinnamon	
	2.2.1 Programmes	Develop and introduce suitable methods for cinnamon cultivation to adapt to the changes in the climate
		Establish high-income-generation model farms at regional level
2.3 Strategy	Develop post-harvest technologies for cinnamon	
	2.3.1 Programmes	Develop appropriate mechanization programme for processing and value addition
		Introduce knowledge and skill development programmes for stakeholders
		Launch programs for selection of best innovations in mechanization process of cinnamon
		Promote establishment of cinnamon processing training centres in major cinnamon growing areas
3. *Thrust area: quality improvement and standardization*		
3.1 Strategy	Promotion of quality and standards of cinnamon products	
	3.1.1 Programmes	Develop programmes for value chain management in cinnamon industry
		Strengthen and introduce new technologies for quality assurance and improvement of cinnamon
		Develop programs to ensure healthy and hygienic production and harvest and post-harvest practices for cinnamon
		Develop and enforce regulatory and monitoring mechanisms for quality control
		Establish and promote good harvesting practices and sound storage systems for cinnamon
		Revise and improve national quality standards and certification systems for Sri Lanka standard system for cinnamon in line with nternational standards
		Establish and promote a sound GMP system
		Develop programmes to promote value chain management
		Develop a ranking system for production houses to ensure the quality of products
		Develop programmes to ensure maintenance of quality cinnamon products at grassroots levels
		Conduct annual quality award ceremony for stakeholders
		Promote stakeholder education and development
		Introduce incentive schemes for stakeholders who achieve the national quality standards in the cinnamon production process
3.2 Strategy	Ensure adoption of standardization procedures for cinnamon production	
	3.2.1 Programmes	Establish regulatory systems to make the use of Ceylon cinnamon logo mandatory for cinnamon exports
		Update the national quality standards and grading criteria for cinnamon products
		Provide infrastructure and equip the national agencies to adopt and issue standard certificates for cinnamon products

(continued)

Table 17.6 (continued)

4. *Thrust area: marketing and promotion*		
4.1 Strategy	Develop, promote and streamline marketing promotion of Ceylon cinnamon at the regional and international levels	
	4.1.1 Programmes	Develop a program to establish the scientific name of Ceylon cinnamon as *Cinnamomum zeylanicum* at the International Botanical Congress through proper scientific guidance and branding of Ceylon cinnamon
		Establish a cess fund for promotional campaign on cinnamon
		Promote Ceylon cinnamon brand products at the international market
		Develop promotional campaign for cinnamon industry through print and electronic media
		Develop and establish a mechanism for cinnamon auction
		Develop and establish sales centres for cinnamon at regional level
		Liaise with relevant international organizations in promoting Ceylon cinnamon
		Conduct market research to identify new markets and expand existing markets for cinnamon
		Develop programs for introduction of new entrepreneurs
		Execute programs to establish niche markets for Ceylon cinnamon
5. *Thrust area: research, development and extension*		
5.1 Strategy	Promote research and development on cinnamon	
	5.1.1 Programmes	Strengthen a national research programme for cinnamon
		Strengthen the National Cinnamon Research and Training Institute
		Develop programmes on research needs analysis for cinnamon
		Establish a research linkage with world-renowned universities and organizations
		Develop novel technologies to improve cinnamon peeling and extraction of cinnamon oil and other plant intrinsic substances
		Promote research on value chain development of cinnamon
		Promote research on marketing aspects of cinnamon
		Promote research on productivity improvement of cinnamon
		Promote research on economic and social aspects of cinnamon industry
		Develop agronomic packages for organically grown cinnamon
		Develop and introduce new cinnamon varieties adaptable to climatic changes
		Develop research programmes on traditional knowledge on agricultural practices and biodynamic farming for cinnamon
		Initiate research on intrinsic properties of cinnamon for human health and well-being

(continued)

Table 17.6 (continued)

5.2 Strategy	Facilitate functioning of sound extension systems	
	5.2.1 Programmes	Establish a national association of cinnamon (Sri Lanka Cinnamon Association)
		Develop programmes to communicate the research findings to stakeholders at all levels
		Establish demonstration units on best practices for cinnamon cultivation and processing at regional level
6. *Thrust area: value addition and by-product development*		
6.1 Promote strategy	Promote value additions and by-product development for cinnamon	
	6.1.1 Programmes	Establish a policy framework for value addition and by product development
		Promote different forms of processing based on the needs of the export markets
		Establish mechanisms and incentive schemes for the producers and exporters to move on to value addition of cinnamon
		Develop multiprocessing centres at regional level
		Develop export model villages
		Develop programmes to invent machinery and equipment necessary for value addition and by-product development
		Develop programmes to establish cottage industries at village level
		Develop and strengthen programmes to motivate stakeholders for high-tech value addition of pharmaceuticals, nutraceuticals, cosmetics and aromatics industry
		Strengthen the laboratory facilities for the testing, quality assurance, value additions and by-products
7. *Thrust area: Mechanization and new technology adoption*		
7.1 Strategy	Promote mechanization and new technology	
	7.1.1 Programmes	Develop programmes for the mechanization of existing product value chain of cinnamon
		Promote and strengthen technical know-how and assistance for the modernization of prevailing industries
		Develop and introduce new high-tech production system into the cinnamon industry
		Develop a programme to simplify production systems and procedures for cinnamon
		Develop a promotion scheme for new inventions in development of systems and machineries
		Develop programmes to identify and re-engineer the production process of cinnamon
		Develop rural-level process technicians

(continued)

Table 17.6 (continued)

7.2 Strategy	Promote adoption of new technologies in the cinnamon industry at all levels	
		Strenthen programmes to adopt new technologies in production and value addition of cinnamon to suit the world demand
		Develop and introduce cinnamon bark distillation unit for SMEs of cinnamon industry
		Develop training programmes for stakeholders to impart knowledge on new technology
		Develop a digital platform to share novel information among stakeholders
8. *Thrust area: stakeholder livelihood enhancement*		
8.1 Strategy	Facilitate livelihood improvement programmes for stakeholders in the cinnamon sector	
	8.1.1 Programmes	Initiate self and lifestyle management programmes for cinnamon farmers and grassroots-level collectors
		Institue and strengthen cinnamon-based community organizations

References

CBSL (2018) Economic and social statistics Sri Lanka, Central Bank of Sri Lanka

DEA (2018) Unpublished report, Department of Export Agriculture, Sri Lanka

DEA (2019) Cinnamon. http://www.exportagridept.gov.lk/web/index.php?Option =comcontent&view=article&id=128&Itemid=159&lang=en. Accessed on 22 Jan 2019

DOA (2018) Annual agricultural statistics 2013–2017. Socio Economics and Planning Center, Department of Agriculture, Sri Lanka

EAS (2013) Annual report of accounts, Exporter's Association of Sri Lanka

EAS (2017–2018) Annual report and accounts, Exporter's Association of Sri Lanka

EAS (2018) Annual report of accounts, Exporter's Association of Sri Lanka

EDB (2018) Export performance indicator 2009–2018, Export Development Board of Sri Lanka

FAOSTAT (various issues) Food and Agriculture Organization of the United Nations. http://www.fao.org/faostat/en/#home . Accessed on 30 Nov 2019

GOSL (2018) Budget speech – 2019. Government of Sri Lanka. http://treasury.gov.lk/documents/10181/684395/BUDGET+++2019+Final.pdf/c603c8ab-5d04-4075-9bea-eea75ae17783. Accessed on 5 Dec 2019

Hirimuthugoda D (2019). http://www.ips.lk/talkingeconomics/. Accessed on 30 Dec 2019

http://en.wikipedia.org/wiki/Cinnamomum_cassia. Accessed on 20 Nov 2019

http://www.dailymirror.lk/business/ features/30034-us-research-study-spells-boom-for-lankas-cinnamonindustry.html. Accessed on 30 Nov 2019

http://www.dailymirror.lk/business/features/30034-us-research-study-spells-boom-for-lankas-cinnamonindustry.html. Accessed on 31 Oct 2019

http://www.exportagridept.gov.lk/web/index.php?option=comcontent&view=article&id=128&Itemid=159&lang=en. Accessed on 30 Nov 2019

http://www.factfish.com/statistic/cinnamon%2C%20yield. Accessed on 30 Nov 2019

http://www.ft.lk/2013/05/23/us-research-shows-massive-potential-for-sri-lankan-cinnamon/. Accessed on 15 Dec 2019

http://www.standardsfacility.org/sites/default/files/STDF_PG_343.pdf. Accessed on 15 Dec 2019
https://cinnamonbenefits.weebly.com/ceylon-vs-cassia.html. Accessed on 15 Dec 2019
IPS (2017) Analysis of cinnamon, pepper, and cardamom value chains in Sri Lanka. Institute of Policy Studies
ISB (2019) Indian spice board. http://www.indianspices.com/monthly-price-international. Accessed on 15 Dec 2019
ITC (2018) International trade center. http://www.intracen.org/. Accessed on 30 Dec 2018
Khan A, Safdar M, Ali Khan MM, Khattak KN, Anderson RA (2003) Cinnamon improves glucose and lipids of people with type 2 diabetes. Diabetes Care 26:15–18
Leelarathne H (2013) US research shows massive potential for Sri Lankan cinnamon. http://www.ft.lk/article/152720/Www.aquavista.lk. Accessed on 10 Jan 2020
MMECP (2014) Development plan for spice and allied crops sector in Sri Lanka (2014–2025). Ministry of Minor Export Crop Promotion. Government of Sri Lanka
SPC (2019) Spice Council of Sri Lanka. https://www.srilankanspices.com/sl_spices_ cinnamon.html. Accessed on 30 Dec 2019

Index

R. Senaratne, R. Pathirana (eds.), *Cinnamon*,
https://doi.org/10.1007/978-3-030-54426-3

The manufacturer's authorised representative in the EU is Springer Nature Customer Service Centre GmbH, Europaplatz 3, 69115 Heidelberg, Germany. If you have any concerns regarding our products, please contact ProductSafety@springernature.com

Printed and bound by CPI Group (UK) Ltd, Croydon, CR0 4YY
30/09/2026
02225287-0016